AF173682

# Simulation Foundations, Methods and Applications

The modeling and simulation community extends over a range of diverse disciplines and this landscape continues to expand at an impressive rate. Over recent years, modeling and simulation has matured to become its own discipline, while continuing to provide support to other disciplines. As such, modeling and simulation provides the necessary conceptual insights as well as computational support which has an established record of significantly enhancing the understanding of dynamic system behavior and improving the system design process, as well as providing the foundations for computational sciences and practical applications, from cyber-physical systems to healthcare. Hybrid methods and combinations with artificial intelligence and machine learning open new possibilities as well. The ever-increasing availability of computational power and the availability of quantum computers make applications feasible that were previously beyond consideration. Simulation is pushing back the boundaries of what it can be applied to and what can be solved in practice. Its relevance and applicability are unconstrained by discipline boundaries.

Simulation Foundations, Methods and Applications hosts high-quality contributions that address the various facets of the modeling and simulation enterprise. These range from fundamental concepts that are strengthening the foundation of the discipline to the exploration of advances and emerging developments in the expanding landscape of application areas. The underlying intent is to facilitate and promote the sharing of creative ideas across discipline boundaries.

As every simulation is rooted in a model, which results from simplifying and abstracting the reference of interest to best answer research questions or support the application domain of interest, we understand the model development phase as a prerequisite for any simulation application. There is an expectation that modeling issues will be appropriately addressed in each presentation. Incorporation of case studies and simulation results will be strongly encouraged.

Titles of this series can span a variety of product types, including but not exclusively, textbooks, expository monographs, contributed volumes, research monographs, professional texts, guidebooks, and other references.

These books will appeal to senior undergraduate and graduate students, and researchers in any of a host of disciplines where modeling and simulation has become (or is becoming) an important problem-solving tool. Some titles will also directly appeal to modeling and simulation professionals and practitioners.

Michael Grieves · Edward Y. Hua
Editors

# Digital Twins, Simulation, and the Metaverse

## Driving Efficiency and Effectiveness in the Physical World through Simulation in the Virtual Worlds

*Editors*
Michael Grieves
Digital Twin Institute
Cocoa Beach, FL, USA

Edward Y. Hua
The MITRE Corporation
McLean, VA, USA

ISSN 2195-2817 ISSN 2195-2825 (electronic)
Simulation Foundations, Methods and Applications
ISBN 978-3-031-69106-5 ISBN 978-3-031-69107-2 (eBook)
https://doi.org/10.1007/978-3-031-69107-2



This Springer imprint is published by the registered company Springer Nature Switzerland AG
The registered company address is: Gewerbestrasse 11, 6330 Cham, Switzerland

# Foreword

Through compelling real-world use cases and case studies, Dr. Michael Grieves, Executive Director of the Digital Twin Institute, an internationally recognized expert in the fields of Digital Twins and Product Lifecycle Management, first introduced the core concept of Digital Twins and further developed and evolved the concept and Dr. Edward Hua, Tech Lead for Digital Twins Simulation Engineering at The MITRE Corporation's Modeling and Analysis Innovation Center, has compiled a comprehensive history of Digital Twin foundational elements, concepts, methodologies, frameworks, and implementations in the chapters comprising *Digital Twins, Simulation, and the Metaverse: Driving Efficiency and Effectiveness in the Physical World through Simulation in the Virtual Worlds.*

Immediately insightful and impactful, this significant body of work, authored by experts in their fields, provides a cross-sectional view where Digital Twins deliver real value and transformative outcomes across diverse market sectors, segments, and industries.

With adoption spanning virtually every primary industry and new applications and use cases continually emerging, readers can understand Digital Twin's growth and adoption, including its near symbiotic relation to the evolving Metaverse landscape. The Metaverse is currently an evolving and not yet clearly defined concept. In the editors' introductory book chapter, the characteristics of Digital Twin Metaverses are outlined to provide a roadmap for the needed future development.

From conceptual methodologies, foundational elements, standardizations, and frameworks, including composability, construction, verification and validation, and other constructs spanning the different lifecycle phases, to exploring AI approaches—including prescriptive to autonomous, such as ML, as described through the application of Reinforcement Learning over the Digital Twin lifecycle—to Generative AI, utilizing co-pilots and multi-agents that deliver increasing value, traditional to intelligent Digital Twin use cases are revealed in highly informative detail.

These include an impressive array of examples that range from Nuclear facilities to Healthcare and Biomedicine, to Smart manufacturing—including the overall value chain along with advances in robotics and battery lifecycle production, to sustainable semiconductor fabrication and the Digital Twin role as "the photorealistic, physics-based, and real-time capable Digital Twin" where in turn simulation is the main ingredient of the Industrial Metaverse.

Examples of Digital Twin applications in Transportation management—Urban Mobility and Distributed AI Modeling and Simulation for Smart Airport Digital Twin with multi-agent transportation management systems, including Metaverse applications, are among several other use cases.

Learnings include the strategic integration of Digital Twins infused with AI and the significant role that reality capture plays in a pioneering journey of a NASA "factory Digital Twin," where the Factory Twin's value is realized and "positioned as a dynamic entity capable of substantial ROI."

For those engaged in or interested in research, this book serves as a compass, helping to guide and provide an understanding of opportunities for both new and existing R&D pursuits. Each chapter includes valuable references for further investigation, pinpointing specific areas of interest.

Through evolving market and business landscape examples, this book further illustrates how industries are progressively innovating as new technologies—encompassing advances in extended reality (XR), AI, 5G, and Edge-Cloud Computing, among other enabling technologies—transform traditional business models and generate new opportunities.

Digital Twin characteristics, oriented to the Metaverse and viewed through a lens into the growth and evolution of this developing landscape, are presented. Historical and current market opportunities are detailed through specific areas of adoption across various APAC regions while also considering the developing future potential.

This compilation examines the development and progression of Digital Twins, including associated opportunities and challenges. The diverse collection of case studies and analyses provides insights into Digital Twin's key role in digital transformation. Fueled by AI-accelerated growth and as exemplified by the use cases described, Digital Twin's evolution and adoption show no signs of slowing, especially when coupled with the emerging Metaverse landscape.

San Francisco, USA
July 2024

Dan Isaacs

# Contents

# Defining, Exploring, and Simulating the Digital Twin Metaverses

**Michael Grieves and Edward Y. Hua**

**Abstract** This chapter presents a brief introduction and history of simulation, Digital Twins and their types and replica twins, and the origin of "metaverse". Underlying all these technologies is the premise that information generated by these technologies is a replacement for the wasted physical resources in human goal-oriented tasks. The chapter then provides the characteristics of a Digital Twin-oriented metaverse. It applies the characteristics to the different DT types. It concludes by discussing the evolution of Digital Twins in replication and prediction that will see Front Running Simulation as our crystal ball into the future. AI is predicted to play a major role in making this evolution possible as an assistance to humans but not a replacement.

**Keywords** Digital twin · Physical twin · Replica twin · Metaverse · Simulation · Front running simulation

## 1 Introduction

Digital Twins (DTs) are a twenty-first-century concept that has enjoyed an exponential growth of interest over the last decade. DTs originated as the underlying component of another twenty-first-century concept, Product Lifecyle Management (PLM). PLM represented a change from a functional-centric approach where each function, engineering, manufacturing, operations, and support had siloed its data and information to a product-centric approach where every function populated and consumed from a common source.

That common source needed to be DTs. DTs took advantage of the exponential increases in information technology to implement this lifecycle-based, product-centric representations of physical products and artifacts. This DT model was

M. Grieves
Digital Twin Institute, Cocoa Beach, FL, USA
e-mail: mgrieves@mwgvp.com

E. Y. Hua (✉)
The MITRE Corporation, McLean, VA, USA
e-mail: ehua@mitre.org



M. Grieves and E. Hua (eds.), *Digital Twins, Simulation, and the Metaverse*, Simulation Foundations, Methods and Applications, https://doi.org/10.1007/978-3-031-69107-2_1

originally proposed for automotive and aerospace. However, DT use has been proposed for almost any physical product or artifact that exists in the physical world, both tangible and intangible. The chapters in this book include DTs for industrial and manufacturing [1–4], nuclear reactors [5], health care [6], airports [7], semiconductors [8], power generation [9], and batteries [10].

Humans have always had physical representations of different scales and fidelity of products and artifacts in the form of Replica Twins as described later. DTs are a digital representation of counterparts that exist or are intended to exist in the physical world. Unlike atom-based objects where all or none of the objects exist, the bit-based data and information of DTs allow for granularity of representation. The fidelity of that granularity and the intervals needed to synchronize the DT and its physical counterpart will depend on the use cases that create value. In order to understand how value is created, it is important to understand what information does and the role information plays with the different types of DTs throughout the product lifecycle.

DTs have gotten a significant amount of attention from both academics and industry. However, equally important is a digital representation of the physical environment that the DTs are or will exist and operate in. While it is necessary and useful to have the data and information about the DT and its physical counterpart, it is equally important to understand the forces of the environment surrounding and affecting the DT and its physical counterpart. This means the creation of digital environments or digital spaces[1] that multiple DTs from different sources can interoperate in. These are what we are calling Digital Twin Metaverses that this chapter will describe. This represents the next wave of evolution for DTs.

While defining and exploring DTs and DT Metaverses are important, this evolution will feature the important capability of simulation. Simulation allows us to predict and anticipate the future, at least probabilistically. We have had simulations for as long as humans have been in existence and could think. What is novel is that we now have the technology to simulate outside of human minds with DTs. This chapter will set the stage for advancing this evolution.

## 2 Simulation

In the computer age, we technologists think that we have invented simulation. However, humans have been doing simulations since the beginning of their existence. Simulation is defined as one process that imitates another process [11]. It's important to note that processes are, by the very nature, time evolved. By that definition, simulation is a foundational aspect of human thinking.

[1] We will use virtual and digital as synonymous here. There actually is no true virtual space. The virtual representation is always instantiated in atom-based physical material. In humans, it's in the carbon-based matter of the brain. In digital computers, it's in the silicon-based matter of digital processor and memory components.

## 2.1 Simulation Through Human History

Since man began to think, he has performed simulations. Man has used simulations from the beginning of human existence for a wide variety of tasks. Simulation has been used for planning, assessments, training, scenario generation, risk assessment, experimentation, and even entertainment.

Take the example of prehistoric man hunting game. Prehistoric man ran through different scenarios in his mind of what he, his fellow hunters, and the animal would do. For example, the hunter mentally simulated what plan would predict running a mammoth off a cliff. He ran through various simulations of which hunters needed to be where, what actions they needed to take, and what reaction the animal would have. He then selected the simulation that he believed would have the highest probability of succeeding.

He shared that simulation with his fellow hunters by tracing his plan in the dirt with a stick. His fellow hunters watched this simulation unfold over time as the hunter traced the stick in the dirt, showing the movement of the hunters and their intended prey. It was crude. It was primitive. It often didn't predict the intended outcomes. However, it was simulation.

The history of the military is intertwined with simulations. Soldiers throughout history were trained in simulation exercises. Tzu [12] writes of convincing an emperor of his ability to train troops by doing a simulation using the emperor's harem. The D-Day invasion of Normandy Beach was planned via simulation in a Scottish harbor that had the characteristics of Normandy Beach [13].

Over history, humans developed more stylized simulations. Early Greek plays were simulations of what would happen over time when certain events took place. Later, in the Middle Ages, written stories were simulations [14]. In the 1900s, movies came about and provided much richer simulations that could be shared by many more people.

The arrival of computers in the last half of the twentieth century advanced the rigor and robustness of simulations. These simulations were mathematically oriented and could be quite complex in terms of calculations. However, these simulations with applications like GPSS were mathematically abstract, and visualization was limited to reams of numbers on paper output [15].

Fast forward to the twenty-first century. We can do very rich detailed simulations that provide photo realistic visualizations of the simulation of physical objects in their environments. These simulations reasonably mirror the changes of their physical counterparts when subject to the same forces.

## 2.2 Simulation Prediction—Causation and Correlation

As noted above, simulations are about predicting possible future outcomes. There are two mechanisms that simulation uses: causation and correlation. From a system's view of causation, we have inputs. We know those inputs cause specific things to happen in our system. We then get outputs. It is deterministic.

In the correlation model, we have inputs. We don't know deterministically what happens in the system, but we get outputs. Even though we don't know what happens within the system, we do know that there is a relationship between the inputs and the outputs. Varying those inputs will result in a varying of the outputs that maintain a correlating relationship. The correlations may be very strong, close to 100% or very weak, close to 0%. We can use strong correlations and ignore weak correlations.

We may have a correlation of 100% or close to that, but we may still not know why or how the inputs result in the outputs. These correlation models can be useful even if we have no idea why one input variable would correlate with an outcome variable [16]. Note that we are very uncomfortable and should be questioning correlations where we cannot theorize some relationship between correlated inputs and outputs.

Simulations can use both methods, causation and correlation. Simulation driven by causation will use formula or algorithms to take inputs and derive outputs. Simulation driven by correlation will take data that they have from previous input/output data and apply that to the input data that they are attempting to determine outputs for.

Causation will give deterministic outputs, while correlations will give probabilistic outcomes. However, it is important to note that for systems of high complexity, there may be other unknown causal variables that will affect the outputs. We may be unaware of these causal variables or that what we believe are causal variables are simply correlated with this unknown causal variable.

Many of the computer-based simulations in the past were driven by formulas or algorithms. The complexity of these simulations was such that this was sufficient for the outputs that were needed. The simplest version of this is $y = f(t)$. This could be an object moving in a straight line with a constant velocity. The variable "$y$" is the location at any time "$t$" produced by the function. While this is not usually thought of as a "simulation", it meets the requirements of the time-evolved definition. These functions or formulae can be very complicated but are causally deterministic.

However, as we get into more and more complex systems with numerous variables, some hidden, we will need to use correlations from big data that we are collecting from products that are in operation. As we collect more and more data from products in the field, Bayesian-based probability models assisted by Artificial Intelligence (AI) become more and more useful [17].

Using causation when we have it is highly preferred. However, it needs to be remembered that causal models are conditional and, for complex systems, may not reflect or predict accurately or capture all causal relationships [18]. Probabilistic correlation models can be highly effective in predicting overall successful outcomes. We can also use a hybrid approach which combines using both causal and big data correlations [19].

Going back to our prehistoric hunter, if he was successful 50% of the time in running a mammoth off a cliff but not knowing exactly why it worked, he and his tribe were well fed.

## 2.3 Front Running Simulation (FRS)

FRS is a specific form of simulation. FRS was introduced by Grieves [20] and is shown in Fig. 2. FRS is a simulation that predicts future states using assumptions based on physics and/or data. However, instead of the initial conditions of the simulation being arbitrarily set, the initial conditions are taken from the state of actual conditions in the physical environment. At every new time zero ($t_0$), FRS simulates future states and attaches probabilities to those future states. The future states that are the concern of FRS are states of adverse events. Adverse events are events that waste resources hindering or preventing us from completing our task goals.

There are two versions of FRS. The first is FRS using inputs from only the physical product itself to predict future states. The second is that FRS uses inputs from the physical product itself and the environment to predict future states. FRS acts as crystal ball into the future.

The specific conditions of FRS are:

- A simulation that contains behavioral assumptions of a corresponding physical product's future states based on physics causality and/or data probabilities, usually Bayesian based.
- The initial conditions of each simulation at $t_0$ are taken from the current state of an object in the physical world and, optionally, the environment that surrounds it.

## 2.4 Value of Simulation

Humans and non-human life, which for wont of a better term we will call "nature", have two different approaches to existence. Nature tries all possible combinations and lets the environment select the winners. Nature can do this because its only goal is survival of the fittest, and it has effectively an unlimited time horizon and resources. Nature also does not care about individual living organisms.

Humans, on the other hand, do care about individual living organisms, especially their own. Humans do not have unlimited time horizons and resources. Humans also have other goals besides survival. The human approach is to be task goal-oriented and to accomplish that task using the minimum of physical resources, time, energy, and materials.

Given that Is the case, the human approach means that for a goal-oriented task, humans employ sophisticated thinking capabilities utilizing data, information, knowledge, and wisdom (DIKW) [21]. This relies on the fact that the expended physical resources used to perform the goal-oriented task can be divided up into two categories. These two categories are shown in the left bar in Fig. 1 [22].[2] The lower part of the bar is the minimum amount of resources that is ideally required to

[2] The figure in the book had a third category, Execution Inefficiencies, where we know what we need to do to eliminate the waste but don't have the technical capabilities yet to do so. That category was dropped in later versions in the interest of simplification.

complete the task. Everything above that is in the second category of wasted physical resources. We can apply a cost function to this in order to value the time, energy, material of the physical resources needed to accomplish the task (Fig. 2).

The right-handed bar shows the role of information in task accomplishment. The amount of physical resources necessary to accomplish the task in the most optimal fashion stays the same. Information cannot replace the minimum amount of physical resources necessary to complete the task. However, information can replace the information inefficiencies or wasted resources ($C_W(x)$) over and above that. For purposes of illustration, this figure shows information as replacing all the wasted resources, but for human endeavors, this will usually not be possible.

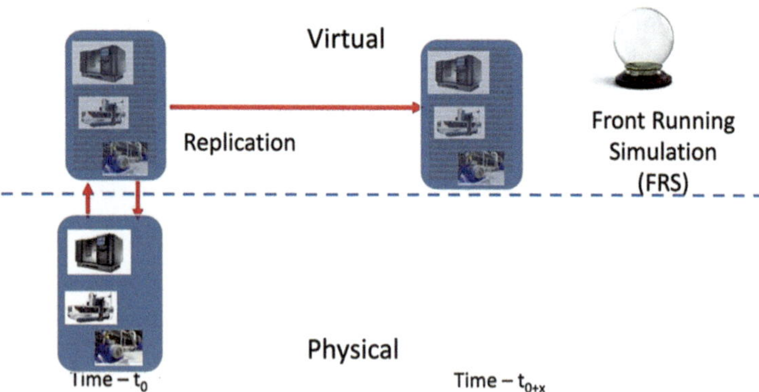


**Fig. 1** Front Running Simulation (FRS)

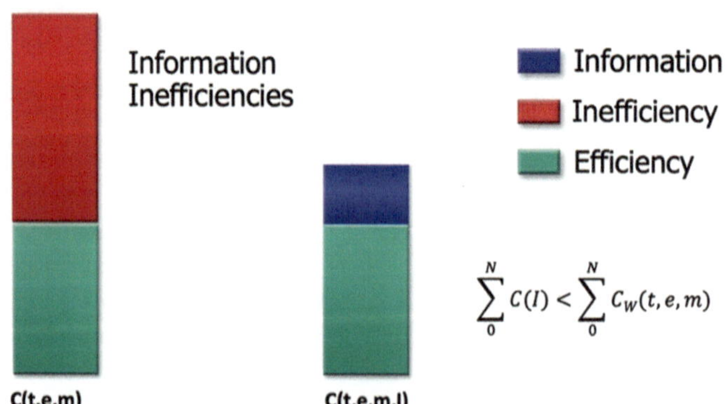


**Fig. 2** Information as time, energy, material trade-off

The issue with this costing function is that the cost of generating information from data[3] does not come in a unit of measure, like physical resources. We can use time and physical resources needed to develop this information as a proxy for a unit of measure. In the past, that proxy has always been human resources and physical materials. Today, the proxy commonly consists of computer hardware, software, and energy in addition to human resources. As represented in the formula, the assumption under which this model holds true is that the cost of developing this information is less than the cost of wasting resources over all the times the task is performed.

Simulation is a method of developing information. Humans have no interest, let alone the resources, in trying all possible combinations and letting the environment dictate which one is successful. Humans want to simulate the possible ways of obtaining their task-oriented goals and then perform the task using the method that minimizes their scarce physical resources of time, energy, and material.

Throughout all of human existence, until very recently, those simulations involved only the computational capabilities of human minds. The development and rapid advancement of digital computing bring a quantum leap in simulation value for human goal-oriented tasks. The more accurately predictive and cheaper that their simulations are, the greater value humans will obtain in completing their goal-oriented tasks while minimizing the use of physical resources.

## 3 Digital Twins

The concept of twins is ontological. We constantly categorize and compare things to see if they are similar, so we can make decisions about how we should interact with them. We expect to find similarities and regularities in our world. If we could not, the world would be a very lethal place [23]. "Twin" is the regularity that something is identical or nearly identical to something else, so that we can apply how one of the twins looks and/or behaves to the other twin.

The only requirements for a "twin" are that it has two key attributes: duality and strong similarity. That is there needs to be two of them, and they need to share significant attributes. There is no ontological or metaphorical requirement for timeline simultaneity, as in human twins, scale similarity, or for the precedence of one type of twin before another type of twin, e.g., a Physical Twin before a Digital Twin.

[3] There is much confusion about what is data and information. We have a functional perspective. Data is a fact or facts about reality and the input to create information: We collect data and process it to create information. Information is a replacement or substitute of wasted physical resources: We use information.

## *3.1 Replica Twins*

There are physical objects that are "twins" in the sense that they are simply independent duplicates. What we are interested in is 3D physical objects that are intended to represent a specific physical object.[4] Since the term "Physical Twin" (PT) refers to the physical object in the dyad of Physical Twin–Digital Twin, we will call these 3D physical objects Replica Twins (RTs). We will use Replica Twins (RTs) in the sense that there is a unique physical object and a replica physical object that can be at different scales and fidelities.

RTs, even if primitive, rudimentary, and abstractly shaped, have been used in human endeavors for all of human existence. RTs have been used at all scales, from small models to full-size replicas and all fidelities, from exact replicas to simple representations. While barely three-dimensional, prehistoric man sketched in the dirt with a stick a representation of a mammoth, the cliff that the mammoth needed to be driven over, and the positions of his fellow hunters. While military sand tables date from the 1800s, equivalents date back to ancient Greece military use and most likely before.

Architecture has used RTs from earliest times all over the world. RT artifacts have been discovered dating from at least 6000 BC. RT model making was prevalent in ancient Greece. The making of RT models to represent actual physical buildings existed throughout the world in all cultures [24].

The RTs were even dynamic and not simply static. Watch any movie about World War II. It will generally feature a table with a geographical map that people will move around representations of military and naval forces. As dispatches come in, people move these representations into different geographical positions so that commanders can assess and plan their next strategic and tactical moves.

But it wasn't simply scale models that were replica Physical Twins. Full-scale RT mockups have been created and used. Full-scale RTs have been used in military preparation as long as military engagements have existed. As noted above, the D-Day preparation included exercises on Scotland beaches that were the replica physical "twin" of Normandy Beach.

As discussed in the simulation section above, we could argue that since very early times plays, and then later, in the twentieth century, movies have used full-scale RTs in a form of simulation. Plays and movies have created exact replicas of existing physical environments and then "simulate" activity within those environments. For example, an exact replica of the US White House Oval Office appears in an innumerable number of movies and television shows.

For equipment and vehicles, RTs were used primarily in development. However, RTs were also used to resolve issues with operational equipment and vehicles. Airplane manufacturers used replica twins to recreate and troubleshoot reported

---

[4] Obviously, there have always been 2D representations of physical objects, such as sketches, drawings, blueprints. However, we wouldn't call them "replicas", as humans must do much mental work to visualize them even poorly as three dimensional. They are more accurately described as abstractions.

problems with their airplanes in the 1930s. When problems were reported with automobiles, it was standard operating procedure for the engineer working on the problem to find an identically configured automobile to try and recreate the problem.

As we developed electronics, we could make these RTs dynamic on their own. The company one of the authors worked for in the 1970s, Lear Siegler Corporation, had an F-16 flight simulator in its Grand Rapids Instrument Division. It was physical, not digital. However, dynamic flight simulators date back to the 1930s with the Link simulator [25]. There have been dynamic replica Physical Twins of nuclear reactor control rooms for training and emergency exercises for over 50 years [26].

The Apollo program is often cited as the first use of Digital Twins. That myth is still being perpetuated today.[5] The common reference is to the Apollo 13 mission, where the myth is that its "digital twin" was used to bring the crew safely back home after an almost catastrophic malfunction.[6]

The reality is that the digital capabilities of the most powerful computer mainframes of the Apollo days were extremely limited compared to today. Main memory of the most powerful mainframes of the era was in the 16 MB range. The Command and Lunar Landing Modules had a miserly 2K of main memory. The extensive troubleshooting on earth was done with a series of RT capsules that had no "digital" aspects.

Replica twins have been in existence throughout humanity's history and are still in use today. Replica twins have been abstracted in such representations as dirt sketches and sand tables. Replica twins have been realistic scale models such as buildings and even cities. Full-scale replica twins have been used to prepare for and track military engagements. Replica twins have existed dynamically as in the Link, F-16, nuclear reactor control rooms, and the Apollo space capsule simulators. The advent and rapid development of digital computers enabled the logical next step of moving "twins" from physical replicas to digital ones.

## 3.2 *The Rise of Digital Twins*

Digital Twins are a twenty-first-century concept. While considered a possibility since the early days of digital computers [27], it isn't until the 2000s that a cohesive model and concept were proposed.

There were two major capabilities that DTs were intended to have: replication and prediction. Replication is the characteristic that the DT would possess all the data of

[5] This can be independently verified by doing a search of academic papers using the keywords "digital twin" and "Apollo".

[6] Apollo 13 might be the most amazing malfunction recovery story ever. One of the authors had the privilege of meeting the Apollo 13 Commander, James Lovell, and hearing first-hand the amazing story of two astronauts sitting on what was basically a couch in Apollo 13, lining up the earth's meridian vertically and horizontally perfectly on a reticule etched on Apollo 13's window so that they didn't burn up or bounce off into space at re-entry. However, the "twin" involved in working the problem on earth was a replica twin capsule simulator, not a digital twin.

its physical counterpart needed for use cases. At its ideal, any data that could be had while in physical proximity of the physical counterpart could be obtained from its DT. The characteristic of prediction is that the DTs would causally or probabilistically predict the future states of their physical counterparts.

The origin of the Digital Twin model is well documented in a multitude of academic papers and industry articles [28]. Figure 3 is the first version of the Digital Twin model that was presented at a Society of Manufacturing Engineering (SME) conference in Troy, Michigan in October 2002. The presentation was entitled Completing the Cycle: Using PLM Information in the Sales and Service Functions [29]. It was about using Digital Twins in the operational and support phase of the product lifecycle when there was both the physical product and its Digital Twin.

The model in Fig. 4 was refined a little later that year to emphasize that products existed in real and virtual spaces. This version was for a meeting of industry executives, automotive software providers, and academics from the University of Michigan. The meeting was to explore setting up the Product Lifecycle Management Development Consortium (PLM DC) at the University of Michigan. Because of the automotive industry attendees, the focus was on different product lifecycle phases than the SME conference, namely engineering and manufacturing.

Both presentations were about the new discipline that was being defined, Product Lifecycle Management (PLM). As a result, the model did not even have a name, as it was simply entitled "Underlying Premise of PLM". It did describe that the model's

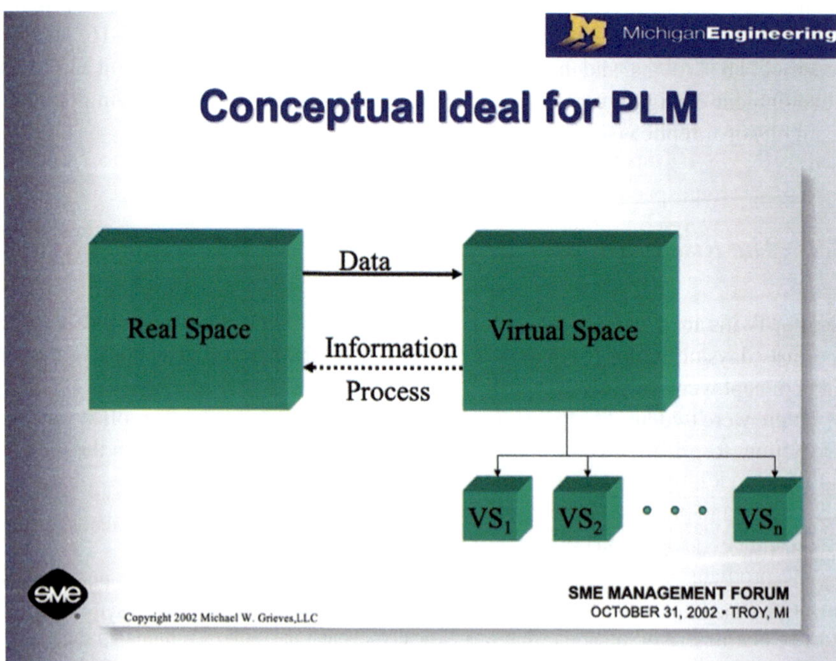


**Fig. 3** Original Digital Twin model

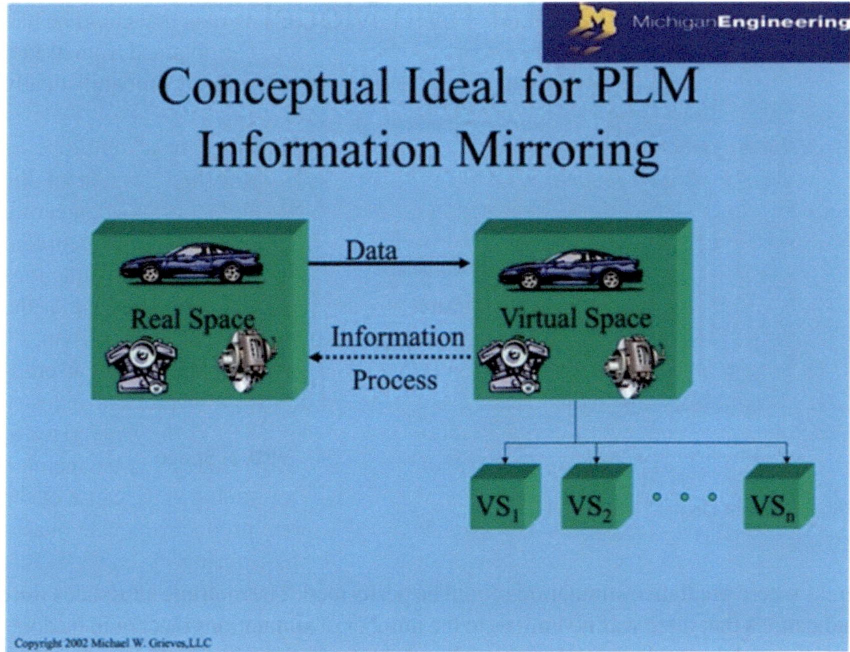


**Fig. 4** Original Digital Twin model version 2

purpose was "Information Mirroring". Even though it was rudimentary, the model contained the major elements of the Digital Twin that exist today.

On the left side were physical products in physical space. These are the Physical Twins (PTs). On the right side were virtual products, which we now refer to as Digital Twins, in virtual space that corresponded to the physical products. The third element is that there were communications between the two spaces and products, with data from physical space and products obtained from sensing and IoT devices populating the virtual space and products, and data and information coming back from virtual space and products to be used in the physical space.

These models also contained the sub-spaces as part of the virtual space, $VS_1$, $VS_2$, $VS_3$… VSn. The idea of virtual spaces was fairly new at the time, so this was to emphasize the fact that while there was only one physical space, there could be an unlimited number of virtual spaces.

This model highlighted that there were two main functions that it implemented: replication and prediction. Replication is the characteristic that the DT would possess all the data of its physical counterpart. The products in physical space were replicated by the products in virtual space. At its ideal, any data that existed in the physical product was replicated in its Digital Twin.

The characteristic of prediction is that the DT would causally or probabilistically predict the future states of the physical counterparts. The subspaces were virtual

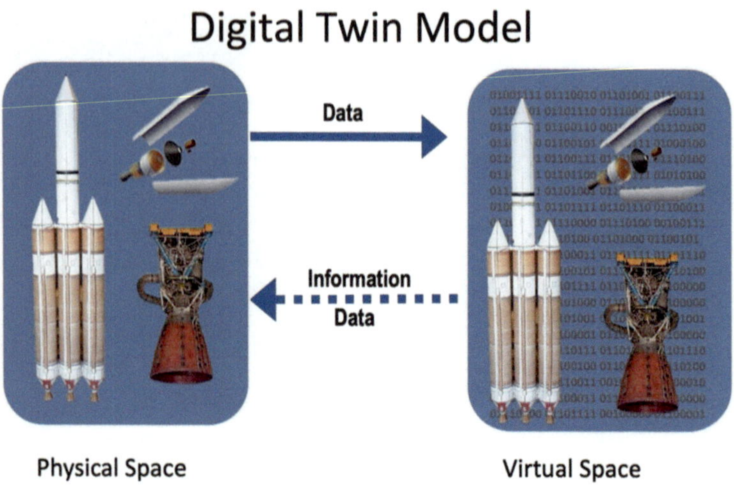

**Fig. 5** Current Digital Twin model

areas where predictive simulations could be performed. The multiple subspaces were indications that there was no limit as to the number of simulations that could be done.

The model was changed during work at NASA. The graphics are better, courtesy of NASA, when the model was used for a Department of Defense (DoD) conference [30]. In addition, the model was also simplified, as shown in Fig. 5. It was felt that sub-virtual spaces unnecessarily complicated the model. However, as we will explain later, the original model better represents a metaverse model.

While the model did not have a name originally, it did acquire some names shortly thereafter. It was called the Mirrored Space Model first, and then shortly thereafter that was changed to Information Mirroring Model. The Information Mirroring Model name remained albeit somewhat obscurely until around the 2010 timeframe. At that time John Vickers of NASA who was working with Grieves suggested the name Digital Twin. The Digital Twin name was a replacement for the relatively strange name that NASA was using, Virtual Digital Fleet Leader.

The Digital Twin name was mentioned in a footnote in Grieves' book on PLM [31], attributing the name to John Vickers. Grieves used the Digital Twin name in one of the seminal and highly cited Digital Twin papers [32], which noted that the Digital Twin name was going to be used for the model from then on. Later, in a short but highly influential piece, Grieves wrote an article for the Economist Magazine in the GE Lookahead section that was subsequently picked up by the World Economic Forum [33]. The Digital Twin was explained to the general audience in that short article.

2015 marks the beginning of an exponential growth in reference to Digital Twins in academic papers, industry white papers, and websites. The uses of DTs were initially proposed for aerospace and automotive. That has exploded to encompass a huge swath of industries and disciplines: power generation, heavy machinery, smart building/cities, oil and gas, ports and airports, archeology, and healthcare, just to name a few. Doing an internet search of "Digital Twins" in 2018 resulted in only one million hits [34]. A search of 2022 results in over 17 million hits. The number of academic papers on Digital Twins shows a similar exponential growth (2015—295 results, 2022—17,100 results).

As of today, the model in Fig. 5 is the accepted model of the Digital Twin. While definition may vary and vary widely, images usually show Digital Twin representations that are fairly consistent in the representation of physical space and products, digital space and products, and a two-way connection between them.

The commonly accepted Digital Twin Model that was introduced in 2002 and simplified to the one as shown in Fig. 5 consists of three main components:

- The physical products (PT) in the real-world environment.
- The Digital Twins (DTs) in a digital environment.
- The two-way connection between the physical and virtual for data and information.

The third component, the connection between the physical and digital has often been referred to as the "digital thread". The connection is a two-way communication connection. Data from physical products and their environment is communicated to the digital environment and populate their DT counterparts for collection, assessment, and response (CAR) [20]. Data and/or information (if action to replace physical resources is proposed) is now available to the physical environment.

The digital environment of the Digital Twin, referred to as the Digital Twin Environment (DTE), requires that it has rules that are identical as possible to our physical environment. We need to be assured that the behavior of the Digital Twin in the DTE mirrors the behavior of its physical counterpart for the use cases we require.

Finally, it is important to remember that when we refer to Digital Twins in a general way, we are implicitly including all three elements of the Digital Twin Model. We are not simply referring to the digital object that represents a specific physical object. While it would be more accurate to use the term, Digital Twin Model, we simply us Digital Twin.

## 3.3 Types of Digital Twins

There are three types of Digital Twins: the Digital Twin Prototype (DTP), the Digital Twin Instance (DTI), and the Digital Twin Aggregate (DTA).

The DTP originates at the creation phase of the product lifecycle. The DTP of a product begins when the decision is made to develop a new product, and work begins doing just that. The DTP consists of the data and information of the product's physical characteristic, proposed performance behaviors, and the manufacturing processes to build it. The DTP should also include the necessary processes and practices to ensure the product is fully supportable and maintainable in the field and to troubleshoot and repair the product effectively and efficiently to keep it operational. As much of this work as possible should take place virtually.

The DTI originates when individual physical products are manufactured. DTIs are the as-builts of the individual products and are connected to their corresponding physical product for the remainder of the physical product's life and even exist beyond that. The DTI implements replication. Much of the DTI can simply be linked to the DTP. For example, the DTI can link back to DTP 3D model and only needs to contain the offset of exact measurements to the designed geometrical dimensioning and tolerance (GD&T).

Pre-production physical product versions that are called physical prototypes will have a DTI since these are actual instantiations of the developmental period. These DTIs should be put to the equivalent digital tests and evaluations as the physical prototype itself. Comparing the digital results to the physical results will increase the confidence on relying on digital testing when the product is moved to full rate production. Where there are significant deviations, digital testing can be improved to converge on producing equivalent results to physical testing, with the goal of dramatically reducing and even eliminating physical tests, except for a final physical validation.

The DTP will contain the manufacturing process, Bill of Process (BoP), and the parameters associated with the BoP. The DTI will contain any variations that occurred in actual production. For example, the DTP process may require heat treating within a temperature range. The DTI will capture the temperature that actually occurred. The data and information that is needed for the DTI will be driven by the use cases of the organization. The digital testing described above will make it possible to test digitally each DTI of its physical counterpart to enable a high level of confidence in the future performance of each individual product [35].

Because the DTI remains connected to its physical counterpart for the rest of that physical product's life, it will also contain the data from its operation. The DTI will contain sensor readings and fault indicators. Based on use cases, the DTI will contain a history of performance of state changes and resulting outcomes.

The DTA is the aggregate of all DTIs. The DTA contains all the data and information about all the products that have been produced. The DTA may or may not be a separate information structure. Some of the DTA data may be processed and stored separately for analysis and prediction purposes. Other DTA data may simply be mined on an ad hoc basis.

The bigger the population of DTIs, the more data that will be available to improve Bayesian-based predictions. The DTA, which consists of the DTIs of physical systems, is subject to model bias for its predictions. However, there are mathematical techniques available for bias identification in DTs [36]. The DTA will also be the source for Artificial Intelligence and Machine Learning (AI/ML) to predict expected performance.

In 2019, Grieves introduced the Intelligent Digital Twin (IDT) to explain the role that AI would have in both assisting Digital Twins in their performance and in dealing with the increasing system complexity and emergent behavior of products themselves [37]. The view here was that AI was not a replacement for humans but an augmentation for humans. IDT specifies four attributes for Intelligent Digital Twins as active, online, goal-seeking, and anticipatory.

The characteristic of anticipatory requires that the Intelligent Digital Twin can be constantly running simulations to look ahead into the future for its PT. That obviously means that FRS is a critical component of Intelligent Digital Twins.

### 3.4 Digital Twins and Simulations

As defined above, simulations are one process that imitates another process. In the case of Digital Twins, we require that the "process" we imitate is our physical universe. Our DT simulations need to have as perfect fidelity to the laws of our universe as we need for our use cases. The characteristics of materials and forces of our physical universe need to be imitated as closely as possible. Simulation is what is needed to implement DT prediction.

The one exception to adherence to the laws of our universe is the cadence of time. We are unconstrained by time [37]. In our physical universe, time is completely out of our control. We cannot go back in time. The only way to go forward in time is to wait for the next tick of the clock. Even then, we go forward only according to the set time. We cannot slow time down nor speed it up.

In digital spaces, we are time unconstrained and can completely control time.[7] We can run our simulations at any clock speed. We can computationally go years and decades into the future. We can also slow down time. We can break down actions that happen in split seconds in our physical universe into microseconds.

In digital space, we can even go back in time which we cannot do in physical space. In the physical world, we employ forensic methods to attempt to determine what happened in the past that resulted in the current present. In simulations with deterministic rules, we can usually reverse the arrow of time in the digital world. If we have been using traceability to capture state changes, we can simply step back through the time frames.

However, another fundamental advantage is prediction, being able to advance the clock to see what's going to happen in the future. This is a crystal ball that sees into the future. With Front Running Simulation (FRS) described previously, we have the ability to do just that.

The assumption is that "simulation time and wall clock time can be kept in sync using conservative and optimistic synchronization protocols" [38]. At every new time $t_0$, we will be able to take the data from the physical world, i.e., replication, and predict at least probabilistically what will happen in the digital world. This is a tremendous opportunity to prevent the waste of physical resources by anticipating

[7] I discovered this firsthand in the early 1970s. Even though I was only a sophomore in college, I was a systems' programmer for a computer timesharing company. We charged by the CPU second. There was a meeting to discuss how to increase revenue. After listening to the staff provide ideas, I simply said, "I can increase what a CPU second is". That worked in increasing revenue until some customers started running benchmarks and complained that their programs were taking more elapsed CPU seconds and therefore were more expensive to run!

and correcting adverse events, especially ones involving the safety of individuals, before adverse events can occur.

### *3.5 Digital Twins, Simulation, and Information*

The purpose and value of DTs are that they provide information that can replace wasted physical resources. One of the three elements of the Digital Twin Model in Fig. 5 is the connection between Physical Space and its Physical Twins and Virtual Space and its Digital Twins. Data is sent from the physical products and optionally the physical environment to the digital DTIs. The information that is created and housed in digital space is used in our physical space. This information is created by processing the data coming from the physical space and, as noted above, by performing simulations.

This information can take a couple of forms. It can be the result of humans doing queries of DTs and creating information by using the result of the queries to take action that will replace wasted resources.

For example, a certain model fuel pump of a helicopter is being recalled and replaced. The traditional method would be to identify the location of every helicopter, send a mechanic to inspect each one, and replace the fuel pumps in the helicopters with the defective ones. With DTs, a query would be run on all the DTIs. Mechanics would only be sent out to only those helicopters with defective fuel pumps. That results in information replacing the wasted physical resources of mechanics' time and expenses for the helicopters with fuel pumps that need no replacing.

Information can also be created by routines that run on a constant basis in digital space looking for specific sensor data patterns that, using physics and DTA data, simulates and predicts adverse events and alerts humans to them. Humans have the responsibility of deciding the actions to take to avoid wasted physical resources by having an adverse event occur. This is a human-in-the-loop version.

The information can also be routines that run in digital space that specify actions to take when they find certain conditions. These actions are coded as commands and sent to the PTs directly, without human intervention. FRS is applicable in both simulation-based situations. All are the result of data coming from the physical world to digital world and data and information returning from digital world to the physical world to replace wasted physical resources.

## 4 Metaverse

"Metaverse" is a portmanteau of "universe" and "meta". "Universe" would commonly be thought of as our one and only ("uni") physical universe. "Meta" is having aspects or capabilities that transcend or are beyond the ones our universe possesses.

Like Digital Twin, "Metaverse" is a twenty-first-century term. Although "metaverse" has its origin in the last decade of the twentieth century as described below, "metaverse" only starts to be prominent in the last few years. The metaverse, virtual spaces, virtual worlds, digital spaces, and digital worlds have been used fairly synonymously. All these terms refer to virtual representations of the physical world. As explained in Product Lifecycle Management: Driving the Next Generation of Lean Thinking [39], humans have had virtual spaces since the beginning of their existence.

Humans have always had the ability to virtually represent the physical world in their own minds. However, those representations suffer from vagueness, impermanence, and an inability to share with other humans. The result was that whenever we developed a mental model or a new idea, it had to take immediate physical form in some fashion: a sketch, a drawing, a blueprint, or a physical model.

## *4.1 Origins of Metaverse*

The term, "metaverse" is almost universally attributed to originating in Neil Stephenson's book, Snow Crash [40]. Stephenson's metaverse is a singular space where an individual as an avatar interacts with other avatars in an immersive environment. The immersive environment was implemented by a Virtual Reality (VR) headset that the user wore.

This type of metaverse was almost exclusively about social interaction. This was described in a contemporaneous definition of "virtual worlds", which defined them as, "A synchronous, persistent network of people, represented as avatars, facilitated by networked computers" [41]. The virtual worlds and Stephenson's metaverse were synchronized with our physical world, could not be paused, and existed for the interactions of people as avatars.

There was some correspondence with our physical world in that it had three dimensions, and inanimate objects had persistence and impermeability. Avatars could move about in this three-dimensional space. There was a monorail that traveled the entire metaverse space and other vehicles were available, although how avatar locomotion itself was accomplished was never really explained [42]. Like Linden's Second Life, users could acquire virtual real estate where they could construct virtual buildings and other virtual artifacts [43].

Stephenson's metaverse had its own laws that loosely, if at all, followed the laws of our physical universe. For example, one can be driving a motorcycle at Mach 1 speed, run into a metaverse building, and the motorcycle and rider simply come to a stop with no damage to anything.

In the years after Stephenson's book and specifically within the last few years, metaverse has become synonymous with almost any and all digital spaces. This has led to the term "metaverse" applying to all digital spaces even if they are in contradiction to each other.

There has been discussion of an "Industrial Metaverse" that is vaguely defined as "the set of metaverse applications designed for industrial users" [44] [45]. The descriptions on an Industrial Metaverse almost always focus on technologies that users can exploit such as Augmented Reality and Virtual Reality. But there is a lack of defining what is required for such a metaverse. This needs to be much more concrete as we will discuss below.

## *4.2 Metaverse and Digital Twins*

The original view of the metaverse is substantially different from what is required for the purposes of Digital Twins. There is no reason that there cannot be all types of metaverses with different rules governing them. The rules can be completely arbitrary and have only a tenuous connection with our physical universe, as was the case for the Stephenson metaverse.

However, we would contend that we can legitimately co-opt the term because our metaverse required for DTs has more fidelity to the meaning of the words, "meta" and "universe". What we require is a digital fidelity to our physical universe with meta capabilities. The metaverse that we require needs to have complete conformance to the selected rules of our unitary, physical universe, although it can have meta capabilities as described below.

In order to move work from the physical world into the digital world successfully, the metaverse of Digital Twins needs to enforce all the laws of our physical universe required by the supported use cases. However, we need a metaverse that is specifically tailored to Digital Twins. From now on, we will refer to this type of metaverse as a "DT Metaverse" (DTM).

As humans perceive it, our universe is a three-dimensional space populated by objects. The DTM will have that same characteristic, a three-dimensional space populated by DTs. This will allow humans to use their ontological understanding of physical space to understand and operate seamlessly in the DTM.

Figure 6 is an evolutionary model of the DT that has been proposed previously by Grieves [27]. The claim is that we are currently in Phase 2, the Conceptual Ad Hoc phase. This is the current state of DT evolution.

In this phase, "the Digital Twin is an entity that we conceptually create from disparate and even fragmented data sources. We use different existing sources to pull data from. We start building correlations and even causations of data source inputs to results. We build different simulation views and determine how well they map to reality. We start to put manual processes in place to pull the data from different sources, even if on an ad hoc basis, to create a Digital Twin view [27].

The next phase in the evolution of DTs is the creation of DT platforms. These platforms are envisioned to support multi-users and multiple DTs. DT systems are possible in Phase 2. However flexible Systems of Systems (SoS) where multiple DTs of completely different functions enter and exit will require platforms that support interoperability.

First will come platforms that support replication, Phase 3. Those platforms will be followed by platforms for prediction, Phase 4. While not referred to as "DT

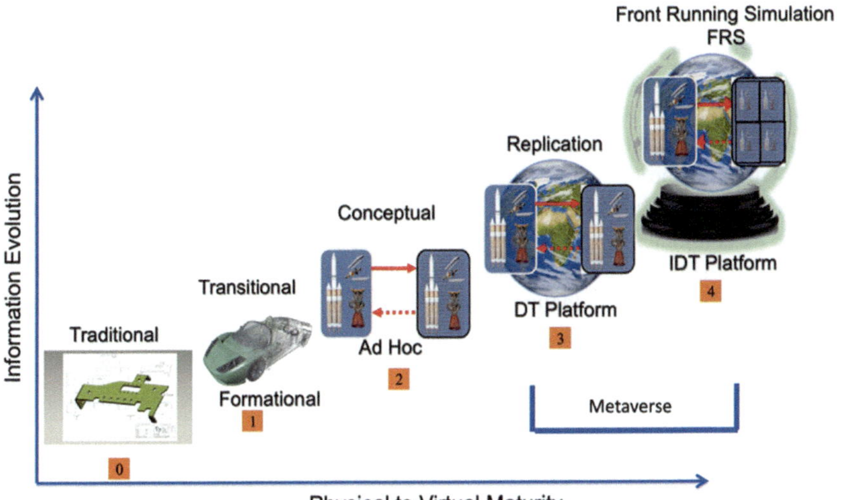


**Fig. 6** Digital twin evolution

Metaverses", that is in essence what these platforms are by our definitions here. These platforms now have a name, DT Metaverse. They are consistent with the long-term vision of DTs.

The DT Metaverse has these key characteristics:

- There are multiple DT Metaverses.
- DT interoperability is a core requirement.
- All laws of the physical universe are implemented and enforced in simulations for all inanimate objects.
- The DT Metaverse supports both replication and prediction.
- Multiple immersive participants as avatars is supported.
- Meta capabilities are allowed for human participants as avatars.
- Time can be synchronous or asynchronous with physical time depending on use case and DT type.
- Cybersecurity is embedded in all aspects of the DT Metaverses.

### 4.2.1 There Are Multiple DT Metaverses

The limitation of our physical universe is that there is only one. If the Many Worlds theory is correct and multiple physical universes exist [46], we will only ever have access to the single physical universe this version of us inhabits. In the digital universes, we can recreate our physical universe at some level. Since we have no physical restrictions on the number of computer-based spaces, we have no restriction or limitation of the number of digital metaverses that will be available to us.

As the original DT model proposed, we have effectively an unlimited N number of sub-virtual spaces. Ironically, the original Digital Twin model in Figs. 3 and 4 is a much more accurate model of the DT Metaverse than the simpler model in Fig. 5 that is evolved from it.

We expect that there will be many different metaverse platforms in the near future. They will be differentiated by different use cases, different DT types, and different phases of the product lifecycle. DT Metaverses can be on a spectrum of completely private to completely public. While we are restricting our discussion to DTs of tangible products, there is no reason that metaverses for intangible or process DTs such as supply nets, manufacturing processes, or financial systems cannot be developed. If we can visualize the data and information in some symbolic fashion, we can have a DT Metaverse for it.

### 4.2.2 DT Interoperability Is a Core Requirement

The concept of a DT Metaverse implies that there will be multiple DTs in it. Otherwise, the DTM would be no different than what we have today, which is a Phase 2 Conceptual Ad Hoc programming space. The ability to easily insert a DT into a digital environment to monitor its interaction and behavior with other DTs will greatly enhance their value and usefulness. Interoperability also implies intra-operability as this will also foster component modularity within DTs.

This means that the DT Metaverses are platforms. Platforms are hardware and software infrastructure that provides underlying tools, services, and governance to accomplish participative specific tasks and interactions [47]. Participative implies that there is intended to be multiple users and multiple DTs. So, this means that the platforms need to enable the interoperability of DTs.

There are numerous organizations working on mechanisms for interoperability. A common perspective for interoperability is to produce standards and ontologies. While this works in the physical world, it is much more difficult in the digital world because of the much finer granularity of data and information. As a result, to address this, we have a multitude of standards, which means that we really don't have "a" standard.

These other mechanisms that platform may employ include defining ad hoc programming conventions, harmonization of programming conventions among software providers, and a platform's own middleware. A promising solution may be to deploy AI. AIs may be able to explore the solution spaces between different DTs and provide mapping and translations. As Fig. 7 illustrates, AI may provide both intra-operability for components of a Digital Twin and interoperability for different DTs in DTMs.

Depending on the DT type and use case, the DT Metaverse will be useful for its ability to support immersion for multiple people assessing the serviceability of a new product. It may also be useful for staging and operating multiple products in simulation from different vendors in virtually commissioning a production line.

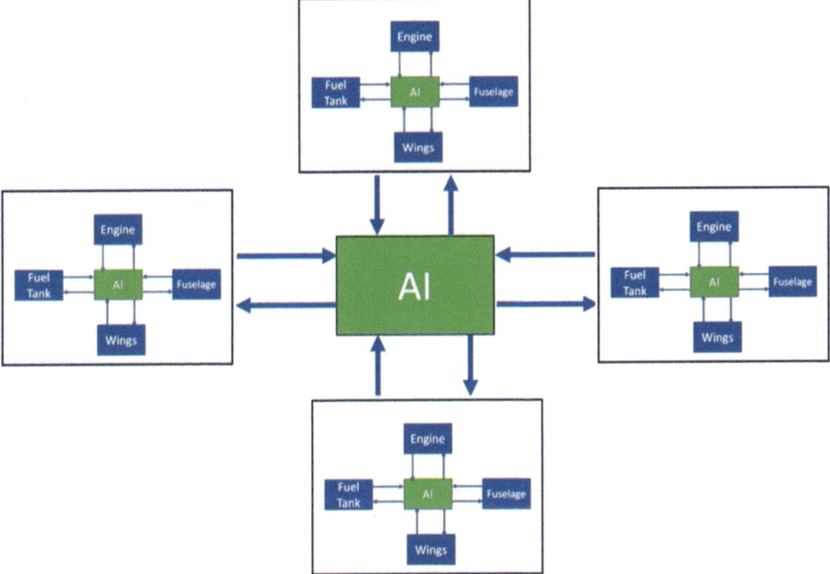


**Fig. 7** AI DTM interoperability at the system of systems level

### 4.2.3 The DT Metaverse Supports Both Replication and Prediction

There are two main characteristics that DTs need. Those characteristics are replication and prediction. Even if a product is being developed in digital space and has not yet taken physical form, the environment that the new product is going to exist in needs to replicate our physical universe for the use cases needed.

When we have DTIs of existing physical products, we need to replicate the data from those PTs that we require for our use cases. The DT Metaverses supporting this will be on a spectrum. On one end, there is simply the DTI in empty space that can be simultaneously and instantaneously interrogated no matter where its PT is in the physical world. On the other end of the spectrum, we can have DTIs and the surrounding environment it currently resides in that an immersive avatar can experience.

We need prediction both when we are developing the product and then when we have a product that is in operation. We need to predict what forces that the product produces will have on its own structure, its operation, and its surrounding environment. We also need to predict the impact that outside forces we'll have on our product. Prediction is done via time-evolved simulations and can be done based on hybrid physics/data probabilities techniques. Again, the DTM will be on a spectrum from simply having the DTI representation of PT itself to having the DTI representation of PT and its current surrounding environment. This gives us the ability to be time and space unconstrained with a probabilistic window into the future.

### 4.2.4 Laws of the Physical Universe Are Implemented and Enforced in Simulations for Inanimate Objects

The primary requirement of the DT Metaverse is that inanimate objects will behave as they would in our physical world. In DTM simulations, if, as described above, a DT motorcycle traveling at Mach 1 slams into a stationary object like a building, that will generate a tremendous amount of force as defined by $F = M * V^2$. That resultant force will deform/destroy the motorcycle, the building, and the surrounding location.

Since the physical world is a tremendously complex place, it may not be feasible to enforce all the physics of our universe. DT Metaverses will need to establish the parameters of their adherence to our universe's physical laws so that users will understand what they can and cannot rely on. DT Metaverses will need to define the use cases that they support.

### 4.2.5 Multiple Immersive Participants as Avatars Is Supported

The DT concept has always been intended that there be a social aspect in terms of having many people collaborate in a shared environment [32]. However, it was originally intended that people be outside observers. When there is immersive capability with people having avatars, we will need to carefully define and design the meta capabilities.

What seems to differentiate metaverse from virtual environments is that metaverses enable immersion and host multiple participants. Immersion can be done first person using avatars and Virtual Reality (VR) and Extended Reality (XR) capabilities. The metaverse is a 3D geographic space that contains visual DTs that can be examined and optionally operated by the avatars. The requirement for avatars is important for knowing what participants are present in the metaverse and to know who is operating the equipment, so multiple avatars don't try and engage with DTs simultaneously.

The DTMs will allow for multiple participants who have independence within the metaverse. Participants are not constrained to a certain view but can move around the metaverse independently of other participants.

This does not mean that it requires an immersive participant to use and obtain value from the DT Metaverse. Authorized outside observers can have access to a DT Metaverse and benefit from both replication and prediction data. Whether or not observers can be active participants in a DT Metaverse will depend on the rules employed by that DT Metaverse.

#### 4.2.6 Meta Capabilities Are Allowed for Human Participants as Avatars

While it is foundational that our DT Metaverse follows the laws of our physical universe, it does not preclude having meta capabilities for our participants of the DT Metaverse. As described above, they can take advantage of the capabilities that simulation and virtuality afford. Participants are unconstrained by location if so allowed, meaning that they can "teleport" from one location to another. Solid structures can become transparent so as to see inside. As long as the inanimate objects are subject to the laws of our physical universe, human participants can have meta capabilities.

If the use case is that participant avatars are merely observers who simply want to experience the contents of the DT Metaverse as would be seen in the physical world such as the layout of a factory and its equipment, we can allow them meta capabilities such as walking through solid objects. However, if the use case involves people interacting with the DT Metaverse physical objects, their avatars need to be subject to the physical rules like every other inanimate object. For use cases such as physical factory walkthroughs or equipment training, avatars should not have meta capabilities. That would defeat the purpose of these types of use cases.

#### 4.2.7 Time Can Be Synchronous or Asynchronous with Physical Time Depending on Use Case and DT Type

Another major difference with DT Metaverses from Stephenson's metaverse is that these DT Metaverses can be asynchronous and non-persistent with the physical world. The original metaverses were synchronous and persistent. The time in the metaverse was the exact same time as in the physical world and moved at exactly the same cadence. The metaverse was persistent in that if people left the metaverse, the metaverse continued without them.

A claim about metaverses is that persistence and synchronicity with the physical world are a requirement for all metaverses [48]. This would be a requirement of a DT Metaverse if the use case is real-time replicative and multiple PTs corresponding to their DTIs exist in the physical world. However, the ability to internally control time is a key advantage and capability of the DT Metaverse for some DT types and use cases that involve prediction of future states.

As discussed above, depending on the DT type and use cases, time can be either synchronous with physical time or asynchronous. However, time must be persistent in either case. This means that time within a DTM must be the same for all entities inside. If time speeds up or slows down, it must speed up or slow down for all inanimate objects. Human participant avatars will simply be brought along as observers with the time as a "fast forward" or simply with a discontinuous jump.

However, as noted above, if we are not time synchronous with the physical world, we are time unconstrained. We cannot only go forward in time. We can reverse time's arrow and go backward. Again, there must be persistence for all DTs with human avatars participating as described above.

#### 4.2.8 Cybersecurity is Embedded in All Aspects of the DT Metaverse

It is not possible to do justice to cybersecurity as a chapter topic, except to highlight its importance. A fortress approach with a remote physical location, impenetrable walls and fences, and armed guards to secure physical products is feasible. A fortress approach is not feasible for DTs and DTMs.

That makes cybersecurity the highest risk for the adoption of DT Metaverse capabilities. The approach to DT Metaverse security needs to have three characteristics: being embedded in all aspects, not just access, continuous monitoring of all participants and interactions, and driven by extreme paranoia. If cybersecurity cannot be guaranteed, DTMs cannot succeed.

## 5 Digital Twins, Simulation, and the DT Metaverse

The spaces below the DT Metaverse, indicated by $VS_1$, $VS_2$, $VS_3$ … $VS_{n,}$ are virtual/digital spaces, "sandboxes" is the term usually used, where different simulations of our DTI could be performed. The initial conditions from the DT Metaverse would all be the same, but other conditions and assumptions would be varied to understand what different outputs would result. The main space and the sub-virtual spaces were what was referred to above as the Digital Twin Environment (DTE) above.

In this section, we will put together the concepts we have described above.

### *5.1 DT Types and Use of Metaverses*

As stated above, the DT Metaverse will not be a singular entity. There can and will be different DT Metaverses, depending on where we are in the product lifecycle, what DT types we are working with, and what use cases we need.

In terms of capabilities, as described above, the DT Metaverse must twin the physical universe consistent with our physical laws. The DT Metaverse needs to reflect what the physical world does. It needs to have persistent physical atom-based material characteristics. This is called coherence that the physical world enforces through atoms but needs to be specifically programmed into digital worlds [22]. For example, if we lengthen a steel beam, it has to not only lengthen the beam, but it must add physical weight.

It must have persistent forces active upon atom-based materials. If a force acts on a particular material, it must do it consistently across all the materials in the DT Metaverse. We cannot have forces acting one way on part of the DT and in a different way on a different part of the DT.

We must have persistent time in each DT Metaverse. All the objects of our DT Metaverse must have the same time basis. We cannot have some DTs speed up in time and some DTs lag in time. There must be a persistent time basis in the DT Metaverse. This requires Digital Twin interoperability. As we move different DTs into a DT Metaverse, they all have to participate as if they were physical objects in the physical universe. Interoperability of different DTs is a major issue that we need to address.

### 5.1.1 Digital Twin Prototype

The DT Metaverse of the DTP will need to support both replication and prediction. In this case, since we are developing a new product, the replication will be of the environment that our product is going to operate in. It will be a synthetic replication that we will put our new product through its paces in. The primary focus of the DT Metaverse in this lifecycle phase will be prediction of behavior over time.

We will want to predict the suitability for purpose and behavior performance of our new product under a variety of conditions. We want to predict our product's performance throughout its entire lifecycle. As has been specified, we will want to create the product virtually, test the product virtually, manufacture the product virtually, and support the product virtually. Only when we get the product as perfected as we possibly can do, we want to move around expensive atoms to build a physical version.

The DT Metaverse will need to adhere to all the laws and rules implemented in our physical universe for the use cases it supports. We want to test our products under a range of conditions to see that they can measure up. We will want to put a new airplane through years of missions that it will encounter. We want our pacemaker to encounter a wide range of patient conditions and react with the required response accordingly.

We want to manufacture a product on a virtual production line. What we want to do here is reflect the movement of materials, machines, and people. We will explore different Bill of Processes (BoPs) to determine the "leanest" version. We will look at the ergonomics of the factory workers to ensure that there are no health and safety issues. As a byproduct, we can do our virtual commissioning of new production facilities during this phase.

In the support/operation phase, we will want to insert problems into our DTP and see how we troubleshoot that particular issue. We want to disassemble and reassemble to ensure that it will be easy and convenient in the field. We want to see how products interface with support equipment such as refueling systems, to make sure that it has been designed appropriately.

We want our DT Metaverse to support interoperability. This will allow us to take components from different manufacturers and make sure that they interface with each other. Interoperability also means that the physical laws are enforced on a coherent basis. When we fill the fuel tank in the wings of an airplane, the plane fuel gauge will register the increase in fuel, the wings will deform appropriately for the amount of new mass they have in it, and the weight of the entire airplane will increase.

Humans will be able to participate via avatars. They will be able to walk around the product, inspect areas of it, open doors and latches, and even operate the product. For an airplane, they can get into the cockpit and fly the airplane through a mission. Using

VR and haptic equipment, they can bring in a refueling hose, connect it to the refueling port, and feed fuel into the airplane. They can participate on the manufacturing line and move parts from one place to another. They can operate and change settings on the digital machines that are producing the parts.

These avatars can have meta capabilities, so they could jump from one place to another or see through outside structures to see what's underneath the surface coverings. Avatar participants will have an indication as to whether time is synchronous or asynchronous, so they can understand whether the simulation is being sped up or slowed down.

### 5.1.2 Digital Twin Instance (DTI) and Digital Twin Aggregate (DTA)

We will consider the DTI and DTA at the same time, since the DTA is made up of DTIs. When we move to looking at the DTI and DTA type, the DT Metaverse requirement changes. The DT Metaverse is now driven by a Physical Twin. The PT is at the mercy of our physical universe clock. PTs can only move forward in time as fast as the clock ticks.

Even though there is a unique DTI for every PT object, there is no reason that the DTI cannot participate in multiple DT Metaverses. There may be different use cases that are involved that the different DT Metaverses are concerned with. What will be true for replication use cases is that in each of the DTMs, the DTI in that DTM will be identical at times that are synchronous to our physical universe to the same DTI in other DT Metaverses.

The DT Metaverses for the DTI and DTA will support both replication and prediction. At this phase in the lifecycle, there is actually a PT to replicate and the physical environment that the PT lives in. Depending on the use cases, we need to populate the DT Metaverse with both the PT data and the environmental data.

Because we have sub-virtual spaces, we can engage in prediction. This is where FRS comes in the play. At every time zero, we can run a simulation in the sub-virtual spaces to predict the future of individual products or even a group of products or products as a whole. We can use the Digital Twin Aggregates to provide Bayesian-based probabilities. For purely deterministic situations, we can simply predict outcomes, such as predicting that unless a ship reduces its speed immediately that it will be unable to stop in time and will crash into its berth. In other cases, we may provide probabilities of different outcomes.

As noted above, for real-time replicative use cases, DT Metaverses will require that time will need to be synchronous and persistent. However, that will not preclude use cases where it may be advantageous to drop out of real time in replication and slow the happening of an event. This will give humans the ability to assess all that is happening at a slower rate to make better decisions. Better decisions mean reducing wasted physical resources to accomplish the goal by selecting the best future actions to take.

Obviously, our DT Metaverse will need to follow the laws and rules of our physical world. While prediction of the PT itself is important, the impact that this PT will have on the external world is also important.

DT interoperability will also be important in this DT Metaverse. In a city-oriented DT Metaverse, DTIs of various vehicles could participate. We would want to have them interoperate with DTIs of roads, traffic signals, and other aspects of the environment. This is so that we could have accurate predictions of what would happen in this particular DT Metaverse as different vehicle DTIs moved about.

We could use multiple participant immersive DT Metaverses. We might have a factory machine that is having problems. Experts from all over the world could congregate in the DT Metaverse of the factory to troubleshoot the problem and recommend a physical fix. These experts could adjust parameters within the DT Metaverse of the DTI machine that would then send those commands to the actual machine itself. This is in essence merging the physical and digital worlds.

Human participants could have meta capabilities. So not only could they be looking at the DTI of the machine, but they could also look within the machine to see certain elements of it so that they had a better understanding of what was transpiring.

Obviously, the machines would need to be instrumented to provide this data, but that is fully within the realm of possibility. While dealing with a PT, there is no other choice but to be synchronous. However, these avatars could step into one of the sub-virtual spaces to speed up the process to see what was going to happen and then transition back into the main DT Metaverse space in order to take action. This looks very much like the original Digital Twin model in Figs. 3 and 4.

## 5.2 *Artificial Intelligence, Simulation, Digital Twins, and Metaverse*

We need to consider the role that Artificial Intelligence (AI) will have in the evolution to a DT Metaverse. It has been asserted that we are moving into an era of Intelligent Digital Twins (IDT) that have the characteristics of being active, online, goal-seeking, and anticipatory [37]. Grieves claim for DTs and IDTs is that AI will be an assistance for humans and not a replacement.

It should be obvious that as we move into DTM platforms with major requirements for replicative and predictive analysis and simulation and ensuing adverse prevention that humans will need help. That help will take the form of AI.

In an above section, we discussed the difference between humans and nature. While humans are task goal oriented while attempting to minimize physical resources, nature tries all possible combinations via genetic mutations, letting the environment select for fitness.

While the habitat of nature is the physical world, the habitat of AI is the digital world or the DT Metaverses. Subject to computational limits and time constraints, AI has the ability to explore the solution space, test it against the criteria for obtaining goal success. AI can then select those solutions that meet or exceed the success criteria.

In order to evolve to Phase 4 with Front Running Simulation, we are going to need AI capability. FRS will require massive amounts of data coming from DTIs on a constant basis. FRS will then require selecting the assumptions and parameters and running simulations of future outcomes. The probabilities of those outcomes will be calculated from the DTA and its causal and Bayesian probabilities.

However, it's simply not enough to have probabilities of adverse events. We will want to take action to prevent them from occurring. From the beginning days of DTs, a key characteristic was cued availability [22]. Cued availability is a digital assistant that assesses what is happening in context and offers information that replaces wasted physical resources. Clearly an adverse event is an event that wastes resources.

Cued availability is a role for AI. Some adverse events will have obvious solutions. It is obvious that the remediation for the prediction that there is a high probability of a component failure in few weeks is to replace the component before the predicted failure. Other predictions of adverse events will not have obvious remediations. An example is a robot glitching momentarily that will result in an assembly bottleneck later that day. AI has the ability to propose alternative preventative solutions for the different adverse events it predicts.

This is clearly beyond human capability, computability, and mental bandwidth. This will require the assistance of an AI agent with an enormous amount of computing capability at its disposal. We may delegate some autonomous operations to AI systems in FRS. However, we should always understand that AI should be an assistant to humans and not a replacement.

## 6 Conclusion

As we have emphasized, we are in the twenty-first century. However, the movement of work into the digital arena as described here is so dramatic that we may want to refer to this as the 3rd millennium to call out how different we are from previous eras that did not possess the computational and communication capability we now have at our disposal. This century/millennium has seen the rise of technological capability that did not exist or was immature prior to 2000. This includes technologies that we are discussing here such as Digital Twins, photorealistic simulations, hybrid physics/big data models, and AI.

While we have always had simulations and Replica Twins, the advances in computing and communication technologies have enabled us to move work into the digital world and use the information developed there to replace wasted physical resources in our goal-oriented tasks. We can use Digital Twins to replicate their physical counterparts so that we have simultaneous and instantaneous access to their data, develop that data into information, and then predict their behavior via simulation.

The idea of a DT Metaverse will allow us to continue to evolve this capability as we move to Phase 3 and 4 DT platforms. However, the DT Metaverse will need the specific characteristics that we've outlined, such as fidelity to our universal laws. Having our products employ Front Running Simulations will be our crystal ball into the future, with the opportunity of using AI to prevent adverse events.

This may seem fantastical until we realize that it was only fifty years ago that computer chip capacity was measured in thousands of bytes and that we recently passed 100 billion bytes on a chip, with more exponential increases to come in the next decade or so. With that in mind, the advances proposed in this chapter seem very reachable and promise to create substantial value.

## References

1. Lu H (2024) A framework for the credibility evaluation of equipment digital twins. In: Grieves M, Hua E (eds) Digital twins, simulation, and metaverse: driving efficiency and effectiveness in the physical world through simulation in the virtual worlds, Forthcoming. Springer, Heidelberg
2. Vickers J (2024) NASA's Michoud assembly facility: developing a factory digital twin. In: Grieves M, Hua E (eds) Digital twins, simulation, and metaverse: driving efficiency and effectiveness in the physical world through simulation in the virtual worlds, Forthcoming. Springer, Heidelberg
3. Malik AA (2024) Digital twins for robot systems in manufacturing. In: Grieves M, Hua E (eds) Digital twins, simulation, and metaverse: driving efficiency and effectiveness in the physical world through simulation in the virtual worlds, Forthcoming. Springer, Heidelberg
4. Boschert S (2024) On the importance of simulation and digital twin for industrial applications. In: Grieves M, Hua E (eds) Digital twins, simulation, and metaverse: driving efficiency and effectiveness in the physical world through simulation in the virtual worlds, Forthcoming. Springer, Heidelberg
5. Darrington J (2024) DThe first real-time digital twin of a nuclear reactor. In: Grieves M, Hua E (eds) Digital twins, simulation, and metaverse: driving efficiency and effectiveness in the physical world through simulation in the virtual worlds, Forthcoming. Springer, Heidelberg
6. Dong Y (2024) Evolution of simulation and digital twin in healthcare: from discovery to design and integration. In: Grieves M, Hua E (eds) Digital twins, simulation, and metaverse: driving efficiency and effectiveness in the physical world through simulation in the virtual worlds, Forthcoming. Springer, Heidelberg
7. Alexandridis K (2024) Distributed AI modeling and simulation for smart airport digital twin applications. In: Grieves M, Hua E (eds) Digital twins, simulation, and metaverse: driving efficiency and effectiveness in the physical world through simulation in the virtual worlds, Forthcoming. Springer, Heidelberg
8. Moradian A (2024) Digital twins for sustainable semiconductor manufacturing. In: Grieves M, Hua E (eds) Digital twins, simulation, and metaverse: driving efficiency and effectiveness in the physical world through simulation in the virtual worlds. Editors, Forthcoming. Springer, Heidelberg.
9. Mondoloni S (2024) The power and application of digital twins: a multidisciplinary perspective. In: Grieves M, Hua E (eds) Digital twins, simulation, and metaverse: driving efficiency and effectiveness in the physical world through simulation in the virtual worlds, Forthcoming. Springer, Heidelberg

10. Singh S (2024) Defining, exploring, and simulating the digital twin metaverses. In: Grieves M, Hua E (eds) Digital twins, simulation, and metaverse: driving efficiency and effectiveness in the physical world through simulation in the virtual worlds, Forthcoming. Springer, Heidelberg
11. Hartmann S (1996) The world as a process: simulations in the natural and social sciences. Modelling and simulation in the social sciences from the philosophy of science point of view. Springer, Heidelberg, pp 77–100
12. Tzu S (2000) Art of War
13. Allison D (2019) Scotland's role in the D-Day landings [cited 2023 August 23]. Available from: https://www.bbc.com/news/uk-scotland-48266481
14. Barzun J (2000) From dawn to decadence: 500 years of cultural triumph and defeat, 1500 to the present. HarperCollins
15. Schriber TJ (1974) Simulation using GPSS. Wiley, New York, xv, 533 p
16. Cass D, Shell K (1983) Do sunspots matter? J Polit Econ 91(2):193–227
17. Korb KB, Nicholson AE (2010) Bayesian artificial intelligence. CRC Press
18. Pearl J (2003) Statistics and causal inference: a review. TEST 12:281–345
19. Guo Y et al (2023) Digital twins for electro-physical, chemical, and photonic processes. CIRP annals manufacturing technology, Forthcoming
20. Grieves M (2019) Virtually intelligent product systems: digital and physical twins. In: Flumerfelt S et al (eds) Complex systems engineering: theory and practice. American Institute of Aeronautics and Astronautics, pp 175–200
21. Grieves M (2024) DIKW as a general and digital twin action framework: data, information, knowledge, and wisdom. Knowledge, Forthcoming
22. Grieves M (2006) Product lifecycle management: driving the next generation of lean thinking. McGraw-Hill, New York
23. Popper K (2014) Conjectures and refutations: the growth of scientific knowledge. Routledge
24. Pillsbury J (2015) Modeling the world: ancient architectural models now on view [cited 2022 September 7, 2022]. Available from: https://www.metmuseum.org/blogs/now-at-the-met/2015/modeling-the-world-ancient-architectural-models
25. New Link Trainer (1936) Aviation. McGraw-Hill Publishing Co, p 3
26. Furet J (1985) New concepts in control-room design: Worldwide, more attention is being given to improving the man-machine interface. In: Nuclear power and electronics
27. Grieves M (2023) Digital twins: past, present, and future. In: Crespi N, Drobot AT, Minerva R (eds) The digital twin. Springer International Publishing, Cham, pp 97–121
28. Holland S (2022)Michael Grieves, Father of Digital Twins, to headline at IOTSWC. In: iot insider
29. Grieves M (2002) Completing the cycle: using PLM information in the sales and service functions [Slides]. In: SME management forum, Troy, MI
30. Caruso P, Dumbacher D, Grieves M (2010) Product lifecycle management and the quest for sustainable space explorations. In: AIAA SPACE 2010 conference & exposition, Anaheim, CA
31. Grieves M (2011) Virtually perfect: driving innovative and lean products through product lifecycle management. Space Coast Press, Cocoa Beach, FL
32. Grieves M (2014) Digital twin: manufacturing excellence through virtual factory replication (White Paper). Available from: https://www.3ds.com/fileadmin/PRODUCTS-SERVICES/DELMIA/PDF/Whitepaper/DELMIA-APRISO-Digital-Twin-Whitepaper.pdf
33. Grieves M (2015) Can the digital twin transform manufacturing. Available from: https://www.weforum.org/agenda/2015/10/can-the-digital-twin-transform-manufacturing/
34. Brown C, Singer T (2019) Defining and utilizing the digital twin. In: Digital enterprise society
35. Grieves M (2023) Digital twin certified: employing virtual testing of digital twins in manufacturing to ensure quality products. Machines 11(808)
36. Andres Arcones D (2023) Evaluation of model bias identification approaches based on Bayesian inference and applications to digital twins
37. Grieves M (2022) Intelligent digital twins and the development and management of complex systems. Digital Twin 2(8)

38. Oren T et al (2023) Simulation as experimentation. In: Oren T, Zeigler BP, Tolk A (eds) Body of knowledge for modeling and simulation: a handbook by the society for modeling and simulation. Springer Nature, pp 77–119
39. Grieves M (2006) Product lifecycle management : driving the next generation of lean thinking. McGraw-Hill, New York, xiii, 319 p
40. Stephenson N (1991) Snow crash: a novel. Bantam Books
41. Bell MW (2008) Toward a definition of. J Virtual Worlds Res 1(1)
42. Joshua J (2017) Information bodies: computational anxiety in neal stephenson's snow crash. Interdisc Literary Stud 19(1):17–47
43. Rymaszewski M (2007) Second life: the official guide. John Wiley & Sons
44. Bohné T, Li K, Triantafyllidis K (2023) Exploring the industrial metaverse: a roadmap to the future. World Economic Forum and Univesity of Cambridge
45. MIT Technology Review Insights (2023) The emergent industrial metaverse
46. Carroll S (2020) Something deeply hidden: Quantum worlds and the emergence of spacetime. Penguin
47. Parker G, Van Alstyne M, Choudary SP (2016) Platform revolution: how networked markets are transforming the economy and how to make them work for you, 1st edn. W.W. NORTON & Company, New York, xiii, 336 pages
48. Richter S, Richter A (2023) What is novel about the Metaverse? Int J Inf Manage 73:102684

# Contrasting Capabilities: A Comparative Analysis of Simulation and Digital Twin Technologies

**Ana Wooley and Daniel F. Silva**

**Abstract** In the realm of advanced manufacturing, confusion often surrounds the distinction between Digital Twin (DT) and simulation due to their complementary nature. While they appear similar, they are fundamentally different concepts. This chapter considers capability-based frameworks to characterize and classify DT and simulation. Specifically, it uses the 4R framework for DT, from the literature, and proposes an analogous one (4S framework) for simulation. By comparing the definitions and levels of capability of DT and simulation using these frameworks, this chapter identifies similarities and differences in capability between these two technologies. Ultimately, this chapter aims to clarify these differences and provide a comprehensive comparison between the two, to prevent future ambiguity on this topic. The conclusion of this study is that differences in capabilities between the two technologies can be observed in data flows, real-time aspects, predictive timelines, optimization approaches, and level of autonomy.

**Keywords** Simulation · Digital Twin · Capability

## 1 Introduction

In the advanced manufacturing literature, there is often confusion surrounding the distinction between Digital Twins (DTs) and simulations. This confusion arises from their complementary nature, leading to the perception that they are similar or interchangeable concepts. However, it is important to recognize that DTs and simulations are fundamentally different. To address this issue, this chapter aims to clarify the differences between the two by exploring their individual capabilities and presenting a comprehensive comparison between simulation and DTs. By providing a clear

A. Wooley (✉)
The University of Alabama in Huntsville, Huntsville, USA
e-mail: ana.wooley@uah.edu

D. F. Silva
Auburn University, Auburn, USA
e-mail: dfs0008@auburn.edu



M. Grieves and E. Hua (eds.), *Digital Twins, Simulation, and the Metaverse*, Simulation Foundations, Methods and Applications, https://doi.org/10.1007/978-3-031-69107-2_2

comparison between DTs and simulations, this chapter emphasizes the synergistic relationship between them while emphasizing their distinct characteristics and capabilities, enabling informed decision-making and effective implementation of these technologies in various industries.

The chapter begins by defining the concept of a DT, introducing the 4R framework [1, 2] to characterize and classify DTs based on levels of capability. Next, it presents the definition of simulation and proposes an analogous framework to characterize simulation based on its levels of capability. This chapter highlights that a simulation can be based on reality, or it can be completely artificial, with the primary objective centered on examining how an object or system evolves and produces outcomes over time, using a predefined set of inputs. On the other hand, a DT fully replicates, in a virtual environment, a physical element that exists or will exist in the physical world, with bidirectional data flow connecting the physical and the virtual environments. The main result presented here is a discussion of the similarities and differences between DT and simulation, based on the capability-based classification frameworks discussed earlier.

In summary, the main conclusion is that simulation and DT share a common foundation of modeling and analysis but diverge in their objectives and advanced features. DTs, unlike simulations, offer real-time adjustments and autonomy, enhancing decision-making and optimizing physical systems.

## 2 Digital Twin Capabilities

The concept of DT has gained significant attention from both academia and industry in recent years [3, 4]. This surge in interest in DT is primarily driven by Industry 4.0 and advancements in technologies, such as the Internet of Things (IoT), Artificial Intelligence (AI), wireless sensor networks, Machine Learning (ML), and big data [5]. These technologies offer opportunities to seamlessly merge physical and virtual environments. The growing interest in DT has also led to an increase of publications about it [6, 7]. However, even academic publications about DT do not always agree on the what the definition, characteristics, and capabilities of DT are, or should be, resulting in a dilution of the DT concept [8–10], and causing confusion regarding its practical applications and advantages [11].

The concept of DT was originally introduced by Michael Grieves in collaboration with John Vickers from the National Aeronautics and Space Administration (NASA) [12]. Grieves's definition of a DT consists of three key elements: a physical space, a virtual space, and the data flow that connects these spaces bidirectionally. Since then, researchers have been trying to establish a unified and universally accepted definition that effectively captures the essence of a DT [6, 10]. Some researchers, as exemplified by [13–16], focus on DTs applied to individual products, where a DT represents a single machine or process. Conversely, others, including [17–21], consider DTs in the context of processes, encompassing either a segment or the entirety of the production environment.

In the community's journey to determine the conclusive definition and capabilities of a DT, some researchers, such as [22, 23], have presented concepts, definitions, and architectures related to DT, while others, such as [24, 25], have delved into emerging technologies relevant to industrial DT applications. Alternatively, [26] proposes that the DT consists of different types (DT Prototype, DT Aggregate, and DT Instance) for the entire product lifecycle.

Although certain authors define DT as a model that integrates both the physical system and its virtual counterpart with a continuous and bidirectional data exchange between them [21], numerous other definitions persist, characterizing DTs solely as accurate representations of the physical environment, disregarding real-time data connections [27]. To achieve a consensus on a DT definition, the specifications of fundamental requirements and capabilities for a DT are necessary. Furthermore, these requirements may change over time because they are linked to advancements in the technologies, such as ML and big data [27].

Several studies have focused on establishing a framework for DTs and explaining the interactions among various DT components [1, 2, 28–31]. Each of these works has attempted to provide a definitive methodology for characterizing DTs. While each of the proposed approaches has advantages, we have chosen to focus on a single taxonomy of DT throughout this chapter to provide a direct comparison between DTs and simulation models.

Specifically, we will use a robust, versatile, and systematic framework for characterizing DTs based on their capability and maturity levels called the 4R framework [1, 2]. This framework comprises four distinct levels of DT: Representation (R1), Replication (R2), Reality (R3), and Relational (R4). Any DT can be classified into one of the 4R levels based on its maturity and capabilities, which increase in each level. Additionally, each level within this framework outlines the developmental process of a DT for a given system.

Representation (R1): In the 4R framework, the first level marks the first stage of building a DT. It involves understanding the operations of the physical system and establishing a mechanism for gathering and storing data derived from the physical surroundings. At this level, data is connected and employed to construct a representation of the physical environment. This data serves the purposes of visualization, validation, and the control of the virtual environment.

Replication (R2): In this level, the emphasis shifts toward replicating the system within a virtual setting using the architecture created during the Representation phase. At this level, the DT has the ability to replicate identical outcomes when provided with identical inputs as those of the physical system.

Reality (R3): The third level is where the DT is deployed to explore what-if scenarios, with the goal of leveraging the insights gained from virtual models to enhance the performance of the physical system. At this level, the DT model possesses the capability to generate results based on a set of inputs, operating independently of the physical system.

Relational (R4): The last level describes a DT with the capabilities of integrating decision-making technologies by using AI or ML. At this level, the physical and

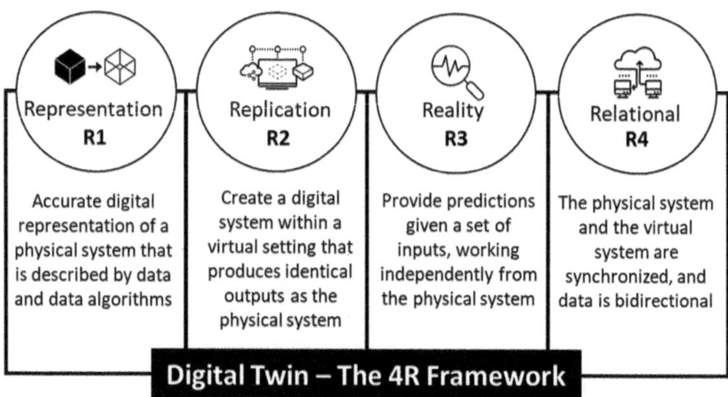


**Fig. 1** Summary of 4R framework. Figure adapted from [1, 2]

virtual systems are connected and the data between them is bi-directional. Additionally, the DT has a level of autonomy that is not present in the preceding capability levels, enabling it to autonomously adapt, identify optimal strategies, and fine-tune itself.

A concise overview of the DT capabilities is depicted in Fig. 1.

## 3 Simulation Capabilities

Simulation is a modeling technique that replicates the operations of a process or system over time, serving the purposes of describing and analyzing the system behavior and conducting what-if scenarios [32]. Simulation offers a host of advantages, providing methodologies and tools for intricately modeling system behaviors, facilitating predictions of future outcomes, and thereby supporting decision-making procedures [33]. It enables the testing of designs without committing resources to physical acquisition, as well as the ability to manipulate and accelerate time, which is relevant in system investigations [34].

Simulation also provides understanding of why things happen the way they do in the system and explores possibilities without the expense and disruption of experimenting with the real system [35]. In essence, a simulation model serves as a precise representation of a system or subsystem that fulfills specific criteria, frequently used to verify and analyze potential improvements prior to their actual implementation in a real-world setting [36].

The utilization of simulation technology is a well-established practice in the field of engineering [37], offering a wide array of simulation paradigms to choose from [38]. The selection of the most appropriate paradigm to apply when simulating a system depends on several factors, including the characteristics of the system, specific requirements that must be met by the simulation, and the desired outcomes of the analysis. The three most prevalent simulation paradigms are Monte Carlo, Agent-based, and Discrete Event Simulation (DES) [39].

In simple terms, a Monte Carlo simulation is employed to assess uncertainties and potential risks by means of probability distributions [40]. Monte Carlo simulation is often used to generate random variates that will be utilized in more complex analysis, such as Monte Carlo Markov Chain models.

In contrast, an Agent-based simulation models individual entities (agents) and their interactions with each other and the environment, facilitating the modeling of intricate systems with numerous interacting components [41]. In Agent-based simulations, agents act in a decentralized manner, following a set of prescribed (possibly randomized) rules that govern their interactions with other entities.

DES, on the other hand, permits the modeler to generate or observe specific events over time and it proves versatile for studying various system types and an extensive range of outcomes [36]. DES stands as the most used approach within the manufacturing domain [42].

Extensive literature exists defining simulation and offering detailed instructions for constructing simulation models. Many authors agree [35, 43–46] that simulation models provide the capability of conducting experiments and analyzing results. However, none of them provide a structured framework for characterizing and classifying simulations that would allow us to distinguish between the different levels of capability or that can be directly compared to other technologies (by comparing capabilities), such as DT.

While numerous frameworks and guidelines exist in the literature for constructing simulations, to the best of our knowledge, there is no framework that precisely delineates the different levels of capability within a simulation model in a manner that is comparable to a framework for DT, such as the 4R framework. To be clear, the capabilities of simulations are thoroughly documented in existing literature; therefore, the objective of developing such a framework is to consolidate these well-established definitions and integrate them into a comprehensive structure that can serve as a roadmap for understanding simulation models based on their capabilities, which is analogous to the existing 4R framework for DTs.

An analogous framework for evaluating simulation capabilities should provide a similar assessment of the level of simulation maturity in an organization. This would enable organizations to gauge their current simulation capabilities, pinpoint areas requiring improvements, and develop a plan for future simulation investments. Furthermore, an analogous framework would facilitate comparisons between simulation and DT capabilities, allowing organizations to better align their simulation and DT strategies.

In this context, we present a framework for simulation, which we called the 4S framework, that was initially introduced in our prior work [47]. It is designed to be analogous to the 4R framework for DTs, allowing for the classification of applications of both DT and simulation and facilitating comparisons between them. It is important to emphasize that there isn't a one-to-one correspondence between the levels in the 4S and 4R frameworks.

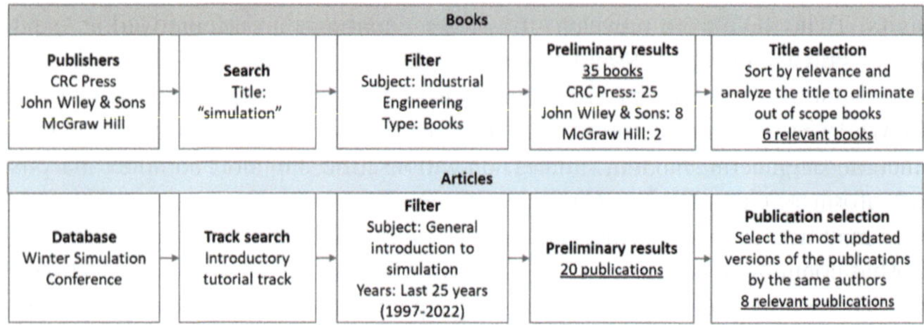


**Fig. 2** Systematic search for textbooks and articles. Figure adapted from [47]

To inform the 4S framework's development, definitions of simulation were gathered from research literature dedicated to simulation and its characteristics. Specifically, a two-pronged literature search was conducted: one for textbooks and another for scholarly articles. The selection process for the textbooks and articles used in this research is outlined in Fig. 2.

To gather pertinent textbooks, three prominent academic publishers were chosen: CRC Press, John Wiley & Sons, and McGraw Hill. These publishers are renowned for their expertise in producing textbooks and reference materials and enjoy a strong reputation for publishing content across a broad spectrum of academic disciplines. They hold a respected position within the academic community, particularly known for their authoritative publications in areas like engineering, technology, and computer science, which are directly relevant to the study of simulation. By focusing on these publishers, the likelihood of identifying relevant books that provide thorough insights and analyses regarding simulation and its attributes is significantly enhanced.

The first step involved conducting a search to identify textbooks featuring the term "simulation" in their titles. To maintain relevance to manufacturing and service system applications, this search was confined to books that listed "industrial engineering" as one of its topic areas. From these databases, CRC Press (25 books), John Wiley & Sons (8 books), and McGraw Hill (2 books), a total of 35 books were retrieved. Subsequently, a thorough and rigorous evaluation process was executed to select the most suitable books for this study. This process entailed sorting the results based on relevance and analyzing the titles of each book, with a focus on those directly providing simulation concepts.

The criteria for inclusion and exclusion were carefully designed to consider only books encompassing the general theory of simulation, such as handbooks, which incorporate definitions and characteristics of simulation. Any books explicitly mentioning a specific simulation type or application field in their titles were excluded to prevent any bias toward a particular definition of simulation associated with a specific simulation paradigm or application area. These criteria were established to ensure that the focus remains solely on the core theory of simulation and its fundamental principles, avoiding any influence from specific simulation types like DES or Agent-based simulation. Following the initial assessment of book titles, 19 books were found to not meet the inclusion criteria and were excluded. As a result, six textbooks [35, 36, 48–51] were identified as pertinent sources for gathering definitions of simulation.

To obtain relevant literature for the article search, the database of choice was the Winter Simulation Conference (WSC) proceedings. This decision was motivated by the prominence of WSC as a conference that consistently features a track dedicated to introductory tutorials, which cover the subject of simulation and its fundamental characteristics. These tutorials are typically authored by leading experts in the field, offering comprehensive insights into the concept and definition of simulation. Such an emphasis on definitions and concepts is not commonly encountered in journal articles, which often assume a pre-existing familiarity with the field and focus only on novel, recent developments.

A filter was applied to identify articles related to a general introduction to simulation, published within the last 25 years in the introductory tutorial track, resulting in the retrieval of 20 publications. It's worth noting that some of these articles were written by the same author in consecutive years. In such cases, the most recent versions were selected, as they often represented refined iterations of what had been presented the previous year. Consequently, eight pertinent papers were obtained [43, 45, 52–57].

The next step involved the aggregation and examination of definitions and attributes of simulation presented in the selected books and articles using text analytics. The definitions were transcribed verbatim into a spreadsheet. Next, each phrase within every definition was isolated into individual cells within the spreadsheet and characterized by identifying the action verb within each phrase.

By deconstructing the definitions into their component parts and clustering them based on common themes, six main characteristics of simulation were identified and labeled as C1 through C6. For instance, many phrases featured the action verb "understand," implying a focus on understanding a specific system or process. By grouping together phrases utilizing similar action verbs, a characteristic denoted as "Understanding how the system works" was established and designated as C1. Similarly, action verbs like "build" and "create" appeared frequently in numerous phrases, indicating an emphasis on simulation's capability to construct a model of a system. These phrases were categorized as C2, labeled as "Represent the physical system in a digital model." A comparable approach was employed to arrive at C3 through C6. A summary of these characteristics is provided in Table 1.

The subsequent step involved revisiting the books and articles to identify which of the characteristics (C1–C6) were explicitly mentioned in their definitions of simulation. A summary of this classification process is presented in Table 2, which illustrates that while certain characteristics are implied, some authors do not expressly articulate them in their definitions, as is the case with C1, C3, and C4. On the other hand, all sources explicitly state that simulation is a digital representation of a system (C2), it is employed to evaluate and predict the behavior of physical systems under varying conditions (C5), and it serves as a source of insights for decision-making (C6). Consequently, after an examination of the individual definitions pertaining to

**Table 1** Summary of the characteristics of simulation

| Characteristics |
|---|
| (C1) Understand how the system works |
| (C2) Represent the physical system in a digital model |
| (C3) Examine and analyze the system and its components |
| (C4) Draw inferences about the system |
| (C5) Evaluate and predict how the physical system will behave under different conditions |
| (C6) Provide insight into the system's optimal operation to support decision-making |

Table adapted from [47]

each characteristic (C1–C6), it was found that certain categories exhibited significant conceptual overlap. As a result, C1 and C2 were consolidated into a single category, and C3 and C4 were likewise combined.

The categories "Understand how the system works" (C1) and "Represent the physical system in a digital model" (C2) were combined into a single category because many authors combine understanding and representation of a system into one step [35, 36, 43, 49–51, 53, 54]. C1 emphasizes the need for understanding a system's behavior to create an accurate simulation model. This encompasses the identification of various system components, their interactions, and their interrelationships [58]. On the other hand, C2 involves the development of a digital representation of a physical system. This process involves constructing a simulation model that faithfully

**Table 2** Simulation definition sources classification

| | Ref | C1 | C2 | C3 | C4 | C5 | C6 |
|---|---|---|---|---|---|---|---|
| Books | [36] | × | × | × | × | × | × |
| | [35] | × | × | × | × | × | × |
| | [48] | | × | × | × | × | × |
| | [49] | × | × | × | × | × | |
| | [50] | × | × | × | | × | × |
| | [51] | × | × | × | | × | × |
| Articles | [52] | | × | | × | × | × |
| | [53] | × | × | × | × | × | × |
| | [54] | × | × | × | | × | × |
| | [43] | × | × | | × | × | × |
| | [45] | | × | × | × | × | × |
| | [55] | | × | × | | × | × |
| | [56] | | × | | × | × | × |
| | [57] | | × | | × | × | × |
| | | S1 | | S2 | | S3 | S4 |

Table ad apted from [47]

mirrors the characteristics and behavior of the physical system. To accomplish this, the simulation modeler must have an in-depth understanding of the behavior of the physical system, its components, and their interactions [33].

The categories "Examine and analyze the system and its components" (C3) and "Draw inferences about the system" (C4) were combined into a single category because both involve the process of analyzing a system to draw conclusions about its behavior. C3 entails the examination and analysis of various system components within the simulation model to understand how they interact with one another. This analysis encompasses the identification of relationships among different components and their impact on the overall system behavior. Similarly, C4 is centered around drawing conclusions about the system's behavior based on the outcomes generated by the simulation model. Upon the completion of the simulation, the modeler can analyze the results and draw conclusions regarding the system's behavior [59].

After combining similar characteristics, the resulting framework comprises four categories. To make it more comparable it with the 4R framework of DT, these categories are defined as the 4S framework of simulation: S1 (combining C1 and C2), S2 (combining C3 and C4), S3 (C5), and S4 (C6). To appropriately label the four categories in the 4S framework of simulation, the most common and relevant terms associated with each category were visually represented in a word cloud. This approach helped to simplify and condense the keywords, making it easier to identify the most important and frequently used terms. The word clouds for each category are shown in Fig. 3.

Following a thorough analysis of the word clouds, names were assigned to each of the four categories in the 4S framework of simulation. To encapsulate the essence of each category as a capability, it was decided to use action verbs when selecting category labels. In the word cloud of the first category (S1), the most frequent terms

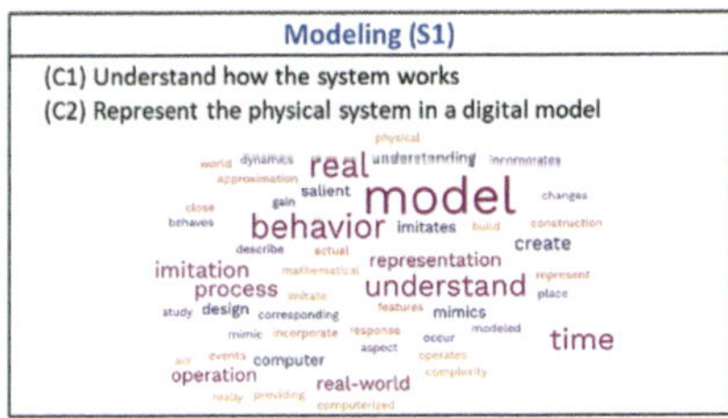


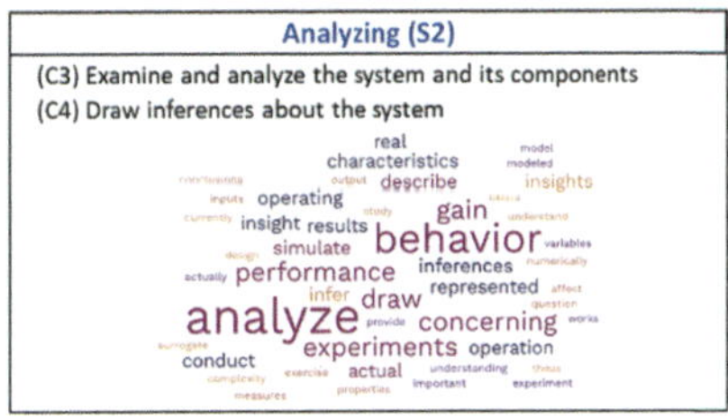


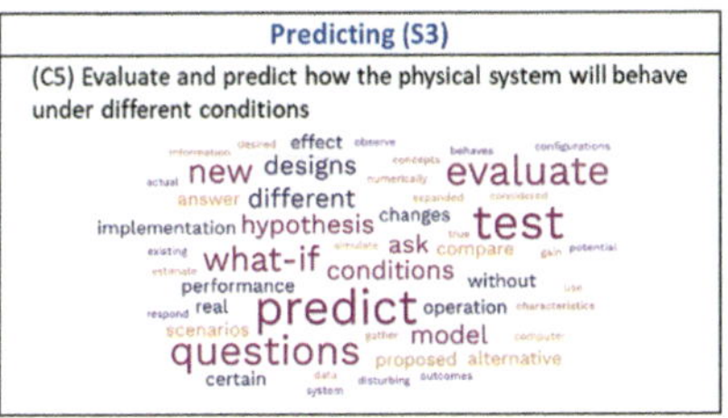


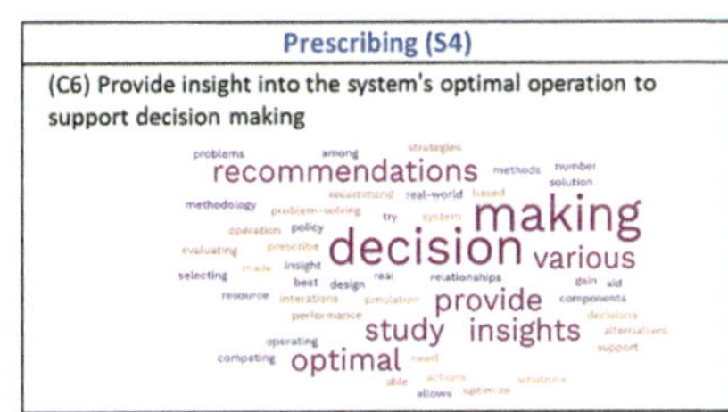


**Fig. 3** Word Clouds for labeling the 4S levels. Figure adapted from [47]

included "model," "behavior," "understand," and "real." "Model" and "behavior" strongly imply a focus on simulating the behavior of a system. Similarly, "understand" and "real" indicate an emphasis on comprehending the workings of the actual system. These terms are closely associated with conceptual and computer models, suggesting that "Modeling" is a fitting title for this category.

In the word cloud for the second category (S2), the most prevalent terms were "analyze" and "behavior". These terms suggest that this category centers on the analysis and assessment of system behavior and performance, making "Analyzing" an appropriate name for this category.

In the word cloud for the third category (S3), the most frequently recurring terms were "predict," "hypothesis," "what-if," "test," and "evaluate". These terms imply that this category concentrates on using simulation to forecast how a system will behave under different conditions and to test and evaluate what-if scenarios. Therefore, "Predicting" serves as a fitting designation for this category.

Lastly, in the word cloud of the fourth category (S4), the most frequent terms included "decision," "making," "provide," "insights," and "recommendations." These terms suggest that this category revolves around employing simulation to provide valuable insights and recommendations that can inform decision-making. The term "prescriptive" is commonly employed in the realm of computing and analytics to refer to a type of analytics that involves recommending specific actions to achieve a desired outcome [60]. Therefore, "Prescribing" is an appropriate label for this category.

Overall, as the level of a simulation increases from S1 to S4, its level of analytical utility and its capabilities both increase [47]. A summary of these levels of capability is shown in Fig. 4.

Modeling (S1): The first level involves creating an abstracted representation of a real-world system through conceptual modeling and computer modeling. Conceptual modeling helps in understanding the system, mapping its processes, and collecting

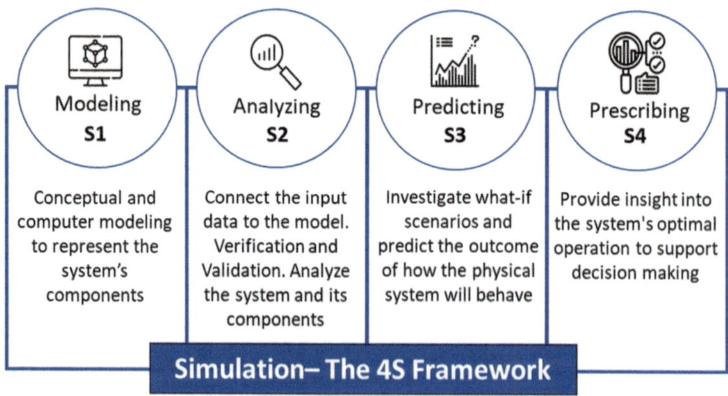


**Fig. 4** Summary of the simulation capabilities (4S Framework). Figure adapted from [47]

requirements. In the computer modeling stage, the system components are represented in a virtual format and the simulation model is created. Proper data collection, parameter selection, and understanding are crucial in this stage.

Analyzing (S2): In this level, the data gathered during modeling is incorporated into the simulation model. Input analysis ensures that the data used approximates real-world behavior. Verification and validation are performed to ensure model accuracy. In this phase, analysis is conducted to draw inferences about the system's behavior.

Predicting (S3): This level is where the simulation has the capability to predict system behavior under various conditions. It allows for the investigation of alternative scenarios, performance measurement, and comparison. Developing accurate scenarios is essential.

Prescribing (S4): This level involves using the simulation model for decision support and optimization. Decision-makers can identify optimal courses of action, and simulation optimization or machine learning methods can be applied to enhance decision-making and system performance.

## 4 What Is the Difference Between Simulation and Digital Twin?

Although traditional simulation and DT models share similar capabilities in replicating physical systems within a virtual environment, they are not the same and should not be referred to as the same technology [61]. The practice of combining simulation and DT can lead to confusion when simulation models are mislabeled as DT and vice versa [62]. While traditional simulation serves a valuable role in various applications, it is not accurate to label it as a DT.

Establishing a clear distinction between the capabilities of a DT and a simulation is essential to maintain consistency in DT implementations. DTs and simulations represent distinct technologies, each offering unique advantages that can yield valuable insights into various problems. Therefore, it is important to accurately categorize them and clarify their differences to prevent misconceptions.

In prior work, we conducted a systematic literature review [63] with the primary objective of determining whether the current body of literature, that claims to combine simulation and DT, is truly applying DT or whether it is using simulation in place of DT. Figure 5 shows the findings from this study. It indicates that one third of the papers (33%) are classified as R0 because they did not achieve any of the DT capabilities according to the 4R framework. These works simply presented simulations and called them DT. This demonstrates a persistent gap between the conceptual understanding of DT and its practical applications as many individuals continue to build models and label them as DTs even though they are essentially simulation models. Even among those who build DTs correctly, they often harness only a fraction of their capabilities, failing to realize their full potential, with no publications among those surveyed reaching the R4 level.

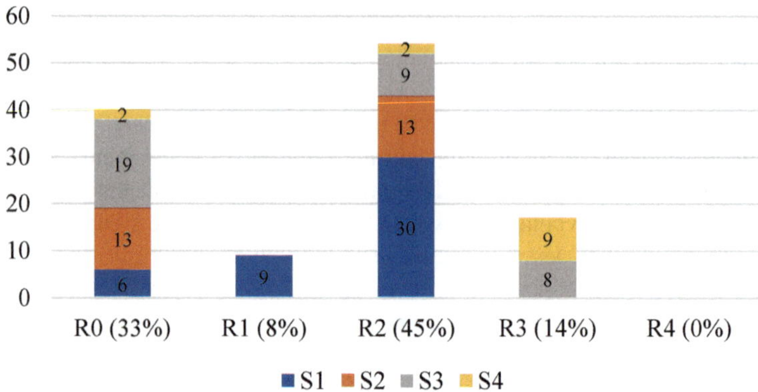


**Fig. 5** Analysis of the systematic literature review. Figure adapted from [63]

With the introduction of the 4S framework for simulation and the existing 4R framework for DTs, we can now proceed with a direct comparison of both technologies. In the following discussion, we will explore the similarities and differences between simulation and DTs.

- **Digital Representation**: Both simulation and DTs involve creating a virtual representation of a physical system by gathering and understanding data from the physical world. The key difference is that a simulation can vary from being very realistic to highly abstracted, whereas DT fully describes an existing physical element or one that will exist in the physical world [26].
- **Real-Time**: Simulations may rely on data-driven approaches, but they usually lack real-time data integration. Even at the most advanced level of simulation capability (S4), real-time data flows are typically absent. Simulation typically relies on historical data and performing input analysis of the data to use probability distributions or rules to represent that system over time, whereas the DT is continuously updated to mirror the real-world counterpart in real time. The real-time capability of a DT happens during the R2 and R3 phase; however, the key parameters and data pipeline of the DT happens during the first level (R1), which requires the establishment of real-time data pipeline integration from sensors, IoT devices, or other sources.
- **Data Flows**: Simulations usually involve one-way data flow, which is established at the first capability level (S1). DTs, on the other hand, commonly feature bidirectional data flow, with the flow from the physical to the virtual domain set up during the representation phase (R1), and the virtual to physical data flow established in the final capability level (R4). Bidirectional data flows are not a typical component of simulations, even at higher capability levels (S2, S3, and S4).
- **Analysis and Visualization**: Both simulation and DTs can be used for analysis and visualization, which involves using the data from the physical system to analyze the system's behavior and replicate it within a virtual environment. Both can replicate identical outcomes when provided with the same inputs as those of the

physical system. The distinction lies in their respective approaches. Simulations often provide a description of the system based on historical data and require extensive pre-analysis to explore how an object or system will change over time under various input conditions. In contrast, DTs place a stronger emphasis on real-time analysis and monitoring, focusing on the current state and performance of the physical system. In other words, DTs can display real-time events as they occur, providing insights into the present situation, past events, and future occurrences. Simulations, on the other hand, can only represent what has happened or what will happen in the future, lacking the ability to capture real-time.

- **Prediction Timeframe**: Both simulation and DTs possess the capability to predict the behavior of a system under various scenarios, allowing for the exploration of "what-if" scenarios. They can generate results based on a specific set of inputs, operating independently from the physical system, and producing outcomes and changes in state that mirror what the physical system would produce within the same timeframes. However, simulations typically place a greater emphasis on medium and long-term-oriented analysis, focusing on how a product evolves as it interacts with external forces and how these forces impact the system over time, which happens during the predicting level of capability of simulation (S3). In contrast, DTs do not exclusively concentrate on predicting the future; they also deliver real-time or near-real-time data concerning the current state and behavior of the physical system, corresponding to the third level of capability of DTs (R3). While simulation can give insight into the distant future and prescribe solutions based on the system's historical state, DTs can adapt to new situations and reveal what is currently happening in the system. If a sudden change in conditions happens, such as the disruption to global supply chains as a result of the Covid-19 pandemic, a simulation is not typically going to reflect it until a long time after the fact, as it is usually fed historical data, which would not immediately register those changes. On the other hand, a DT, which is focused on reproducing the physical system in real-time, would quickly incorporate the effects of such a disruption and be capable to extrapolate those impacts into the future.
- **Autonomy**: The predictive (S3) and prescriptive (S4) capabilities of simulation are closely linked to the reality level of DTs (R3). Both simulation and DTs are utilized to support decision-making and optimize processes. However, the key distinction lies in the fact that simulations often depend on human operators to make decisions and implement them based on the results generated by the simulation. In contrast, DTs operate autonomously, reducing the need for direct human intervention. In fact, at the R4 level, when data flows from the virtual world to the physical world have been established, DTs can perform prescriptive tasks and send commands directly to the physical system to implement those decisions. In the context of DTs, they can provide prescriptions at the R3 level, and when progressing to R4, automation is introduced.
- **Optimization**: Because DTs appeared at the same time as an explosion of research and development of both ML and AI, DT architecture has been developed to allow (at the R4 level) ML and AI to be directly integrated into DTs. This feature is absent in most commercial simulation software. By incorporating advanced

decision-making technologies such as AI and ML, jointly with bi-directional data flows, DTs become capable of adapting, identifying optimal strategies, and autonomously improving. By contrast, the prescriptive capabilities of simulation (at the S4 level) are often done through calling external optimization software or using iterative procedures between simulation and external tools. At best, most commercial simulation software only provides the ability to fine-tune parameters through enumerative approaches.

In summary, both simulation and DT start with modeling or representation and progress through stages of analysis and replication. However, the key differences lie in their data flows, the approach to real-time data adjustments, integration of advanced technologies such as AI and ML, and autonomy.

## 5 Conclusion

In conclusion, this chapter addressed the prevalent confusion in the literature regarding the distinctions between DTs and simulations, often misinterpreted as interchangeable concepts due to their complementary nature. It highlighted the critical need to differentiate between DTs and simulations, emphasizing their fundamental differences in capabilities. The chapter started by introducing an existing framework for characterizing DTs based on their capability and maturity levels, called the 4R framework. Next, the chapter presented the creation of an analogous framework to characterize simulation based on its levels of capability, the 4S framework, allowing for the classification of applications of both DT and simulation and facilitating direct comparisons between their capabilities.

This chapter delineated the similarities and differences in capabilities between simulations and DTs. The choice between simulations and DTs is not straightforward. Nonetheless, DTs serve a broader array of purposes beyond running scenarios. While both simulations and DTs employ digital models to replicate products and processes within a virtual environment, several crucial distinctions set them apart.

The most significant difference between DT and simulation lies in the real-time data integration and bidirectional data exchange in DTs, connecting the virtual environment with sensors for continuous data collection. In contrast, simulations rely on historical data and typically extensive pre-analysis, lacking real-time integration. This enhanced connectivity in DTs enhances the precision of predictive analytical models, offering a deeper understanding for product, policy, and procedure management and monitoring. Furthermore, DTs emphasize bidirectional data flows, a feature uncommon in simulations, and exhibit greater autonomy and advanced decision-making technologies such as AI and ML, reducing the need for human intervention. DTs can even autonomously command physical systems and boast superior optimization capabilities compared to most commercial simulation software.

## References

1. Hyre A, Harris G, Osho J, Pantelidakis M, Mykoniatis K, Liu J (2022) Digital twins: Representation, Replication, Reality, and Relational (4Rs). Manuf Lett 31:20–23. https://doi.org/10.1016/j.mfglet.2021.12.004
2. Osho J, Hyre A, Pantelidakis M, Ledford A, Harris G, Liu J et al (2022) Four Rs framework for the development of a digital twin: the implementation of representation with a FDM manufacturing machine. J Manuf Syst 63:370–380. https://doi.org/10.1016/j.jmsy.2022.04.014
3. Liu M, Fang S, Dong H, Xu C (2021) Review of digital twin about concepts, technologies, and industrial applications. J Manuf Syst 58:346–361. https://doi.org/10.1016/j.jmsy.2020.06.017
4. Ludwig S, Karrenbauer M, Fellan A, Schotten HD, Buhr H, Seetaraman S et al (2018) A 5G architecture for the factory of the future 2018
5. Syafrudin M, Alfian G, Fitriyani NL, Rhee J (2018) Performance analysis of IoT-based sensor, big data processing, and machine learning model for real-time monitoring system in automotive manufacturing. Sensors (Switzerland) 18. https://doi.org/10.3390/s18092946
6. Tao F, Zhang H, Liu A, Nee AYC (2019) Digital twin in industry: state-of-the-art. IEEE Trans Industr Inform 15:2405–2415. https://doi.org/10.1109/TII.2018.2873186
7. Melesse TY, Di Pasquale V, Riemma S (2021) Digital Twin models in industrial operations: state-of-the-art and future research directions. IET Collaborative Intell Manuf 3:37–47
8. VanDerHorn E, Mahadevan S (2021) Digital Twin: generalization, characterization and implementation. Decis Support Syst 145:113524. https://doi.org/10.1016/j.dss.2021.113524
9. Walton RB, Ciarallo FW (2022) A framework to manage systems based on digital twin. In: IIE annual conference. Proceedings, Institute of Industrial and Systems Engineers (IISE), pp 1–6
10. Jones D, Snider C, Nassehi A, Yon J, Hicks B (2020) Characterising the digital twin: a systematic literature review. CIRP J Manuf Sci Technol 29:36–52. https://doi.org/10.1016/j.cirpj.2020.02.002
11. Boschert S, Heinrich C, Rosen R (2018) Next generation digital twin. In: Proceedings of TMCE, Las Palmas de Gran Canaria, Spain, pp 7–11
12. Grieves M, Vickers J (2017) Digital Twin: mitigating unpredictable, undesirable emergent behavior in complex systems. In: Kahlen F-J, Flumerfelt S, Alves A (eds) Transdisciplinary perspectives on complex systems: new findings and approaches. Springer International Publishing, Cham, pp 85–113. https://doi.org/10.1007/978-3-319-38756-7_4
13. Abramovici M, Göbel JC, Savarino P (2016) Virtual twins as integrative components of smart products. In: Harik R, Rivest L, Bernard A, Eynard B, Bouras A (eds) Product lifecycle management for digital transformation of industries. Springer International Publishing, Cham, pp 217 226
14. Schroeder GN, Steinmetz C, Pereira CE, Espindola DB (2016) Digital Twin data modeling with automation ML and a communication methodology for data exchange. IFAC-PapersOnLine 49:12–17. https://doi.org/10.1016/j.ifacol.2016.11.115
15. Shao G, Kibira D (2018) Digital manufacturing: requirements and challenges for implementing digital surrogates. In: 2018 Winter Simulation Conference (WSC), IEEE, pp 1226–1237
16. Tao F, Zhang M (2017) Digital Twin shop-floor: a new shop-floor paradigm towards smart manufacturing. IEEE Access 5:20418–20427. https://doi.org/10.1109/ACCESS.2017.2756069
17. Gabor T, Belzner L, Kiermeier M, Beck MT, Neitz A (2016) A simulation-based architecture for smart cyber-physical systems. IEEE Int Conf Autonomic Comput (ICAC) 2016:374–379
18. Rosen R, von Wichert G, Lo G, Bettenhausen KD (2015) About the importance of autonomy and digital twins for the future of manufacturing. IFAC-PapersOnLine 48:567–572. https://doi.org/10.1016/j.ifacol.2015.06.141
19. Coronado PDU, Lynn R, Louhichi W, Parto M, Wescoat E, Kurfess T (2018) Part data integration in the shop floor Digital Twin: mobile and cloud technologies to enable a manufacturing execution system. J Manuf Syst 48:25–33
20. Negri E, Fumagalli L, Macchi M (2017) A review of the roles of digital twin in CPS-based production systems. Proc Manuf 11:939–948

21. Kritzinger W, Karner M, Traar G, Henjes J, Sihn W (2018) Digital Twin in manufacturing: a categorical literature review and classification. IFAC-PapersOnLine 51:1016–1022. https://doi.org/10.1016/j.ifacol.2018.08.474
22. Singh M, Fuenmayor E, Hinchy EP, Qiao Y, Murray N, Devine D (2021) Digital twin: origin to future. Appl Syst Innov 4. https://doi.org/10.3390/asi4020036
23. Juarez MG, Botti VJ, Giret AS (2021) Digital twins: review and challenges. J Comput Inf Sci Eng 21. https://doi.org/10.1115/1.4050244
24. Lim KYH, Zheng P, Chen CH (2020) A state-of-the-art survey of Digital Twin: techniques, engineering product lifecycle management and business innovation perspectives. J Intell Manuf 31:1313–1337. https://doi.org/10.1007/s10845-019-01512-w
25. Rathore MM, Shah SA, Shukla D, Bentafat E, Bakiras S (2021) The role of AI, machine learning, and big data in digital twinning: a systematic literature review, challenges, and opportunities. IEEE Access 9:32030–32052. https://doi.org/10.1109/ACCESS.2021.3060863
26. Grieves MW (2023) Digital Twins: past, present, and future. The Digital Twin. Springer, Heidelberg, pp 97–121
27. Barricelli BR, Casiraghi E, Fogli D (2019) A survey on Digital Twin: definitions, characteristics, applications, and design implications. IEEE Access 7:167653–167671. https://doi.org/10.1109/ACCESS.2019.2953499
28. Fan Y, Yang J, Chen J, Hu P, Wang X, Xu J et al (2021) A digital-twin visualized architecture for flexible manufacturing system. J Manuf Syst 60:176–201. https://doi.org/10.1016/j.jmsy.2021.05.010
29. Wu C, Zhou Y, Pereia Pessôa MV, Peng Q, Tan R (2021) Conceptual digital twin modeling based on an integrated five-dimensional framework and TRIZ function model. J Manuf Syst 58:79–93. https://doi.org/10.1016/j.jmsy.2020.07.006
30. Zhang H, Qi Q, Tao F (2022) A multi-scale modeling method for digital twin shop-floor. J Manuf Syst 62:417–428. https://doi.org/10.1016/j.jmsy.2021.12.011
31. Jones D, Snider C, Nassehi A, Yon J, Hicks B (2020) Characterising the Digital Twin: a systematic literature review. CIRP J Manuf Sci Technol 29:36–52. https://doi.org/10.1016/j.cirpj.2020.02.002
32. Morgan CB (1984) Discrete-event system simulation
33. Beck A (2008) Simulation: the practice of model development and use. J Simul 2:67. https://doi.org/10.1057/palgrave.jos.4250031
34. Yin C, McKay A (2018) Introduction to modeling and simulation techniques
35. Banks J (1998) Principles of simulation. Handbook of simulation: principles, methodology, advances, applications, and practice, pp 3–30
36. Law AM, Kelton WD, Kelton WD (2007) Simulation modeling and analysis, vol 3. Mcgraw-Hill New York
37. Schluse M, Rossmann J (2016) From simulation to experimentable digital twins: simulation-based development and operation of complex technical systems. In: ISSE 2016—2016 international symposium on systems engineering—Proceedings Papers, Institute of Electrical and Electronics Engineers Inc. https://doi.org/10.1109/SysEng.2016.7753162
38. McHaney R (1991) Computer simulation: a practical perspective. Academic Press
39. MOSIMTEC (2022) 4 Types of simulation models to leverage in your business 2022. https://mosimtec.com/types-of-simulation-models/. Accessed 17 Dec 2022
40. Harrison RL (2010) Introduction to monte carlo simulation. In: AIP conference proceedings, vol 1204. American Institute of Physics, pp 17–21
41. Davidsson P, Holmgren J, Kyhlbäck H, Mengistu D, Persson M (2006) Applications of agent based simulation. In: International workshop on multi-agent systems and agent-based simulation. Springer, Heidelberg, pp 15–27
42. Moreira ACM, Silva DF (2021) Applying discrete-event simulation and value stream mapping to reduce waste in an automotive engine manufacturing plant. Winter Simul Conf (WSC) 2021:1–12. https://doi.org/10.1109/WSC52266.2021.9715477
43. Banks J (1999) Introduction to simulation. In: Proceedings of the 31st conference on Winter simulation: simulation—a bridge to the future, vol 1, pp 7–13

44. White KP, Ingalls RG (2015) Introduction to simulation. Winter Simul Conf (WSC) 2015:1741–1755. https://doi.org/10.1109/WSC.2015.7408292
45. Ingalls RG (2011) Introduction to simulation. In: Proceedings of the 2011 Winter Simulation Conference (WSC), pp 1374–88. https://doi.org/10.1109/WSC.2011.6147858
46. Bennett BS (1995) Simulation fundamentals. Prentice Hall International (UK) Ltd
47. Wooley A (2023) Simulation and Digital Twin: contrasting the capabilities and bridging the gap 2023
48. Chung CA (2003) Simulation modeling handbook: a practical approach. CRC Press
49. Sokolowski JA, Banks CM (2011) Principles of modeling and simulation: a multidisciplinary approach. John Wiley & Sons
50. Bossel H (2013) Modeling and simulation. Springer, Heidelberg
51. Bandyopadhyay S, Bhattacharya R (2014) Discrete and continuous simulation: theory and practice. CRC Press
52. Maria A (1997) Introduction to modeling and simulation. In: Proceedings of the 29th conference on Winter simulation, pp 7–13
53. Shannon RE (1998) Introduction to the art and science of simulation. In: 1998 winter simulation conference. Proceedings (cat. no. 98ch36274), vol 1, IEEE, pp 7–14
54. Carson JS (2005) Introduction to modeling and simulation. In: Proceedings of the Winter simulation conference, IEEE, p 8
55. Goldsman D (2007) Introduction to simulation. In: 2007 Winter simulation conference, IEEE, pp 26–37
56. Sanchez PJ (2006) As simple as possible, but no simpler: a gentle introduction to simulation modeling. In: Proceedings of the 2006 Winter simulation conference, IEEE, pp 2–10
57. White KP, Ingalls RG (2018) The basics of simulation. In: 2018 Winter Simulation Conference (WSC), IEEE, pp 147–61
58. Robinson S, Arbez G, Birta LG, Tolk A, Wagner G (2015) Conceptual modeling: definition, purpose and benefits. Winter Simul Conf (WSC) 2015:2812–2826. https://doi.org/10.1109/WSC.2015.7408386
59. Smith JS, Sturrock DT (2021) Simio and simulation—modeling, analysis, applications, 6th edn
60. Davenport TH, Harris JG, Morison R (2010) Analytics at work: smarter decisions, better results. Harvard Business Press
61. Wright L, Davidson S (2020) How to tell the difference between a model and a digital twin. Adv Model Simul Eng Sci 7. https://doi.org/10.1186/s40323-020-00147-4
62. Ante L (2021) Digital twin technology for smart manufacturing and industry 4.0: a bibliometric analysis of the intellectual structure of the research discourse. Manuf Lett 27:96–102. https://doi.org/10.1016/j.mfglet.2021.01.003
63. Wooley A, Silva DF, Bitencourt J (2023) When is a simulation a digital twin? A systematic literature review. Manuf Lett 35:940–951

# Reinforcement Learning for Digital Twins

**Deena Francis, Jonas Friederich, Adelinde Uhrmacher, and Sanja Lazarova-Molnar**

**Abstract** Digital Twins (DTs) aim to ongoingly replicate complex systems through data acquisition, simulation, and analysis to monitor, optimize, and/or experiment to achieve systems' goals. Typically, systems adapt and evolve during their lifetimes, which requires updating simulation models and analysis as new data arrives or conditions change. The dynamics of environments under which DTs commonly operate necessitate using a paradigm that can deal with the uncertainties and unprecedented scenarios that may arise throughout its operation. Reinforcement Learning (RL) is a learning paradigm that provides tools to do precisely this. It is concerned with sequential decision-making in dynamic, uncertain environments. In this work, we discuss the current and potential role of RL in the context of DTs, motivate its usage through a concrete case study, and finally discuss the opportunities and challenges.

## 1 Introduction

A Digital Twin (DT) is a high-fidelity representation of an intended or current physical counterpart in terms of a computational model, often in the form of machine learning models or simulation models [1]. The digital and physical twins are typically

D. Francis (✉)
Department of Engineering Technology, Technical University of Denmark, Ballerup, Denmark
e-mail: dfra@dtu.dk

J. Friederich
University of Southern Denmark, Odense, Denmark
e-mail: jofr@mmmi.sdu.dk

A. Uhrmacher
University of Rostock, Rostock, Germany
e-mail: adelinde.uhrmacher@uni-rostock.de

S. Lazarova-Molnar
Karlsruhe Institute of Technology, Germany and University of Southern Denmark, Odense, Denmark
e-mail: lazarova-molnar@kit.edu

 
M. Grieves and E. Hua (eds.), *Digital Twins, Simulation, and the Metaverse*, Simulation Foundations, Methods and Applications, https://doi.org/10.1007/978-3-031-69107-2_3

connected in near-real-time to exchange information and data, allowing a frequent "twinning" between the virtual and the physical. DTs are inherently data-driven; therefore, supervised and unsupervised machine learning plays an important role in their realization [2]. In addition, due to the twinning, the digital and physical twins are often updated frequently, bringing us to another class of algorithms that learns from interacting with their environment, i.e., reinforcement learning. Reinforcement learning is concerned with sequential decision-making in dynamic, uncertain environments.

In the following, we explore how the development and deployment of DTs in a dynamic environment can be supported by reinforcement learning. Our focus is on DTs based on discrete event simulation models. In particular, in our case study, we discuss a DT in which not only parameters of a simulation model are adapted, but the simulation model itself is learned from data [3]. Discrete decisions are plentiful within the DT life cycle (as shown in Fig. 1). They may refer to what data to collect, what methods to use for validating the data, what methods to extract knowledge from the data, how to develop the model (e.g., which model components to compose), what methods to use for calibrating or validating the model, and what actions to send back to the environment. We present several applications of reinforcement learning in the context of DTs to gain insights about which questions in developing and deploying DTs can be effectively supported by reinforcement learning.

## 2 Reinforcement Learning

Reinforcement learning is about sequential decision-making; the agent learns how to map situations to actions to maximize a cumulative reward from the environment. Reinforcement learning assumes a closed loop in which the learning system's actions will influence its later inputs. The agent discovers which actions yield the most reward by trial and error. Markov Decision Processes (MDPs) are mathematical models used to formalize and study sequential decision-making in reinforcement learning [4]. Table 1 gives a short introduction to reinforcement learning concepts [4]. The notation used in this chapter is mostly from the third chapter of [4].

Reinforcement learning appears intuitive and, as such, attractive to solve problems, but its application is not without challenges. One concern is the complexity of the environment; real-world environments can be highly complex, making it difficult to model accurately. Generally, RL algorithms struggle when faced with high-dimensional state and action spaces or when dealing with continuous or stochastic environments, so it is crucial to generate the state space for RL carefully. The underlying hypothesis of applying RL is that the goal can be described as maximizing the expected cumulative rewards, which might be difficult for some applications. RL algorithms typically require a large number of interactions with the environment to learn effective policies. This can be prohibitively expensive, time-consuming, or impractical in some applications. Exploration and exploitation need to be care-

**Table 1** Central concepts of reinforcement learning

| Concept | Description |
|---|---|
| Agents | Reinforcement learning concerns at least two active entities, i.e., the environment and one or more agent(s). An agent receives observation and reward from the environment and sends actions to the environment; the agent employs reinforcement learning to maximize its rewards |
| Environment | The environment is also considered an active entity with dynamics of its own; it receives the actions $A_1$, ..., $A_i$, ... and sends observations $O_1$, ..., $O_i$, ... and rewards $R_1$, ..., $R_i$, ... to the agent |
| States | An agent uses its state to select its next action, forming the reinforcement algorithm's basis. Agent's state $S_t$ is typically a function of the history of encounters between the agent and its environment, i.e., $S_t = f(H_t)$ with $H_t = O_1, R_1, A_1, \ldots, O_{t-1}, R_{t-1}$, and is assumed to be Markov, i.e., that the future is independent of the past, given the present. Often, the environmental state is not fully observable by the agent, which makes it formally a Partially Observable Markov Decision Process (POMDP) |
| Actions | An agent selects actions to maximize its cumulative reward, $\mathcal{A}$ is the set of actions an agent selects from |
| Rewards | A reward is a scalar feedback signal that indicates how well the agent is doing at step $t$, |
| Discount factor | The discount factor $\gamma \in [0, 1]$ influences how much an agent values future rewards |
| Value function | The action-value function $q_\pi(s, a)$ is the expected return starting from state $s$, taking action $a$, and following policy $\pi$. The action-value function can be decomposed into immediate reward and the discounted return from the successor state, $q_\pi(s, a) = \mathbb{E}_\pi[R_{t+1} + \gamma q_\pi(S_{t+1}, A_{t+1}) \| S_t = s, A_t = a]$ |
| Model | Reinforcement learning approaches are distinguished based on whether they construct an explicit model of the environment's dynamics, typically in the form of a transition function $\mathcal{P}^a_{ss'} = \mathbb{P}[S_{t+1} = s' \| S_t = s, A_t = a]$, or whether they are model-free |
| Policy | A policy $\pi$ maps the states to actions and is defined as a distribution over actions given the states, $\pi(a\|s) = \mathbb{P}[A_t = a\|S_t = s]$ The goal of RL is to learn the optimal/near-optimal policy to maximize the action-value function for all policies $\pi$, $q_*(s, a) = \max_\pi q_\pi(s, a)$ |

fully balanced. Both exploration (trying out new actions to discover their effects) and exploitation (leveraging known information to maximize rewards) are crucial in RL. Finding the right balance is challenging, especially in environments with sparse rewards or deceptive dynamics [4].

## 3 Application of Reinforcement Learning in Digital Twins

In [5], data collection, data validation, knowledge extraction, model development, and model validation were identified as constituent parts of DTs. Exemplarily, the

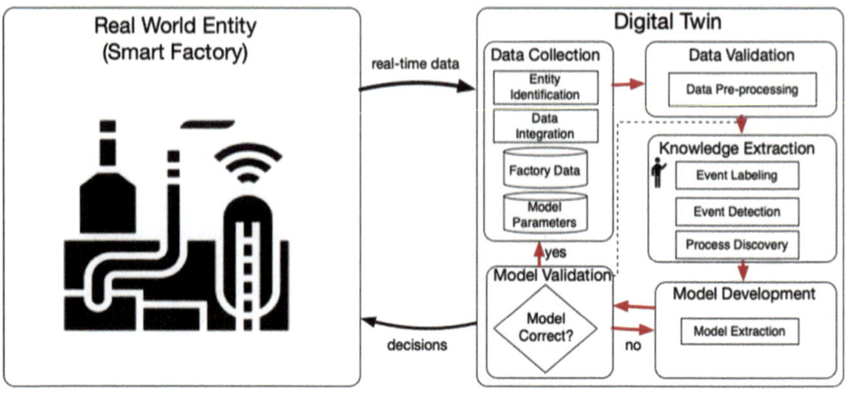


**Fig. 1** The framework for a DT for a factory, adapted from [5]

**Table 2** An example of the use of RL in the data collection of a data-driven DT

| | Data collection |
|---|---|
| Main task(s) | Recommend strategies to the PT for data collection |
| Simulation environment | The physical twin for a telecommunication network with multiple physical counterparts like mobile devices and network nodes, which are the agents. There are $K$ agents in this environment |
| Type of algorithm | Model-based Bayesian multi-agent reinforcement learning (MARL) |
| States | Network-related artifacts like traffic load conditions, queue length of packets at the devices denoted as $s_t^k$, where $k = 1, 2, ..., K$ |
| Actions | Decisions related to the network such as send a packet, do not send the packet denoted as $a_t$ |
| Policy | Data collection policy for the PT, $\pi(a_t^k \mid h_t^k)$, which assigns a probability to each action $a_t^k$, given the action-observation history $h_t^k = (o_1^k, a_1^k, ..., o_t^k, a_t^k, ...)$. Here $o_t^k$ is an observable function of the state $s_t$ such as the traffic load of a base station |
| Reward | Low model uncertainty, which happens when there are only a few disagreements between the various model predictions |
| Reference | [6] |

DT framework in the context of a smart factory is shown in Fig. 1. In this section, we explore each part of the framework in detail, identifying prospects for how RL can enhance the related processes. We, furthermore, identify promising examples to illustrate.

## 3.1 *Data Collection*

Data collection is the DT element concerned with real-time data gathering and storage from the physical counterpart of the DT. This data is used to generate and update

**Table 3** An example of the use of RL in the data validation

| | Data validation |
|---|---|
| Main task(s) | Determine value of data |
| Environment | Predictor model training and evaluation |
| Type of algorithm | REINFORCE [14] (a neural network-based RL) algorithm adapted for data value determining problem |
| States | Samples of data |
| Actions | Decision to select data instance (yes or no) |
| Policy | Data value determining policy $\pi(\mathcal{D}, \mathbf{s})$, where $\mathcal{D}$ is the training dataset of $N$ samples, $\mathbf{s} \in \{0, 1\}^N$, a data selection vector |
| Reward | Performance of the predictor on a validation set |
| Reference | [15] |

DTs. In the design of DTs, the interest has been centered on the technologies for data collection, such as Internet of Things (IoT), sensors, etc. [7], but in a DT, there are other aspects of data collection such as *how much* and *when* to collect (frequency) that are also important.

In order to answer the question of how much data to collect, there are some previous works that have used pre-training in RL to improve the performance in downstream learning tasks while learning the most useful features [8]. It has been found that when the data is collected passively in finite state episodic Markov Decision Processes (MDP), to obtain an $\epsilon$-optimal policy, one would need an exponential number of samples. More formally, to obtain an $\epsilon$-optimal policy $\Omega(A^{\min(S-1,H)}\epsilon^{-2})$ episodes would be needed [9]. The question of when to collect the data can be explored by looking at an example of RL being employed in the data collection is described in Table 2. RL has been used in developing optimal data collection strategies in DTs for wireless communication systems with notable benefits [6]. Such strategies inform when and what data to collect to refine the RL models.

Their RL-based data collection method was evaluated against other data collection schemes in an experimental study, and it was found that their method performed as well as an oracle-aided baseline method as the number of rounds (more data samples) increased. Gleeson et al. [10] explored the possibility of reducing the data collection time as well as computation time by utilizing the parallelization offered by GPUs. They demonstrate a 13.4x faster implementation of a deep RL framework. Reducing the state space has been done by including sensor selection as part of the learning task [11]. Long periods of training and/or data collection are often needed to obtain success in real-world learning problems [12]. Inspired by the success of data augmentation in supervised learning, [13] proposed using data augmentation for RL for pixel-based (image) and state-based data. Their results demonstrated that simple RL methods can outperform state-of-the-art on several tasks. This approach could possibly benefit RL for DTs in that the training/data collection time can be reduced while providing performance boosts.

**Table 4** An example of the use of RL in the knowledge extraction phase of a data-driven DT

| | Knowledge extraction |
|---|---|
| Main task(s) | Event detection |
| Environment | Battery-less energy harvesting sensors |
| Type of algorithm | Deep RL, Proximal Policy Optimization (PPO) |
| States | Energy storage, time (day, hour, minutes), light intensity |
| Actions | Turn on or off the Passive Infrared (PIR), humidity, temperature, light, and pressure sensors |
| Policy | Event detection policy $\pi(a\|s)$ |
| Reward | For every event that it catches, +1 is assigned; for every node that depletes its energy, −1 is assigned |
| Reference | [19] |

## *3.2 Data Validation*

The data validation of a digital twin consists of tasks, such as imputation, anomaly detection, etc., which can be solved by supervised and unsupervised learning, respectively [16]. RL has also been used for anomaly detection in sequential and time series data [17, 18]. In addition, RL can also be used to determine the value of data [15]. This method determines whether the data in hand is useful from a learning/modeling perspective. Their results indicate that an RL-based approach performed better than other approaches. Determining the value of data has also been used effectively in selecting training samples that are similar to the data in a given validation set, which is quite beneficial in the realm of domain adaptation [15]. The details are shown in Table 3.

Often, the collected data contain missing values. A strategy to combat this issue is to perform data imputation. Data imputation has been approached from unsupervised and supervised perspectives [20]. RL-based approaches for imputing data have also been studied with some success [21, 22]. Data measured through sensors have measurement noise. Handling noisy data appropriately is a crucial task. [23] proposed a noise-robust RL method that demonstrated reliable control in a network control system. Their method consists of an extended Kalman filter with RL.

## *3.3 Knowledge Extraction*

Knowledge extraction is another essential part of the DT framework that involves event detection and process mining, as well as mining other application- and goal-relevant elements of the model. The dynamic nature of events in the physical counterpart motivates the use of a learning paradigm such as RL. Event detection is an important element of DTs, and RL can be effectively used for event detection when no historical data is available, as demonstrated by Fraternali et al. [19]. They used a self-supervised strategy to collect baseline data. An example of RL applied to knowledge extraction is described in Table 4.

### 3.4 Model Development and Deployment

In the DT framework outlined in Fig. 1, the model development part is where the simulation model is generated from the data and extracted knowledge. The DT is based on this underlying simulation *model* of the physical counterpart, reflecting its dynamics and functionalities. The choice of the modeling formalism depends on the application domain and the goal(s) of the DT. The model's parameters are updated at pre-defined intervals to ensure high fidelity. The dynamic nature of the DT makes adopting an RL model beneficial in this phase. Deployment of the model involves the model providing insights for decision-making from the PT's perspective.

### 3.5 In-situ Learning of Methods to Apply for Digital Twins

Various simulation methods are used in developing, analyzing, and deploying the digital twin. To make the possible role of RL for selecting and configuring suitable methods for a DT more concrete, let us look at the selection and configuration of simulation algorithms. The efficiency of model execution determines how thoroughly a model can be analyzed and validated. In addition, as the DT is used to inform its physical twin, the execution efficiency is also decisive for this information to reach the physical twin in time. Also, the DT needs to reflect changes within its physical counterpart and shall be functional over a longer period of time; therefore, adaptations appear as an intrinsic ingredient of the DT. Depending on the simulation model and the available hardware structure, the efficiency of simulation algorithms varies [24], implying that for the DT to be effective, simulation algorithms must be selected and configured automatically. Various approaches have been developed to select and configure simulation algorithms automatically [25, 26].

The problem of selecting among the different simulators and their configurations, given the model and the available hardware, can be posed as a learning problem, either as a supervised learning problem [27] or as a reinforcement learning problem. To interpret it as a reinforcement learning problem, i.e., choosing iteratively between different options with previously unknown rewards, we assume that we can use part of the simulation, e.g., either some part of a long simulation run or a few of the replications, to identify a setup that performs best with respect to a user-defined criterion, e.g., simulation speed. The approach to tackle algorithm selection problems [28] with reinforcement learning is not unusual [29], even to apply it to reinforcement learning as a meta learner [30]. To illustrate the challenges, we will shortly sketch the approach developed in [31], which aims at and analyses automatic, frequent

**Table 5** Reinforcement learning to select the most efficient simulator configurations online [31]

| Concept | Description |
|---|---|
| Agents | Adaptive simulator |
| Environment | Execution environment, simulation model |
| States | Model information, e.g., variable values or coupling scheme of components, information about the simulator's state, e.g., event queue length, and information about the hardware, e.g., available cores |
| Actions | Set of simulation algorithms (and components that might be configured) |
| Rewards | Rate of event computations per second |
| Discount factor | 0 |
| Value function | Q-learning, $Q[s, a] :=$ $Q[s, a] + \alpha(N[s, a]) \cdot (r + \gamma \cdot max_{a'} Q[s', a'] - Q[s, a])$ |
| Model | *model-free* |
| Policy | f(s,Q,N) |

adaptations of simulators during simulation runs via reinforcement learning. The approach is based on Q-learning. Q-values record the value of taking an action $a \in A$ in a state $s \in S$. These values are unknown at first but are updated step by step by applying the following function: $Q[s, a] := Q[s, a] + \alpha(N[s, a]) \cdot (r + \gamma \cdot max_{a'} Q[s', a'] - Q[s, a])$ (Table 5)

Q-learning takes the received reward $r$ (of applying $a$ in state $s$) and the observed state $s'$ (as a result of taking action $a$ in state $s$) into account to update the q-value. $\gamma$ is the discount factor, which determines the influence of possible future utilities on the current q-value, and the function $\alpha$ determines the learning rate, i.e., the impact an update has on the q-value, which depends on how often the action $a$ has already been chosen in a state $s$, $N[s, a]$. The policy $\pi : S \times \mathbb{R}^{|S| \times |A|} \times \mathbb{N}^{|S| \times |A|} \rightarrow A$ is responsible for action selection. The policy, selecting the next action, balances exploitation (based on $Q$, choose actions that promise a high utility) and exploration (based on $N$, choose actions that have been rarely chosen in the current state).

In our case, the set of options to select from is defined by the different simulation algorithms (and components that might be configured) for the models to be executed. Therefore, a pre-requisite is a clear separation of concern between model and simulator [32] so that simulators can be exchanged and composed even during simulation runs [33]. A suitable reward (as the goal is to minimize the time needed for a simulation run) can easily be identified and calculated, e.g., the rate of event computations per second. To define the state space for reinforcement learning, information about the model state in terms of variable values or coupling scheme of components, information about the state of the simulator in terms of event numbers per minute or event queue length, and information about the environment in terms of available cores might be valuable. The discount factor $\gamma$ was set to zero, as experiments had shown that, in the benchmark cases used, a higher value of $\gamma$ had no significant impact on the results [31]. This may be different when using stochastic simulation algorithms and data structures with longer warm-up phases. With $\gamma = 0$, the reinforcement learning problem reduces to searching (i.e., it is assumed that actions do not affect future situations and rewards). To serve as a basis of reinforcement learning, this complex, highly-dimensional information needs to be suitably aggregated into more simple states. Therefore, dynamic, flexible approaches are required [34], which might benefit from applying learning methods [35]. In any case, a suitable partitioning and aggregation of the state space is essential for an effective reinforcement learning of simulator selection and configuration [31].

Also, the question of when to trigger a reinforcement learning step needs to be answered: it could depend simply on the time passed since the last learning step or be triggered by drastic changes in the dynamics. To identify such changes, information about the current and previous states of the model, simulator, or hardware and the current performance must be interpreted to identify suitable points for adapting (respectively when invoking the reinforcement step) [36]. Generally, change point detection [37] is an important ingredient to support decisions when dealing with dynamic systems. Due to the frequent feedback (performance data) between the agent (adaptive simulator) and environment (execution environment including model, chosen simulator configuration, hardware), suitable simulator configurations can be learned effectively [31] and facilitate leveraging algorithms and data structures tailored to specific requirements and application contexts.

One could apply reinforcement learning to support other tasks in developing the DT by method selection and configuration. For example, reinforcement learning has also been proposed to learn optimization algorithms [38, 39]. The approach uses guided policy search at the meta-level to train the optimizers and showed good performance in optimizing neural networks.

The agent and environment must frequently interact for reinforcement learning to be effective. Therefore, sample efficiency is critical for the in-situ learning of methods in the context of DT. Other conditions are that an immediate reward can be easily calculated and is due to the action (credit assignment problem). Also, it should be noted that reinforcement learning is only one possibility to solve the algorithm selection problem [40], even in the context of machine learning approaches, and thus to support developing or deploying the DT effectively.

## 4 Case Study in Reliability-Focused Digital Twins of Production Systems

In this section, we present an illustrative case study of a reliability-focused DT to demonstrate the application potential of RL in DTs. A reliability-focused DT aims to support decision-making processes related to equipment purchases (i.e., whether to invest in a more reliable piece of equipment), maintenance scheduling and staffing, and other system configuration decisions that impact the systems' reliability. When considering the life cycle of a DT in the context of a smart factory in Fig. 1, we focus

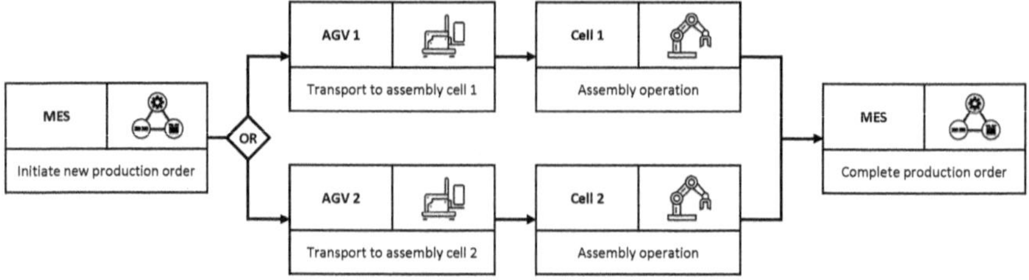


**Fig. 2** Overview of the case study system

solely on the knowledge extraction, model development, and model validation phase in this case study.

The following subsections describe the case study system, followed by the DT extraction procedure. We then examine how an extracted DT model is validated. Finally, we reflect and discuss the different application areas of RL for this case study.

## 4.1 Case Study System: A Flow Production Line with Concurrency

Figure 2 provides an overview of the case study system, which is a flow production line commonly found in manufacturing systems. The production line is fully automated and consists of five resource components: a manufacturing execution system (MES), two automated guided vehicles (AGVs), and two assembly cells. Both assembly cells work concurrently, performing the same assembly operation. The MES controls the production process by initiating new production orders, routing them to either assembly cell 1 or 2, and marking orders as completed. When a new production order is initiated and assigned to one of the assembly cells, the AGVs transport the raw material to the designated cell.

The AGVs and the assembly cells are susceptible to failures, while the MES is always fully operational. In the event of a production resource failure, the resource stops operating, and a repair crew is dispatched to repair the malfunctioning resource. The maintenance policy in place is purely reactive. The AGV has an unlimited buffer and a capacity of one, while both assembly cells 1 and 2 have finite buffers and capacities of one.

We assume that several types of data can be collected from the case study system, including event logs, which capture production processes, and state logs, which record operational state changes of production resources. These two log types are necessary for the DT extraction process, described in the next section.

## 4.2 Digital Twin Model Extraction for the Case Study System

To extract a reliability-focused DT of the case study system, we employ our methodology described in [3]. Our proposed methodology uses advanced process mining (PM) and statistical techniques to extract and parameterize a stochastic Petri net (SPN), which serves as a conceptual model for the DT of the physical system. In the following, we describe the extraction and parameterization process in further detail and present the resulting DT for the case study system.

The class of SPNs that we use to describe the conceptual model for a DT of a physical system is defined as follows:

$$SPN = (P, T, A, m_0)$$

where:

- $P = \{P_1, P_2, .., P_p\}$ is the set of places, drawn as circles,
- $T = \{T_1, T_2, .., T_q\}$ is the set of transitions along with their distribution functions or weights, drawn as bars,
- $A = A^I \cup A^O \cup A^H$ is the set of arcs, where $A^O$ is the set of output arcs, $A^I$ is the set of input arcs, and $A^H$ is the set of inhibitor arcs, and each of the arcs has a multiplicity assigned to it,
- $m_0$ is the initial marking of the Petri net.

Each transition $T_i$ can be either timed or immediate. A timed transition is drawn as a hollow bar and is assigned a probability distribution function that describes the firing time of the corresponding activity. An immediate transition is drawn as a filled-out bar and is assigned a weight that describes the firing probability of the associated event.

To extract the reliability-focused DT, we first extract a manufacturing process model and then integrate it with fault models for production resources.

The extraction of the manufacturing process model utilizes the event log and involves the following four steps:

1. Identification of a Petri net of the material flow within a production line using process discovery algorithms. The material flow is the path that production orders follow through the system [41].
2. Determination of transition types (i.e., timed or immediate). Timed transitions correspond to the arrival of new production orders or to resource activities.
3. Estimation of probability distributions for timed transitions and extraction of weights for immediate transitions.
4. Extraction of resource capacities and buffers.

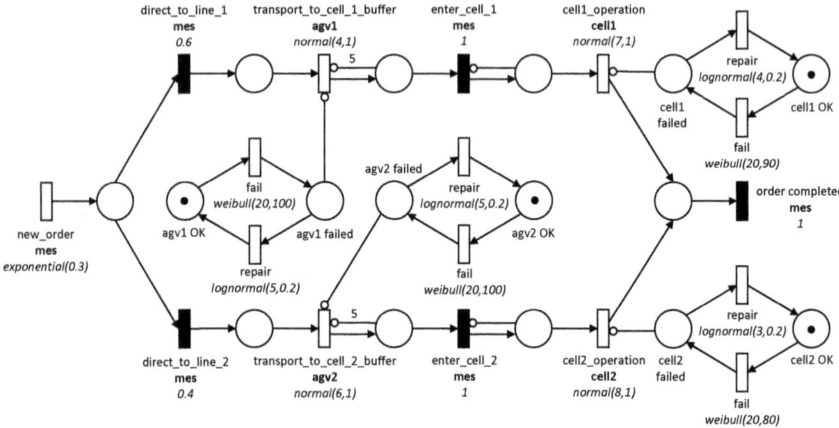


**Fig. 3** Extracted DT of the case study system

The extraction of the fault models for production resources utilizes the information captured by the state log and involves the following two steps:

1. Creation of necessary places and transitions from a fault model template.
2. Estimation of resource failure and repair distributions.

Finally, the fault models are integrated into the manufacturing process model using inhibitor arcs, preventing the corresponding resource activity timed transition from firing.

Figure 3 depicts the extracted DT obtained through the described model extraction process. The figure shows the extracted and parameterized manufacturing process model as well as the resource fault models. For each timed transition, the corresponding distribution function, including parameters, is shown. For each immediate transition, the corresponding weight is displayed. Furthermore, the capacity of one for both assembly cells, as well as their finite buffer sizes, have also been extracted.

### *4.3 Validation of Extracted Digital Twins*

Validation is an essential step to ensure that the extracted DT accurately mimics the real-world system. In [42], we propose to validate extracted models in two phases: *validation of initial model* and *validation of model at run-time*. In the first phase, the validity of a newly extracted DT model is evaluated to ensure that the model is safe to deploy in a production environment to support decisions. In the second phase, the deployed DT model is validated to ensure continuous validity.

**Table 6** Applying RL for data collection

| Concept | Description |
|---|---|
| Goal | Determine data streams and amounts of data needed for model extraction |
| Environment | Cyber-physical system with numerous IoT devices |
| States | Set of data streams and sizes of data considered for extraction, structure of the PN, including model validation results |
| Actions | Adding or removing data streams, Extending and reducing size of data, Decreasing and increasing frequency of data |
| Rewards | Model validation results for the extracted PN |

## 4.4 Application of Reinforcement Learning to the Case Study

In the following, we investigate and illustrate the application of RL to the case study, considering all phases of the DT life cycle. Furthermore, we highlight how the case study system can be optimized using RL and the extracted reliability-focused DT.

### Data Collection

The data collection process for reliability-focused DTs can be supported by using RL to determine the right data streams and the amounts of data needed for model extraction (Table 6). By adjusting data streams, their size, and frequency of collection, the impact on the model validation results of the extracted PN model can be observed and fed back to the agent as a reward. The main challenge in this phase is the possible large size of the state space, which can lead to high time complexity [43].

### Data Validation

RL can be used to to learn and identify patterns, anomalies, or errors within datasets such as event logs and state logs used for model extraction (Table 7). To do so, specific metrics or criteria that define the quality and accuracy of the dataset need to be established. These could include measures such as data consistency, completeness, accuracy, and adherence to defined standards. The data value determination method of [15] could serve as a possible solution to selecting useful or error-free data.

### Model Extraction

One possible application of RL in the DT model extraction phase could be for material flow identification (Table 8). RL can be used to interactively extract the material flow model using process discovery algorithms where the system learns from user feedback. For instance, when a user corrects or validates certain parts of the material flow model, the RL system can learn which types of structures are preferred or which common mistakes to avoid in the following process discovery iterations.

**Table 7** Applying RL for data validation

| Concept | Description |
|---|---|
| Goal | Identify and address anomalies, errors, and inconsistencies within a dataset |
| Environment | The dataset |
| States | Subsets of data, specific data points, or statistical summaries of the dataset |
| Actions | Flagging data points as potential anomalies, suggesting corrections or imputations for missing values, or triggering automated data cleaning processes |
| Rewards | Positive rewards for improvements in data quality and negative rewards for data quality degradation |

**Table 8** Applying RL for the material flow identification

| Concept | Description |
|---|---|
| Goal | Extract an accurate material flow model |
| Environment | Material flow Petri net (PN) model |
| States | Structure of the PN |
| Actions | Adding or removing places, transitions, and arcs |
| Rewards | User validation of correct and incorrect PN modeling elements |

Similarly, RL can be applied to improve the final extracted DT model interactively. Here, the actions would not just include adding or removing places, transitions, and arcs but also adjusting the token distribution, transition types, arc multiplicity, distribution functions, and weights. A possible issue is the requirement for a large number of steps in order to find the optimal policy. In one study, several deep RL methods were employed for the task of optimization in a manufacturing setup with some success. Their problem had a state vector with 12 variables and three possible actions per state. In this setting, the Proximal Policy Optimization (PPO) algorithm was found to converge toward the optimal policy in around 160k steps.

### Model Validation

RL can support the model validation process of an extracted model by determining the right validation strategy as well as the time intervals at which validation should take place (Table 9). For the latter, an RL agent could adjust the validation intervals and receive a reward based on the validation results. The work of [44] is an example of RL being used to successfully validate a model in three varied engineering systems.

### System optimization

RL can be used in several ways to optimize the real system by making adjustment to the DT and feeding the gained knowledge back to the real system (Table 10). This

**Table 9** Applying RL for model validation

| Concept | Description |
|---|---|
| Goal | Determine the right time intervals for model validation |
| Environment | The extracted DT and the physical system it mimics |
| States | Validation intervals |
| Actions | Adjusting validation intervals |
| Rewards | Validation results |

**Table 10** Applying RL for joint model and system optimization

| Concept | Description |
|---|---|
| Goal | Enable automated decision support for the real system |
| Environment | The extracted DT and the real system |
| States | Production process states |
| Actions | Adjustments to Buffer sizes, logistic routes, schedules, etc. |
| Rewards | KPI changes |

could eventually lead to automated decision support in the real system, unleashing the full potential of RL. Some of the activities RL could help optimize the real system are:

- Find optimal parameters to increase/improve a KPI (e.g., production volume, resource downtime).
- Use an RL agent to determine routing decisions instead of hard-coded routing probabilities (i.e., *direct_to_line_1* and *direct_to_line_2* transitions).
- Order scheduling: Use an RL agent to determine production order schedule, instead of using manually defined order schedules (i.e., *new_order* transition).
- Buffer size allocation: Use an RL agent to dynamically adjust buffer sizes based on system state (i.e., buffer for cell 1 and cell 2).
- Optimal repair scheduling: Use RL agent to schedule and dispatch repair crews.

A deep Q-network model was used in the regulation of a subsystem of the Fermilab booster accelerator complex with notable success in a test scenario [45][46]. RL has also been demonstrated to be effective model optimization in a sheet metal assembly setup [47].

## 5 Discussion and Conclusion

In our study, we investigated how Digital Twins can be enhanced using Reinforcement Learning at the different steps of Digital Twins' lifecycle. Reinforcement Learning is suited to the scenarios found in Digital Twins where sequential decision-making needs to be done based on interactions with an environment that is dynamic. We, furthermore, provided a case study to illustrate the different opportunities. There are several aspects of utilizing Reinforcement Learning in Digital Twins that lead to opportunities and challenges, which we briefly describe in the following.

- *Quality of data*: Large quantities of data are generated from the physical counterpart of the Digital Twin setup, which aids the Reinforcement Learning model to learn optimal policies. The downside is, however, that not all data that has been collected is relevant to building a reliable model. Determining the quality or value of data has been studied by [15] by formulating this problem as an RL problem.
- *Curse of dimensionality*: When the state-action space is large, reinforcement learning methods will suffer from high time complexity. The Digital Twin scenario will typically have a large state-action space as discussed in the case study (Sect. 4). Several works have attempted to tackle this issue from various perspectives [48, 49].
- *Pre-training*: In systems with sensors that collect videos or audio signals for long periods of time, large amounts of data are available. It has been found that using this data to learn a good representation of the state improves the performance of the downstream Reinforcement Learning task [8]. It would be interesting to see if such a strategy would also benefit a Digital Twin scenario.
- *Uncertainty quantification*: The Digital Twin model needs to be calibrated to account for the uncertainties with respect to what it knows about its physical counterpart. Failure to do so would result in the physical counterpart receiving poor decisions, which can lead to failures. Quantifying of the uncertainty can be done using a Bayesian perspective [6].

To summarize, we can state that there remains much to be done and investigated in the application of Reinforcement Learning in Digital Twins. Our initial investigation provides initial directions and opportunities that are worth further investigation.

## References


1. Grieves M, Vickers J (2017) Digital twin: Mitigating unpredictable, undesirable emergent behavior in complex systems. Transdisciplinary Perspect Complex Syst New Find Approaches 85–113
2. Huang Z, Shen Y, Li J, Fey M, Brecher C (2021) A survey on ai-driven digital twins in industry 4.0: smart manufacturing and advanced robotics. Sensors 21(19):6340
3. Friederich J, Lazarova-Molnar S (2022) Data-driven reliability modeling of smart manufacturing systems using process mining. In: 2022 winter simulation conference (WSC) pp 2534–2545, Dec 2022. https://doi.org/10.1109/WSC57314.2022.10015301. ISSN: 1558-4305
4. Sutton RS, Barto AG (2018) Reinforcement learning: an introduction. MIT Press
5. Friederich J, Francis DP, Lazarova-Molnar S, Mohamed N (2022) A framework for data-driven digital twins of smart manufacturing systems. Comput Ind 136:103586
6. Ruah C, Simeone O, Al-Hashimi B (2023) A bayesian framework for digital twin-based control, monitoring, and data collection in wireless systems. IEEE J Sel Areas Commun

7. Liu Mengnan, Fang Shuiliang, Dong Huiyue, Cunzhi Xu (2021) Review of digital twin about concepts, technologies, and industrial applications. J Manufact Syst 58:346–361
8. Ghosh D, Bhateja CA, Levine S (2023) Reinforcement learning from passive data via latent intentions. Int Conf Mach Learn 11321–11339. PMLR
9. Xiao C, Lee I, Dai B, Schuurmans D, Szepesvari C (2022) The curse of passive data collection in batch reinforcement learning. In: International conference on artificial intelligence and statistics. PMLR, pp 8413–8438
10. Gleeson J, Snider D, Yang Y, Gabel M, de Lara E, Pekhimenko G (2022) Optimizing data collection in deep reinforcement learning. arXiv preprint arXiv:2207.07736
11. Kishima Yasutaka, Kurashige Kentarou (2013) Reduction of state space in reinforcement learning by sensor selection. Artif Life Rob 18:7–14
12. Kalashnikov D, Irpan A, Pastor P, Ibarz J, Herzog A, Jang E, Quillen D, Holly E, Kalakrishnan M, Vanhoucke V et al (2018) Qt-opt: Scalable deep reinforcement learning for vision-based robotic manipulation. arXiv preprint arXiv:1806.10293
13. Laskin Misha, Lee Kimin, Stooke Adam, Pinto Lerrel, Abbeel Pieter, Srinivas Aravind (2020) Reinforcement learning with augmented data. Adv Neural Inf Process Syst 33:19884–19895
14. Williams RJ (1992) Simple statistical gradient-following algorithms for connectionist reinforcement learning. Mach Learn 8:229–256
15. Yoon J, Arik S, Pfister T (2020) Data valuation using reinforcement learning. Int Conf Mach Learn 10842–10851. PMLR
16. Chandola Varun, Banerjee Arindam, Kumar Vipin (2009) Anomaly detection: a survey. ACM Comput Surv (CSUR) 41(3):1–58
17. Oh M-H, Iyengar G (2019) Sequential anomaly detection using inverse reinforcement learning. In: Proceedings of the 25th ACM SIGKDD international conference on knowledge discovery and data mining, pp 1480–1490
18. Mengran Yu, Sun Shiliang (2020) Policy-based reinforcement learning for time series anomaly detection. Eng Appl Artif Intell 95:103919
19. Fraternali F, Balaji B, Sengupta D, Hong D, Gupta RK (2020) Ember: energy management of batteryless event detection sensors with deep reinforcement learning. In: Proceedings of the 18th conference on embedded networked sensor systems, pp 503–516
20. Emmanuel Tlamelo, Maupong Thabiso, Mpoeleng Dimane, Semong Thabo, Mphago Banyatsang, Tabona Oteng (2021) A survey on missing data in machine learning. J Big Data 8(1):1–37
21. Mei H, Li J, Shi B, Wei H (2023) Reinforcement learning approaches for traffic signal control under missing data. arXiv preprint arXiv:2304.10722
22. Yamaguchi N, Fukuda O, Okumura H (2020) Model-based reinforcement learning with missing data. In: 2020 eighth international symposium on computing and networking workshops (CANDARW). IEEE, pp 168–171
23. Bui V-P, Raj Pandey S, Ana PM, Popovski P (2023) Value-based reinforcement learning for digital twins in cloud computing. arXiv preprint arXiv:2311.15985
24. Jeschke M, Ewald R (2008) Large-scale design space exploration of ssa. In: Computational methods in systems biology: 6th international conference CMSB 2008, Rostock, Germany, October 12-15, 2008. Springer, Berlin, pp 211–230
25. Ewald R (2012) Automatic algorithm selection for complex simulation problems. Vieweg+Teubner
26. Gratl FA, Seckler S, Tchipev N, Bungartz H-J, Neumann P (2019) Autopas: Auto-tuning for particle simulations. In 2019 IEEE international parallel and distributed processing symposium workshops (IPDPSW). IEEE, pp 748–757
27. Ewald R, Himmelspach J, Uhrmacher AM (2008) An algorithm selection approach for simulation systems. In: 2008 22nd workshop on principles of advanced and distributed simulation. IEEE, pp 91–98
28. Rice JR (1976) The algorithm selection problem. In: Advances in computers, Vol. 15, pp 65–118. Elsevier
29. Lagoudakis MG, Littman ML et al (2000) Algorithm selection using reinforcement learning. In: ICML, pp 511–518

30. Afshar RR, Zhang Y, Vanschoren J, Kaymak U (2022) Automated reinforcement learning: An overview. arXiv preprint arXiv:2201.05000
31. Helms T, Ewald R, Rybacki S, Uhrmacher AM (2015a) Automatic runtime adaptation for component-based simulation algorithms. ACM Trans Model Comput Simul (TOMACS) 26(1):1–24
32. Zeigler BP, Praehofer H, Kim TG (2000) Theory of modeling and simulation. Academic Press
33. Himmelspach J, Uhrmacher AM (2007) Plug'n simulate. In: 40th annual simulation symposium (ANSS'07). IEEE, pp 137–143
34. Helms T, Mentel S, Uhrmacher A (2016) Dynamic state space partitioning for adaptive simulation algorithms. In: Proceedings of the 9th EAI international conference on performance evaluation methodologies and tools, pp 149–152
35. Asadi K, Abel D, Littman ML (2020) Learning state abstractions for transfer in continuous control. arXiv preprint arXiv:2002.05518
36. Helms T, Reinhardt O, Uhrmacher AM (2015b) Bayesian changepoint detection for generic adaptive simulation algorithms. In: Proceedings of the 48th annual simulation symposium, pp 62–69
37. Aminikhanghahi S, Cook DJ (2017) A survey of methods for time series change point detection. Knowl Inf Syst 51(2):339–367
38. Li K, Malik J (2016) Learning to optimize. arXiv preprint arXiv:1606.01885
39. Chen T, Chen X, Chen W, Heaton H, Liu J, Wang Z, Yin W (2022) Learning to optimize: a primer and a benchmark. J Mach Learn Res 23(189):1–59. http://jmlr.org/papers/v23/21-0308.html
40. Kerschke P, Hoos HH, Neumann F, Trautmann H (2019) Automated algorithm selection: Survey and perspectives. Evol Comput 27(1):3–45
41. Lugaresi G, Matta A (2021) Automated manufacturing system discovery and digital twin generation. J Manufac Syst 59:51–66, Apr 2021. ISSN 0278-612 https://doi.org/10.1016/j.jmsy.2021.01.005. https://www.sciencedirect.com/science/article/pii/S0278612521000054
42. Friederich J, Lazarova-Molnar S (2023) A framework for validating data-driven discrete-event simulation models of cyber-physical production systems. In: Proceedings of the 2023 winter simulation conference. San Antonio, USA, IEEE
43. Dietterich TG (2000) An overview of maxq hierarchical reinforcement learning. In: International symposium on abstraction, reformulation, and approximation. Springer, Berlin, pp 26–44
44. Schena L, Marques P, Poletti R, Ahizi S, Van den Berghe J, Mendez MA (2023) Reinforcement twinning: from digital twins to model-based reinforcement learning. arXiv preprint arXiv:2311.03628
45. Kafkes D, Schram M (2021) Developing robust digital twins and reinforcement learning for accelerator control systems at the fermilab booster. arXiv preprint arXiv:2105.12847
46. John JS, Herwig C, Kafkes D, Mitrevski J, Pellico WA, Perdue GN, Quintero-Parra A, Schupbach BA, Seiya K, Tran N et al (2021) Real-time artificial intelligence for accelerator control: A study at the fermilab booster. Phys Rev Accelerators Beams 24(10):104601
47. Cronrath C, Aderiani AR, Lennartson B (2019) Enhancing digital twins through reinforcement learning. In: 2019 IEEE 15th international conference on automation science and engineering (CASE). IEEE, pp 293–298
48. Yang L, Wang M (2020) Reinforcement learning in feature space: matrix bandit, kernels, and regret bound. In: International conference on machine learning. PMLR, pp 10746–10756
49. Osband I, Van Roy B (2014) Near-optimal reinforcement learning in factored mdps. Adv Neural Inf Process Syst 27

# A Framework for the Credibility Evaluation of Digital Twins

**Han Lu, Lin Zhang, Kunyu Wang, Zejun Huang, Hongbo Cheng, and Jin Cui**

**Abstract** The widespread applications of Digital Twins are facing a significant challenge due to the lack of systematic and effective credibility evaluation methods. To address this issue, this chapter analyzes the dynamic evolution, virtual-real interactivity, and other key features of Digital Twins. Subsequently, a credibility evaluation framework for Digital Twins is proposed, encompassing the concept of Digital Twins credibility, a multi-dimensional and multi-level credibility evaluation index system, and a credibility evaluation methodology. The evaluation process is elucidated using the robotic arm Digital Twin as an example, thereby providing valuable guidance for the evaluation and construction of Digital Twins.

**Keywords** Digital twin · Credibility evaluation · Dynamic evolution · Index system · Modeling and simulation

## 1 Introduction

A Digital Twin can be taken as a digital model of a physical object that undergoes real-time self-evolution through data integration from its physical counterpart, ensuring consistent synchronization throughout its entire lifecycle [1]. Simulations based on the model can help to optimize and make decisions on the physical object [1, 2]. The physical object can be any type of physical system, such as a transportation system, a factory, a production line, or an equipment. The Digital Twins in this chapter mainly focus on equipment. The equipment can be the product of the manufacturing industry or the equipment used in the manufacturing process, e.g.,

H. Lu · L. Zhang (✉) · K. Wang · Z. Huang · H. Cheng · J. Cui
Beihang University, Beijing 100191, China
e-mail: johnlin9999@163.com

Engineering Research Center of Complex Product Advanced Manufacturing System, Ministry of Education, Beijing 100191, China

State Key Laboratory of Intelligent Manufacturing System Technology, Beijing 100191, China

L. Zhang
Hangzhou International Innovation Institute of Beihang University, Hangzhou 311115, China

 
M. Grieves and E. Hua (eds.), *Digital Twins, Simulation, and the Metaverse*, Simulation Foundations, Methods and Applications, https://doi.org/10.1007/978-3-031-69107-2_4

the aircraft, car, machine tool, or robot. The adoption of Digital Twins in the equipment's lifecycle offers numerous benefits, including enhanced equipment innovation and design, reduced development cycles, cost reductions, and improved autonomy in equipment operation and maintenance [3–8]. During the research and design stage of equipment development, the utilization of modeling and simulation-based system engineering techniques enables activities to be conducted within the digital realm [9, 10]. Consequently, design schemes can be thoroughly tested and pre-demonstrated, leading to significant reductions in physical experiment investments.

Credibility is a metric to measure the "correctness" of a model and the model's results for a specific use or purpose [11], which is the degree of user's confidence in the correctness of a model [8]. The concept of Verification, Validation, and Accreditation (VV&A) was proposed by the US Department of Defense (DoD) [12] to guarantee the credibility of modeling and simulation [13]. How to evaluate the credibility of a model is a critical task and foundation of VV&A. As a self-evolving and dynamic model, Digital Twin's credibility is much more difficult to be evaluated. In practice, lacking of credibility evaluation for Digital Twins is one of the critical bottlenecks in the widespread utilization of Digital Twins, as well as Digital Twins. Only a credible Digital Twin can effectively capture the equipment's characteristics and status in a timely manner, thereby assisting decision-makers in making accurate judgments. Considering that the Digital Twin maintains real-time data interaction with its physical counterpart, the usage of an untrustworthy Digital Twin may lead to failed tasks and even catastrophic consequences.

The research on credibility evaluation of Digital Twins is currently limited worldwide. Consequently, the widespread adoption and extensive application of Digital Twins face significant barriers. Existing credibility evaluation methods for Digital Twins primarily focus on visual resemblance and simulation output results, which are insufficiently comprehensive and scientific for complex Digital Twins.

Firstly, it is essential to recognize that a digital model cannot perfectly replicate a physical object. The excessive pursuit of realism or high accuracy may result in unnecessary complexity and introduce additional uncertainty, thereby reducing model credibility and usability. It is crucial to strike a balance between accuracy and complexity to facilitate efficient model computation and utilization. Secondly, evaluating credibility solely based on the consistency of output data collected at a certain time interval between the Digital Twin and its physical counterpart is insufficient. While external performance similarity is valuable, it does not guarantee that the internal mechanistic structure of the Digital Twin matches that of the physical object. Thirdly, relying on data samples from a specific time period fails to represent all the potential output variations of the model. Therefore, the credibility of Digital Twins cannot be solely determined by the visual resemblance or simulation output results.

In summary, the credibility evaluation of Digital Twins demands a comprehensive and systematic approach. There is an urgent need for the development of a comprehensive evaluation theory and methodology. Such methodologies can offer practical solutions for assessing the credibility of Digital Twins. Moreover, they can provide guidance on constructing credible Digital Twins, thus ensuring their credibility for various applications.

## 2 Related Work

During the past decade of concept development related to DTs, most of the literature focused on discussing the concept itself and its underlying connotations [14–17]. However, there remains a notable dearth of a systematic credibility evaluation framework for reference within the field. Evaluating a DT entails assessing both the DT model itself and the real-time data utilized for its construction and evolution. While theoretically, data changes should be incorporated and reflected in the model, a separate evaluation of the credibility of the data can help prevent the introduction of untrustworthy data into the DT, thereby enhancing DT credibility. Despite the limited resources specifically addressing credibility evaluation for DTs, there is considerable research available in the field of M&S that may serve as a valuable resource to develop credibility evaluation methods for DTs.

### 2.1 *Data Credibility Evaluation*

The credibility of data used in a DT significantly influences the credibility of the DT. These data encompass a large number of sensor data and non-sensor data related to human activities.

In the case of sensor data, the reliability can be compromised if the sensor is damaged or experiences failure [18]. Additionally, if a sensor fails to function properly within a sensor network, the data from that sensor node may be considered unreliable [19]. Even under normal circumstances, sensor data can be influenced by various factors such as the working environment [20], as well as the installation distance and angle of the sensors [21]. To evaluate the credibility of sensor data itself, Liao et al. [22] utilized a cumulative residual Chi-square check to compare current data with historical data and determine their credibility.

Digital Twin involves human participation during both the construction and operation decision-making stages. The credibility of the data in Digital Twin systems has been evaluated from various perspectives. For instance, the influence of ambiguity and authority on data credibility has been studied through psychological evaluations [23]. Additionally, the credibility of human-related propositions has been assessed based on a priori knowledge and a posteriori knowledge [24].

### 2.2 *Model Credibility Evaluation*

The existing methods for model credibility evaluation mainly come from the M&S. The specific methods for assessing model credibility can be categorized into three groups: qualitative analysis methods, quantitative analysis methods, and comprehensive analysis methods.

As an important means of ensuring credibility, VV&A-related technologies has been widely studied. Research on VV&A of weapon equipment system simulation has been carried out by the research team on system simulation at the National University of Defense Technology [25, 26]. Additionally, the teams at Harbin Institute of Technology have achieved significant progress in VV&A and credibility evaluation research on distributed interactive simulation systems [27]. In the field of aerospace simulation systems, Kim et al. [28] have proposed a trust evaluation framework for the entire life cycle of modeling and simulation development. The trust evaluation theory of complex weapon equipment models has been discussed and elucidated in references by Zhou et al. [29], Zhang and Ye [30], Zhen and Hu [31], and Sim and Lee [32].

More specifically, Beydoun et al. [33] proposed a set of model evaluation methods based on expert scoring, simulation requirements, and the simulation environment. Acar [34] suggested that the predictive ability of meta-modeling can be enhanced by integrating various types of models using weighted average integration. Li et al. [35] investigated the credibility evaluation technology of complex simulation models using the multi-agent interactive network method. Addressing the issue of high complexity in simulation models, Ferson and Oberkampf [36] developed the u-pooling region indicator. Li et al. [37] introduced the multi-variate probability integral transformation (PIT). Dornheim and Brazauska [38] proposed a hybrid linear expectation model for determining the credibility of complex systems in an automated and efficient manner. Liang et al. [39] put forward a credibility measurement method based on dynamic Bayesian networks. Hu [40] proposed a dynamic data-driven simulation method framework that utilizes Monte Carlo simulation to model real-time wildfire scenarios. Wang et al. [41] devised an incremental external attention temporal convolution network (IExATCN) model to establish a dynamic evolution framework for black-box Digital Twins, efficiently balancing computation accuracy and efficiency. Wang et al. [42] developed a lifelong learning method based on event-triggered online frozen-Elastic Weight Consolidation (EWC) transformer encoder for the dynamic evolution of Digital Twins, effectively managing computation costs and dynamic evolution performance.

However, traditional M&S primarily focuses on offline simulations. DT, on the other hand, is specifically designed for online simulations, incorporating interactive and evolutionary characteristics based on traditional offline simulations. Therefore, existing credibility concepts and evaluation methods from M&S could serve as references. However, considering the unique features of Digital Twins, new principles and methods must be proposed to implement effective credibility evaluation of DTs.

## 3 Concept of Digital Twin Credibility

The current evaluation of credibility for complex simulation systems primarily focuses on correctness based on user requirements. However, the analysis for dynamically changing models is often overlooked or not given enough emphasis. As a result, existing methods cannot be directly applied to assess the credibility of Digital Twins. Digital Twins are highly interactive and evolve in real time, making it crucial to examine whether they can provide correct simulation results in a timely manner during interaction and evolution. Furthermore, the credibility requirements for equipment vary greatly throughout its life cycle, leading to different focal points in credibility evaluation.

The primary purpose of credibility evaluation is to thoroughly analyze the quality of simulation results from the perspective of the model user's interests. This allows users to gain a full understanding of the limitations of the results and their impact on the decision-making process. As the understanding of credibility and evaluation methods is still in the exploratory stage, the quantified credibility values may be somewhat incomplete. Therefore, it is necessary to include a description of this incompleteness in the credibility analysis to provide decision-makers with a more comprehensive understanding of the credibility of Digital Twins.

The credibility value serves as a quantitative measure of credibility. In this chapter, the concept of credibility for Digital Twins is defined as follows: Based on specific user requirements, and during continuous virtual-real interactions and dynamic evolutions, the credibility of a Digital Twin refers to the correctness and timeliness of the twin model, the evolution process, and the simulation results, as well as the uncertainty analysis in each stage of the entire life cycle.

Figure 1 illustrates the process of assessing the credibility of Digital Twins based on user requirements. A Digital Twin can be considered credible when the difference between its behavior and that of the physical object falls within the tolerance of the user's requirements.

The fulfillment of user requirements is measured through two main aspects: correctness and timeliness. Correctness and timeliness are assessed by evaluating the mapping model between the Digital Twin and physical objects, the evolution process of the Digital Twin, and the simulation results generated using the Digital Twin. Timeliness refers to the ability to provide the necessary simulation results within an acceptable time period upon request. The specific evaluation elements for correctness and timeliness are defined in the index system presented in Sect. 5.

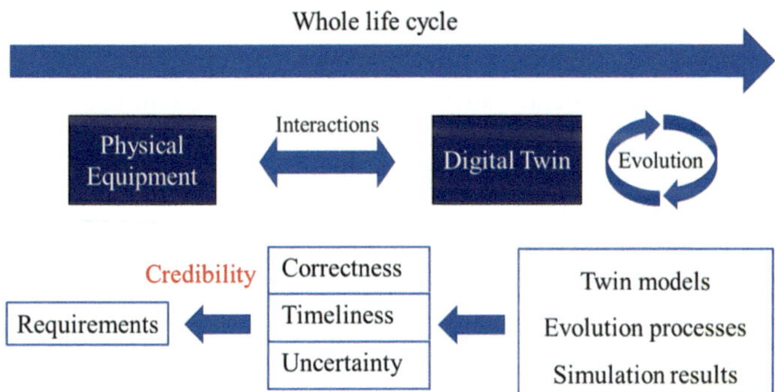


**Fig. 1** Credibility of DT

Uncertainty analysis plays a crucial role in credibility evaluation. Uncertainty exists in various aspects of deriving quantitative values for correctness and timeliness. The evaluation methodology outlined below provides a methodology for analyzing different types of known uncertainties.

The parameters, structure, and mechanisms of the twin model undergo changes and evolution over time. This ongoing process of assessing the credibility of Digital Twins distinguishes it from the evaluation of traditional basic models, which solely focus on providing simulations without considering real-time data and evolution mechanisms.

## 4 Features of DT Credibility Evaluation

The credibility evaluation of DT is a complex systematic process. The model for DT is profoundly intricate owing to the multi-disciplinary and multi-level attributes of the equipment itself. It possesses two fundamental elements of a Digital Twin: dynamic evolution and real-time interaction. In comparison to traditional models, these characteristics lend distinct features to the credibility evaluation process of Digital Twins.

### *4.1 Multi-dimensional Evaluation*

Equipment is characterized by multi-disciplinary, multi-level, and long life cycle. The Digital Twin of equipment is a complex model that involves multiple disciplines, hierarchies, and model integrations. The credibility evaluation of Digital Twins needs to be carried out from six perspectives, as shown in Fig. 2: model life cycle, model form, model transparency, multi-disciplinary domain, model granularity, and model scale.

(1) Model life cycle: A Digital Twin is developed around mapping the whole life cycle of equipment (design, processing, testing, operation, and maintenance). Meanwhile, the Digital Twin model also has a complete life cycle for any life cycle stage of equipment, including design, construction, use, evolution, and management. Each stage has an independent model form and function focus. Therefore, it is necessary to investigate the life cycle stage of the twin model and select the appropriate essential credibility metrics and evaluation methods according to the user requirements and model characteristics.
(2) Model form: Different life stages correspond to various model forms, such as conceptual models, mathematical models, and simulation models. The conceptual model is a formal interpretation of requirements. The mathematical models are divided into mechanism models and non-mechanism models. Simulation models are the code expression of conceptual models and mathematical models,

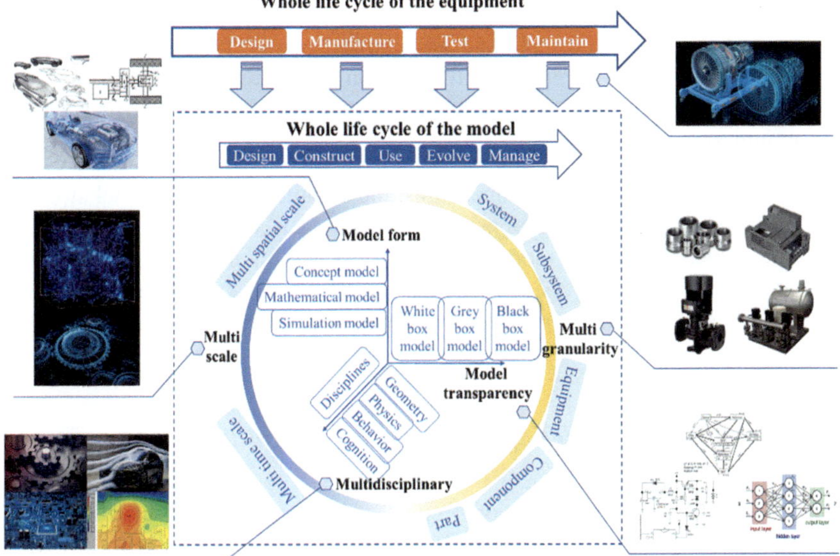


**Fig. 2** Multi-dimensional perspectives of equipment models

so the inspection of its logic and standardization is an important content of credibility evaluation.

(3) Model transparency: According to understanding of a model mechanism, there are three types of transparency: black box, white box, and gray box. The mechanism of the white-box model is completely clear. The black-box model fits only the input and output data without using the mechanism. The gray-box model has an unclear mechanism, mainly presented by empirical formulas. Its evaluation should consider both the mechanism and the output.

(4) Multi-disciplinary domain: Equipment models may include single or multiple domains like electromechanical, hydraulic, pneumatic, thermal, and control, which are presented in the form of continuous models and discrete models to describe geometric, physical, behavioral, and cognitive models. For models in different disciplines, specific theories and tools should be applied to evaluate the relevant ability of the model according to the domain knowledge.

(5) Model granularity: A DT model of an equipment usually has a multi-granularity structure including parts, components, equipment, subsystems, and systems. Each granularity model type has different evaluation emphases according to its unique characteristics. Compared with the parts or components, systems have strong emergence and complex functions. The evaluation of systems should be closely combined with the needs and be conducted hierarchically.

(6) Model scale: The scale of equipment includes temporal scale and spatial scale. The scale measures the scope of the investigation, while the granularity measures

the size of basic constituent units within the scope. There is a strong correlation between spatial scale and granularity. Due to the limitation of evaluation resources, the smaller the investigation granularity, the smaller the corresponding spatial scale, and the larger the spatial scale, the larger the investigation granularity.

In addition to the above six dimensions, the complexity of the model structure, the model composition, the relationship between models, the model interoperability, and the interaction between models and real-time data should be considered in the evaluation. Therefore, it is necessary to have a set of index system involving multi-dimensional elements and corresponding hierarchical evaluation methods.

## *4.2 Coupling of Modeling–Application–Evaluation*

The real-time evolutionary nature of Digital Twins couples the construction process and application process of twin models. Unlike the traditional basic model where there is a clear boundary between construction and application, the twin model keeps changing along with the changes of physical equipment during its applications. Thus, the evaluation of the construction and application should be regarded as part of the whole process, which can be called as a "modeling-application-evaluation cycle". As shown in Fig. 3, the key output and state data of the running equipment are collected in real time. According to the difference between the real and virtual data, the synchronized twin model identifies the biased components and conducts online evolutionary calibration. Thus, the consistency between the model and the equipment in terms of mechanism, structure, and parameters could be maintained. Once the credibility of the twin model meets the requirements, various simulation-based applications are allowed to be executed on the equipment.

The effect of the dynamic evolution process is shown in the credible evolution box in Fig. 3, where the blue line refers to the states and outputs of the physical equipment and the red line shows those of the twin model. The physical equipment is influenced by the environment and teams including the equipment suppliers, the DT users, and the evaluators. Compared with the equipment, the initial twin model still has difference which will increase with the equipment running. The trigger module calculates the incoming difference in real time and triggers the evolution process according to the preset threshold. Then, the evolution module receives the messages and starts assimilating the data.

As the number of loop iterations increases, the output error of the twin model will be gradually reduced and eventually converged with the equipment output. In this process, the evolution method will be optimally adjusted according to the obtained credibility. Optimization aspects include the precision, over-advancement, and light weighting, so that the twin model can continuously achieve higher credibility under the given resource conditions.

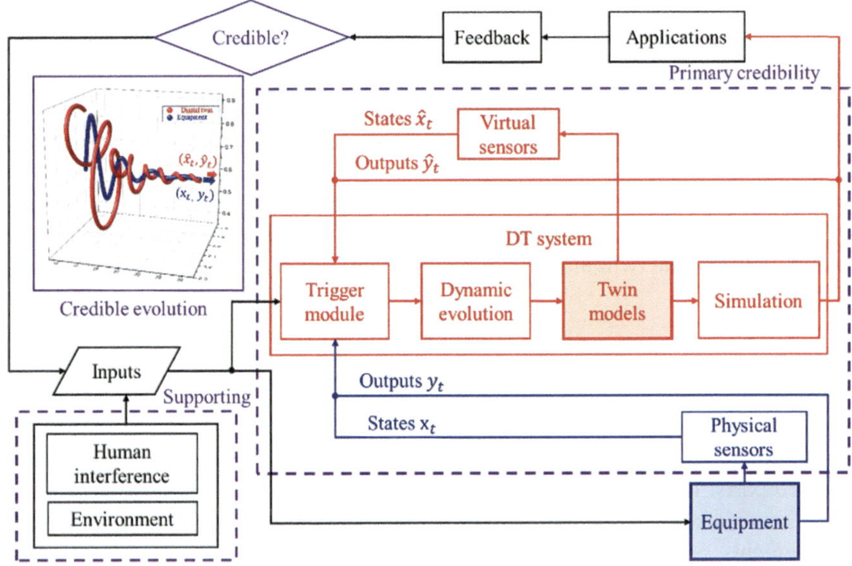


**Fig. 3** Coupling of modeling–application–evaluation in DTs

From the above analysis, DT is a combination of modeling–application–evaluation processes. The credibility evaluation of the DT is an online evaluation of the dynamic model. For a given evaluation time window, the DT has one credibility value. As the evaluation time window slides on the real-time axis, there is a corresponding series of changing credibility values. The trend of the credibility value is more indicative of the actual credibility than the value under any specific time window.

### 4.3 *Integrated Evaluation of Data and Models*

The real-time interaction between the equipment and the twin model is the premise of DT dynamic evolution, which includes forward calibration and feedback optimization. Forward calibration transfers data from the equipment to the Digital Twin, including real-time collection of equipment operating information, data generated from various software, manually updated data like text drawings. Feedback optimization delivers optimized control signals and scheduling planning from the Digital Twin to the equipment, to promote the iterative updating of the equipment. As the equipment is in operation, the user's application requirements for the equipment may change from time to time, which will deeply affect the credibility connotation of the relevant Digital Twin. So, it is necessary to recognize the changes in the user's requirements and update the method of credibility evaluation in time.

The two directions of real-time interactions are relatively independent but closely related. Forward calibration is the foundation of credibility insurance. The quality and timeliness of the data in this process will deeply affect the credibility of subsequent predictions and simulations. Feedback optimization is the purpose of DT. If the performance of the system is degraded in this step, it is likely to lead to a vicious cycle. So actual feedback should be implemented cautiously with sufficient simulation demonstration and risk prevention measures. During the evaluation, the independence and relevance of the two directions should be fully considered from the perspectives of both data and the twin model.

## 5 Overall Framework for Credibility Evaluation

As shown in Fig. 4, the credibility evaluation of the Digital Twin is mainly composed of three parts: the construction of the index system, the basic evaluation, and the evolution evaluation. The Digital Twin is a simulation model constructed around the requirements. So, the multi-dimensional and multi-level index system required for the evaluation should be obtained by analyzing the user requirements for the model. The appropriate evaluation methods are then constructed to form a methodological framework, by combining the characteristics of the model and investigation results of the requirements.

On a given evaluation time window, the twin model is composed of a sequence of basic models arranged on the time axis. The model evolution which occurs at a time point generates a basic model for a new time point according to the previous basic model; thus, an update of the twin model is completed. The basic evaluation involves every basic model in the model sequence. The first basic evaluation needs to be conducted in a comprehensive and detailed manner. The subsequent evaluations are more lightweight and could only focus on the changed components. The

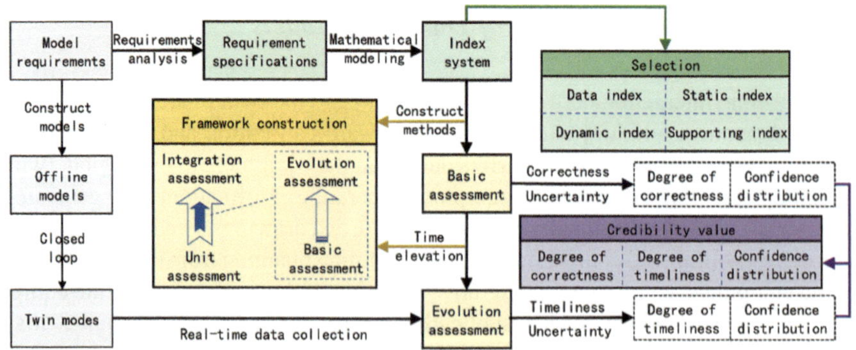


**Fig. 4** Framework of DT credibility evaluation

evolution evaluation examines the changes in the basic model sequence. The correctness of each basic model and the timeliness of the overall basic model sequence are influenced by each other. In general, the pursuit of higher correctness would lead to lower timeliness. The examination of uncertainty gives the confidence distribution for each quantitative value. The overall combination of correctness, timeliness, and confidence distribution forms the credibility of the twin model.

When a closed loop is formed by connecting the basic model with the equipment, the evaluator should continuously conduct credibility evaluation throughout the life cycle of the model according to the basic characteristics of each stage. The following section will focus on the framework of the index system.

## *5.1 The Index System*

Although simulation technology has been developed for more than 70 years, there is no unified index system about credibility.

Due to the complexity of the object, the credibility of the twin model needs to be examined from multiple aspects. Synthesizing the various types of index systems on credibility in different fields, and considering the characteristics of Digital Twins, this chapter proposes a framework of credibility evaluation index system for Digital Twins (Fig. 5). The superscript of the index symbols is the classification to which it belongs, and the subscripts is the meaning of the indexes. The system consists of four major parts, which can be further refined and expanded according to the meanings of the indexes below, combined with the needs and characteristics of the evaluation object.

(1) Data Indexes $I^D$

Data is the foundation of Digital Twins construction and has a significant impact on DT credibility. DT data is often characterized by real-time, high concurrency, and multi-source heterogeneity. Data evaluation examines both the source that generates the data and the data itself. The indexes are shown in Table 1.

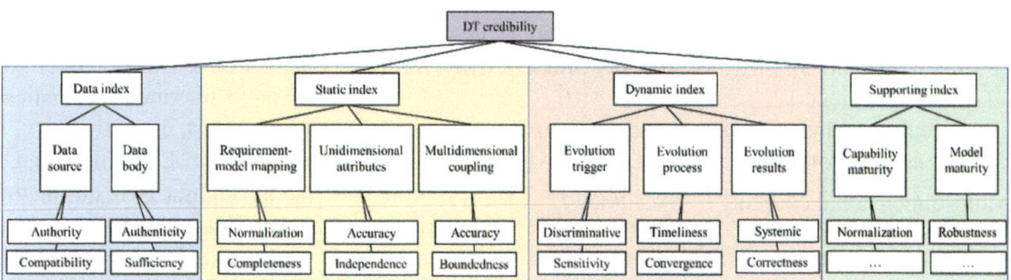


**Fig. 5** Framework of DT credibility evaluation indicator system

**Table 1** Data indexes

| Object | Index | Meaning |
|---|---|---|
| Data source | Compatibility $I_{c}^{D}$ | Generate, collect, and transmit data that meets performance requirements |
| | Authority $I_{ao}^{D}$ | Equipment manufacturers, personnel organizations, and software are authoritative or reliable |
| Data body | Authenticity $I_{ae}^{D}$ | The data arrives at the system in time and is consistent with the reality |
| | Sufficiency $I_{s}^{D}$ | The data fully contains the required information and is standardized and easy to access |

(2) Static Indexes $I^S$

Evolution of a twin model is the generation of a new model by making parametric, structural, or mechanistic changes to a basic model. Both the basic models before and after evolution can be considered static. Static indexes are for the consistency of the virtual-real mapping of the basic model and are important for subsequent dynamic index. The static index is shown in Table 2 which considers the following areas: analysis and correlation of requirements and models, single-dimensional key attributes, and cross-dimensional coupling effects.

**Table 2** Static indexes

| Object | Index | Meaning |
|---|---|---|
| Requirement-model mapping | Normalization $I_{n}^{S}$ QUOTE $Z_{gf}^{J}$ | Using standardized formal methods to characterize requirements and models. |
| | Completeness $I_{c}^{S}$ QUOTE $Z_{wz}^{J}$ | The requirements are systematically and completely correlated and mapped to a multi-dimensional model. |
| Unidimensional attributes | Accuracy $I_{ua}^{S}$ QUOTE $Z_{zq1}^{J}$ | All unidimensional mechanisms and parameters in Fig. 2 accurately reflect the equipment attributes. |
| | Independence $I_{i}^{S}$ QUOTE $Z_{dl}^{J}$ | Each unidimensional model, as well as its internal attributes and functions, is relatively independent. |
| Multi-dimensional coupling | Accuracy $I_{ma}^{S}$ QUOTE $Z_{zq2}^{J}$ | The description of coupling effects is concise and consistent with actual results. |
| | Boundedness $I_{b}^{S}$ QUOTE $Z_{yj}^{J}$ | The coupling process has reasonable time or value boundary constraints. |

(3) Dynamic Indexes $I^{Dy}$

When the basic model has formed a closed loop with the physical equipment, the model parameters, structure, and mechanism will be updated based on the data collected from the physical system. Model evolution is an important feature which separates Digital Twin models from traditional models. As shown in Table 3, the dynamic indexes are for the evolution process, which mainly examine whether the virtual model correctly follows the physical object, whether the response to various evolutionary events is timely, and whether the whole evolution trend converges. Based on the multi-dimensional correlation analysis of the model in the preparation of Sect. 5.2.1, the dynamic changes in the model structure and mechanism can be parameterized for consistency evaluation.

(4) Supporting indexes $I^{Su}$

The above three indexes are primary, while the supporting indexes are often used as references. These include maturity of capabilities and maturity of models [43], which are more difficult to obtain and has indirect impact on model credibility (Table 4).

**Table 3** Dynamic indexes

| Object | Index | Meaning |
|---|---|---|
| Evolution trigger | Discriminative $I_{d}^{Dy}$ | Be able to correctly identify the changed part of the model when the physical system undergoes changes |
| | Sensitivity $I_{se}^{Dy}$ | The minimum data fluctuation required to achieve identification of changing parts |
| Evolution process | Timeliness $I_{t}^{Dy}$ | The ratio of evolution complexity to the time spent on corresponding evolution |
| | Convergence $I_{cv}^{Dy}$ | The convergence of differences between model and equipment under the influence of evolution and uncertainty |
| Evolution results | Systemic $I_{sy}^{Dy}$ | Evolution participation level of parts that are related to the trigger one and affect the requirements |
| | Correctness $I_{cr}^{Dy}$ | The degree of difference between the multi-time window output of the measurable object and the actual output |

**Table 4** Supporting indexes

| Object | Meaning |
|---|---|
| Capability maturity | Including team collaboration, management standardization, team authority, etc. |
| Model maturity | Including reusability, security, robustness, sustainability, etc. |

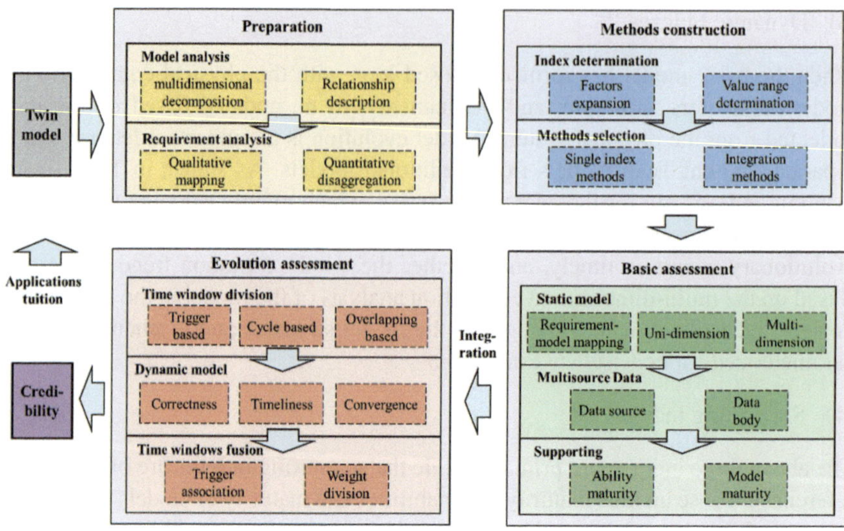


**Fig. 6** Framework of DT credibility evaluation methods

Based on the quantitative mapping relationship between the requirements and the model, the degree to which each index value is within the acceptable range of the requirements is the degree of correctness. For the index values involving any processes, the degree to which the time consumption is within the acceptable time range is the degree of timeliness.

## 5.2 *Methodology Framework for Credibility Evaluation*

The methodology framework for credibility evaluation of Digital Twins is shown in Fig. 6. The evaluation process is mainly composed of four segments: preparation, method construction, basic evaluation, and evolution evaluation. The core idea and focus of each segment of the evaluation work are described below. At the end of the evaluation, a dynamic multi-dimensional distribution of credibility values considering correctness, timeliness, and uncertainty will be formed. The integration of values is based on the demand-model mapping. The obtained credibility values could provide an important reference for further optimization of the construction and decision-making application of Digital Twins.

### 5.2.1 Preparation

(1) Analyze the model composition of the evaluation object. According to Fig. 2, a multi-dimensional decomposition diagram of the complex model (Fig. 7) should

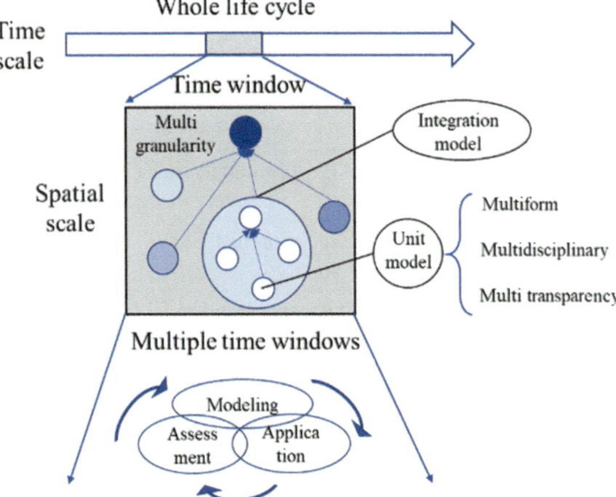


**Fig. 7** Multi-dimensional decomposition diagram of complex models

be constructed from six perspectives. Also the cross-dimensional associations and the associations among the same dimensions should be described by the characterization methods, such as directed graphs. Based on this description, the geometric structure variation index or integrated analysis methods, such as AHP and Bayesian network, can be used to parametrically characterize the changes of the structures and mechanisms.

(2) Analyze the requirements for the evaluation object. In this process, methods like formal analysis and mathematical modeling are applied. The fuzzy requirements described in natural language are transformed into semi-formalized descriptions, further specified qualitatively or quantitatively. Finally, each requirement is fully decomposed and mapped to the six-dimensional perspectives.

### 5.2.2 Methodology Construction

(1) Determine the specific index system. The specific indexes are further refined by considering both the characteristics of the model and the requirement analysis. A complete multi-dimensional hierarchical index system would be formed based on the index system shown in Fig. 5. Furthermore, the acceptable value range of each index is determined under the guidance of quantitative requirement analysis, which would serve as a benchmark for the correctness of the subsequent evaluation.

(2) Select appropriate methods. Eighty-one existing credibility evaluation methods are sorted out and organized into in eight categories (Table 5). Different indexes can be quantified and assessed using the appropriate methods.

For each single index, it is necessary to select the appropriate method or combination of methods from the list by considering the life cycle stage where the model serves, model characteristics, and user requirements. There is still a lack of evaluation methods for some indexes, which is also an important part of future study. In

**Table 5** List of credibility evaluation methods

| Top level | Secondary level | Top level | Secondary level |
|---|---|---|---|
| Software evaluation F1 | Code inspection F11 | Results comparison F5 | Sensitivity analysis F51 |
| | Algorithm analysis F12 | | Probability statistics F52 |
| | Execution test F13 | | Deterministic data analysis F53 |
| | Dataset validation F14 | | Evaluation of error effects F54 |
| Subjective methods F2 | Manual review F21 | Integrated analysis F6 | Analytic hierarchy process F61 |
| | Turing Test F22 | | Bayesian network F62 |
| | Expert scoring F23 | | Gray correlation analysis F63 |
| | Empirical comparison F24 | | Complex network analysis F64 |
| Semi-formal methods F3 | Consistency check F31 | Evolution analysis F7 | Timeliness analysis F71 |
| | Graphical analysis F32 | | Accuracy analysis F72 |
| Formal method F4 | Inference by induction F41 | | Smoothness analysis F73 |
| | Assertion check F42 | | Systematic analysis F74 |
| | Theory of evidence F43 | Data examination F8 | Data certification F81 |
| | Formal review F44 | | Data validation F82 |
| | Logical inference F45 | | Data source analysis F83 |

addition, appropriate integration methods need to be chosen according to primarily three types of situations: multi-disciplinary coupling, multi-dimensional integration, and multi-temporal fusion.

### 5.2.3 Basic Evaluation

(1) Evaluation of Static Models

1. Requirement-model mapping evaluation. According to the analysis in Sect. 5.2.1, examine whether the user requirements are correctly analyzed or normatively characterized for the normalization index $I_n^S$. Afterward,

examine whether the system architecture and functional design of the model sufficiently support the requirements for the completeness index $I_c^S$. Subjective methods such as F21 and F23 are mostly used.

2. Unidimensional evaluation. The static model is initially divided into specific evaluation objects of single life cycle and single granularity under the overall spatial and temporal scales. The model is then gradually decoupled into unit models from three perspectives: model forms, disciplinary domains, and transparency of mechanisms. Using a combination of F1, F3, F4, and F5 methods, the indexes under each single dimension are assessed for accuracy index $I_{ua}^S$ and independence index $I_i^S$. This process is the focus of the basic evaluation, which requires multi-disciplinary expertise and analytical tools.
3. Coupling evaluation: F13, F32, F5, and other methods are mainly used to examine the coupling effect between multiple disciplinaries and multiple dimensions. The accuracy index $I_{ma}^S$ QUOTE $Z_{zq2}^J$ of the coupling effect characterization can be examined from the unity of the conserved quantity, the coverage of the associated factors, and the consistency of the feedback effect. The boundedness index QUOTE $Z_{yj}^J$ $I_b^S$ of the coupling can be examined from the consistency of the boundary conditions and the convergence of the coupling.

   It is also necessary to integrate the results of multi-dimensional and multi-disciplinary single-index evaluations into a coherent hierarchical network of credibility values according to the intrinsic correlations of the system. In this case, multiple integration analysis methods in F6 are used.

(2) A credibility evaluation of the multi-source heterogeneous data (Fig. 8) is performed to ensure that the basement for subsequent model evolution is correct. This step considers both the data source and data body. The examined data source includes collection devices, human-related organizations, and software applications. The main purpose is to examine the performance of the related equipment, the rationality of the layout, the interference resistance, and the data standardization, etc. The key is to examine the compatibility index $I_c^D$ with methods such as F81, F83, and to take the authority index $I_{ao}^D$ of the data source as a reference. Data body refers to the data that arrives at the system. The key is to evaluate its authenticity index $I_{ae}^D$ through data comparison, cross-validation, and to examine its sufficiency index $I_s^D$ in combination with traceability, value consistency. Methods such as F5 and F82 are mainly used.

(3) Evaluation of supporting elements. For the maturity of capabilities and models, a comprehensive examination of the process of their evaluations and improvements under multiple application scenarios and different needs is performed. These experiences could serve as references for credibility estimation of unpracticed scenarios under the same scope, which can be referred to in the literature [43].

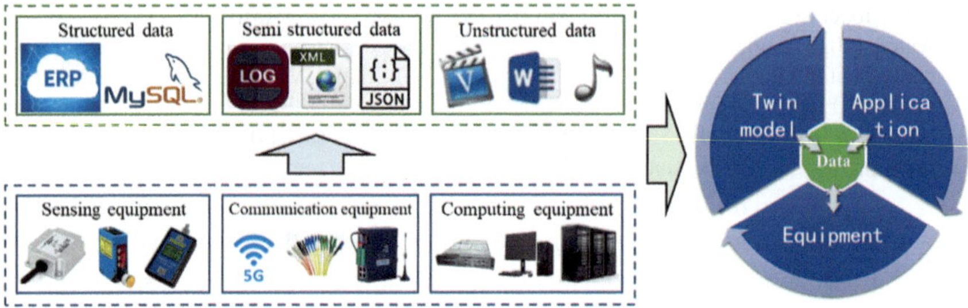


**Fig. 8** Multi-source heterogeneous data of DT

### 5.2.4 Evolution Evaluation

Evolution evaluation focuses on the evolution process of the simulation model. There are some similarities between the dynamic characteristic analysis of the traditional basic model and the evolution evaluation of the twin model. However, they are significantly different. On the one hand, the evolution of the model makes the dynamic characteristic curves change continuously. So the evolution evaluation cannot only analyze the curves at a specific time point, but rather, it has to comprehensively consider the entire sequence of curves on a period of time window. The key element to be examined during evolution evaluation is the change process of the basic model sequence. On the other hand, the twin model shows the property of modeling–application–evaluation cycle as mentioned in Sect. 4.2. Thus, during an evolution process, the object keeps being assessed and the twin model may change accordingly. Then, the evaluation criteria should be changed as well.

In general, there are three types of credibility evaluation triggering methods corresponding to the changes of the object: time-driven, event-driven, and hybrid-driven. There are also four types of time windows, which are shown in Fig. 9. In the scenario of online dynamic evaluation of Digital Twins, it is necessary to divide the time windows according to the actual triggering mechanism, periodicity, and overlapping situation. Generally, the evolution-driven discriminative index $I_{\mathrm{d}}^{\mathrm{Dy}}$ is evaluated using F23, F31, and the sensitivity index $I_{\mathrm{se}}^{\mathrm{Dy}}$ is analyzed using methods such as F51. For a single time window evolution process, its timeliness index $I_{\mathrm{t}}^{\mathrm{Dy}}$ can be checked using methods such as F71, and its convergence index $I_{\mathrm{cv}}^{\mathrm{Dy}}$ can be analyzed using methods such as F73. For the corresponding evolution result, the systemic index $I_{\mathrm{sy}}^{\mathrm{Dy}}$ of the evolution can be assessed using methods such as F74, and the correctness index $I_{\mathrm{cv}}^{\mathrm{Dy}}$ of the evolutionary result can be analyzed using methods such as F72.

The dynamic integration method is based on multiple time windows. The intrinsic correlation between multiple triggering mechanisms is analyzed. Then, the dispersion of the weights on the multiple time windows are obtained, so that the dynamic index values could be continuously synthesized from the basic index values.

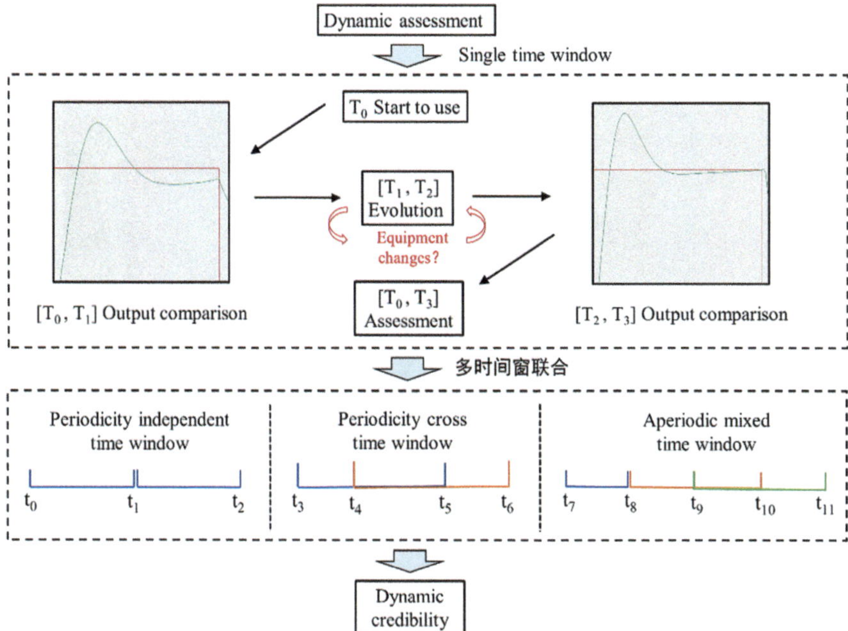


**Fig. 9** Multiple time windows' based credibility evaluation

Uncertainty is examined at the last step. For each index value, the data, evaluation methods, and management methods involved in the acquisition process should be examined. Both the inherent uncertainty, including systematic parameter errors, environmental perturbations, and the cognitive uncertainty, including inappropriate model forms and empirical parameter errors are included. The failure probability of the assessed value brought about by the uncertainty is then quantified and uniformly mapped to the distribution of confidence intervals. By combining this distribution with its degree of correctness and timeliness, the degree to which the model actually meets the demand, which is the degree of credibility, could be obtained. Based on the requirement-model mapping, the single-index value, multi-dimensional integration value, and dynamic evaluation value are transformed into the final dynamic multi-dimensional distribution of the credibility.

## 6 Example of Credibility Evaluation of a Digital Twin

The process of implementing the credibility evaluation methodology framework is illustrated based on the Digital Twin of the ROKAE industrial robotic arm constructed by the research team. The experimental environment is shown in Fig. 10. The hardware used is the XB4 robotic arm, the Mech-mind Nano depth camera, the Dahuan

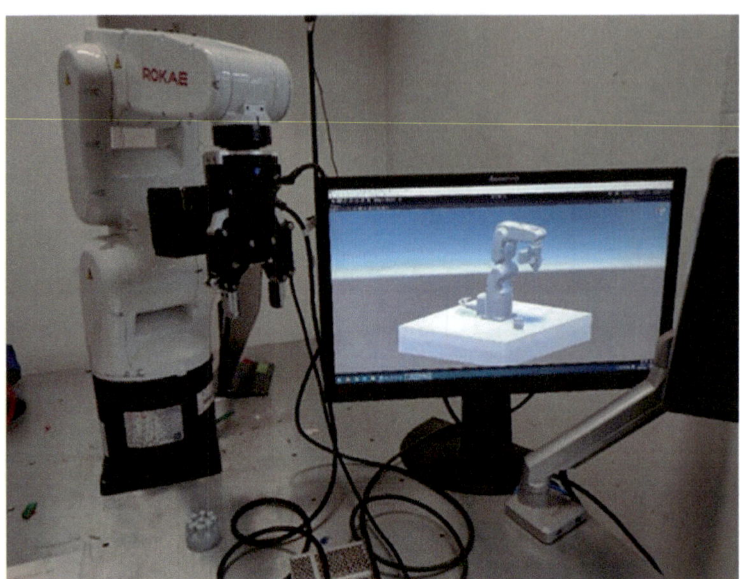


**Fig. 10** DT environment of a robotic arm

AG95 claw, sensors such as thermometers, hall sensors, voltage and current sensors, and communication devices such as routers and edge gateways. The software used is Unity, ANSYS Workbench, C#.

The credibility evaluation started with a multi-dimensional decomposition of the robotic arm DT model by the modeling experts. The process is followed by analyzing the requirements for "grasping a target and placing it in a specified area". The two key inspection points are then identified: "stable grasping with appropriate force" and "timely routing and obstacle avoidance". Afterward, the model is decomposed step by step from qualitative to quantitative to obtain the specific index system and the acceptable range of requirements. In this step, the normalization index $I_{\mathrm{n}}^{\mathrm{S}}$ and completeness index $I_{\mathrm{c}}^{\mathrm{S}}$ of the requirements-model mapping are assessed.

A unidimensional evaluation focusing on multi-disciplinary domains and simulation forms was then conducted. Among the multi-disciplinary domains, geometry, kinematics, mechanics, control, electricity, and magnetism have a strong influence on the grasping requirements, so the correctness of their disciplinary models is the focus of the evaluation. Meanwhile, the correctness of the system structure, code and interface specification, and logical correctness of the simulation models are the main investigations of the polymorphic dimension.

Thereafter, the compatibility index $I_{\mathrm{c}}^{\mathrm{D}}$ of the acquisition communication device and the authority index $I_{\mathrm{ao}}^{\mathrm{D}}$ of data sources such as vendors and built-in software of the robotic arm were evaluated using formal review and expert scoring methods. The authenticity index $I_{\mathrm{ae}}^{\mathrm{D}}$ of historical modeling data, real-time status data, and requirement data was evaluated in depth using a variety of data validation methods.

The overall technology maturity was assessed using subjective and formal methods in conjunction with historical data related to model development and application. The uncertainty of each evaluation component was analyzed using dataset validation and result comparison methods.

After completing the basic evaluation, a multi-disciplinary and multi-dimensional integration was performed using a modified DT theory of evidence. The integration was based on the following Dempster synthesis rule:

$$m(A) = \begin{cases} 0, (A = \Phi) \\ \frac{\sum_{A_i \cap B_j = A} m_1(A_i) m_2(B_j)}{1-K}, (A \neq \Phi) \end{cases},$$

where $K$ is the conflict factor which meets $K = \sum_{A_i \cap B_j = A} m_1(A_i) m_2(B_j) < 1$ and reflects the degree of conflict between the various pieces of evidence.

On this basis, the two key evolution processes of the robotic arm attitude change and real-time deformation of the gripped object are assessed. The process and results of the calibration of each parameter and the geometrical structure change are systematically examined. The time-frequency domain analysis and the convergence analysis are comprehensively used to obtain the six index values of the whole evolution process. The quantitative value of the model credibility is given by combining with the uncertainty analysis.

In the credibility synthesis process, the weights of each index are given mainly based on the analysis of scene requirements and experts' experience. The weights assigned to the data indexes are relatively small because the experimental environment is relatively simple and stable, and the robotic arm as well as the related acquisition equipment has high reliability. High weights are assigned to the indexes related to unidimensional and coupled evaluation due to the high impact of the correctness of the discipline model on the grasping requirements. Because the real-time calibration of the robotic arm attitude changes and the deformation of the grasped object has a greater impact on the grasping and obstacle avoidance needs, high weights are also assigned to the evolution-related indexes. The other indexes have less influence on the application requirements of the robotic arm twin model and are uniformly assigned smaller weights. A multi-objective optimization algorithm will be used in the subsequent study to determine the optimal weight allocation for each index.

To validate the credibility evaluation methods, four types of DT models with different credibility are constructed to form a basic model benchmark, as shown in Fig. 11. The grasping effects of the four types of DT models are shown in Fig. 12.

The horizontal axis is the time axis, and the vertical axis is the distance between the grasped object and the end point. The evaluation calculation results give the order of model credibility as Model 3 > Model 4 > Model 2 > Model 1, which is consistent with the experimental performance results. Thus, the effectiveness of the credibility evaluation framework proposed in this chapter is preliminarily validated. It is necessary to expand the library of benchmark models in terms of equipment type, model mechanism, and complexity of scenario requirements, so that the credibility evaluation framework could be developed and validated in a wider scope.

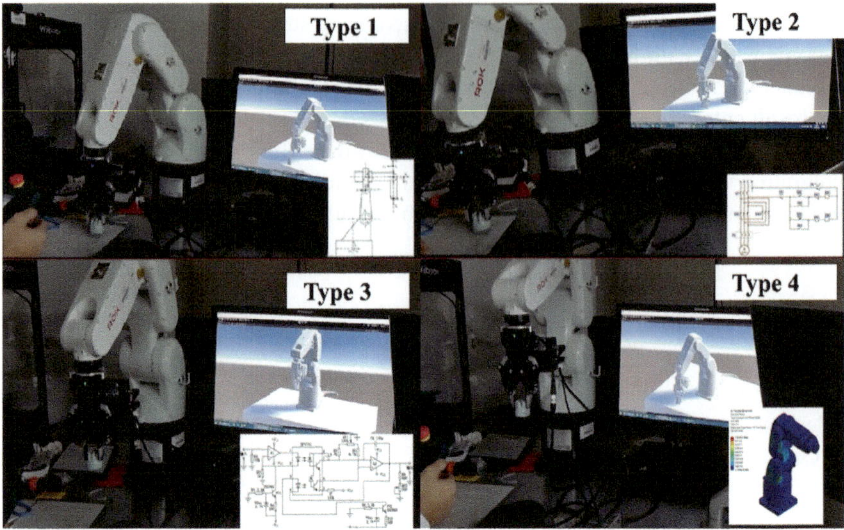


**Fig. 11** Grasping experiments on four types of models with different credibility

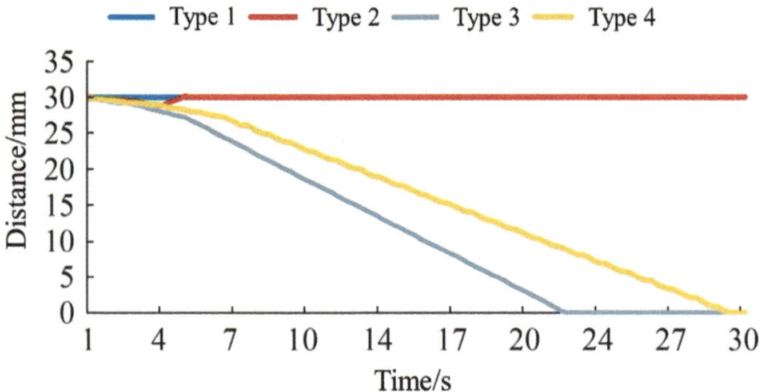


**Fig. 12** Experiment results of four types of models with different credibility

## 7 Conclusions

This chapter analyzed three typical characteristics of Digital Twins: dynamic evolution, real-time interaction, and system complexity. On this basis, the definition of credibility of Digital Twin is given. Then, a set of credibility evaluation index system is proposed based on previous researches and practical experiences. The general process of DT credibility evaluation is described afterward. The credibility evaluation framework is validated through experiments.

In the future research, the evaluation index system will be further extended and refined so that credibility can be described more comprehensively and accurately. Moreover, some single-index evaluation methods will be improved or supplemented, such as uncertainty quantification method and credibility integration method. In addition, more effective quantification methods will be developed for the credibility evaluation of the evolution model. The refinement and combination of evaluation methods would be further explored according to the commonality and difference of the geometric appearance model, the system structure model, and the mechanism model. The benchmark library of Digital Twin models would be extended to better validate the generalization capability of the credibility evaluation framework. An evaluation software system would be developed based on the framework to support the automatic credibility evaluation to a certain extent.

## References


1. Zhang L, Zhou L, Horn BKP (2021) Building a right digital twin with model engineering. J Manuf Syst 59:151–164
2. Shafto M, Conroy M, Doyle R et al (2010) Technology area 11: modeling, simulation, information technology and processing roadmap. NASA Office of Chief Technologist, November 2010
3. Boschert S, Rosen R (2016) Digital Twin-the simulation aspect. Springer International Publishing, Cham, pp 59–74
4. Bielefeldt B, Hochhalter J, Hartl D (2015) Computationally efficient analysis of SMA sensory particles embedded in complex aerostructures using a substructure approach. In: ASME 2015 conference on smart materials, adaptive structures and intelligent systems. ASME, New York, USA, V001T02A007
5. Liu S, Bao J, Lu Y et al (2021) Digital Twin modeling method based on biomimicry for machining aerospace components. J Manuf Syst 58(Part B):180–195
6. Zhang L (2020) Cold thinking about the digital twin and the modeling and simulation techniques behind it. J Syst Simul 32(4):1–10
7. Zhang L, Lu H (2021) Discussing digital twin from of modeling and simulation. J Syst Simul 33(5):995–1007
8. Tuegel E J (2012) The airframe digital twin: some challenges to realization. In: 53rd AIAA/ASME/ASCE/AHS/ASC structures, structural dynamics and materials conference. 20th AIAA/ASME/AHS adaptive structures conference 14th AIAA, 1812
9. Huang J, Gheorghe A, Handley H et al (2020) Towards digital engineering: the advent of digital systems engineering. Int J Syst Syst Eng 10(3):234–261
10. Zhang L, Wang K, Laili Y et al (2022) Modeling & simulation based system of systems engineering. J Syst Simul 34(2):179–190
11. Rabeau R (2013) Credibility in modeling and simulation. In: Simulation and modeling of systems of systems, pp 99–157
12. Department of Defense (2007) DoD Modeling and Simulation (M&S) Management. Directive Number 5000.59, 8 Aug 2007
13. Sargent RG (2015) Model verification and validation. In: Modeling and simulation in the systems engineering life cycle: core concepts and accompanying lectures. Springer London, pp 57–65
14. Grieves M (2019) Virtually intelligent product systems: digital and physical twins. In: Flumerfelt S et al (eds) Complex systems engineering: theory and practice. American Institute of Aeronautics and Astronautics, pp 175–200

15. Lim KYH, Zheng P, Chen CH (2020) A state-of-the-art survey of Digital Twin: techniques, engineering product lifecycle management and business innovation perspectives. J Intell Manuf 31(6):1313–1337
16. Liu M, Fang S, Dong H et al (2021) Review of digital twin about concepts, technologies, and industrial applications. J Manuf Syst 58:346–361
17. Lu Y, Liu C, Kevin I et al (2020) Digital Twin-driven smart manufacturing: connotation, reference model, applications and research issues. Robot Comput Integr Manuf 61:101837
18. Sun Y, Luo H, Das SK (2012) A trust-based framework for fault-tolerant data aggregation in wireless multimedia sensor networks. IEEE Trans Dependable Secure Comput 9(6):785–797
19. Jiang J, Han G, Wang F et al (2014) An efficient distributed trust model for wireless sensor networks. IEEE Trans Parallel Distrib Syst 26(5):1228–1237
20. Miao CL (2014) Research on trustworthiness problems in wireless sensor networks. University of Science and Technology of China
21. Mao HM (2016) Research of ultrasonic sensor detection model based on credibility evaluation. Shenyang Jianzhu University
22. Liao M, Liu J, Meng Z et al (2021) A sins/sar/gps fusion positioning system based on sensor credibility evaluations. Remote Sens 13(21):4463
23. Levelt P, Caramazza A (2007) The Oxford handbook of psycholinguistics. Oxford University Press, USA
24. Barber A, Robert JS (2010) Concise encyclopedia of philosophy of language and linguistics. Elsevier
25. Tang JB (2009) Research on the credibility of warfare simulation system. Graduate School of National University of Defense Technology
26. Liao Y, Feng XJ, Zhang J et al (2003) Research on calibration, verification and validation (VVA) technology for strategic missile system modeling and simulation. In: Proceedings of the 2003 National Academic conference on system simulation, pp 187–192
27. Yang M, Zhang B, Wang ZC et al (1999) VV&A and credibility evaluation of distributed interactive simulation system. J Harbin Inst Technol 05:48–51
28. Kim CJ, Lee D, Hur S et al (2017) M&S VV&A for aerospace system development: M&S life-cycle model and development paradigm. J Korean Soc Aeronautical Space Sci 45(6):508–516
29. Zhou JH, Xue JJ, Li HY et al (2020) Thinking on digital twin for weapon system. J Syst Simul 32(04):539–552
30. Zhang SL, Ye MC (2006) Simulation credibility evaluation method for missile weapon system. Comput Simul 23:48–51
31. Zhen BT, Hu Y (2015) Research on foreign weapon and equipment modeling and simulation credibility assessment methods. Qual Reliab 2:6–9
32. Sim SH, Lee JC (2014) On improving the verification, validation and accreditation process by including safety requirements in M&S-based development of weapon systems. J Korea Safety Manage Sci 16(4):123–131
33. Beydoun G, Low G, Bogg P (2013) Suitability assessment framework of agent-based software architectures. Inf Softw Technol 55(4):673–689
34. Acar E (2015) Effect of error metrics on optimum weight factor selection for ensemble of metamodels. Expert Syst Appl 42(5):2703–2709
35. Li N, Dong L, Zhao L et al (2021) A credibility evaluation method for complex simulation systems based on interactive network analysis. Simul Model Pract Theory 110:102289
36. Ferson S, Oberkampf WL (2009) Validation of imprecise probability models. Int J Reliab Saf 3(1–3):3–22
37. Li W, Chen W, Jiang Z et al (2014) New validation metrics for models with multiple correlated responses. Reliab Eng Syst Saf 127:1–11
38. Dornheim H, Brazauskas V (2011) Robust–efficient credibility models with heavy-tailed claims: a mixed linear models perspective. Insurance Math Econ 48(1):72–84
39. Liang J, Bai Y, Bi C et al (2013) Adaptive routing based on Bayesian network and fuzzy decision algorithm in delay-tolerant network. In: 2013 IEEE 10th international conference on high performance computing and communications & 2013 IEEE international conference on embedded and ubiquitous computing. IEEE, pp 690–697

40. Hu XL (2011) Dynamic data driven simulation. SCS M&S Mag 5:16–22
41. Wang K, Zhang L, Jia Z et al (2023) A framework and method for equipment digital twin dynamic evolution based on IExATCN. J Intell Manuf 1–13
42. Wang K, Zhang L, Cheng H et al (2023) A lifelong learning method based on event-triggered online frozen-EWC transformer encoder for equipment digital twin dynamic evolution. IEEE Internet Things J
43. Zhang L, Liu Y, Laili Y et al (2020) Model maturity towards modeling and simulation: concepts, index system framework and evaluation method. Int J Modeling Simul Sci Comput 11(03):2040001

# On the Importance of Simulation and Digital Twin for Industrial Applications

**Stefan Boschert, Christoph Heinrich, Vincent Malik, Roland Rosen, and Uliana Soellner**

**Abstract** The term Digital Twin has become ubiquitous recently as virtual representation of a physical asset to leverage business opportunities, i.e., with a dedicated purpose. It consists of very heterogeneous data and can take very different shapes, from the collection of datasheet information together with a visual reproduction of the real object via the compilation of operational data to the application of physics-based simulations or predictions based on artificial intelligence (AI). Obviously, any combination thereof is also possible. The applications of Digital Twins are spread over the whole lifecycle of products and systems including the interaction of Digital Twins at different hierarchical levels. In addition, Digital Twins can create a connection between different value chains. An example is the exchange of Digital Twins along the supply chain with a link to the shop floor, which then enables the execution of cooperative simulations for different purposes. In our book chapter, we investigate the role of simulation for a Digital Twin with respect to value chains, the lifecycle phases, and their business value. Furthermore, in the age of the Industrial Metaverse the importance of simulation for Digital Twins is getting even more important, as immersion, interaction, and collaboration in real-time play a greater role, making it necessary to have fast and good enough predictions of the behavior of industrial assets. Novel algorithms and machine learning (ML) methods enhance the capabilities of the traditionally engineering-focused simulation tools. The mentioned aspects will be illustrated with several examples of real-life industrial applications.

**Keywords** Digital Twin · Simulation · Industrial Metaverse

## 1 Introduction

The efficient development and optimal operation of automated systems are well-known, but still challenging. Automated systems are mostly complex technical systems that are controlled by (automation) software and increasingly networked with

S. Boschert (✉) · C. Heinrich · V. Malik · R. Rosen · U. Soellner
Technology, Siemens AG, 85748 Garching, Germany
e-mail: stefan.boschert@siemens.com

 
M. Grieves and E. Hua (eds.), *Digital Twins, Simulation, and the Metaverse*, Simulation Foundations, Methods and Applications, https://doi.org/10.1007/978-3-031-69107-2_5

one another, which transforms technical systems into cyber-physical systems (CPS). Concrete examples are production systems, their sub-systems, and components in the process and manufacturing industry.

In addition to the consideration of quality, time, and cost targets, the flexibility of these systems in operation is increasingly coming to the fore. At the same time, the variety of product types is increasing and the batch sizes are becoming smaller. Short-term changes in orders as well as changes in supplies are necessarily to be taken into account. Instead of supplementing classic automation, Digital Twins are increasingly being named and realized to fulfill these functions. In this chapter, we consider Digital Twin applications that have simulation-based and data-driven function modules and can improve operation in various ways (cf. [1, 2]).

The term Digital Twin has grown historically, introduced in the context of PLM as to collect the information about a product across its lifecycle [3] and first mentioned as Digital Twin in a NASA Roadmap in 2010 [4]. Nowadays, it contains all information about an asset for the entire life cycle and becomes even a synonym for the asset administration shell in the Industrie 4.0 context [5]. Unfortunately, no consistent definition is used in practice. It is mostly shaped by the perspective and use of the respective user.

In the following, we discuss different lifecycle aspects of Digital Twins and motivate the concept of bundling best practices for simulation-based Digital Twins in the context of engineering and automation. Then we discuss several novel applications of the Digital Twin and give concrete examples of their realization. We conclude with a discussion on the future development of Digital Twins.

## 2 Digital Twin Aspects and Goals

In recent years, we have seen an increasing digitization and the triumph of Internet technologies. Most of the products and systems are created in a digitally supported development process. Standardized interfaces and data formats are the foundation of integrated toolchains for the creation and execution of models during all lifecycle phases [6]. In operation phases, Internet technologies for data acquisition, data transformation, and data analysis (e.g., by artificial intelligence (AI) methods) have become cheap, easy to install and use, and have been commoditized. The pervasion of virtual and augmented reality (VR/AR) into daily-life applications, called Metaverse, is currently creating a demand for interactive and intrusive solutions. All this leads to an increased interest and awareness of Digital Twins and their purpose and goal.

## 2.1 Digital Twin in the Lifecycle

Digital representations are found everywhere during the lifecycle of a system or product. Already during the initial conceptual phase, first ideas, mainly in the form of requirements, are collected mostly in a semi-formal way. In a more systematic way, the requirement specification is created, to describe the basic behavior of the final system – a first comprehensive description. During the following design and engineering steps this description is continuously modified and complemented with new information and data. First (system) architectures, CAD representations of sub-components, and simulation models to virtually test the components against the given requirements are developed, see Fig. 1. Often different versions and variants of each of those digital artifacts must be considered as well. Even if the real system or product is not yet built, this set of information gives a reasonable picture of the final product and its behavior—at least under certain pre-considered situations. They form a virtual representation of the knowledge about the physical asset—an initial Digital Twin [7, 8].

The Digital Twin does not only contain descriptive information like CAD models, but also executable models in the form of simulation models, based on physical laws or heuristics like previous experience. This allows us to calculate the behavior of the physical asset for a pre-defined situation. It must be noted that these models are specific for their intended use, and it must always be checked if the model is suitable (i.e., it satisfies its prerequisites) for the current situation under consideration. If this is not the case, a new model or a modification of an existing similar (simulation) model must be created to answer the specific design question or to validate specific properties. This newly created model becomes then part of the Digital Twin as well. As such, the Digital Twin becomes more and more comprehensive.

Another aspect to consider is that most systems are not completely developed in all details by a single stakeholder. We must consider the supply chain as well. Partners and suppliers are involved who supply components or sub-systems. These suppliers in turn have suppliers and both use Digital Twins to describe their system. This results in a linked chain and an exchange of digital artifacts respectively Digital Twins. This finally forms a Digital Twin ecosystem.

Looking at the manufacturing of a product, Digital Twins can support it as well. On the one hand, the compatibility of the product with the production line can be tested in advance. This can be done for example by checking different geometrical or

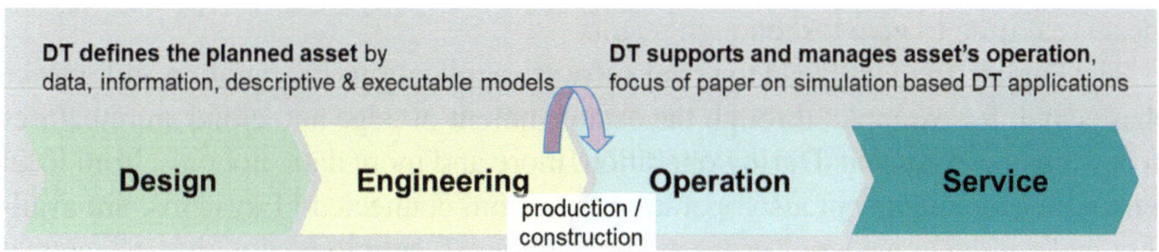


**Fig. 1** Digital Twin along the life cycle

functional parameters of the product with the capabilities of the production system. On the other hand, the Digital Twin of the production line can be used to test out different operating/manufacturing strategies. There is also a certain analogy in the process industry. Here, too, the behavior of the process plant can be tested and validated.

The production, however, is a stage in the lifecycle of the product where the characteristics of the Digital Twin change significantly. Before the production of the physical asset, only the "theoretical"/desired behavior of the product in a generalized environment is considered assuming estimated operating conditions. As soon as the product is realized, individual aspects of each of the products are added to its Digital Twin. This could be the manufacturing history like "manufactured on machine x with quality y", "used parts/substances from supplier z" or "batch id number and environment conditions like temperature or humidity". Especially in the case of mass products, each of the products carries along its own production and operational history, even if the basic product description is initially the same. In [9], this transformation is called instantiation of the Digital Twin: The Digital Twin prototype (DTP) that was developed during the design becomes a Digital Twin instance (DTI) of each individual product. Consequently, the content of the initial DTP is also available as a common set of information in all instances. This common set of knowledge is complemented by individual aspects of each instance, like the individual and specific environment conditions or operational modes. Using these individual data, individualized predictions for each asset can be made. This requires on the other side that the data for each asset must be treated individually, which can be done locally—on the edge—or in a more centralized way or in a mixture of both extremes, depending on the availability of computing resources.

Based on this set of information/data, executable models can be created to describe a specific behavior aspect of the asset. Usually, these models are either numerical simulation models based on a mathematical description of the relevant physics, or data-based models that are trained with available sensor data. The desired accuracy of these models widely depends on the necessities of the specific use case. As these requirements are often opposite to each other, one has to compromise, e.g., execution speed versus accuracy, resulting in a "good enough" simulation model for the specific purpose. Therefore, it could also happen that models of the same physical effect are part of the Digital Twin in different model fidelity.

Further challenges arise as the model must cope with real sensor data. Measurements include uncertainties that should be considered in the simulation model as well. And more importantly, the simulation must process the sensor information in (near) real time to give fast enough results.

The Digital Twin will lead to novel software applications in operation and service phases and, for example, through the establishment of edge and cloud architectures to new business models. During operation, more and more data, not only from local sensor measurements but also operating data from connected IT systems, are available and can be considered in these applications. Technologically, this means that simulation models are also combined with data-based methods and AI techniques.

Typical applications of such simulation-based Digital Twin applications support systems for operation, where for example operation alternatives can be tested in advance on the virtual system so that the operator can choose the best option. In some situations, even a fully automatic optimization of the operation is possible, if the models and data provide a sufficient accuracy and execution speed. Another typical application could be the support of maintenance actions. Running a simulation in parallel to the real asset offers the possibility either to detect derivations from the expected (i.e., simulation) and the real behavior (obtained from sensor readings) or to improve operational aspects. Concrete applications are virtual sensors, calculations of remaining useful lifetime, failure detection, and identification or improved spare part management.

One fascinating idea with Digital Twins is that different types of digital information like descriptive and executable models can be used in a linked way. For a machining tool, a service application can be built which calculates the optimal process time and process parameters (=executable model), depending on the state of the machine, tool, and orders (=descriptive data). Another example is an online production planning application, which optimizes the production steps of each product unit in a line based on the actual order times, current machine states, and related maintenance plans.

Looking at the supply chain perspective, the linked digital artifacts along all tiers form a root network of data. This makes it possible to trace this information and answer questions such as "Which parts does my system consist of?", "Which suppliers manufactured it?", and "Of what quality is it?", and "Which footprint does it have?". When considering these topics, often the term "Digital Thread" is used, which emphasizes the temporal and logical connection of the information. Digital Threads help to navigate through them in order to solve specific challenges like product improvements swiftly and reliably.

## 2.2 *Bundling Best Practices for a Novel Simulation-Based DT Approach*

From a research point of view and to derive technical ideas for the preliminary development, we propose a novel simulation-based Digital Twin concept which extends traditional simulation approaches and improves most likely the usage of Digital Twins. This vision of the future has developed further in recent years (see [5, 10, 11]) and manifests itself in the following characterization:

The (in previous papers called next generation) Digital Twin refers to a description of a component, product, system, infrastructure, or process by a set of well-aligned, descriptive, and executable models. It is a semantically linked collection of all relevant digital artifacts, including design and engineering data, operational data, and behavioral descriptions.

The Digital Twin has several characteristic features:

- It evolves along the lifecycle—use and content should be pre-planned but is open for additional, un-planned artifacts.
- It integrates the currently available and commonly required information and knowledge.
- It addresses (proto)type and instance aspects.
- It is synchronized with the physical world—the type of synchronization changes over the different lifecycle phases.
- It offers an efficient way to generate new services and applications—besides extended solutions in development new offerings in operation and service promise new business opportunities.

With respect to the focus of this paper on simulation-based Digital Twin application, we see new or at least increased business value in a realization of this approach even if several individual aspects need to be considered. This will be illustrated in the following section.

## 3 Novel Simulation-Based Digital Twin Applications

### *3.1 Digital Twins in Networked Value Chains*

The development and use of technical systems almost always take place in a network of development partners, suppliers of parts, components, and products, and of customers who use these systems or in turn integrate them into their systems and sell them to their customers. At the end of this chain is the end customer who uses or operates this system. An example illustrates these connections of partners, e.g., the component could be an electric drive designed for a control valve. This product is manufactured in the manufacturing plant and installed in a production plant. Or the valve is used by system manufacturer to manufacture a complete process module. It is an open, dynamically changing network. After all, information is already exchanged between partners or suppliers and customers today (and in the past). The Digital Twin approach will significantly increase the effectiveness and efficiency of information exchange, as more complex knowledge can be exchanged using models. The artifacts exchanged can be Digital Twins or, in terms of the description given above ("a set of well-aligned, descriptive, and executable models"), a subset of digital artifacts. Regarding the simulation part, the previously mentioned modularization is essential here, as it improves reuse and thus increases cost-effectiveness. To put it briefly, the previous fragmented exchange will be replaced by comprehensive puzzle pieces in the future. In implementation, their unambiguous description of their content and their interfaces is important to be able to use them as independent simulation-based Digital Twins or to connect them with other twins (puzzle pieces). In this way, these Digital Twins are transformed into independent products that are subject to specific

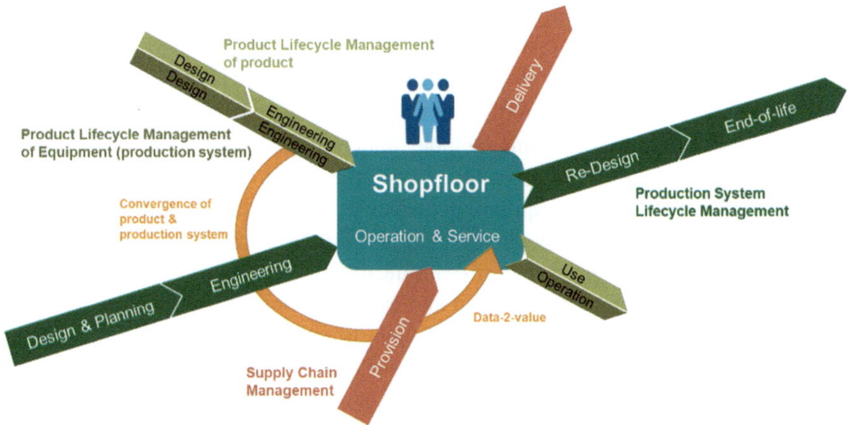


**Fig. 2** Crossing of value chains in production, adapted from [14]

quality requirements [12, 13]. Encapsulation in modules also improves maintainability, because if an update like the change of a physical asset is necessary, the corresponding module can be updated or replaced in a targeted manner.

One application focus for Digital Twins is the creation, operation, and modernization of production facilities. Figure 2 (adapted from [14]) shows the value chains of the product, the production system, and the supply chain that meets on the shop floor. Each value chain represents related views and tasks [15]:

- Product Lifecycle Management of the product (PLM): This is the systematic process of managing the product over its whole lifecycle. This includes ideation, design, engineering tasks as well as its production and service up to disposal of the product. This product is the output of the production system under consideration. A second type of products is the means of production (PLM of equipment). These are machines, actors, sensors, automation systems, and so on which are the elements of the production system itself.
- Production System Lifecycle Management (PSLM): It handles all processes which are relevant for the production system and its context like the factory building or plant area. The PSLM contains the phases design, engineering, construction and commissioning, operation (including service and maintenance), and end-of-life aspects.
- Supply Chain Management (SCM): SCM is the management of the flow of goods and services. An important aspect for the production is the logistics of raw material and supply parts as well as warehouse aspects and delivering of finished products.

To achieve economically successful and sustainable production, the combination of information from these value chains is required. The Digital Twin is an important success factor here. It contains various information and thus transports knowledge across the lifecycle phases and, analogous to the representation in Fig. 2, it enables a linking of the information, e.g., between product development and production.

A digital convergence of product and production system expands the opportunities in the planning and development of production systems as well as in their (virtual) commissioning and operation. Potential benefits of (simulation-based) DT applications lie in the construction, modification (e.g., product-specific reconfiguration), and expansion of production. This includes matching product requirements and the capabilities of production resources (e.g., machines and process units) together with the electrical and automation equipment properties of the plant in planning and validation tasks. In case of short-term production disruptions or changes in the supply chain (e.g., changed deliveries in terms of time, quantity or quality as well as changed orders), alternatives can be determined and evaluated online. This requires a media discontinuity-free transfer of all Digital Twin artifacts which are required to implement simulation-based applications. To achieve this goal, practical implementations for the above-mentioned modularization concept must be achieved. An increased use of semantic models will help to link all information.

Increasing the economic benefit from the further use of measurement and operating data is summarized by the buzzword data-to-value. The simulation-based Digital Twin will expand the (so far) often purely data- and AI-based methods by including simulation models from the development phases of the production equipment and the production system. This extends the performance capabilities of diagnostic and service applications and allows online controls for operation optimization [10, 11]. The synchronization of the Digital Twin with its physical asset is decisive for the technical feasibility. The physical asset's status must be identified adequately and in a purpose-oriented way by the input data coming from the real asset. The technical implementation needs to consider the specific required update cycles and transfer rates.

Such a Digital Twin, combining information from different domains, will continue to reinforce the development and operation of cyber-physical systems, which are increasingly moving toward autonomous systems. These aspects are presented in the next section.

## *3.2 Digital Twin Contributions Toward System Integration*

Autonomous systems have become popular for several years, especially through assistance systems in vehicles. In addition to the application in the automotive industry, this development direction is also relevant for other complex technical systems, e.g., in manufacturing and process industries.

According to Bekey's definition [16] "Autonomy refers to systems capable of operating in the real-world environment without any form of external control for extended periods of time", autonomous systems are intelligent machines (or systems again) that execute high-level tasks without detailed programming and without human control. They know their capabilities and skills, and their internal state, and they use sensors to perceive their environment and the current situation. They can make decisions based on scenario analysis on how to proceed. In order to make this happen, the autonomous systems will need access to models that are as fine-grained as the use case requires and their own behavior in interaction with their environment in the real world—these are typical tasks of a Digital Twin [2]. This once again proves the relevance of the topic.

Full autonomous behavior of systems will remain a vision for years, but we will continuously approach this goal. The concept of Digital Twins will be very useful for this. An intermediate step is the realization of cyber-physical systems (CPS) which communicate with their environment and can fulfill more complex tasks. The economic realization of such CPS will require new development methods, and the Digital Twin approach will support it.

A typical approach for system development is to start from requirements and define functional blocks to fulfill them. These functions are further sub-divided into logical systems which are finally realized in their physical representation. Mechatronic systems become more and more cyber-physical, which means that the communication aspect either within the system but also between different systems becomes more and more important. Therefore, the system under consideration becomes larger and finally harder and harder to understand in all details. The quest for the optimal system design becomes challenging. One approach could be to split the full system into several sub-systems, e.g., functional blocks and optimize their operation. Especially in classical mechatronic systems like automotive industry this approach delivers good results, e.g., by sub-dividing the system "car" into many different components which could be finally produced by a large ecosystem of suppliers, each one specializing on some components and optimizing the components by themselves. Only by integration of the components to the full system are the interactions (physical or communication) between the separate components realized in a pre-defined way—mostly determined by the initial functional and logical sub-division of the full system.

However, as the communication aspect becomes more dominant in cyber-physical systems, it also becomes more challenging to identify components that are mutually related and could benefit from each other. This becomes obvious at the example of a logistics system, where on the one side of a conveyor belt the belt is loaded from different sources, and on the other side, the material must be sorted into different buckets. If on the one side the sources are running at maximum speed (are optimized for maximum performance), the sink side could run into trouble as several buckets get overloaded—even if the sink side runs on maximum as well as some buckets are full and must be emptied. In this situation, it would be helpful to slow down the performance of the source side—or modify the mixture in such a way that the sinks are more evenly accessed—to maximize the overall performance of the system. Here source and sink form a symbiotic pair by coordinating their behavior to the benefit of both components and finally the overall system.

A central part for the realization of such symbiotic effects is to establish a communication link between the components. This, however, requires that the relevant components are already identified. Approaches to identify such symbiotic pairs can be found, e.g., in [17]. The identification of the symbiotic pairs should have an impact on the design of the control system as well, as a direct communication is established more easily if the components are controlled by the same controller. Obviously, an initial partition (probably dominated by geometric reasons) of the system must be adjusted after the identification of symbiotic interdependencies.

From the benefit point of view, the operation of systems that can solve complex tasks independently is attractive. A common approach for production (equally in the manufacturing and process industries) is that production systems become more modular. These modular units will act more independently. To do this, they need to exchange information with each other. This is one of the basic principles of Industrie 4.0. With the definition of the Asset Administration Shell (AAS) [6], vendor-independent exchange of information is being driven forward by corresponding standardizations. This information relates, among other things, to machines and their components as well as other information such as order data. Since Digital Twins are required for decision making in autonomous systems, the AAS must also provide services for using the Digital Twins. One aspect of the simulation of sub-models is that it makes intensive use of co-simulation concepts [18].

The development of CPS and the use of Digital Twins are an economically relevant and a technologically interesting intermediate step. In production, the aim is to create cyber-physical production systems that are structured more modular and enable better autonomous operation of the production units. Their benefit is the increasing flexibility. Changes in the production process and intralogistics can be implemented ad hoc. Like cars, the technical possibilities and the trust in the decisions of these systems will grow step by step. Digital plant companions which leave the final decision to humans represent an intermediate step here [19].

## *3.3 Simulation in the Metaverse Era*

Only recently has the Industrial Metaverse (IMV) as evolution of the Digital Twin era emerged. Whereas the term Metaverse is already used in the gaming market, it is developing in three additional fields. In the consumer market, it represents an avatar-centric virtual world for socializing and entertainment, whereas in the enterprise market, it is used in connection with real-time collaboration to facilitate office tasks employing large language models in the future. In contrast to that, the IMV as the third field aims at solving real-world problems, increasing productivity, shortening time-to-market and, thus, adding value to industry. Prognosis of market segmentation even sees the IMV at around 100 billion US$, outcompeting the other two fields by a factor of more than 2–3 [20].

Concretely, the IMV is a space to experience Digital Twins of industrial assets with the key features immersion, collaboration, and interaction where the combination goes beyond Digital Twins. The visualization of the Digital Twin in its context and realistic environment leads to new insights. Meetings between different stakeholders to jointly review Digital Twins facilitate problem solving where changes can be directly applied. Moreover, interactive simulation of the Digital Twin behavior and faster-than-real-time prediction leverage predictive maintenance to a new level. The connection to the physical assets to monitor and analyze in a closed loop is a major differentiator of the IMV.

We see the IMV as a convergence of various key technologies to create industrial virtual worlds, e.g., Internet of Things, edge and cloud computing, artificial intelligence machine learning, user experience, networks (5G/6G), and blockchain. But the key enabler of the IMV is the photorealistic, physics-based, and real-time capable Digital Twin where in turn simulation is the main ingredient, see Fig. 3. Simulation based on the conservation of physical laws, like, e.g., the finite element method (FEM) for structure mechanics, is physically correct by construction and thus makes it possible to get the accuracy needed for the respective application. This drawback of AI-based methods is currently a strong research field with the goal to fulfill physical laws also for, e.g., neural networks.

The IMV creates added value along the lifecycle and connects the different phases. In design and engineering, it helps to perform interactive simulations with collaborative review and evaluation. Photorealistic environments for synthetic data generation speed up virtual testing and validation tremendously. In production, new assets and control software can be tested in the virtual worlds and deployed to the real hardware after successful virtual commissioning. Remote control of assets and their service reduce downtimes in operation, saves costs, and overcomes the shortage of skilled workers by intuitive visualization of data and interaction with software and hardware as well as remote training.

Although the IMV rather represents a vision where foundations are still being laid, first implementations already show its great potential. By using simulation and Digital Twin to plan a new factory in Nanjing the productivity could be increased up to 20%

**Fig. 3** Digital Twin of a production line based on a simulation model and its photorealistic visualization [21]

compared to conventional and comparable factories [20]. This, in turn, has impact on space efficiency, required material, and energy consumption. Consequently, the IMV can be also seen as a key enabler for sustainability.

Another example will be explained in the example section.

# 4 Digital Twin Realizations

## *4.1 A Digital Twin for Design and Operation of a Complex Chemical Reactor*

The progress in science and technology leads to improvements in manufacturing processes. One of the modern methods of manufacturing is additive manufacturing (AM). AM allows to manufacture complex functional structures. This in turn makes it possible to adapt geometries with respect to physical processes and realize new advanced applications with high performance. Examples of such applications could be cooling systems or chemical reactors, like the methanol synthesis reactor described in the following. Conventional reactor geometries for methanol synthesis are quite simple, but they have a significant size and are mainly used in industrial chemical plants. However, methanol synthesis can be applied to store electrical energy from renewable energy sources, so-called Power-to-X reactors [22]. In this case, the reactors work under transient conditions, are significantly smaller, and therefore require additional studies for a novel reactor design. AM is a promising method which allows compact design of reactors thus saving material and as a result improving the sustainability of the power-to-chemical process. The challenges of reactor design for methanol synthesis are operation requirements and boundary conditions: high pressure, temperature control, condensation of chemical reactions products. The Digital Twin allows to understand the physical processes in the reactor in more detail and to find the optimal design.

Figure 4 shows the development chain of optimal design. At the beginning, an initial design concept is developed, which is based on the analysis of existing designs and a comprehensive understanding of physical processes. Once the geometry for the first prototype is ready, its characteristics can be studied by physical test or simulations. In many cases, the numerical modeling can estimate the properties of component prototype with less time and material efforts than physical tests. This is especially important at early stages of design development. And even if simulations are not able to describe all multi-physics processes without simplifications, they allow us to estimate the geometry-functionality dependency. In a chemical reactor for the methanol synthesis, we have the following processes: chemical reactions on the catalyst surface, phase transition of reaction products methanol and water, and temperature conditioning. Since the system is under high pressure, the stiffness of the structure must be considered during developing of the reactor design. All these processes can be described as multi-scale and multi-physics processes. Including all

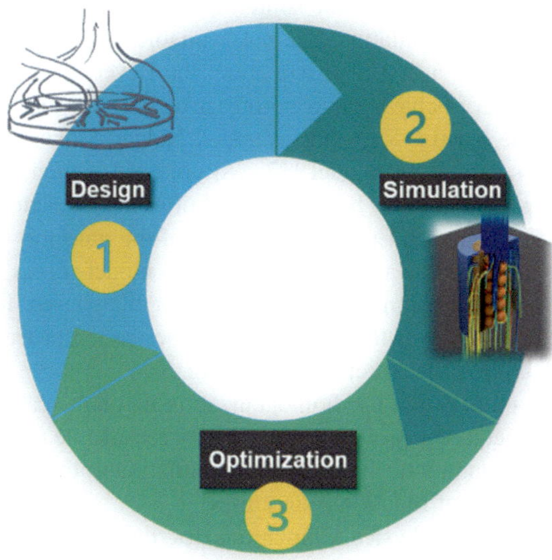


**Fig. 4** Development circuit for the functional design

physical and chemical processes in detail in a single simulation model will lead to high complexity and computational costs. Therefore, it is reasonable to use simplification methods like homogenized models. To ensure that the selected models are applicable and correct for describing a specific process, it is necessary to validate these models with physical tests or high-resolution multi-physics simulations (if high-fidelity simulations are available for the respective process). For the model verification, a physical test or high-resolution simulations can be performed for a representative volume or simplified geometry. The validated models are then used to simulate and optimize the complex structures and geometries of functional components. As such, already during the early design phase a set of specific simulation models is created that form the initial Digital Twin of the reactor.

Let us look at the workflow from detailed representations to homogenized models for 3D simulations of methanol synthesis reactor. The methanol synthesis is a complex chemical process that can be described by means of kinetic models [23] received from fitting the measurement results. In the high-resolution multi-physical simulations, the geometry of the catalyst is reproduced in detail. Therefore, the properties such as pressure drop or effective thermal conductivity of the catalyst bed for a homogeneous model are calculated. Similarly, the basic properties of the porous metallic structures in the reactor are defined. Using the high-resolution simulations and tests the porous materials were analyzed, taking into account the geometry including all pores and collected information. Then this information is used to find out the parameters for the homogenized models like the Darcy-Forchheimer model for pressure drop [24, 25]. The difficulty so far lies in the calculation of the dynamic capillary pressure with detailed pore geometry, since this simulation is transient, requires a high computational effort, and depends on the contact angle, which does not always remain constant during the solid–liquid interaction. Currently, the high-resolution multi-physics simulations are not available for the porous structures in our chemical reactor. As a first approximation, the Leverett model [26] was used to describe the capillary behavior in a porous structure. For the description of the turbulent flow in the reactor, Reynolds-averaged Navier–Stokes equations (RANS equations) with $k$-$\varepsilon$ turbulence model [27, 28] are used, which provide results of sufficient accuracy for given flow boundary conditions in the reactor. To describe the phase transition a model with Eulerian averaging of the transport equations for the additional phase interaction was used [29, 30]. This model solves the conservation equations and assumes a common pressure field for the phases. It is important to mention that the main goal of the models is to describe the process quantitatively and not the micro-scale process reproduction (Fig. 5).

The simulation models that use homogenization of small-scale phenomena and parametrization of geometry provide results during reasonable time for the industry and allow to build an automatic circuit "CAD-simulation-optimization-CAD" to get an optimal novel design of functional structures.

The created 3D multi-physics simulations for the understanding of detailed processes and geometry optimization can be also applied to determine the missing input data and boundary conditions like pressure drop in the catalyst bed and heat transfer coefficients for one-dimensional simulations (1D simulations). The 1D simulations allow not only to calculate the chemical transformation in a reactor under given conditions but also can consider the reactor as a part of a chemical plant and define the operation points for an entire facility. The local 3D simulation of the reactor can be included in an overall system model to optimize the operation of the facility. On the other hand, the system model provides more accurate boundary conditions

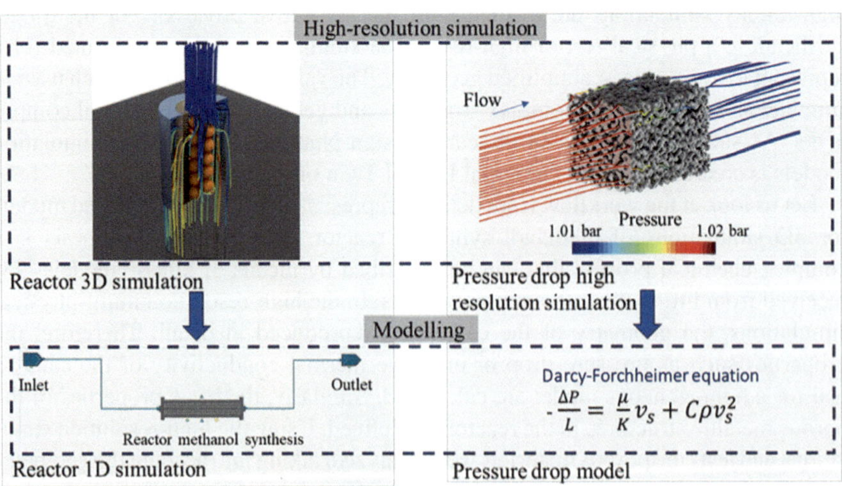


**Fig. 5** Example of the connection between high-resolution simulations and modeling for methanol synthesis reactors

for the local model. Therefore, the example of the methanol synthesis shows that the development of design for a functional component is an iterative process using Digital Twin models of varying degrees of scale. These models improve during this iterative process and are then the base for a proper operation of the whole plant.

## 4.2 *Health Monitoring for Drive Train Components*

For a health monitoring application, a specific aspect of Digital Twin is utilized that focuses primarily on modeling and simulating the behavior of a physical system or process (Behavior Twin). This can be used to predict how the system will behave in different scenarios and to support decisions or optimizations. This approach is applied at different stages of the development to ensure the functionality and reliability of complex systems, particularly in operation and service for health monitoring of critical components.

The knowledge of the health status of electrical drive train components, such as converter and motor, is important for safe operation and for planning the service activities. To calculate the aging, knowledge of the temperature of critical components (converters electronic, bearing, stator insulation) during the operation is essential. Using numerical simulation models, the calculation of health and remaining useful life by processing of operational data and estimation of temperatures of critical components becomes possible. It eliminates the need for direct measurements, which are often unfeasible due to limited access to the components and the huge volume of operational and sensor data generated.

The workflow for monitoring the health of the drive train is shown in Fig. 6. This process consists of several steps which contribute to the development and deployment of a comprehensive health monitoring service. The first step involves the creation of detailed models for each critical component of the drive train: (a) the application, (b) the motor, and (c) the converter. These models describe components' behavior and consist of various digital artifacts, including system architectures, CAD representations of sub-components, materials, and control models. In the next phase (system integration), the component models are combined to a model of the entire drive train system. This step allows to simulate the interactions between the different components, evaluate the system's performance, and calculate the temperature of critical components. To create a robust health monitoring service, the aging models are integrated into the system model. These models consider the wear that drive train components undergo over time and allow to estimate the components' health and remaining useful life.

It is important to know that these models are developed for a specific application and need to be precisely evaluated to ensure that they are applicable for another scenario. If a model does not meet the defined requirements, it is necessary to create a new model or modify an existing model.

The execution of the developed service usually takes place on edge computers (d) close to the data source. The most important advantages are low transmission

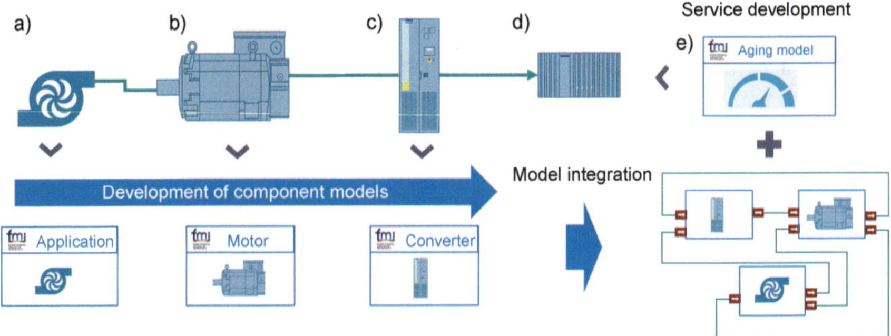


**Fig. 6** Workflow for the health monitoring of the drive train

latency, reliability, and security. The other option is to run it as a cloud service. The main reasons for this are scalability, flexibility in updates and maintenance.

In summary, the Digital Twin technology provides an innovative and physics-based solution for the calculation of stress, temperature, and damage levels in components without direct measurements. By applying the correlation between component use, characteristics, and operating conditions, these predictive maintenance models provide the way for safer and more efficient operations as well as the development of more robust components.

## 4.3 *Manufacturing-X and Catena-X*

Manufacturing-X is a German initiative, started by business, politics, and academia to digitalize supply chains in the industry [31] and to realize an open and global data ecosystem which allows companies to share data across production and supply chains. The general goal is to implement a federated, decentralized, and collaborative data ecosystem for smart manufacturing. Concrete aims are

- "the reorganization of value networks so that they can react quickly to incidents (resilience),
- new business models, a circular economy, and increased efficiency (sustainability), and
- digital innovation that will secure and expand the global leadership of German industry (competitiveness)" (see [32]).

The initiative pursues a cross-sectoral solution approach (see Fig. 7) and drives the internationalization by implementing global standards and setting up international communities, e.g., the International Manufacturing-X Council.

A blueprint for Manufacturing-X is the Catena-X flagship project, which is publicly funded by the Federal Ministry for Economic Affairs and Climate Action (BMWK) and was started in August 2021 with a duration of 3 years [33]. Goal of

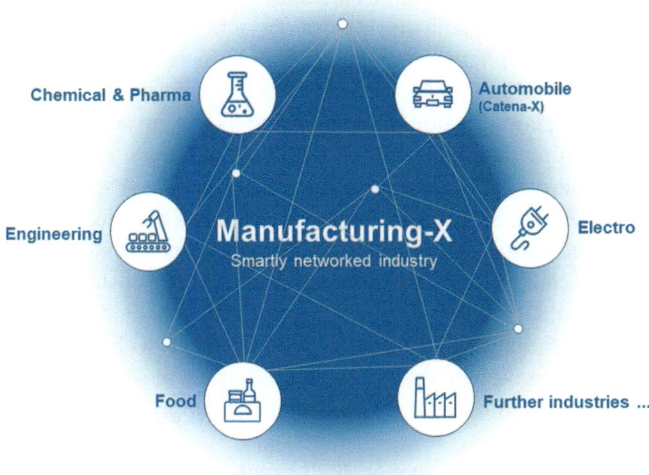


**Fig. 7** Cross-sectoral approach to realize a networked industry [31]

the project is to create a collaborative and open data ecosystem for the automotive industry. Digital Twins are a core element to realize the provision and exchange of data [34].

The information required for the exchange between partners in the data ecosystem is described as semantic data models. For transmission, they are filled with the specific data to be exchanged, i.e., instantiated. This principle is implemented via asset administration shells [35]. In Industrial Digital Twin Association (IDTA) terminology, these data models correspond to sub-models. These sub-models are supplemented by information about the object to which this data relates. Both parts together form the asset administration shell. In the understanding of IDTA, the asset administration shell and Digital Twin are synonyms [6]. Digital Twins are therefore the means of transport for software solutions in data ecosystems. They are used, for example, to exchange information along the supply chain, as the next example illustrates.

### *4.4 Online Control and Simulation*

Online Control and Simulation (OSim) is the name of a use case in the German public-funded project Catena-X. The goal of the research development is a software solution which allows a distributed, collaborative material flow simulation of production and logistics. The strategic goal is the increased resilience of supply chains by faster, more precise reactions in case of short-term disruptions [36]. OSim is one of five use cases in the Catena-X Business Domain Resiliency [37].

Supply chains are subject to various disruptive factors, and it is important to react flexibly and make effective adjustments. An obstacle to this is the often insufficient exchange of information between the individual partners in the chain. Even if individual logistics or production partners carry out their planning locally optimally, without overarching coordination this can lead to inefficient supply chains.

The OSim solution approach aims to simulate the behavior of the entire supply chain, including the effects of short-term disruptions or changes, at the material flow modeling level [38]. The simulation is not carried out in a holistic, monolithic model, but as a distributed simulation in which each partner simulates their processes and sequences. The exchange of information with suppliers and customers necessary for the respective simulation takes place via standardized data models. It maintains the data sovereignty of the partners and guarantees compliance with antitrust regulations.

Through collaborative simulation, it becomes possible to react to changes in advance and improve efficiency across the entire supply chain by the increased observation horizon.

All partners involved in the supply chain simulate their respective processes and actions (production or logistical sequences) based on current information from the shop floor and its IT systems (e.g., manufacturing execution systems). The results of the simulation, essentially delivery data, and quantities of parts produced or transported, are passed on to the immediate successor in the supply chain, who in turn carries out a simulation of their material flow processes based on this possibly changed information. In OSim, a "horizontal" exchange of information across the supply chain is combined with a "vertical" exchange of the material flow simulation of its production or its logistics processes (e.g., transport times). Through iterative simulations by all partners and information exchange upstream and downstream, improved efficiency of the entire supply chain is achieved in a collaborative way. The core of this approach is an OSim manager, which is operated by every partner involved. The OSim manager is a software application that records all the information that is necessary for the local execution of a material flow simulation and supplies it to the locally executed simulation. After the simulation has been carried out, the OSim manager provides the results of the simulation to the partners again. For the secure and data-sovereign transmission of information, the Eclipse Data Space Connector (EDC) is used in accordance with the Catena-X regulations [39].

The executed OSim manager with its connected local simulations forms a Digital Twin of the material flow of the entire supply chain. Also, every exchange of information between the partners uses (in the IDTA understanding) a Digital Twin, which essentially contains the results of the simulation.

## 4.5 The Industrial Metaverse for Battery Production

Use cases for the Industrial Metaverse are numerous and manifold. Although it still represents a vision, Siemens and Freyr have already showcased in [40] how

**Fig. 8** Digital Twin of a battery plant including production and assembly lines [40]

manufacturers can work in the future based on the example of a battery production site, see Fig. 8.

By creating the Digital Twin including an Industrial Metaverse experience of a factory that hasn't been built yet, confidence can be built among stakeholders including investors. Being immersed into the future production site, seeing the production lines virtually commissioned, and having transparency on the performance of the whole plant creates trust in the capability of the corresponding company and reduces the risk of wrong investments significantly.

Moreover, the IMV provides the means to not only display KPIs from the whole plant down to the component level but also to optimize them. Here, simulation plays a crucial role. For example, in the critical mixing process (see Fig. 9), the mixing degree can be simulated by means of CFD (computational fluid dynamics) and particle simulation to finally optimize the process time impacting the overall performance of the plant. If this is not enough, then even the machine itself or machine components can be optimized. Simulation and design optimization enable dedicated improvements like the geometry of the stirrer. Thus, the IMV allows for virtually testing different parameters before the plant is built or without compromising the actual production of an existing plant. This becomes especially important in the operational phase of a plant when anomalies are detected, and a root-cause analysis can be performed in the IMV. Here, different stakeholders from different locations can virtually meet in the plant, analyze operational data, and test corrective actions before applying them in the real plant. With this approach, the costly downtimes of a plant can be reduced to a minimum.

There are many more use cases where the IMV creates an added value in a factory. In the assembly (see Fig. 9 on right) health and safety issues can be evaluated such as the interplay between robots and human workers or test the ergonomics by simulation. As energy consumption and the corresponding $CO_2$ equivalent became a major KPI in production, the IMV of a plant can also be used to create transparency and to optimize scheduling to additionally save costs by making use of changing energy prices.

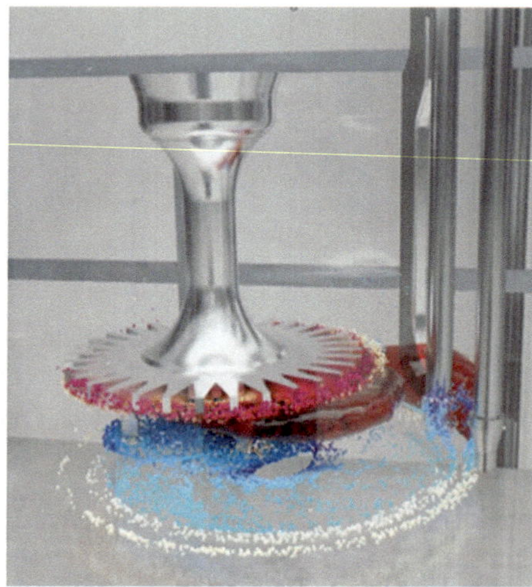

**Fig. 9** Simulated mixing process allows for optimization of mixing time and mixing degree [20]

## 5 Commenting on the Future Development of the Digital Twin Approach

The term Digital Twin is now being used for more than ten years. Its roots lie in simulation, other influences came and continue to come from areas such as data analytics, the Internet of Things (IoT), and Industry 4.0. The latter claims to equate the terms Digital Twin and asset administration shell [6].

A comparison between the definition given here by the authors and the Asset Administration Shell shows great similarities. Information and data from different sources are used and integrated. Knowledge capturing takes place; data and simulation-based models play a central role here. Both approaches address the entire lifecycle of assets, from design and development to operation and maintenance. Furthermore, there is a clear relationship to the counterpart of the Digital Twin—the real twin. These similarities justify using both terms synonymously and further convergence will increase as further sub-models of the AAS are defined. To evaluate the future development of the Digital Twin approach, it makes sense to take a focused look at the core properties of both terms and the approaches behind them. The Asset Administration Shell is a standardized framework that provides a digital representation of a physical asset. It serves as a comprehensive information model, which includes in particular administrative and business-related information on the physical asset. The framework defines a common language designed for seamless integration between different assets and interoperable communication between systems and devices. The Digital Twin approach emphasizes the creation process of the digital representation. This representation includes relatively detailed and dynamic behavioral models consisting of detailed analysis, optimization, and automated support of users in their different tasks in all life cycle phases. A Digital Twin refers not only to physical assets, but also to planned real objects (from small components to large infrastructures), technical and business processes and people.

The term Digital Twin is associated with two aspects that must be viewed in a differentiated manner. On the one hand, this is the representation of the real counterpart itself and, on the other hand, the use and application of the Digital Twin. The first aspect includes elements of a scientific discipline that pursues technical improvement of the models used and works on further developing the methods required for this. The second aspect addresses the creation of software applications that directly provide the Digital Twin as a concrete solution for a task. The types of solutions range from simple tools to fully automated IT solutions.

This leads to a future vision of the Digital Twin that is characterized by several elements. The importance of the models used will continue to increase. These must be easier to combine in their usage and therefore more modular in their methodological basis. The models must meet typical product characteristics. These are explicit descriptions of the model contents and model limitations as well as assurances about model quality in terms of maturity and stability. In order to ensure economic use, these models must be easier to develop and maintain. This refers on the one hand to the further use of the models or the Digital Twin for variants of the planned or real counterpart and on the other hand to the successor generation of the real twin. This in turn requires better integration into development tools and existing or new software solutions for operation and service, which is also a task in the Digital Twin technology area. These more scientific and technical future aspects of the Digital Twin continue approaches that were started in the 1980s with the CAx methods and are now finding a future with the Industrial Metaverse. On the application perspective of Digital Twins, another important element is the efficient generation of mostly software applications which allow the direct usage of a Digital Twin for concrete tasks. The Asset Administration Shell framework plays a crucial role in implementing these goals. It enables the management of Digital Twins and their integration into various applications. However, other objectives require improvements on technical and scientific elements. On the Digital Twin application perspective, it is expected that specific application solutions besides the AAS will be continuously developed which are independent and directly marketable products. The Digital Twin and its further development offer numerous opportunities and potential for companies in various industries. By using models that have modularity and clear product descriptions as well as integrating appropriate tools and applications, new uses can be opened up and efficiency increased. The Digital Twin is therefore an important technology for the future of industry to realize better insights and decision making for the planned or existing real counterpart.

## References

1. Kuehn W (2018) Digital twins for decision making in complex production and logistic enterprises. Int J Des Nat Ecodynamics 13(3):260–271. https://doi.org/10.2495/DNE-V13-N3-260-271
2. Rosen R, von Wichert G, Lo G, Bettenhausen KD (2015) About the importance of autonomy and digital twins for the future of manufacturing. IFAC-Pap 48(3):567–572
3. Grieves M (2002) Completing the cycle: using PLM information in the sales and service functions [slides]. SME Management Forum, Troy, MI
4. Piascik R, Vickers J, Lowry D et al (2010) Technology area 12: materials, structures, mechanical systems, and manufacturing road map. NASA Office of Chief Technologist
5. Wagner C, Grothoff J, Epple U, Drath R, Malakuti S, Grüner S, Hoffmeister M, Zimmermann P (2017) The role of the Industry 4.0 Asset Administration Shell and the Digital Twin during the life cycle of a plant. In: Proceedings of the 2017 IEEE conference on Emerging Technologies Factory Automation (ETFA 2017), Limassol, Cyprus
6. Rosen R et al (2021) Die Rolle der Simulation im Kontext des Digitalen Zwillings: Virtuelle Inbetriebnahme-Modelle als Mittler zwischen Phasen und Anwendungen? atp magazin, Bd 63, Nr 04, S. 82–89
7. Boschert S, Rosen R (2016) Digital Twin—the simulation aspect. In: Hehenberger P, Bradley D (eds) Mechatronic futures. Springer International Publishing, Cham, pp 59–74
8. Drath R, Malakuti S, Grüner S, Grothoff J, Wagner C, Epple U, Hoffmeister M, Zimmermann P (2017) Die Rolle der Industrie 4.0 „Verwaltungsschale" und des „digitalen Zwillings" im Lebenszyklus einer Anlage - Navigationshilfe, Begriffsbestimmung und Abgrenzung. In: Tagungsband zur Automation 2017. Langfassung auf Tagungs-CD (12 Seiten), VDI-Verlag, Baden-Baden. ISBN 978-3-18-092293-5
9. Grieves M, Vickers J (2017) Digital Twin: mitigating unpredictable, undesirable emergent behavior in complex systems. In: Kahlen FJ et al (eds) Transdisciplinary perspectives on complex systems. Springer, Heidelberg
10. Boschert S, Rosen R, Heinrich C (2018) Next generation digital twin. In: Horvath I et al (eds) Proceedings of TMCE 2018, S 209–218
11. Rosen R, Fischer J, Boschert S (2019) Next generation digital twin: an ecosystem for mechatronic systems? IFAC-Pap 52(15):265–270. https://doi.org/10.1016/j.ifacol.2019.11.685
12. Barth M, Kübler K, Heinzerling T, Rosen R, Jäkel J (2020) Eine systematische Bewertung der Qualität von Simulationsmodellen: Identifikation und Clustering von Qualitätskriterien. atp magazin (06–07), S 68–75
13. Barth M et al (2021) Eine systematische Bewertung der Qualität von Simulationsmodellen, Teil 2: Bewertungsmetrik und Validierung, atp magazin, Bd 63, Nr 04, S 56–63
14. Kalhoff J, Borchers E, Rosen R (2021) „Diskussionspapier: Production System Lifecycle Management im Wandel der Fahrzeugindustrie", Plattform Industrie 4.0, available at https://www.plattform-i40.de/PI40/Redaktion/DE/Downloads/Publikation/Diskussionspapier_kopa35c.html. Last access 04 June 2021
15. Bröcker S et al (2021) Prozesssimulation – Fit für die Zukunft?, Positionspapier der DECHEMA und VDI. Available at: https://dechema.de/Positionspapier_Prozesssimulation-path-123212,124930.html. Last access 04 June 2021
16. Bekey GA (2005) Autonomous robots: from biological inspiration to implementation and control. The MIT Press. ISBN 9780262025782
17. Zeman K et al (2020) Symbiotic mechatronics: an alternative view on complex systems. In: Proceedings of TMCE 2020, 11–15 May 2020, Dublin
18. SG AAS Teilmodell Simulation, Plattform Industrie 4.0 (2021) „Interoperabilität hoch zwei für Simulationsmodelle. Available at https://www.plattform-i40.de/PI40/Redaktion/DE/Downloads/Publikation/Interoperabilit%C3%A4t-hoch-zwei.html. Last access 04 June 2021
19. Fischer J et al (2021) Assistenzsysteme der nächsten Generation für Produktionssysteme, VDI conference Automation 2021, Paper accepted

20. MIT Technology Review Insights (2023) The emergent industrial metaverse. https://assets.new.siemens.com/siemens/assets/api/uuid:260d1ead-250b-45fb-b32f-caa9924c1ddd/The-Emergent-Industrial-Metaverse.pdf
21. https://www.siemens.com/global/en/company/digital-transformation/industrial-metaverse/what-is-the-industrial-metaverse-and-why-should-i-care.html
22. Kalz KF, Kraehnert R, Dvoyashkin M, Dittmeyer R, Gläser R, Krewer U, Reuter K, Grunwaldt J-D (2017) Future challenges in heterogeneous catalysis: understanding catalysts under dynamic reaction conditions. ChemCatChem 9:17–29
23. Lacerda de Oliveira Campos B, Herrera Delgado K, Pitter S, Sauer J (2021) Development of consistent kinetic models derived from a microkinetic model of the methanol synthesis. Ind Eng Chem Res 15074
24. Darcy H (1856) Dalmont V (ed) Les Fontaines Publiques de la Ville de Dijon: Exposition et Application des Principes à Suivre et des Formules à Employer dans les Questions de Distribution d'Eau (Paris)
25. Forchheimer P (1901) Wasserbewegung durch Boden. Z Ver Dtsch Ing 45:1781–1788
26. Leverett MC (1941) Capillary behavior in porous solids. Trans AIME 142:151
27. Lien FS, Chen WL, Leschziner MA (1996) Low-Reynolds number eddy-viscosity modelling based on non-linear stress-strain/vorticity relations. In: Proceedings 3rd symposium on engineering turbulence modelling and measurements, 27–29 May, Crete, Greece
28. Teli SM, Mathpati CS (2020) Computational fluid dynamics of rectangular external loop airlift reactor. Int J Chem Reactor Eng 18(5–6):20200009
29. Drew DA, Passman S (1998) Theory of multicomponent fluids. Springer, New York
30. Ishii M (1975) Thermo-fluid dynamic, theory of two-phase, eyrolles
31. Plattform Industrie 4.0. Manufacturing-X. Available at https://www.plattform-i40.de/IP/Navigation/EN/Manufacturing-X/Manufacturing-X.html. Last access 19 Oct 2023
32. Plattform Industrie 4.0. White Paper on Manufacturing-X. Available at https://www.plattform-i40.de/IP/Redaktion/EN/Downloads/Publikation/Manufacturing-X_long.html. Last access 19 Oct 2023
33. Federal Ministry for Economic Affairs and Climate Action (BMWK). Projektsteckbriefe. Available at https://www.bmwk.de/Redaktion/DE/Downloads/P-R/projektsteckbriefe.pdf?__blob=publicationFile&v=1. Last access 19 Oct 2023
34. Catena-X (203) Semantic layer/digital twins. https://catena-x.net/en/offers-standards/semantic-layer. Last access 19 Oct 2023
35. IDTA, Industrial Digital Twin Association, information available at https://industrialdigitaltwin.org/en/technology. Last access 19 Oct 2023
36. Eclipse Tractus-X. Osim Kit. Available at https://eclipse-tractusx.github.io/docs-kits/kits/OSim%20Kit/Adoption%20View%20OSim%20Kit. Last access 19 Oct 2023
37. Eclipse Tractus-X. Business domain resiliency. Available at https://eclipse-tractusx.github.io/docs-kits/next/Resiliency/. Last access 19 Oct 2023
38. Wolfrum P, Haager D, Isinger T, Rosen R, Kutzler T (2023) Kollaborative Materialflusssimulationen zur Steuerung der Verknüpfung von Supply Chain und Shopfloor - ein Catena-X Use-Case. In: Automation, Baden-Baden
39. Catena-X, EDC—The Central Component. Available at https://catena-x.net/en/offers/edc-the-central-component-1. Last access 19 Oct 2023
40. Siemens FREYR Industrial Metaverse showcase. YouTube. https://www.youtube.com/watch?v=LQyKc6SI5u8

# A Formal Framework for Digital Twin Modeling, Verification, and Validation

**Mamadou Kaba Traore, Simon Gorecki, and Yves Ducq**

**Abstract** In today's technological landscape, Digital Twins have emerged as pivotal technologies, offering unprecedented opportunities for innovation and optimization across various industries. Despite numerous on-going research and development initiatives, the verification and validation (V&V) of the Digital Twin (DT) remains a major scientific obstacle. In order to disambiguate the DT concept (as various interpretations exist), this chapter first formalizes it, using the Discrete Event System Specification (DEVS) approach. Then, following a review of the state-of-the-art in V&V, it examines three avenues for developing methodologies for V&V of DTs: software engineering-based V&V, simulation-based V&V, and formal method-based V&V. That way, this chapter provides a framework for DT understanding and a roadmap for future research on DT V&V.

**Keywords** Digital Twins · Verification · Validation · DEVS · Formal Framework · Smart Systems

## 1 Introduction to Digital Twin Concepts

The notion of "smart everything" is surfacing alongside the continuous digitalization of society, industries, health, educational, and urbanization sectors. As a result, new production systems are emerging, characterized by the prominent integration of data and virtual technologies. These systems are so complex that they need model-based management approaches.

M. K. Traore · S. Gorecki (✉) · Y. Ducq
University of Bordeaux, CNRS, Bordeaux INP, IMS, UMR 5218, 33400 Talence, France
e-mail: simon.gorecki@u-bordeaux.fr

M. K. Traore
e-mail: mamadou-kaba.traore@u-bordeaux.fr

Y. Ducq
e-mail: yves.ducq@u-bordeaux.fr

 
M. Grieves and E. Hua (eds.), *Digital Twins, Simulation, and the Metaverse*, Simulation Foundations, Methods and Applications, https://doi.org/10.1007/978-3-031-69107-2_6

The Digital Twin (DT) concept has risen as a pivotal strategy, landing in top strategic technology trends. It revolves around the concept of mobilizing a model in place of a system, ensuring continuous synchronization between the model and the actual system. This synchronization serves to reflect real-world events on the model, enabling the assessment of management initiatives on this ever-updated artifact before implementation. Hence, the model is more than a simple representation, evolving into a digital counterpart intricately linked to the specific system in question. NASA pioneered the system-pairing approach, simulating situations in its spacecraft from distant command centers to guide astronauts in interventions. Notably, the famous Apollo 13 mission in 1970 relied on a system-pairing technique but didn't involve a DT; instead, it utilized two Physical Twins (one in space and the other on land).

The term "Digital Twin" was introduced by [1], acknowledging its underlying principle foreseen by [2]. In the context of product lifecycle management, the concept evolved through various stages: Mirrored Spaces Model [3], Information Mirroring Model [4], and finally as Digital Twin [5]. The concept is defined as "a set of virtual information constructs that fully describe a potential or actual physical manufactured product, from the micro-atomic to the macro-geometric level" [6]. This data-centric view shifted to behavioral aspects in [7], defining a DT as "an integrated multi-physics, multiscale, probabilistic simulation of an as-built vehicle or system that uses the best available physical models, sensor updates, fleet history, etc., to mirror the life of its corresponding flying twin". From the simulation perspective, this approach is disruptive, as simulation experiments are based on current information from the system rather than assumptions. Simulation experiments rely on the current information available from the system rather than assumptions [8, 9]. In this application, DT not only serves a representational function but also proves effective in predicting the expected behavior of the system [10, 11]. Consequently, a DT is not merely a comprehensive model of the system it represents but a collection of integrated sub-models reflecting different aspects of the system [12]. Additional dimensions have emerged, including the use of DT for prognostic and diagnostic activities [13, 14], as well as real-time optimization [15, 16].

Nowadays, DT applications are diverse and extend across various industries, including automotive [17], avionics [18], aerospace [7], energy [19], manufacturing [8], health care [20], and services [21]. Within industrial contexts, DTs play a crucial role in predictive maintenance of equipment, enhancing asset reliability and safety, and optimizing both product design and process operations. In health care, the DT approach holds the potential to revolutionize medical treatments by tailoring them to the individual patient's needs rather than relying on generalized approaches. Furthermore, DTs contribute to the servitization trend by enabling companies to monitor their products during customer usage, enhancing the overall customer experience.

DTs can be divided into three [22] main types: Digital Twin Prototype (DTP) oriented for Product Lifecycle Management, Digital Twin Instance (DTI) oriented for individual product that replicates one single physical or numerical ToI, and Digital Twin Aggregate (DTA) that is an aggregate of several DTIs. In this chapter, we will

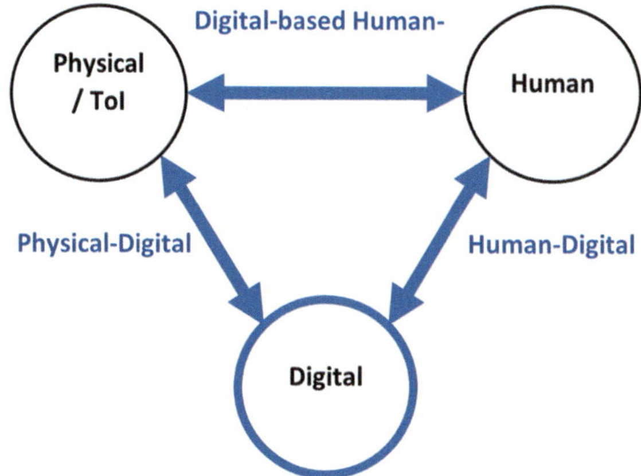


**Fig. 1** High-level view of Digital Twin in interaction with both human and physical sides

mainly focus on DTI type, but contributions can be applied to both DTP, DTI, and DTA.

Despite numerous research and development initiatives, the verification and validation (V&V) of DT remains a significant scientific challenge. To address this obstacle, there is a pressing need for a comprehensive framework. This chapter responds to this need by introducing a potential framework designed to achieve effective DT verification and validation. As a second contribution, we propose three different approaches to reach the goal of DT V&V which can be applied to all DT types: simulation based, software engineering based, and formal method based. This approach extends to modeling, verifying, and validating the interoperability middleware, specifically the Internet of Things (IoT) connection between the system and its corresponding DT.

Thus, we can observe two main interacting elements: a physical or numerical model of a system of interest, sometimes called Physical Twins, or Replica Twins [6] in literature; however, we prefer to refer to it as the Twin of Interest (ToI), enabling us to incorporate both physical models and non-physical ones, such as numerical models. Perpetual synchronization between the TOI and the Digital Twin is maintained, based on data tracked in the TOI. Considering that humans are in most cases present in systems as stakeholders or decision-makers, e.g., health care [20], smart cities [23], services [21], cyber physical and human systems [24], we can add a third interacting element, the human.

Figure 1 provides an overview of the value chain, emphasizing key components necessary for deploying an effective Digital Twin (highlighted in blue). The triangle comprises three primary stakeholders: the physical (component to be replicated), the digital (core Digital Twin), and the human (users and managers). On the physical side, representing a cyber-physical system, the integration of cyber and physical components allows operations executed by actuators, with data collected by sensors and transmitted through a network. The digital side hosts the Digital Twin, receiving data into models, enabling decisions sent back to the system or used by the user or manager for further management decisions. The human side accommodates major stakeholders, including users and managers, utilizing the Digital Twin to explore use cases and develop strategies.

Digital–physical interactions focus on the symbiotic relationship between the physical part and the Digital Twin infrastructure, addressing scientific and technological challenges. These include designing the end-to-end data circuit, conceptualizing data needs, integrating heterogeneous data sources, and addressing requirements for IoT liability and efficiency. Additionally, edge computing is employed for faster computations related to real-time data streams, such as algorithms transforming raw data for the Digital Twin's use.

Digital–human interactions involve applying advanced technologies like Virtual/Augmented Reality and Web/Mobile approaches to bridge the reality gap in interfacing humans with the Digital Twin. This addresses issues like using Metaverse-type technologies for the last mile to end-users, enhancing decision-making, and formalizing "cognitive interoperability" in the Metaverse through social interactions. Digital-based human–physical interactions explore how Digital Twin models and services impact the relationship between the Twin of Interest/physical subject and the human subject: user (behavioral changes) or decision-maker.

The blue component of Fig. 1 (the Digital Twin part) can be zoomed in and illustrated as Fig. 2 according to an understanding framework highlighting the tree main nodes-graphs common to all Digital Twins. The intertwining of these three nodes is commonly referred to as a DT. However, a DT can be assimilated at different scales: a process perspective (macro) or a product perspective (micro). For instance, in the case of a production system, the DT at a process level could represent an entire industrial production line. The feedback from the twin would include information about the flows moving through the factory, with the twin providing guidance for managing this chain. In the same context, at a different scale, from a "product" perspective, the twin could be the DT of an engine in the production line. This distinction can be illustrated with an example from a service system. The twin of a process (macro) might be the DT of a smart city, while the twin of a product (micro) could be that of a patient in the healthcare domain. This flexibility in scale allows the concept of Digital Twins to be applied broadly, capturing both the holistic view of entire systems and the detailed perspective of individual components or products within those systems.

## 2 Digital Twin Understanding Framework

In this chapter, we present a diagram illustrating the core DT. Figure 2 illustrates its composition as three nodes: data, models, and services.

Figure 2 is the understanding framework, the symbiotic association of data, models, and services. In the Digital Twin, data stored and used are not only real-time information from sensors but also draw from various legacy sources, aggregating historical data, user equipment (e.g., smartphones, embedded cameras), open data, and Internet-based APIs (e.g., Google Maps). Digital Twin models are structured into modules, each focused to specific objectives and designed to address particular ques-

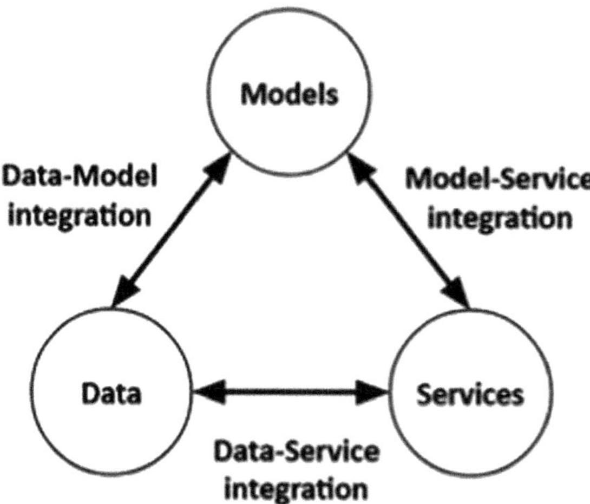


**Fig. 2** High-level architecture for Digital Twin

tions (e.g., a physical system evolution in its environment under specific conditions, the impact of adding/modifying infrastructures, identification of upcoming security vulnerabilities, etc.). Modules are developed across specific domains depending on the subject of interest.

As we can see on Fig. 3, one of the modules is the services provided by the Digital Twin, also defined considering the needs and the area of the subject twinned. They can be separated in four main categories: (1) monitoring data according to geometric (2D/3D) or non-geometric models, or directly from data [25], (2) diagnosis is based on regression, statistical, and stochastic models to identify and prevent events [26], (3) prognosis capability [27] allows forecast or predicts future states, conditions, or events based on the analysis of current and historical data and simulation models [7, 13], and (4) prediction ability to anticipate or forecast future outcomes, behaviors, or trends based on simulation and optimization models [16].

Service node component is based on the model node which contains a collection of integrated sub-models reflecting different aspects of the system twinned. Those models can be geometric or non-geometric models for monitoring, statistical or stochastic models for diagnosis services, simulation models and optimization models for prognosis and prediction services. The use of these models fed by data enables service execution. Moreover, these models can have the ability to update through data–model interaction. Data node component serves as the foundational pillar that includes all relevant information generated by real-time sensors, legacy data sources, and data results from other services. The data node is crucial for fueling models; thus, the quality and diversity of these data determine the reliability of predictions, diagnostics, and analyses performed by the Digital Twin. By integrating varied data, ranging from real-time conditions to historical data, the "Data" node provides a robust foundation for prediction, diagnostics, and optimization services, contributing to an informed and effective decision-making process.

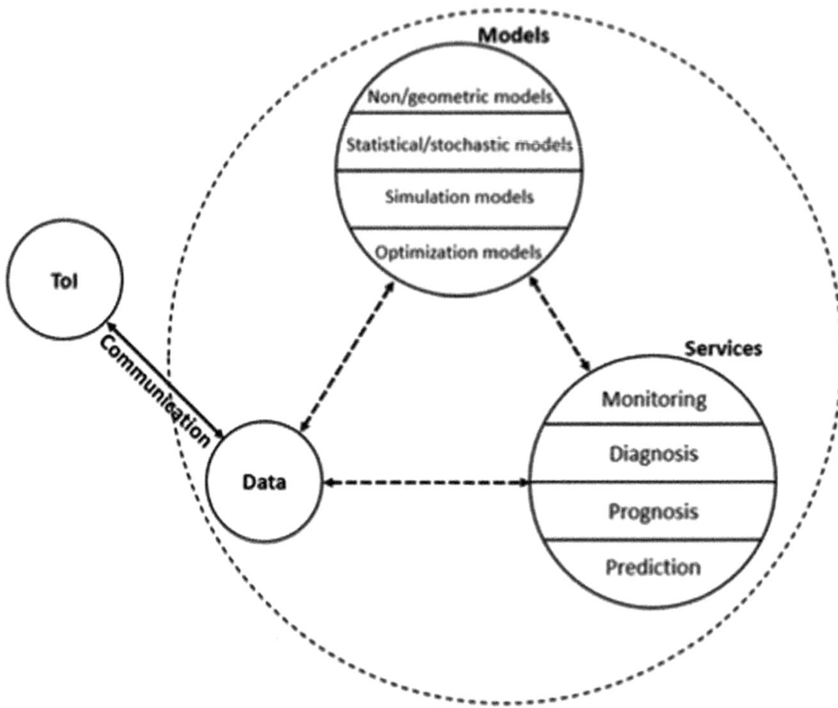


**Fig. 3** DT models and services

Data–model integration stands out as a central challenge in the domain of Digital Twin technology, representing a critical effort in ensuring the seamless adaptation of data to dynamically detect changes in the real system. This integration aims at creating a responsive environment where real-time data are assimilated on-the-fly to reflect alterations of the physical system, enabling continuous updates to the corresponding models [28]. The challenge takes place in establishing a harmonious relationship between the evolving data stream and the digital models, ensuring that the virtual representation remains accurate and aligned with the real-world counterpart.

Data–service integration is focused on services that can be directly derived from raw data, bypassing the need for an intricate model. This involves organized access to historical data for mainly monitoring purposes, allowing stakeholders to visualize valuable information without relying on complex modeling structures. This approach is an efficient way to access information for immediate decision-making and monitoring activities.

Model–service integration takes a different approach by focusing on how models are employed and potentially combined to feed services. This integration is pivotal in optimizing the utility of digital models, exploring their versatility in providing the largest spectrum of services possible. By strategically aligning models with specific services, this integration ensures that the Digital Twin leverages its modeling capabil-

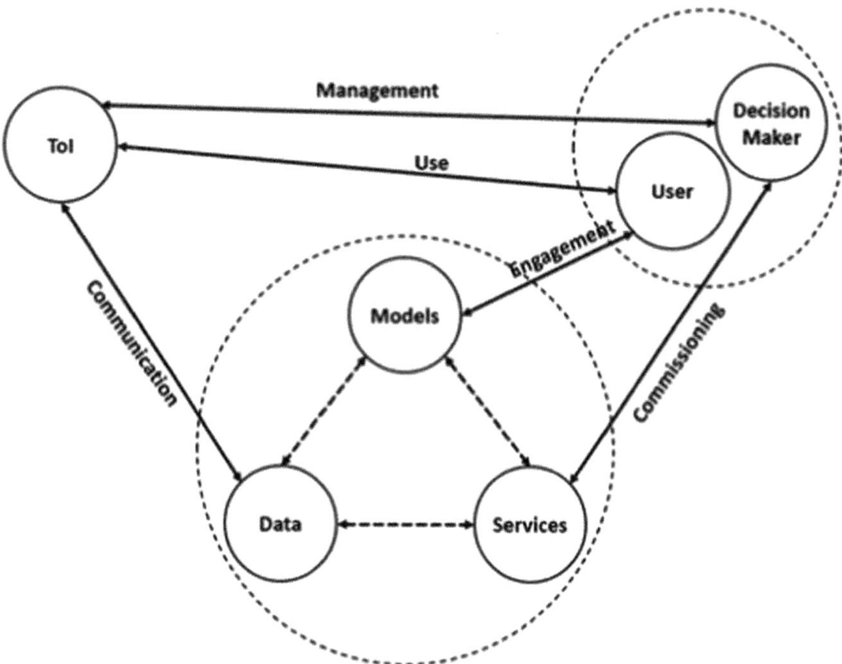


**Fig. 4** Digital, physical, and human interoperations

ities to offer a diverse range of functionalities, contributing to its overall effectiveness in decision support and system optimization.

The three main nodes composing the DT depicted in Fig. 3 will be in continuous interaction with external entities through various means, protocols, and at varying frequencies. Figure 4 provides a detailed breakdown of the elements interacting with the DT and the nature of the interactions it will exchange with them. The interconnectedness is established through a network of dynamic exchanges, emphasizing the diverse and multifaceted engagements the DT maintains with external entities.

Figure 4 serves as a comprehensive synthesis of insights from both Figs. 1 and 2. It introduces a nuanced perspective on the human side, distinguishing between the user and the decision-maker. To illustrate the intricacies of this figure, we'll delve into a case study focusing on a smart city. The case of smart cities illustrates this high-level architecture; however, it's just an example that can be swapped with other Digital Twins where humans are in the loop (e.g., DT health care). In this smart city scenario, the Twin of Interest (ToI) is the city itself, intricately twinned with a DT (hairy circle at the bottom of Fig. 3). This digital representation engages in continuous interaction with the "Human" system, segmented into two pivotal sub-systems: the decision-maker, embodying the smart city manager, and the user, encompassing all other inhabitants and stakeholders within the smart city, such as citizens.

Moreover, the complexity shown in the figure includes various layers of interactions involved in the Digital Twin's overall process. The data node of the Digital Twin emerges as the gateway facilitating seamless communication with the physical realm. Concurrently, the services' node acts as the portal enabling the commissioning (activation and utilization) of the Digital Twin by the urban decision-maker. On the other hand, citizen engagement finds its avenue through the models' node, shaping a dynamic and interactive ecosystem within the smart city's digital representation. This intricate network of nodes and interactions highlights the sophisticated web of relationships and functionalities within the Digital Twin framework.

Having explored the structure and high-level architecture of the Digital Twin in this chapter, the next logical step is to dive into the modeling and formal specification aspects. Understanding the composition and intricacies of the Digital Twin sets the stage for a more detailed examination of its modeling and standardization in the following chapter. This transition is pivotal as it emphasizes the need for a systematic and formalized approach to enable the subsequent phases of verification and validation. With a comprehensive understanding of the Digital Twin's composition, the focus now shifts toward establishing robust models and standardized practices, paving the way for a thorough examination of its capabilities and functionalities.

## 3 Digital Twin Modeling

Formal specification brings valuable advantages. Firstly, it clarifies the intended concept, making it less open to different interpretations. Secondly, it allows the creation of systematic approaches to generate models that can be executed. Thirdly, it provides a possibility for symbolic manipulation to formally check the consistency of the specification and its alignment with certain requirements: verification and validation (V&V). Thus, the need for methods and tools for modeling and specifying Digital Twins is a crucial point addressed in this document. Without a formal description of these twins, it becomes impossible to perform formal verification and specification to ensure their alignment with the case study, regardless of the application domain.

The scientific community agrees that there is a significant difference between traditional models and digital twins. In [29–32], authors provided summaries of different definitions of Digital Twins and proposed generalized definitions and distinctive characteristics to differentiate Digital Twins from other models.

However, it is challenging to find traces of research that focus not on modeling Digital Twin models, but on modeling the Digital Twin itself. Some attempts have been made to achieve this goal, but they are often not sufficiently thorough. In [33], authors identified and categorized different models within Digital Twins into six categories: application domain, hierarchy, discipline, dimension, universality, and functionality. However, this contribution mainly focuses on analyzing models within Digital Twins, without addressing the overall model of Digital Twins itself. The same author, in [34], proposes to deconstruct and investigate the digital models into six distinct aspects: model construction, model assembly, model fusion, model verification, model modification, model management, but once again, this work focuses on the models inside the Digital Twin, not the general model of the DT itself. A similar work has been done in [35] where author in this paper reviews and analyzes several methods and modeling techniques used in DT domain in order to classify modeling methods in groups, e.g., physics-informed ML; data-driven modeling; system modeling; physics-based modeling; geometric modeling, etc. We will see that this approach brings an interesting perspective because depending on the application domain, modeling methods differ, which implies variations in verification and validation methods. In [22], a DT is presented as a composition of basic components that provide on one hand, basic functionalities (identification, storage, communication, security, etc.), and on the other hand, an aggregated DT is defined as a hierarchical composition of other DTs, which is a first trace in the literature of a model not internal to the DT, but of the DT itself. The article proposes a reference architecture of DT using Automation ML through model-driven engineering approach.

Based on the state of the art, we can observe that DT modeling domain covers two main aspects. First aspect is "modeling for DT", which means the design of the model within the Digital Twin, or the architecture of the Digital Twin. The second aspect is "modeling of the DT itself", where there are very few articles, if any, that talk about modeling the Digital Twin itself.

Formal specification methods offer significant advantages for Digital Twin modeling [36], regardless of the application context. However, as we can see, there is no standardized approach for the formal modeling of Digital Twins. Consequently, facilitating the verification and validation (V&V) of Digital Twins is not a straightforward task. This gap in standardization shows the need for a systematic and formal framework that can be universally applied to ensure the accuracy and reliability of Digital Twins across diverse domains. Establishing such a standard would contribute to advancing the field, providing a common ground for researchers and practitioners to enhance the V&V processes and overall robustness of Digital Twin implementations.

To formally specify a DT, we use a system-theoretic approach introduced in 1976. In this approach [37], the DT model is considered a black box with inputs and outputs that connect it to the environment through sensors and actuators. Zeigler's paradigm includes fundamental entities in the modeling and simulation field. These entities are represented by Fig. 5 left side, consisting of the source system (the system under study), the context, the model, and the experimental frame (EF). Please note that the simulator is not shown here. The source system provides behavioral data, the context defines the observation conditions, the model represents the system abstractly using rules or equations, and the EF, acting as an abstract representation of the context, couples with the system model to generate data under specific conditions. The simulator, although not illustrated, is the automaton capable of executing instructions from the resulting coupled model.

In contrast to a conventional model (left side in Fig. 5), which is built as an abstract representation of a source system for experimentation under specific conditions defined by an Experimental Frame (EF), a DT model (right side in Fig. 5) is continuously refreshed with data from the source system. This ongoing update neces-

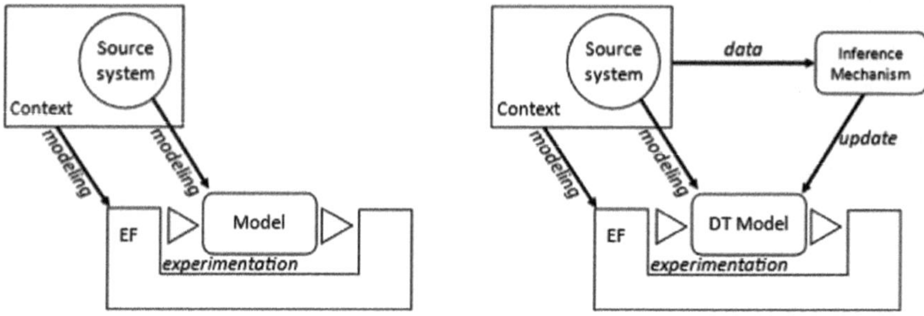


**Fig. 5** DT system-theoretic approach

sitates an inference mechanism (see data–model integration of Fig. 2) to maintain the model's validity within the same EF. This formalization is crucial for ensuring the accuracy and reliability of the DTs through the inference-model mechanism, but raises issues related to verification & validation techniques.

Introducing the Discrete Event System Specification (DEVS) paradigm, [37] defines a model as a mathematical and logical object which serves as a way of specifying a dynamic system, with the following elements:

- State: A state can either be a "hold state" or a "passive state". A hold state is one that the model will stay in for a certain amount of time before automatically changing to another state (via an internal transition). A passive state is one that the model will remain in indefinitely (or until it receives a message that triggers an external transition). One state in the model must be designated as the initial state from which all interactions with the external word commence.
- Time Advance: Every state has a time advance value which specifies the amount of time that expires before it automatically changes to another state (via an internal transition). The time advance for a "hold" state is a finite real value. The time advance for a "passive" state is infinity.
- Internal Transitions and Output: Every hold state in the model has one internal transition defined in order to specify the state to which the model should transition after the specified amount of time. Any state that has an internal transition can also have one or several output messages that are generated before that internal transition occurs.
- Input and External Transitions: Any state can have one or more external transitions defined. An external transition defines an input message that the model might receive when in a given state and the state to which the model should transition in reaction to that input message.

Our framework extends these elements with the following:

- Phase: In the case of a discrete event system, a phase is a state, i.e., its properties are defined by a vector of discrete values. In the case of a hybrid (i.e., combined discrete-continuous) system, the phase properties are defined by equations (linear, differential, etc.). Examples of phase for a vehicle are "accelerating", "cruise-controlling", "decelerating", "stopping", etc., where each phase can be defined by some equations binding the position, speed, acceleration rate, etc., of the vehicle. A phase can be seen as a set of states logically grouped together (and formalized by the properties defining that phase).

- Semantic domain: A phase is defined by its properties, which are expressed by means of state variables. These variables form the semantic domain of the vehicle phases. For example, position, speed acceleration rate, etc., form the semantic domain of the vehicle previously mentioned.
- Parameter: A parameter is an abstraction of some assumption or constraint on the reality represented by the model. Usually, the value of a parameter doesn't change during a simulation experiment, as the experiment is performed under some circumstances that the value of the parameter captures (partially or entirely). The definition of this value for each of the parameters of a model is what is called model calibration (or tuning). However, simulation experiments with a DT model must be possible under varying circumstances, which requires the dynamic change of value of a parameter (i.e., dynamic calibration) be possible during the same experiment.
- Activity: An activity is a set of computations performed by a model during a phase or during the transition from one phase to another. The duration of any activity is zero in the simulated time (it is an instantaneous operation in the real world). Activities are formalized with logical predicates.

We then specify a DT as a 7-uplets $M =< \Lambda, X, Y, \Phi, \Sigma, \Delta, \Delta* >$, where:

- $\Lambda$ is the parameter set: It models all assumptions made on the context in which the DT model is used. $\forall\, \lambda \in \Lambda,\ dom(\lambda)$ is the set of all admissible values for $\lambda$. $\Lambda$ is defined by variables which values can be updated by the context sensors pairing the DT with the ToI.
- $X$ is the input set: Its models the influences received from the DT environment $\forall\, x \in X,\ dom(x)$ is the set of all admissible values for x.
- $Y$ is the output set; it models the DT influences on the environment. $\forall y \in Y,\ dom(y)$ is the set of all admissible values for y. X and Y are defined by variables whose values can be updated by the interface sensors pairing the DT with the ToI.
- $\Phi$ is the phase set; it models the stable discrete/continuous steps of the DT model. Each phase is defined by its properties, as well as the amount of time that expires before it changes to another phase (via an internal transition), and the activities performed during the phase.
- $\forall\, \varphi \in \Phi, \Phi_\phi =< \pi_\varphi, \theta_\varphi, \tau_\varphi >$ is the phase's definition, with:
  - $\pi_\varphi$ is the phase properties, a predicate on $\Sigma$ (noted $\pi_\varphi \in \wp(\Sigma) \cup NIL$) giving the semantics of $\pi$ in $\Sigma$.
  - $\theta_\varphi$ is the phase activitu, a predicate on $\Sigma$ modeling the computations performed during $\varphi$.
  - $\tau_\varphi$ is the phase's time advance, a delay (i.e., expected lifetime of $\varphi$; $\tau_\varphi \in \mathbb{R}_+$) before the model changes from to another phase.

- $\Sigma$ is the semantic domain: This is the set of variables, which the phases are mapped onto. $\forall\, \sigma \in \Sigma$, $dom(\sigma)$ is the set of all admissible values for $\sigma$. $\Sigma$ is defined by variables which values can be updated by the system sensors pairing the DT with the ToI.
- $\Delta = \{\Delta_{i?}^{j}, \Delta_{i}^{j!}, \Delta_{i?}^{j!}, i \in \Phi, j \in \Phi\}$ is the phase-to-phase transition set. $\Delta_{i?}^{j}$, $\Delta_{i}^{j!}$, and $\Delta_{i?}^{j!}$ are respectively external, internal, and confluent transitions, from phase to phase (an internal transition occurs when the delay of the current phase elapses; an external transition is triggered by the receipt of an input; a confluent transition occurs when both internal and external transition conditions occur simultaneously), with:

  - $\Delta_{i?}^{j} =< \omega_{i?,j}, \varepsilon_{i?,j}, \theta_{i?,j} >$ where:

    - $\omega_{i?,j} \in \wp(X)$ is the condition of input receipt.
    - $\varepsilon_{i?,j} \in \wp(\Sigma) \cup \{NIL\}$ is the condition of transition.
    - $\theta_{i?,j} \in \wp(\Sigma) \cup \{NIL\}$ is the action performed in transition.

  - $\Delta_{i}^{j!} =< \varepsilon_{i,j!}, \rho_{i,j!}, \theta_{i?,j} >$ where :

    - $\varepsilon_{i,j!} \in \wp(\Sigma) \cup \{NIL\}$ is the condition of transition.
    - $\rho_{i,j!} \in \wp(\Sigma) \cup \{NIL\}$ is the output sending predicate.
    - $\theta_{i,j!} \in \wp(\Sigma) \cup \{NIL\}$ is the action performed in transition.

  - $\Delta_{i?}^{j!} =< \omega_{i?,j!}, \varepsilon_{i?,j!}, \rho_{i?,j!}, \theta_{i?,j!} >$ where:

    - $\varepsilon_{i?,j!} \in \wp(X)$ is the condition of input receipt.
    - $\varepsilon_{i?,j!} \in \wp(\Sigma) \cup \{NIL\}$ is the condition of transition.
    - $\rho_{i?,j!} \in \wp(\Sigma) \cup \{NIL\}$ is the output predicate.
    - $\theta_{i?,j!} \in \wp(\Sigma) \cup \{NIL\}$ is the action done in transition.

  - $\Delta^{*} =< i^{*}, \theta^{*}, \tau^{*} >$ is the initialization, giving the phase and action prior to the DT execution, where :

    - $i^{*}$ the initial phase.
    - $\theta^{*}$ is the activity performed in initializing the model.
    - $\tau^{*}$ is the time already elapsed in the initial phase.

As an example, let's consider the DT of a smart city which operates a ring road management system (Fig. 6). The DT collects data on weather and congestion through road sensors and controls speed limit of the highway. Figure 4 illustrates the components involved in the engineering of this DT (Fig. 7).

The DT model of a ring road management system presents three phases: stopped (i.e., accident or roadwork), regular (i.e., good weather conditions), and free-flowing

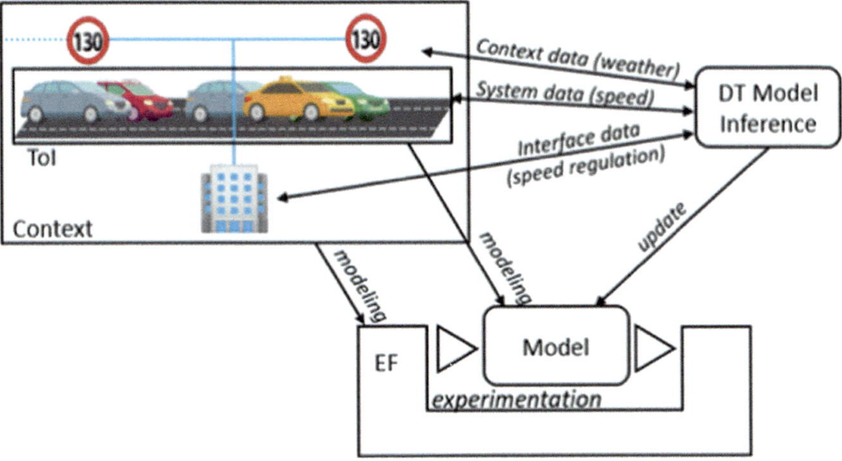


**Fig. 6** Digital Twin of a smart city—road management system

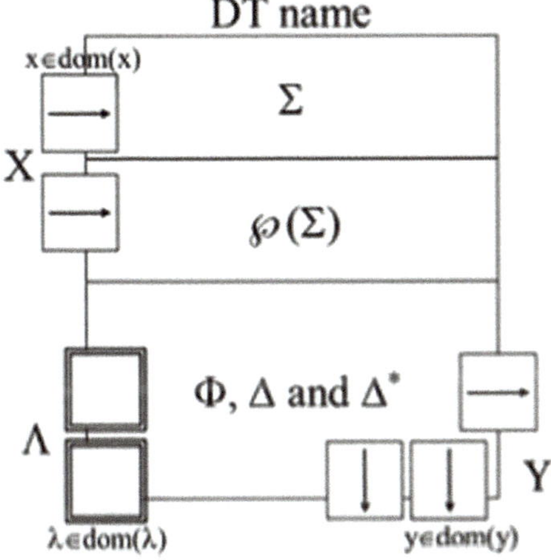


**Fig. 7** DT formal specification metamodel

traffic. The quantity of vehicle (vehicle flow) will be noted F and the speed limit (determined by the system) will be noted "S". Roads' number will be "R" and number of vehicles "n". Weather factor W), and reduced speed limit (i.e., due to bad weather conditions or traffic jams).

The formal specification of this model is illustrated by 8 and given by:

DT Road Management = $< \Lambda, X, Y, \Phi, \Sigma, \Delta, \Delta* >$, where:

- $\Lambda = \{n, W\}$, $n$ and $W$ can be updated with context data (number of vehicles and weather factor).
- $X = \{in, warning\}$ and $Y = \{F\}$, in can be updated with interface data (i.e., modifying speed limit), and the speed limit set is sent through the output pin.

- $\Sigma = \{R, F\}$, $R$ (number of roads) can be updated with system data (i.e., object fallen on a road), and $F$ (speed limit) is calculated, but the model is in regular or reduced phases.
- $\Phi = \{stopped, regular, reduced\}$, with
  - $\Phi stopped =< F = 0, NIL, +\infty >$, the road will remain stopped until switched on (speed limit positive). No activity on this phase, as indicated by NIL.
  - $\Phi regular =< F = 130, NIL, +\infty >$, the road is open, and speed limit is cruising for a infinity time.
  - $\Phi reduced =< F = f(n, W) \times 110, NIL, +\infty >$, the road is open but speed limit is reduced by 25% for a infinity time.
- $\Delta = \{\Delta^{regular}_{stopped?}, \Delta^{stopped}_{regular?}, \Delta^{reduced}_{regular?}, \Delta^{regular}_{reduced?}$ with:
  - $\Delta^{regular}_{stopped?} =< in?on, NIL, NIL >$, the external transition from stopped phase to regular phase arises when road manager switches it on, without any other condition (as indicated by the first NIL) or activity (as indicated by the second NIL).
  - $\Delta^{stopped}_{regular?} =< in?off, NIL, NIL >$, the external transition from regular phase to stopped phase arises when the road manager switches it off.
  - $\Delta^{reduced}_{regular?} =< warning?off, NIL, I = W >$, the internal transition from regular phase to reduced phase arises when the weather factor (W) crosses a threshold.
  - $\Delta^{regular}_{reduced?} =< warning?on, NIL, I = W >$, the internal transition from reduced to regular phase arises when the weather factor (W) crosses again the threshold.
- $\Delta^* =< regular, NIL, 0 >$, the speed limit is initially regular (before any simulation experiment).

The adoption of this formal approach allows the DT developers to clarify the intended concept behind each model that composes DT and allows possibility for symbolic manipulation in order to formally check the consistency through several verification and validation methods. The DEVs' model we have just described provides a formal framework for structuring the "model" component (as depicted in Fig. 2) of the Digital Twin. This formalism can be extended to rigorously describe the various services that may interact with both the model and data layers. Adopting a consistent approach for specifying and understanding the intricate dynamics of services within the Digital Twin ecosystem is essential for ensuring seamless interoperability and functionality. The utilization of such formalism will play a pivotal role in achieving a comprehensive and well-defined representation of the entire Digital Twin framework. This enables a more detailed analysis and validation of the interconnected components, fostering a deeper understanding of the intricate relationships between services, models, and data layers within the Digital Twin paradigm.

However, the "DT Model Inference" layer is not explicitly addressed in the formalism proposed by Zeigler [37] (Fig. 5 left side). It remains an open challenge for the scientific community to establish a method for formalizing the data inference over model within the Digital Twin framework. A DEVS model can adeptly represent and design various types of models, such as multi-agents, discrete events, process flows, and cellular automata. Moreover, a DEVS model can also represent several services, such as prediction, diagnosis, and optimization services. The scientific community must now engage in defining a formalism that seamlessly accommodates the representation of a data lake, regardless of its form (whether it be a database, compilation of data sources, files, a centralized data entry point, ontology, etc.). This objective necessitates a versatile approach that aligns with the diverse nature of data representation in different Digital Twin applications, thereby ensuring a comprehensive and adaptable framework for handling the intricacies of data management within the Digital Twin paradigm.

## 4 Digital Twin Verification and Validation Methodology

Verification and validation is indispensable in ensuring the accuracy and reliability of Digital Twins. It must be based on a clear and precise specification formalism [38], especially in the context of increasingly complex systems such as smart cities. In the literature, many works mention the verification and validation of various systems, but there are very few documents that delve into this issue regarding Digital Twins [39]. Among these papers, various approaches have been used in recent research. These approaches can be broadly classified into two main categories: offline and online. Offline validation of a Digital Twin model involves providing it with historical data to evaluate its accuracy and calibrate its parameters accordingly. In contrast, online validation continuously feeds real-time data from the physical system to the Digital Twin model, allowing for dynamic calibration of the Digital Twin model's parameters to maintain its accuracy and adaptability. Research on the verification and validation of Digital Twins is still in its early stages, and only a limited number of comprehensive studies have been conducted. Most existing studies have focused on offline validation methods to evaluate Digital Twin models: Khan et al. [40] proposed an offline model-based testing approach to validate the model. The proposed approach uses a modeling tool to create a specification model and generate test cases for a production system. Hasan et al. [41] proposed a predictive Digital Twin model for autonomous surface vessels to diagnose faults. Argota Sánchez-Vaquerizo [42] addressed the issue of balancing heavy traffic flow in a complex urban network using a Digital Twin. Finally, other works exist [43] providing a state-of-the-art review of V&V of cyber-physical systems by classifying existing V&V methods in this field, while emphasizing that they are currently insufficient.

Based on what we just observe, there is no notable work on the verification & validation of Digital Twins. However, we see three ways of approaching the DT V & V issue:

- V&V the Digital Twin infrastructure by **software engineering** [44]. Considered as a system. Verification and validation methods from systems engineering are applied. V&V involves unit validation, followed by integration validation, and then system validation: Sect. 4.1.
- V&V by **Modeling and Simulation** [45]. This involves constructing a simulation model of the DT using the DEVS approach: Sect. 4.2.
- DT analysis by **formal method** [46] allows V&V if the DT has certain properties and checking if it does not have certain properties, e.g., constructing the possibility tree of the DT, or use formal simulation techniques: Sect. 4.3.

## 4.1 V&V by Software Engineering

The mechanisms of V&V for complex systems like Digital Twins can draw inspiration from methods within the field of software engineering. It is well known that software engineering uses several levels of tests to verify and validate any software system [47]. These methods can be applied on each types of DT (DTP, DTI, and DTA) dependent on the test level chosen. From a software engineering point of view, V&V can be broken down into clearly identified stages that we propose to map with DT architecture and described in Fig. 8: iterative **unit tests** (I), initially focusing at the lowest level of testing. Here, each sub-components (including the three main entities I.1-I.2-I.4 and inference model interaction I.3) are tested locally. The goal of this testing level is to insure that the component being tested is conforming to its specifications and ready to be integrated with other components of the DT. The second level is **integration testing** (II of Fig. 8) which consists in insuring that the interfaces between components are correct and the sub-components combined to execute the DT's functionality correctly. It goal is to assess the communication between Twin of Interest (II.1) and the entire Digital Twin (II.2). Finally, the last level is **system testing**, which is the process of verifying that the entire DT meets its specified requirements (the all structure in Fig. 8).

### 4.1.1 Unit Test

Unit tests involve verifying and validating the smallest building blocks of the system. The aim is to meet the requirements of each sub-component of the twin. In the framework case, this involves addressing the data, models, and services' sub-components of Fig. 8. These unit tests are at the microscopic scale, thus corresponding to the tests of DTI (Digital Twin Instance) type.

- Methods for **Models'** verification and validation (I.1 of Fig. 8) are essential to ensure the accuracy and reliability of models within DTs. Different types of models exist for different fields of application [48] (physical, mathematical, statistical,

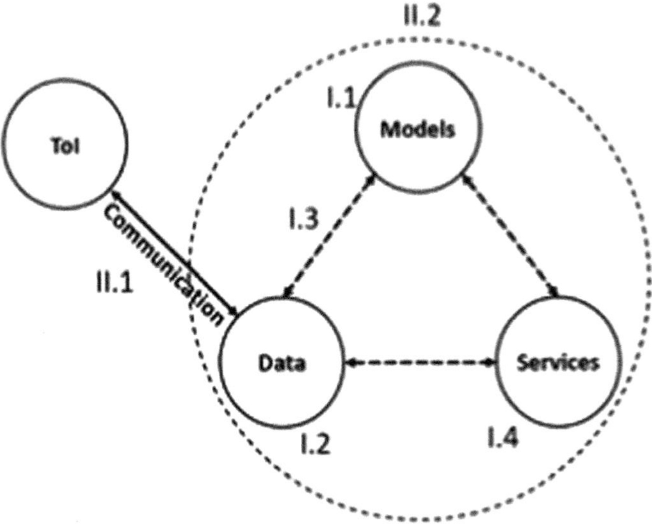


**Fig. 8** V&V unit tests (I) and integration tests (II)

conceptual, 3D graphical, IT, economic, simulation, data, process, systemic models, etc.). There are as many V&V methods as there are formalisms [49]. Many works exist in the literature about model V&V methods, of all possible types [49]. Most known methods used for it are: residual method [50], cross-validation [51], statistical analysis (regression analysis [52], variance analysis [53], hypothesis testing [54]), stability test [55], cross-model validation [56], metric evaluation [57], mean square error [58], etc.). By combining these methods and techniques according to the context of the targeted Digital Twin, it is possible to guarantee the adequacy of the models with the Physical Twin, and therefore, with the physical object.

- Methods for **Data** verification and validation (I.2 of Fig. 8) are essential to ensure the ability to carry a given type of traffic under the right conditions, in terms of availability, throughput, transmission times, quality, completeness, value, availability, etc. [59]. The importance of data quality has been highlighted in various aspects of operating processes [60], decision-making activities [61], and inter-organizational cooperation requirements [62]. Both public and private sectors have initiated numerous efforts where data quality plays a central role. For instance, the UC government passed the Data Quality Act in 2002 [63], while the Government of USA introduced the Data Quality Initiative Framework in 2004 [64]. Therefore, formal methods exist in the literature [65] to verify that a data model adequately meets the requirements set for it based on its application domain: Quality Of Services (QOS). According to the literature, most of these methods can be classified into different techniques: data-driven techniques (e.g., [66] or [67]); process-driven techniques (e.g., [68] or [69]).

Those techniques of verifying the QOS of a dataset can be applied on a DT as unit testing for the data lake layer.

- Methods for **Services'** verification and validation (I.3 of Fig. 8). Before discussing V&V of services, it is important to establish boundaries on what we will call "Services". In any domains (even outside of DTs), each service relies on both data and models. This is also the case in our context of Digital Twins. There is a wide variety of services that can vary greatly depending on the application domains. In our case study, and to encompass the broadest spectrum of Digital Twins possible, we define several categories of services:

- **Monitoring** services, for data visualization: relies on statistical models or visualization models. Two types of monitoring services can be distinguished; e.g., 2/3D geometric representation or non-geometric (e.g., statistical models);
- **Diagnostic** services (limited to "diagnosis models"): relies on regression, statistical, or stochastic models;
- **Prognostic/predictive** services, which rely on simulation models (discrete event, continuous, or hybrid), regression models, optimization (discrete or continuous);
- **Prescription** services: based on simulation or optimization models to enable decision-making, control, or feedback from a Digital Twin.
- **Control** services, which allow to continuously maintain system in a desired state by providing adapted responses.

Based on literature, V&V of DTs services is not a significant problem as numerous works exist to address this issue. Services within the Digital Twin can be V&V by classical QOS methods [70]. We also can take inspiration from V&V of Services Oriented Architecture [71], or from web-services framework V&V applications [72].

#### 4.1.2 Integration Tests

In software engineering, integration testing is a testing phase aimed at verifying that the various modules or components of a system interact correctly when combined to form subsystems or the complete system. Integration testing is at the macro-level of software engineering test, and it can be associated to Digital Twin aggregate level V&V. The main objective of integration testing is to detect and resolve any potential interaction problems between modules, such as interferences, incompatibilities, or communication errors. Integration tests can be conducted at different levels, including the level of individual components, subsystems, or the entire system.

- Interaction between layers within the Digital Twin (I.3 of Fig. 8) is part of the interaction testing. The interactions between the layers of the DT (model, services, data) must also be part of the unit testing and V&V process. For this, the difficulty will be unequal. Communication between services and models is a trivial task as this configuration is common in many cases. Similarly, communication between data and services is straightforward. However, a much more complicated part lies in the interactions between data and models. Here, the context of the Digital Twin imposes a very different treatment compared to other contexts. Thus, a data inference mechanism takes place in the DT. Data can, on the fly, influence the structures of the models [73]. It is therefore important to note that this part of the V&V opens up prospects for evolution in the scientific community.

- Integration testing between the ToI and data layer (II.1 of Fig. 8) is crucial to ensure seamless interaction and functionality. This integration testing phase focuses on verifying that the ToI effectively communicates with the data layer and that the data exchanged between them are accurate, consistent, and appropriately utilized. Testing scenarios may include validating data input/output mechanisms, ensuring compatibility between data formats and structures, and verifying the synchronization of data updates between the ToI and the Digital Twin's data layer. Additionally, integration testing evaluates the responsiveness of the Digital Twin to queries and commands from the ToI, ensuring that the desired actions are executed correctly and in a timely manner. Through rigorous integration testing, developers can identify and address any potential issues or discrepancies in the interaction between the Twin of Interest and the data layer, ultimately enhancing the reliability and effectiveness of the Digital Twin system.
- Entire Digital Twin (II.2 of Fig. 8) testing phase focuses on validating the integration points between the three layers to ensure that they work harmoniously to achieve the intended objectives of the Digital Twin. Testing scenarios may include verifying the flow of data between the data layer and the model layer, ensuring that data inputs are accurately processed by the models and that the outputs are consistent with expected results. Additionally, integration testing evaluates the interaction between the model layer and the services layer, ensuring that services can effectively leverage the models to provide desired functionalities. Moreover, integration testing examines the integration points between the data layer and the services layer, validating that services can access and utilize data from the data layer efficiently. By thoroughly testing these integration points, developers can identify and address any potential issues or inconsistencies in the interactions between the layers of the Digital Twin, ultimately ensuring the reliability and effectiveness of the entire system.

### 4.1.3 System Test

System testing serves as a crucial phase aimed at evaluating the overall functionality, performance, and behavior of the Digital Twin as a cohesive system. System testing for a Digital Twin involves assessing its ability to fulfill its intended purpose and objectives across various scenarios and use cases. This scale level can be associated to Digital Twin Prototype. The comprehensive testing process examines the interaction and integration of all components within the Digital Twin, including the data layer, model layer, and services layer. It verifies the accuracy of data inputs, the effectiveness of models in simulating real-world behavior, and the reliability of services in leveraging the Digital Twin for decision-making and optimization. System testing also encompasses testing under different environmental conditions and stress levels to ensure the robustness and resilience of the Digital Twin. By conducting systematic and rigorous system testing, developers can identify any potential weaknesses, errors, or performance bottlenecks in the Digital Twin and take corrective actions to enhance its overall quality and reliability.

### 4.2 V&V by Modeling and Simulation

Simulation-based verification and validation is an approach used to assess and ensure the quality, reliability, and accuracy of simulation models. This method relies on the use of computer simulations to verify and validate models by comparing them to real data or reference standards [74]. In the context of simulation-based V&V, the real system is replaced by a simulation model. Thus, the Digital Twin is subjected to various tests and scenarios to evaluate its performance and behavior under varied conditions, replicating cases that may occur in reality. This may include sensitivity tests to assess the model's response to variations in input parameters, robustness tests to evaluate its ability to produce consistent results under varied conditions, and validation tests to compare simulation results to experimental data or theoretical results. The primary objective of this method (Model Continuity) is to ensure that the Digital Twin (both DTI and DTA) is capable of adapting and accurately reproducing the behavior of the real system it represents and providing reliable and meaningful results to support decision-making and understanding of the studied phenomenon [45].

### 4.3 V&V by Formal Method

Formal methods represent a powerful alternative for carrying out the V&V of Digital Twin Instance. They provide structured and systematic techniques to ensure the compliance of Digital Twins with the specifications and requirements of the real-world system.

Formal methods are approaches [75] based on mathematical and logical concepts to specify, design, and verify software or hardware systems. They rely on formal languages and formal analysis techniques to guarantee the correctness and safety of systems. In the context of Digital Twins, formal methods can be used to specify expected behaviors, detect design errors, and validate the consistency between the virtual model and the real system.

The first step in V&V of Digital Twins using formal methods is to develop a precise formal modeling of the system. This modeling can be carried out using formal languages such as Petri nets [76], automata [77], or formal specification languages. By formally describing the system's behavior and its interactions with the environment, this modeling provides a solid foundation for analysis and verification.

Once the formal model is established, various techniques of formal verification can be used to ensure its correctness and compliance with specifications. Among these techniques, model checking [78] involves automatically verifying if the model satisfies certain specified properties by traversing the possibility tree to determine the absence of deadlock or system termination. Theorem proving techniques [79], on the other hand, use mathematical reasoning methods to demonstrate the correctness of the model against a predefined set of properties.

In addition to formal verification, formal validation of Digital Twins aims to ensure that the virtual model accurately reflects the behavior of the real system. This step may involve the use of formal simulation techniques [80], where the formal model is executed in a simulation environment to evaluate its behavior under different conditions. Comparing the results of formal simulation with real data validates the accuracy and reliability of the Digital Twin.

Formal methods offer a systematic and rigorous approach to the verification and validation of systems, characteristics that can be beneficial for the Digital Twin domain. By using formal languages, advanced analysis techniques, and specialized tools, it is possible to ensure the reliability and accuracy of these virtual models, paving the way for their use in critical applications such as industry, health care, and transportation.

## 5 Conclusion

In this paper, our first contribution lies in proposing the design of an architecture for understanding the Digital Twin, enabling high-level modeling. This architectural understanding framework provides a structured approach to conceptualizing and representing Digital Twins, facilitating their effective modeling and analysis through three main levels: data level, model level, and service level.

Building upon this foundation, our subsequent analysis delves into the question of verification and validation of Digital Twins, a topic that has received limited attention in the existing literature despite the increasing prominence of this technology. Recognizing this research gap, we have conducted an in-depth analysis to explore the potential of existing methods in the state of the art to establish the conceptual and methodological foundations for ensuring the accuracy and reliability of Digital Twins. Our investigation has identified three primary axes: software engineering, modeling and simulation, and formal methods, each offering valuable insights into addressing the verification and validation challenges associated with Digital Twins. This study serves as both a comprehensive review of the current state of research and a roadmap for future endeavors in this emerging field. By consolidating existing knowledge and pointing areas requiring further investigation, our work aims to facilitate significant progress in the verification and validation of Digital Twins, fostering their widespread adoption across diverse application domains. Through collaborative efforts and continued research, we envision a future where Digital Twins play an increasingly integral role in enhancing decision-making processes and driving innovation across various industries.

## References

1. Piascik RS et al (2010) DRAFT materials, structures, mechanical systems and manufacturing roadmap-technology Area 12". In: Obtained from: http://www.nasa.gov/pdf/501625mainTA12-MSMSM-DRAFT-Nov2010-A.pdf
2. Gelernter D (1993) Mirror worlds: or the day software puts the universe in a shoebox... How it will happen and what it will mean. Oxford University Press
3. Grieves MW (2005) Product lifecycle management: the new paradigm for enterprises. Int J Product Dev 2(1–2):71–84
4. Grieves M (2006) Product lifecycle management: driving the next generation of lean thinking
5. Grieves M (2011) Virtually perfect: driving innovative and lean products through product lifecycle management, vol 11. Space Coast Press Cocoa Beach
6. Grieves M, Vickers J (2017) Digital twin: mitigating unpredictable, undesirable emergent behavior in complex systems. In: Transdisciplinary perspectives on complex systems: new findings and approaches, pp 85–113
7. Glaessgen E, Stargel D (2012) The digital twin paradigm for future NASA and US Air Force vehicles. In: 53rd AIAA/ASME/ASCE/AHS/ASC structures, structural dynamics and materials conference 20th AIAA/ASME/AHS adaptive structures conference 14th AIAA, p 1818
8. Rosen R et al (2015) About the importance of autonomy and digital twins for the future of manufacturing. Ifac-papersonline 48(3):567–572
9. Grieves MW (2019) Virtually intelligent product systems: digital and physical twins
10. Boschert S, Rosen R (2016) Digital twin—the simulation aspect. In: Mechatronic futures: challenges and solutions for mechatronic systems and their designers, pp 59–74
11. Michael S, Juergen R (2016) From simulation to experimentable digital twins: simulation-based development and operation of complex technical systems. In: IEEE international symposium on systems engineering (ISSE). IEEE, pp 1–6
12. Elisa N, Luca F, Marco M (2017) A review of the roles of digital twin in CPS-based production systems. Proc Manuf 11:939–948
13. Reifsnider K, Majumdar P (2013) Multiphysics stimulated simulation digital twin methods for fleet management. In: 54th AIAA/ASME/ASCE/AHS/ASC structures, structural dynamics, and materials conference, p 1578
14. Fei T et al (2018) Digital twin-driven product design, manufacturing and service with big data. Int J Adv Manuf Technol 94:3563–3576
15. Söderberg R et al (2017) Toward a Digital Twin for real-time geometry assurance in individualized production. CIRP Ann 66(1):137–140
16. Hao Z et al (2017) A digital twin-based approach for designing and multi-objective optimization of hollow glass production line. IEEE Access 5:26901–26911
17. Violeta Damjanovic-Behrendt (2018) A digital twin-based privacy enhancement mechanism for the automotive industry. In: 2018 international conference on Intelligent Systems (IS). IEEE, pp 272–279
18. Tuegel EJ et al (2011) Reengineering aircraft structural life prediction using a digital twin. Int J Aerosp Eng
19. Zhang M, Zuo Y, Tao F (2018) Equipment energy consumption management in digital twin shop-floor: a framework and potential applications. In: 2018 IEEE 15th International Conference on Networking, Sensing and Control (ICNSC). IEEE, pp 1–5
20. Bramlet M et al (2016) Virtual reality visualization of patient specific heart model. J Cardiovasc Magnetic Resonance 18(1):1–2

21. Bolton RN et al (2018) Customer experience challenges: bringing together digital, physical and social realms. J Service Manage 29(5):776–808
22. Schroeder GN et al (2021) A methodology for digital twin modeling and deployment for industry 4.0. In: Proceedings of the IEEE, vol 109, issue 4, pp 556–567. https://doi.org/10.1109/JPROC.2020.3032444.
23. White G et al (2021) A digital twin smart city for citizen feedback. Cities 110:103064
24. Barnabas J, Raj P (2020) The human body: a digital twin of the cyber physical systems. Adv Comput 117(1):219–246
25. Bailenson JN, Segovia KY (2010) Virtual doppelgangers: psychological effects of avatars who ignore their owners. In: Online worlds: convergence of the real and the virtual, pp 175–186
26. Jain P et al (2019) A digital twin approach for fault diagnosis in distributed photovoltaic systems. IEEE Trans Power Electronics 35(1):940–956
27. Zhou X et al (2023) A fuzzy-set-based joint distribution adaptation method for regression and its application to online damage quantification for structural digital twin. Mech Syst Signal Process 191:110164
28. Mama D, Kaba TM (2023) Formal approach to digital twin specification. In: Annual Modeling and Simulation Conference (ANNSIM). IEEE, pp 233–244
29. Barricelli BR, Casiraghi E, Fogli D (2019) A survey on digital twin: definitions, characteristics, applications, and design implications. IEEE Access 7:167653–167671
30. David J et al (2020) Characterising the digital twin: a systematic literature review. CIRP J Manuf Sci Technol 29:36–52
31. VanDerHorn E, Mahadevan S (2021) Digital twin: generalization, characterization and implementation. Decis Support Syst 145:113524
32. Mengnan L et al (2021) Review of digital twin about concepts, technologies, and industrial applications. J Manuf Syst 58:346–361
33. Tao F et al (2022) Digital twin modeling. J Manuf Syst 64:372–389. ISSN: 0278-6125. https://doi.org/10.1016/j.jmsy.2022.06.015. https://www.sciencedirect.com/science/article/pii/S0278612522001108
34. Tao F et al (2021) Theory of digital twin modeling and its application. Comput Integr Manuf Syst 27:1–15
35. Thelen A et al (2022) A comprehensive review of digital twin–part 1: modeling and twinning enabling technologies. Struct Multidisc Optimization 65(12):354
36. Cronrath C, Ekström L, Lennartson B (2020) Formal properties of the digital twin—implications for learning, optimization, and control. In: 2020 IEEE 16th international Conference on Automation Science and Engineering (CASE). IEEE, pp 679–684
37. Zeigler BP (1976) The hierarchy of system specifications and the problem of structural inference. In: PSA: proceedings of the biennial meeting of the philosophy of science association, vol 1976, 1. Cambridge University Press, pp 226–239
38. Worden K et al (2020) On digital twins, mirrors, and virtualizations: frameworks for model verification and validation. In: ASCE-ASME J Risk Uncertainty Eng Syst Part B: Mech Eng 6(3):030902
39. Zheng X et al (2017) Perceptions on the state of the art in verification and validation in cyber-physical systems. IEEE Syst J 11(4):2614–2627. https://doi.org/10.1109/JSYST.2015.2496293
40. Khan A et al (2018) Digital Twin for legacy systems: simulation model testing and validation. In: 2018 IEEE 14th international Conference on Automation Science and Engineering (CASE), pp 421–426. https://doi.org/10.1109/COASE.2018.8560338.
41. Hasan A et al (2023) Predictive digital twins for autonomous surface vessels. In: Ocean engineering, vol 288, p 116046. ISSN: 0029-8018. https://doi.org/10.1016/j.oceaneng.2023.116046. https://www.sciencedirect.com/science/article/pii/S0029801823024307. A formal framework for digital twin modeling, verification and validation 25
42. Argota Sánchez-Vaquerizo J (2022) Getting real: the challenge of building and validating a large-scale digital twin of Barcelona & rsquo;s traffic with empirical data. ISPRS Int J Geo-Information 11(1). ISSN: 2220-9964. https://doi.org/10.3390/ijgi11010024. https://www.mdpi.com/2220-9964/11/1/24

43. Zheng X, Julien C (2015) Verification and validation in cyber physical systems: research challenges and a way forward. In: 2015 IEEE/ACM 1st international workshop on software engineering for smart cyber-physical systems, pp 15–18. https://doi.org/10.1109/SEsCPS.2015.11
44. Wallace DR, Fujii RU (1989) Software verification and validation: an overview. IEEE Softw 6(3):10–17
45. Traore MK, Gorecki S, Ducq Y (2022) A simulation based approach to digital twin's interoperability verification & validation. In: "Workshop interoperability challenges and solutions within industrial networks" co-located with 11th international conference on Interoperability for Enterprise Systems and Applications (I-ESA 2022), vol 3214
46. Gore R, Diallo S (2013) The need for usable formal methods in verification and validation. In: 2013 Winter Simulations Conference (WSC), pp 1257–1268. https://doi.org/10.1109/WSC.2013.6721513
47. Collofello JS (1988) Introduction to software verification and validation. Carnegie Mellon University, Software Engineering Institute
48. Fishwick PA, Miller JA (2004) Ontologies for modeling and simulation: issues and approaches. In: Proceedings of the 2004 Winter Simulation Conference, vol 1. IEEE
49. Thacker BH et al (2004) Concepts of model verification and validation
50. Wang C et al (2018) Detecting aberrant behavior and item preknowledge: a comparison of mixture modeling method and residual method. J Educ Behav Stat 43(4):469–501
51. Rasmus B et al (2008) Cross-validation of component models: a critical look at current methods. Analy Bioanalytical Chem 390:1241–1251
52. Darlington RB, Hayes AF (2016) Regression analysis and linear models: concepts, applications, and implementation. Guilford Publications
53. Yue J (2009) Variance analysis in software fault prediction models. In: 20th international symposium on software reliability engineering. IEEE, pp 99–108
54. Koch K-R (2013) Parameter estimation and hypothesis testing in linear models. Springer Science & Business Media
55. Kim WC, Ahn SC, Kwon WH (1995) Stability analysis and stabilization of fuzzy state space models. Fuzzy Sets Syst 71(1):131–142. 26 Mamadou Kaba Traore Simon Gorecki and Yves Ducq
56. Gidskehaug L, Anderssen E, Alsberg BK (2008) Cross model validation and optimisation of bilinear regression models. Chemometrics Intell Lab Syst 93(1):1–10
57. Confalonieri R et al (2009) Multi-metric evaluation of the models WARM, CropSyst, and WOFOST for rice. Ecol Modelling 220(11):1395–1410
58. Willmott CJ, Matsuura K (2005) Advantages of the mean absolute error (MAE) over the root mean square error (RMSE) in assessing average model performance. Climate Res 30(1):79–82
59. Murat K, Arjan D (2017) Quality of service (QoS) in software defined networking (SDN): a survey. J Netw Comput Appl 80:200–218
60. Data Warehousing Institute. Data quality and the bottom line: achieving business success through a commitment to high quality data. http://www.dw-institute.com/ (visited on 02/08/2024)
61. Chengalur-Smith ISN, Ballou DP, Pazer HL (1999) The impact of data quality information on decision making: an exploratory analysis. IEEE Trans Knowl Data Eng 11(6):853–864
62. Monica S (2006) Data quality: concepts. Data-Centric Systems and Applications, Springer, Methodologies and Techniques
63. Office of Management and Budget. Information quality guidelines for ensuring and maximizing the quality, objectivity, utility, and integrity of information disseminated by agencies. http://www.whitehouse.gov/omb/fedreg/reproducible.html (visited on 01/17/2006)
64. DQI (2004) Data quality initiative framework. Project report. https://nces.ed.gov/fcsm/pdf/FCSM.20.04_A_Framework_for_Data_Quality.pdf (visited on 2004)
65. Batini C et al (2009) Methodologies for data quality assessment and improvement. ACM Comput Surv (CSUR) 41(3):1–52

66. Bertolazzi P, De Santis L, Scannapieco M (2003) Automatic record matching in cooperative information systems. In: Proceedings of the international workshop on Data Quality in Cooperative Information Systsems (DQCIS), p 9
67. Hernández MA, Stolfo SJ (1998) Real-world data is dirty: data cleansing and the merge/purge problem. Data Mining Knowl Discovery 2:9–37
68. Muthu S, Whitman L, Hossein Cheraghi S (1999) Business process reengineering: a consolidated methodology. In: Proceedings of the 4th annual international conference on industrial engineering theory, applications and practice. University of Texas San Antonio TX, pp 17–20
69. Hammer M (1990) Reengineering work: don't automate, obliterate. Harvard Bus Rev 68(4):104–112
70. Calinescu R et al (2015) Formal verification with confidence intervals to establish quality of service properties of software systems. IEEE Trans Reliab 65(1):107–125
71. Tsai W-T, Chen Y, Paul R (2005) Specification-based verification and validation of web services and service-oriented operating systems. In: 10th IEEE international workshop on object-oriented real-time dependable systems. IEEE, pp 139–147
72. Bai X et al (2007) A framework for contract-based collaborative verification and validation of web services. In: Component-based software engineering: 10th international symposium, CBSE 2007, Medford, MA, USA, July 9–11. Proceedings 10. Springer, Heidelebrg, pp 258–273
73. Diakité M, Traoré MK (2024) Formalizing a framework of inference capabilities for Digital Twin engineering. In: Simulation, p 00375497241228281
74. Dahmen U, Roßmann J (2018) Simulation-based verification with experimentable digital twins in virtual testbeds. Tagungsband des 3. Kongresses Montage Handhabung Industrieroboter. Springer, Heidelberg, pp 139–147
75. Clarke EM, Wing JM (1996) Formal methods: state of the art and future directions. ACM Comput Surv (CSUR) 28(4):626–643
76. Cortes LA, Eles P, Peng Z (2003) Modeling and formal verification of embedded systems based on a Petri net representation. J Syst Archit 49(12–15):571–598
77. Carroll J, Long D (1989) Theory of finite automata with an introduction to formal languages
78. Rushby JM (1995) Model checking and other ways of automating formal methods. In: Position paper for panel on model checking for concurrent programs, Software Quality Week, San Francisco
79. Goguen JA (2021) Theorem proving and algebra. In: arXiv preprint arXiv:2101.02690
80. Abarbanel Y et al (2000) Focs–automatic generation of simulation checkers from formal specifications. In: Computer aided verification: 12th international conference, CAV 2000, Chicago, IL, USA, July 15–19, 2000. Proceedings 12. Springer, Heidelberg, pp 538–542

# Digital Twins for Advanced Manufacturing: The Standardized Approach

**Guodong Shao, Deogratias Kibira, and Simon Frechette**

**Abstract** Digital Twins are becoming more prevalent in a wide range of industries such as manufacturing, construction, smart city, and healthcare for various purposes, including observing, predicting, optimizing, and controlling. Digital Twins are in the early adoption stage. Currently, few standards directly address Digital Twins and a commercial ecosystem of Digital Twins has not been well established. Developing and implementing Digital Twins present significant challenges. Most current Digital Twin applications are customized solutions, which are expensive to create and difficult to integrate with other systems. Foundational work is needed to support an open marketplace for Digital Twin developers, users, and technology service providers. This includes the development of standardized frameworks, reference models, and interfaces to provide a solid foundation for ensuring interoperability, reliability, validity, security, and trust. This Chapter identifies implementation challenges for Digital Twins for manufacturing, reviews relevant standardization efforts, introduces the ISO Digital Twin framework standard for manufacturing, presents use cases, and discusses potential research topics and future standardization directions.

**Keywords** Digital twin · Standards · Advanced manufacturing · Robot workcell

G. Shao (✉) · D. Kibira · S. Frechette
Engineering Laboratory, National Institute of Standards and Technology, 100 Bureau Drive, Gaithersburg, MD 20899, USA
e-mail: guodong.shao@nist.gov

D. Kibira
e-mail: deogratias.kibira@nist.gov

S. Frechette
e-mail: simon.frechette@nist.gov



M. Grieves and E. Hua (eds.), *Digital Twins, Simulation, and the Metaverse*, Simulation Foundations, Methods and Applications, https://doi.org/10.1007/978-3-031-69107-2_7

# 1 Introduction

The idea of developing physical mockups of planned objects, infrastructure, or scenes has been practiced by humans for millennia. NASA provides one of the notable recent applications of twinning during the planning and execution of space missions. Mission planners created two similar space vehicles. One was sent on a mission while its physical "twin" remained on Earth. The vehicle on Earth emulated the state of the flying twin and could be used to test possible solutions to problems encountered with the flying twin [1]. As computer technology advanced, physical models gave way to digital models. The Digital Twin concept gained recognition in 2002 after Dr. Grieves presented his vision of real space, virtual space, and information flowing between real and virtual spaces [2]. The key idea is the state synchronization of the real space and virtual space. However, it was not until recently that this concept became one of the top strategic technology trends. The advancement of technologies such as the Internet of Things (IoT), smart sensors, Artificial Intelligence (AI), and cloud computing facilitate the realization of Digital Twins.

Digital Twins involve highly complex functional subsystems, including data collection, data processing, communication, information modeling, data analytics, visualization, simulation, optimization, and control. Some of these could be distributed systems. This complex system of systems presents significant challenges for manufacturers to understand and seamlessly integrate these diverse functional subsystems. Many companies already have some forms of digital transformation efforts underway. Others may already have multiple versions of digital solutions from various vendors. Some vendors claim that they have complete solutions for Digital Twin development. However, it is impossible to have all companies discard existing digital systems and source new replacements from the same vendor because not all manufacturers are technically able or can afford to redesign and replace their entire plants. Therefore, existing systems and technologies must be integrated when implementing new Digital Twins. Interoperability standards are needed to support the communication and integration between (1) the physical and virtual systems, (2) multiple Digital Twins, and (3) Digital Twins and legacy systems such as Manufacturing Execution Systems (MES), Enterprise Resource Planning (ERP), and Product Life Cycle Management (PLM).

A Digital Twin comprises three main components—the physical system, the virtual representation of the physical system, and bi-directional communication between the physical and virtual worlds. Recent studies have introduced two more dimensions: data and service [3]. It has then been noted that a framework is needed to realize a Digital Twin, which comprises different components so developers can use it for various applications and domains. While each Digital Twin may differ in composition details, a high to medium-level general framework can help reduce the Digital Twin development effort. Researchers have developed frameworks for various aspects of Digital Twin development. Frameworks that emphasize the two-way communication between the physical twin and the Digital Twin have been proposed [4, 5]. Many Digital Twins today are implemented for real-time state monitoring. The aerospace industry has used Digital Twins to monitor jet engines for many years. These types of Digital Twins use one-way synchronization. They receive data from the physical object but do not provide control feedback.

Galli et al. [6] analyzed a variety of proposed frameworks for building Digital Twins. This research focused on the structure of the frameworks to find correlations in terms of form and conceptualization. The results showed three types of architectures used for the Digital Twin. The first is the traditional and is based on the original framework proposed by Grieves and Vickers [7]. This architecture supports the Digital Twin in paralleling the real system, discussing interfaces and interoperability with operations management systems. It also ensures synchronization between the physical system and its corresponding digital twin. The second is the "service-oriented" type, which consists of four components: a physical shop floor, a virtual model of the shop floor, a service system, and the shop floor Digital Twin data. The third architecture is the fractal, where local or specialized Digital Twins make up the global Digital Twin, each with a similar structure. Khan et al. proposed a six-dimensional framework to include (1) a physical asset, (2) a digital duplicate of the physical asset, (3) data generated by the physical asset, (4) programs that improve product performance and make the production process more efficient, (5) spiral-rings like iterations that generate a more optimized product or efficient production process, and (6) synchronization between the physical asset and the digital twin [4].

Despite these efforts, the Digital Twin technology is still in its early stages. There is a lack of universal definitions, implementation frameworks, and protocols. There is also a lack of comprehensive and in-depth analysis of Digital Twins from the perspective of concepts, technologies, and industrial applications research [8]. Further, the ecosystem of Digital Twins is not well established, and most Digital Twins are developed using customized solutions. Customized solutions do not support reuse and are costly and time-consuming. A systematic approach is needed to characterize and manage Digital Twin subsystems to ensure cross-disciplinary interoperability and credibility of the Digital Twin. Standards are the solutions to enable such interoperability. However, only a few standards have been developed for Digital Twins so far.

According to Accenture research, companies are not taking full advantage of Digital Twins because most Digital Twin applications are standalone for single functions, which focus on functional optimization instead of enterprise optimization, and have no comprehensive strategy for data integration and sharing [9]. Additionally, if a Digital Twin is the current representation of a physical asset, the associated Digital Twin data at any life cycle stage could be helpful for future asset management. The architecture or mechanism supporting the flow of information about a product's performance and use from design, production, use, disposal, and recycling is referred to as the digital thread [10]. Using the digital thread will enable the traceability of Digital Twins from requirements to the retirement of the physical asset. Interoperability among Digital Twins for different life cycle stages through digital threads should help overcome these challenges. Standardized data representation will help avoid customized Digital Twin development and the duplication of efforts. Digital threads can also provide an integrated view of the physical asset for Digital Twin development, avoiding redundancy during information exchange.

To make the best use of Digital Twins, manufacturers need to apply interoperability standards from both systems of systems and life cycle perspectives. In addition, standards on vocabulary, reference architecture, and trustworthiness can help ensure the interoperability, value, and credibility of the Digital Twins. These standards will enable manufacturers to build, manage, and deploy their Digital Twins more efficiently.

This Chapter focuses on a standardized approach to building Digital Twins for advanced manufacturing. It identifies current challenges for manufacturers to implement their Digital Twins, reviews relevant standards efforts, introduces the ISO Digital Twin framework standard, presents use-case scenarios, and discusses some potential standards development research directions. The rest of the chapter is organized as follows: Sect. 2 discusses various applications of Digital Twins, Sect. 3 presents how standards help address the challenges of implementing and adopting Digital Twins that manufacturers face, Sect. 4 introduces the ISO standard, ISO 23247—Digital Twin Framework for Manufacturing, Sect. 5 discusses additional relevant standards, Sect. 6 describes a case study of applying standards for building a Digital Twin of a robot work cell, Sect. 7 discusses potential topics and research directions toward standardization, and Sect. 8 summarizes the chapter and presents future work.

## 2 Applications and Enabling Technologies for Digital Twins

Digital Twins are becoming increasingly prevalent in a wide variety of industries, including manufacturing, construction, smart cities, and healthcare. The applications of Digital Twins include system monitoring, anomaly detection, prediction, optimization, and control. The major general classifications of Digital Twins, shown in Fig. 1, are descriptive, diagnostic, predictive, prescriptive, and intelligent Digital Twins. From left to right, Fig. 1 shows the different Digital Twin functionalities, with increasing complexity from monitoring to intelligent control, to support appropriate decision-making and automation. The bulletized items in each box are examples of enabling technologies for that category of Digital Twins. Each category is described in the following [11].

- *Descriptive Digital Twins* observe their physical counterparts to identify what has happened or is happening. These kinds of Digital Twins can generate different views of the data collected using smart sensors based on the purpose of the Digital Twin. Based on stakeholders' requirements, data and parameters can be visualized in the form of text, tables, and charts. For example, one of these important data is the cycle time of each product type during production. Enabling technologies may include real-time data streaming, database queries, and dashboard reporting.

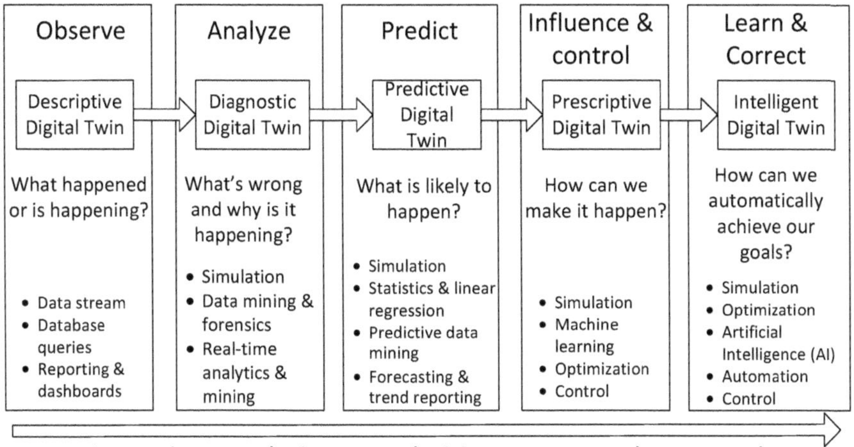


**Fig. 1** Various applications of digital twins with examples of supporting technologies

- *Diagnostic Digital Twins* analyze what has gone wrong and why it has happened or is happening to their physical counterparts. Diagnosis includes analyses of the impact of a data input and an operational strategy on key performance indicators (KPIs). For example, a product's cycle time increase may be caused by machine breakdowns or bad scheduling decisions. Diagnostics are supported by technologies and methodologies such as simulation, data mining, machine learning, and data analytics.
- *Predictive Digital Twins* predict what will happen and when it will happen. These kinds of Digital Twins can be used to estimate when a machine tool or robot deterioration will likely reach a point of failure based on past performance patterns. They can also pinpoint the cause or source of failure. Enabling technologies may include modeling and simulation, predictive data mining, and parameter tracking. For example, Digital Twins integrated with machine learning can be used to predict cycle times for incoming manufacturing orders [5].
- *Prescriptive Digital Twins* provide the influence and control of the physical counterparts and decide how to make it happen based on the objectives of the Digital Twins. These kinds of Digital Twins can help identify the strategies and inputs leading to optimal performance. For example, prescriptive analytics provide the best possible input parameters and methods that enable cycle-time reduction and increase throughput. Enabling technologies may include simulation, optimization, and control.
- *Intelligent Digital Twins* are envisaged to control their physical counterparts based on the strategies and parameters identified by the prescriptive Digital Twins. These kinds of Digital Twins can learn new strategies based on the data collected and take actions accordingly, which could include dynamically adjusting themselves based on the changes in their physical counterparts to keep up with them or based

on the Digital Twin objectives to ensure the physical counterparts operate optimally. Artificial Intelligence (AI), including large language models, is the primary modeling technique used for modeling the intelligent Digital Twin. Other enabling technologies may include simulation, optimization, automation, and control.

The kind of Digital Twin to be implemented depends on the use case and is driven by the objective and scope of the Digital Twin.

## 3 How Standards Support Digital Twin Development and Integration

As discussed in Sect. 1, there are still significant challenges for manufacturers, especially Small and Medium-sized Enterprises (SMEs), to implement their Digital Twin applications efficiently and effectively. Current implementations mostly use ad hoc, customized approaches. Customized solutions not only increase the development time and cost but also make it challenging to integrate with other systems and do not support reuse. Standards are needed for manufacturers to go beyond custom, expensive Digital Twins to an affordable marketplace of products and tools for Digital Twins. Standards, such as frameworks, reference models, and interfaces, will provide a solid foundation for Digital Twin developers, users, and technology and service providers to ensure interoperability, reliability, validity, security, and trust.

Standards for Digital Twins will facilitate the composition and integration of Digital Twins by providing guidelines, methodologies, common terminologies, architectures, and interface specifications. These standards can help make the creation, integration, update, and validation of Digital Twins more accurate and consistent. Standards can also help formalize requirements for Digital Twin projects, enable the use of building blocks for Digital Twin implementations, analyze Digital Twin performance, communicate between suppliers, partners, and customers, secure Digital Twin information and protect privacy, and facilitate the verification and validation of Digital Twins according to stakeholders' requirements. Ultimately, standards will help achieve "plug and play," i.e., enabling interoperability between Digital Twins and among software and hardware from various vendors. In different areas, standards support the following:

- Defining a common language for data representation, communication protocols, and Application Programming Interfaces (APIs) to ensure Digital Twin systems can understand and interact with each other's data. Examples of APIs include those for data query, data update, and data synchronization.
- Developing shared metadata and ontologies to describe the properties, attributes, and relationships within Digital Twins for easy mapping and translating data between Digital Twins.

- Developing tools or middleware that can map data from one Digital Twin's format to another using techniques such as data transformation, data normalization, or data translation.
- Implementing robust security measures to ensure only authorized Digital Twins can access and interact with each other using techniques such as authentication, encryption, and access control.

## 4 Digital Twin Framework for Manufacturing

The ISO standard, ISO 23247—Digital Twin Framework for Manufacturing, was created to facilitate the implementation of Digital Twins in manufacturing. The standard defines a "Digital Twin in Manufacturing" as a "fit for purpose digital representation of an observable manufacturing element with synchronization between the element and its digital representation [12]." It provides a generic guideline, a reference architecture, and a framework for Digital Twin applications in manufacturing. The standard also provides examples of data collection, data communication, integration, modeling, and applications of relevant standards [12]. The standard provides procedures for manufacturers and solution providers to analyze Digital Twin requirements, define scope and objectives, use common terminologies, comply with a generic reference architecture, and integrate multiple existing standards for various purposes. The framework includes the sub-entities and components as building blocks for manufacturers to pick and choose for their own case-specific Digital Twin development. It helps manufacturers systematically identify and determine subsystems and components, their relationships, and the characteristics of their interactions from which appropriate standards can be selected for interoperability.

A key feature of ISO 23247 is that it enables the deployment of the digital thread, implying that model-based engineering standards for various stages of a product life cycle can be included in the framework. For example, for the Digital Twins developed to support a product at different stages, including design, manufacturing, and inspection, relevant standards such as Standard for the Exchange of Product Model Data (STEP) [13], MTConnect [14], and Quality Information Framework (QIF) [15] can be applied. Therefore, the standard supports Digital Twins' compatibility and interoperability throughout the life cycle stages, allowing information reuse and traceability.

The standard series includes four parts: (1) overview and general principles, (2) reference architecture, (3) digital representation, and (4) information exchange. The reference architecture in the standard includes a reference model with domains and entities. There are four domains (layers), each with a logical set of tasks and functions performed by functional entities. Figure 2 shows the entity-based reference model and an illustration of the four domains and their interactions [16]. Each domain is briefed as follows:

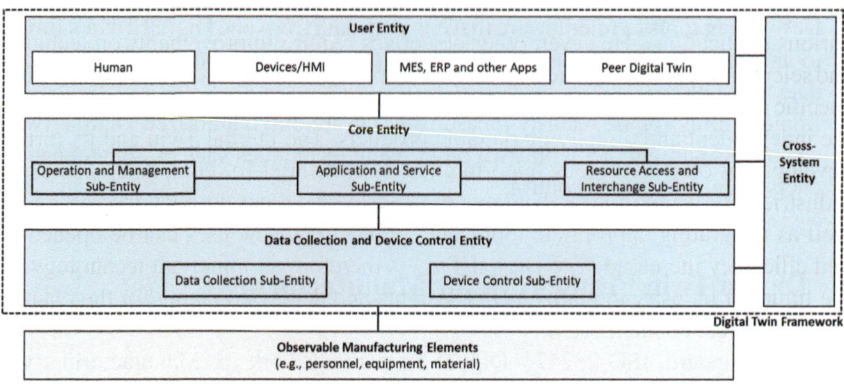


**Fig. 2** Functional view of the digital twin reference model for manufacturing

- The observable manufacturing domain: contains the Observable Manufacturing Elements (OMEs), including any physical artifact, process, or behavior such as personnel, equipment, material, processes, facilities, assets, and systems on the factory floor. These OMEs can be represented by Digital Twins.
- The device communication domain: is a layer between OMEs and their Digital Twins to support data exchange and synchronization. OMEs are monitored, and real-time data are collected using device sensors and standard protocols in the OME domain. This domain is also responsible for transferring commands and signals for control and actuation of the OMEs.
- The Digital Twin domain, or core domain: is responsible for the modeling, operating, and managing Digital Twins. This domain hosts models, applications, and services such as data analytics, simulation, and optimization to support provisioning, monitoring, analysis, and synchronization. It also interacts with users and systems, including other Digital Twins. As indicated in the definition, Digital Twins are built "fit-for-purpose" because each has its own objectives, is context-dependent, and only requires relevant data and models. The purpose of the digital twin dictates its information content, model fidelity, and frequency of synchronization.
- The user domain: includes users or systems such as a human, a device, an application, or a system that uses applications and services provided by the Digital Twin domain.

The cross-system entity in Fig. 2 is an entity that resides across domains to provide common functionalities such as data translation, data assurance, and security support. Digital Twins can be developed based on the Digital Twin Framework depicted within the dotted line in Fig. 2. The framework supports the applications of IoT infrastructure for data collection, communication protocols for data transmission, and information flows between entities of different domains—OMEs, Data Collection and Device Control, Digital Twin Core, and User layers.

Developers tested the Standard by implementing several industry use cases to demonstrate and validate the Standard. The use-cases included a Digital Twin for robot drill and fill to increase equipment utilization, a Digital Twin that optimizes the sizes of aircraft fasteners, and a Digital Twin to optimize Computer Numerical Control (CNC) cutting tool life. These use-case implementations provided valuable feedback to standards developers and demonstrated the viability of the standard framework for constructing Digital Twins [17].

## 5 Additional Relevant Standards

Digital Twin standards, such as architectural frameworks, can provide implementation guidelines for developing Digital Twins. Other existing standards can support specific Digital Twin functionalities such as data collection, data communication, information modeling, systems integration, simulation modeling, and automation and control. This section identifies some examples of manufacturing standards that could apply to various aspects of the Digital Twin development. Standards related to information security, data assurance, and trustworthiness are also necessary, but not listed in this Section.

### *5.1 Frameworks and Architectures*

- IEC 62832-1: 2020, Digital Factory Framework defines a framework to establish and maintain the digital representations of production systems throughout their life cycle. This framework supports a consistent exchange of information between all processes and partners. Information becomes understandable, reusable, and exchangeable throughout the production system life cycle [18].
- IEEE P2806:2019, System Architecture of Digital Representation for Physical Objects in Factory Environments, supports the development of digital factories. It describes the objective, components, data sources required, and procedure of digital representation in factory environments [19].
- IEC 63278-1: Asset Administration Shell for industrial applications—Part 1: Asset Administration Shell (AAS) structure [20]. AAS aims to enable one or more software applications to exchange information and use that information in a trusted and secure way. It specifies the connector between the real and virtual worlds and includes a model of the shell covering the fundamental concepts: Asset, Submodel, and Concept Description. Identifiers are defined for all elements in the model, concept descriptions, and property definitions of external repositories such as ECLASS and IEC CDD. Mappings of the AAS model are specified for several widely used information models such as XML, JSON, RDF, OPC-UA, and AutomationML.

- ISO/IEC 30141:2018, Internet of Things (IoT)—Reference Architecture provides a common vocabulary, reusable designs, and industry best practices. It starts with collecting the essential characteristics of IoT, abstracting them into a generic IoT conceptual model, and deriving a high-level systematic reference with subsequent dissection of that model into five architectural views [21].
- ISO/IEC 21823-1:2019, Internet of things (IoT)—Interoperability for IoT systems—Part 1: Framework provides an overview of the interoperability of IoT systems and the various entities within them. It enables IoT systems to be built so the entities of the IoT system can exchange information efficiently. It also supports peer-to-peer interoperability between IoT systems [22].
- ISO/IEC/IEEE 15288:2015, Systems and software engineering—System life cycle processes establish a common framework of process descriptions for describing the life cycle of systems. It defines a set of processes and associated terminology that can be applied at any level in a system's structure. They can manage the stages of a system's life cycle. It also provides processes that support the definition, control, and improvement of the system life cycle processes [23].
- Microsoft Digital Twin Definition Language (DTDL)—A language for describing models and interfaces for IoT Digital Twins. DTDL is based on JSON-LD and is programming language independent. DTDL is used in different Microsoft services such as IoT Hub, IoT Central, and Azure Digital Twins; it is also used to represent device data in other IoT services such as IoT Plug and Play. DTDL covers the resource description and not resource discovery and access. Resources (interfaces) contain telemetry, properties, commands, relationships, and components [24].
- The High-Level Architecture (HLA) defines an architecture for distributed simulation, its components, and the rules that outline the responsibilities of HLA federates and federations for a consistent implementation. The standard also supports maintaining the information model of each simulation to retain its meaning and purpose but enables data communication and time synchronization of the distributed simulation systems [25].

## 5.2 *Data Collection, Data Modeling, and Data Exchange*

- MTConnect supports interoperability by providing a vocabulary for manufacturing equipment, making structured contextualized data possible, and avoiding proprietary formats. Data sources of MTConnect include equipment, sensor packages, and other factory floor hardware [14].
- OPC-Unified Architecture (UA) is a platform-independent standard used to send messages between clients and servers over diverse networks with syntactic interoperability [26].
- The MTConnect-OPC-UA Companion Specification supports interoperability and consistency between MTConnect specifications and the OPC-UA specifications, as well as devices and software that implement those standards [27].

- ISO/IEC 20922, Information Technology—Message Queuing Telemetry Transport (MQTT) v3.1.1, is a data protocol that supports client and server messaging transport for publishing and subscribing. It is open and simple in design and suitable for use in machine-to-machine (M2M) communication and IoT contexts [28].
- ISO/IEC 17826, Information Technology—Cloud Data Management Interface (CDMI), specifies how to access and manage stored cloud data [29].
- ISO 13374 series, Condition Monitoring and Diagnostics of Machines—Data Processing, Communication, and Presentation, provides basic requirements for open software specification [30].
- ISO/IEC 30161:2020, Internet of Things (IoT)—Requirements of IoT data exchange platform for various IoT services, specifies requirements for an IoT data exchange platform for: (1) middleware components of communication networks allowing the co-existence of IoT services with legacy services, (2) end-points performance across the communication networks among IoT and legacy services, (3) IoT-specific functions allowing the efficient deployment of IoT services, (4) IoT service communication networks' framework and infrastructure, and (5) IoT service implementation guideline for IoT data exchange platforms [31].
- Automation Markup Language (AutomationML) is a neutral data format based on XML that can store and exchange information for plant engineering, connecting heterogeneous modern engineering tools. Disciplines include mechanical plant engineering, electrical design, human–machine interface development, Programmable Logic Controller (PLC), and robot control [32].
- The Core Manufacturing Simulation Data (CMSD) standard provides an information model schema to support data representation and exchange between simulation and other manufacturing applications [33].
- ISO 15531, Industrial Automation Systems and Integration—Industrial Manufacturing Management Data supports information exchange between software applications in production activities, including planning, scheduling, simulation, control, and execution [34].
- ISO 14649-201; Industrial Automation Systems and Integration—Physical Device Control—Data Model for Computer Numerical Control Controllers—Part 201: Machine Tool Data for Cutting Processes describes technology specific to machine tools. It defines data elements for manufacturing and machine characteristics [35].
- ISO/IEC 21823-2:2020, Internet of things (IoT)—Interoperability for IoT systems—Part 2: Transport interoperability, specifies a framework and requirements for transport interoperability to enable the construction of IoT systems with information exchange, peer-to-peer connectivity, and seamless communication between different IoT systems and among entities within an IoT system [36].
- ISO/IEC 21823-3:2021, Internet of Things (IoT)—Interoperability for IoT systems—Part 3: Semantic interoperability, provides: (1) basic concepts for IoT systems including requirements of the core ontologies for semantic interoperability, (2) best practices and guidance on how to use ontologies to develop

domain-specific applications, (3) relevant IoT ontologies along with comparative study of the characteristics and approaches in terms of modularity, extensibility, reusability, scalability, interoperability with upper ontologies, and (4) use cases and service scenarios that exhibit necessities and requirements of semantic interoperability [37].
- ISO/IEC 21823-4:2022, Internet of things (IoT)—Interoperability for IoT systems—Part 4: Syntactic interoperability, describes five facets of IoT interoperability: transport, semantic, syntactic, behavioral, and policy. It includes specifications on how to achieve syntactic interoperability among IoT devices and a framework for processes for developing information exchange rules related to IoT devices [38].

## 5.3 Digital Representation

- The American Society of Mechanical Engineers (ASME) Y14.26M, Digital Representation for Communication of Product Definition Data, focuses on the representation and communication of data to define products. It supports product data exchange developed in computer-aided manufacturing (CAM) systems [39].
- ISO 10303, Automation Systems and Integration—Product Data Representation and Exchange, also known as Standard for the Exchange of Product model data (STEP), supports the exchange of product manufacturing information [13].
- ISO 10303-IR 105, Automation Systems and Integration—Product Data Representation and Exchange—Part 105: Integrated Application Resource: Kinematics focuses on the representation of kinematics information of a mechanical product [40].
- Quality Information Framework (QIF)—This framework standard enables the capture, use, and reuse of metrology-related information throughout the Product Life Cycle Management (PLM) and Product Data Management (PDM) domains. It supports the creation of digital threads. It applies to product design, manufacturing, and quality inspection. It relies on the Extensible Markup Language (XML) standard and contains a library of XML schemas. It supports data integrity and interoperability in implementing model-based enterprise and IoT [15].
- ASME B5.59-2, Information Technology for Machine Tools Part 2, defines the properties needed to describe machine tools used for milling and turning [41].
- ISO 13399, Cutting Tool Data Representation and Exchange, provides a model and a reference dictionary to represent cutting tools. EXPRESS schema is used for the product description, and product files can be generated according to the schema [42].
- ISO 16400, Equipment Behaviors Catalogue (EBC) defines a template and rules for describing equipment behaviors, such as state transition and time series of operation results, that are produced because of machine activities to be registered in the common repository. It specifies the methodology to construct catalogs of equipment behavior to plan and analyze production system performance [43].

- Predictive Model Markup Language (PMML) is used to develop predictive and descriptive models and to represent pre- and post-processed data. PMML is based on XML and supports the representation of statistical and data-mining models and the model sharing between PMML-compliant applications. Examples of models include neural networks, decision trees, Gaussian progress, and Bayesian networks [44].

With these relevant standards from various functional categories, users can select those applicable to their Digital Twin implementations.

## 6 A Use Case of Developing Digital Twins with a Standardized Approach

In this section, a robotic work cell Digital Twin is discussed to exemplify the application of relevant standards. The work cell consists of collaborative robot arms for material handling and machine tending, a CNC machine tool for machining, and a Coordinate Measuring Machine (CMM) for product geometry measurements and quality control. Figure 3 shows the workflow and equipment for the use-case scenario in the work cell. The workflow includes receiving parts, loading parts to the CNC (by ROBOT #1), cutting the parts, unloading parts from the CNC (by ROBOT #1), loading parts to the CMM (by ROBOT #2), inspecting the parts, and offloading parts from the CMM (by ROBOT #2). Parts that fail the inspection will be sent to the rework buffer by ROBOT#2. The cell has a single input location and a single output location. The robots are fitted with a 2F-85 gripper to handle components with different geometries effectively.

One of the research objectives is to build a process digital twin that is a composite of the Digital Twins of each individual component.

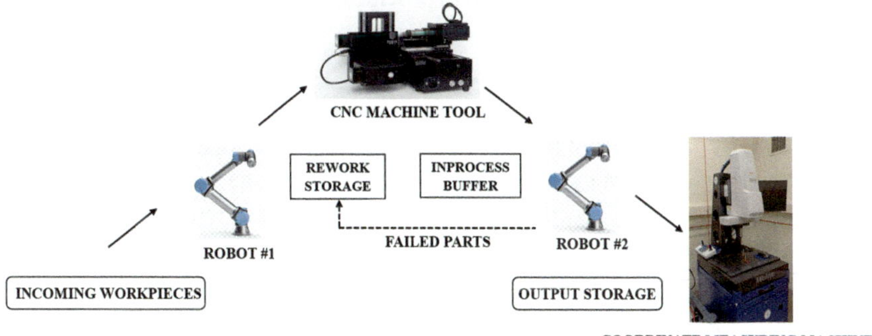


**Fig. 3** Workflow through machines and equipment in the robot work cell

The standards used in implementing the use-case include ISO 23247, MTConnect, STEP, QIF, and ASME Verification, Validation, and Uncertainty Quantification (VVUQ).

### *6.1 ISO 23247*

Based on the Digital Twin framework introduced in Sect. 3, each piece of equipment in the work cell is treated as an OME, for which data need to be collected, and Digital Twins for various scenarios must be developed. In this section, we focus on one robot arm in the work cell to showcase the method of Digital Twin development.

The ISO Digital Twin framework standard is instantiated for the robot arm (UR5e), whose operations include picking and loading a workpiece to a CNC machine tool. Figure 4 illustrates the instantiation of the framework for the robot arm. The figure shows data being collected from the robot arm through a MTConnect adapter. The Digital Twin entity comprises the simulation model of the robot arm and analytical models that manipulate real-time data to support decision-making. The modeling method and environment support the three-dimensional geometry of the robot arm components, including the robot base, links, joints, end effector, and workpiece. The user entity includes the developer and user of the Digital Twin, production software systems, or other Digital Twins. Similarly, the standard can be instantiated for the cutting part, the CNC machine, and the CMM.

### *6.2 MTConnect*

The process of acquiring data and building a scalable data pipeline for the work cell involves multiple activities: (1) collecting real-time operational data from the robot, (2) leveraging the MTConnect standard to provide defined machine data and make it available in a standard format, and (3) developing a set of tools to enable client-side use of the MTConnect agent.

**Operational Data Collection**: To collect the operational data from the UR5e robot arm, we utilized the Universal Robot's Real-Time Data Exchange (UR-RTDE) interface with APIs supporting data collection. Several data items collected include angular position, velocity, acceleration, torque, current, and temperature for each of the six joints of the robot arm. Universal Robots also shows the data items (and the corresponding units) that the vendor offers through the UR-RTDE interface.

**MTConnect Interface**: Implementing the MTConnect standard requires an adapter and an agent. The adapter serves as a data collection element from the equipment while the agent collects data from the adapter. Many machine vendors provide a preinstalled adapter but not the UR5e robot. The adapter was installed on an interface connected to the robot arm. The adapter packages data into a format that is readable

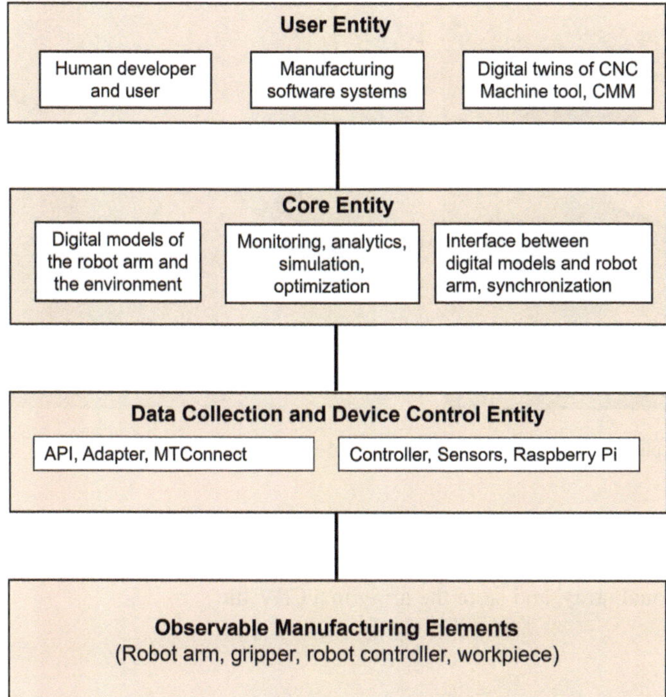


**Fig. 4** Implementation method of building a Digital Twin for the robot based on ISO 23247

by the agent. The agent provides an application interface to retrieve the MTConnect data gathered from the adapter. Figure 5 shows the data flow from a physical device (UR5e) to the Digital Twin. A semantic structure was provided for the physical data generated by the UR5e robot arm through MTConnect. This semantic structure includes data tags and units based on the MTConnect 2.0 Standard [42]. Using the Python UR-RTDE API, a socket-based adapter that sends MTConnect-compliant data to the agent was developed. An instance of the MTConnect agent receives and serves the data in a machine-readable format.

**Client-Side Integration**: The MTConnect agent is implemented in C++ and displays data in XML format on a Hypertext Transfer Protocol (HTTP) server. The tool used for modeling the robot work cell is the Simulink/Simscape, which can input comma-separated values (CSV) where multiple data items over a time interval are

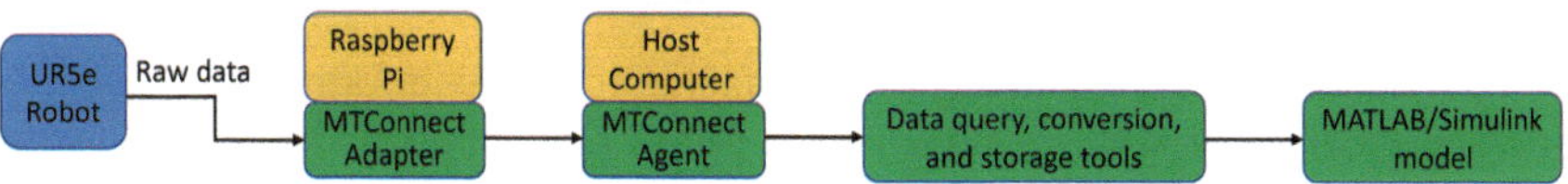


**Fig. 5** Data pipeline for the UR5e robot Digital Twin

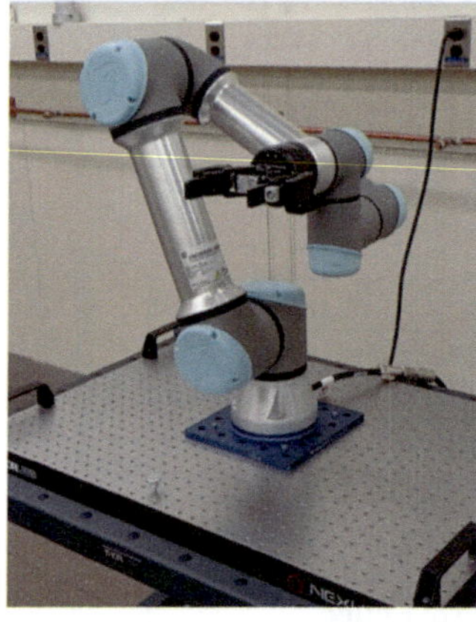

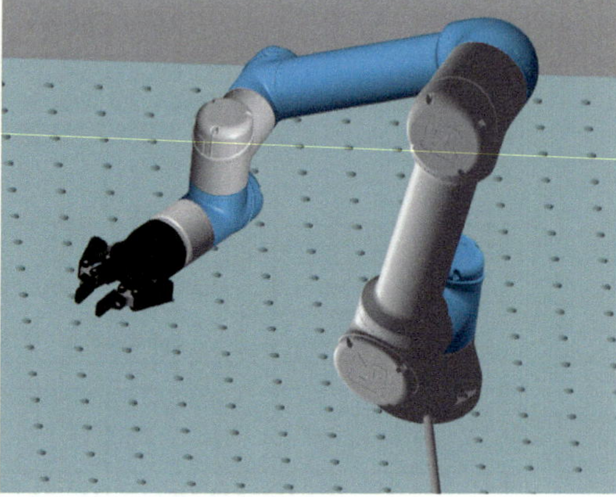

**Fig. 6** The physical and virtual model of the UR5e robot arm with attached gripper

recorded and synced up to their respective timestamps (UTC format). The tools have been developed to parse the XML output from the MTConnect Agent, populate a 2-dimensional array, and store the array in a CSV file.

## *6.3 Step*

To develop a physical model of the robot arm, Computer Aided Design (CAD) models of its components and those of the work environment are fundamental. These models include the description of the geometry of the links and how they are connected to the robot arm. These CAD models are imported into the Digital Twin environment to create a physical model of the UR5e robot arm. The CAD models for the gripper are "assembled" into the end effector in the AutoCAD Inventor environment, exported to the Digital Twin environment, and "attached" to the robot arm model. Figure 6 shows the physical and virtual models of the robot arm with the attached end effector.

The CAD models of the part, CNC, and CMM can be in STEP standard format, which allows easy exchange and visualization. The design part model in STEP will also be used to compare the measurements of the finished product.

## *6.4 Quality Information Framework (QIF)*

QIF provides an integrated model for manufacturing quality information. The CMM provides measurement information about a product regarding conformance to specifications per Geometric Dimensioning and Tolerancing (GD&T) design. Design tolerances are defined by the amount a feature is allowed to vary from the nominal.

Assigning GD&T to a designed part is a way to consider process variation within manufacturing. This product data is obtained directly from the design process and should be available throughout the product life cycle. Being able to map this data back to a single source—the native CAD, QIF enables model-based workflows that are part of the digital manufacturing transformation. The Digital Twin of the CMM will receive measurement data from the actual machine and compare it with the design data. QIF facilitates this data exchange between the real CMM and its Digital Twin.

## 6.5 *Verification and Validation*

Digital Twins are complex systems that are sometimes used as virtual testbeds for verification and validation of systems, especially in cases where actual tests are complex to perform. However, the quality of decisions made with Digital Twins depends on the validity of the underlying models. A valid Digital Twin should accurately describe the system that changes over time. Thus, the development of the Digital Twins needs to be validated before its use in supporting decision-making. Verification, Validation, and Uncertainty Quantification (VVUQ) standards need to be followed to ensure that the Digital Twin is built correctly and that the right Digital Twin has been built. Zhang et al. [45] discuss verification and validation methods for Digital Twins, which are categorized into qualitative, quantitative, and integrated methods. Both qualitative and quantitative methods require metrics for Digital Twin validation. These metrics include credibility/fidelity, complexity, standardization, and capability maturity of model construction.

Hua et al. [46] summarized general strategies to validate a Digital Twin. These include visual inspection of the Digital Twin for correctness using established standards, testing properties of the Digital Twin, model-based testing using methods such as input–output conformance testing, and machine learning or artificial intelligence-based testing. Kibira and Weiss [47] used a model-based approach to validate the Digital Twin model of a robot arm. Joint position and orientation data, velocity data, and acceleration data were collected from the physical twin for validation under the model-based approach.

Verification and validation standards include:

- ASME V&V 10: Standard for Verification and Validation in Computational Solid Mechanics [48].
- ASME V&V 20: Standard for Verification and Validation in Computational Fluid Dynamics and Heat Transfer [49].
- ASME V&V 40: Assessing Credibility of Computational Modeling Through Verification and Validation: Application to Medical Devices [50].
- Other ASME V&V 50, 60, 70, and 80 standards are under development. V&V 50 standards are for advanced manufacturing, and V&V 70 standards are for data-driven models.

# 7 Future Research Topics for Standardization

The current four parts of the ISO 23247 series provide a fundamental generic Digital Twin framework for manufacturing. The framework can be extended to industries that employ specialized manufacturing processes and technologies. Future work on this standard may result in new additions supporting the development and validation of Digital Twins. The new research topics for extension of the standard include (1) digital thread for Digital Twins, (2) Digital Twin composition, (3) ontologies of the Digital Twin framework to clarify the entities and relationships, (4) building Digital Twins from reusable components to increase the consistency and reduce the development time, (5) credibility assessment of Digital Twins to increase the trustworthiness and value for decision-making, (6) Digital Twins and the metaverse to provide guidelines that enable the integration between Digital Twins and industrial metaverse, (7) plug and play Digital Twin integration by standardizing interfaces with customers' environment and application platforms, and (8) extending the framework to specific sectors, e.g., semiconductor manufacturing, biomanufacturing, and additive manufacturing to address domain-specific needs [51]. The following subsections discuss the potential new parts of the standard that could enable better and easier Digital Twin development.

## *7.1 Digital Thread for Digital Twin (ISO 23247-5)*

When performing digital transformation, standalone Digital Twins require a lot of duplicated efforts. Therefore, a life cycle approach needs to be taken. However, effectively bringing the life cycle approach is challenging. Guidelines and methodologies for supporting Digital Twin development using a digital thread of the product life cycle will be needed to access all product life cycle information, including design, manufacturing, inspection, and use data, and enable information traceability.

A new part of ISO 23247 on this topic will specify how the digital thread enables the creation, connectivity, management, and maintenance of Digital Twins across the product life cycle. This will involve defining the principles to follow before embarking on a journey for digital transformation, describing methodologies, and providing use-case examples. This part will describe how the digital thread supports the generation, implementation, and transformation of Digital Twins in manufacturing. Information in the digital thread enables the Digital Twin to express the changes in a product throughout its life cycle, which can be used to improve future iterations of the product. The digital thread ensures this product life cycle information is readily accessible, traceable, reliable, and secure.

### 7.2 Digital Twin Composition (ISO 23247-6)

Digital Twin composition implies that multiple Digital Twins are developed and integrated with the support of a digital thread. For example, the Digital Twin of a part and that of the machine that manufactures the part can interact dynamically and seamlessly. When the Digital Twin of a cutting tool, a machine tool, and a part interact, they can be used to determine the tool wear, tolerance conformance, and the machine's health. Digital Twins of multiple partners coordinate and communicate in real time in a supply chain. However, it is challenging to aggregate, compose, and integrate multiple Digital Twins and applications to achieve a new goal. Standard-based methods and guidelines will help achieve this, reduce development time, and mitigate risks for such undertakings.

This part of ISO 23247 will provide guidelines on enabling multiple Digital Twins to communicate and interoperate effectively. The new part could provide generic methodologies, principles, and examples to help users understand the purpose of the Digital Twin and develop the appropriate Digital Twin to address the identified problem(s). Relevant standards and technologies could be selected and applied to demonstrate the integration.

This part of ISO 23247 will also specify Digital Twin composition by defining principles, showing methodologies, and providing use-case examples of Digital Twin configuration, communication, aggregation, composition, integration, and collaboration during manufacturing.

### 7.3 Ontologies for Digital Twin Framework

The current four parts of the ISO 23247 standard define the terms, relationships, components, and processes necessary for developing a Digital Twin and provide guidelines for Digital Twin implementation. However, logical formalism does not support it, which may lead to inconsistent implementation. The ontologies for the Digital Twin framework will enable the definitions of terms that are both human-understandable and computer-processable, which result in an unambiguous representation of a particular construct and consistent interpretation, regardless of the initial data source. It also enables explicit representation of the connections between different terms; different connections permit a consistent presence and representation of the required metadata. A potential new part of ISO 23247 on this topic could provide an ontology for the Digital Twin framework.

## 7.4 Building Digital Twins from Reusable Components

Digital Twins could be developed for different control levels depending on the application, including equipment, work cells, production lines, factories, and supply chains. While some approaches exist to support model component reuse [52], most of them are not explicitly designed for Digital Twins. Therefore, almost all Digital Twins are built from scratch, which makes implementations time-consuming and costly. Customized designs also make a Digital Twin challenging to modify, extend, and reuse. Manufacturing knowledge, information attributes, and use-case configurations are often developed using different specialized abstractions for each application. The reusability of Digital Twin components in a Digital Twin library could considerably reduce the development cost, time, and the required level of expertise.

A potential new part of ISO 23247 on this topic could provide guidelines on building component libraries and creating templates for organizing data, information, and models. Reusable Digital Twin components may include templates for data collection, common information attributes, and modular models. Digital Twin development would be supported by enabling technologies and relevant standards for various Digital Twin functions. The new part could provide generic methodologies, architectures, frameworks, knowledge bases, and examples for building and using Digital Twin component libraries.

## 7.5 Credibility Assessment of Digital Twins

The current four parts of ISO 23247 do not cover VVUQ and testing on Digital Twins. Given the potential use of Digital Twins in critical decision-making for various manufacturing applications, the results generated by Digital Twins must be trustworthy for real manufacturing needs. Model credibility assessment, including VVUQ techniques, must be applied throughout the life cycle of Digital Twins. VVUQ should be embedded in the design, creation, and deployment of Digital Twins to establish trust in the model and its outputs [53]. Verification and Validation (V&V) activities are necessary to ensure that a Digital Twin meets its intended purpose and design goals. Uncertainty Quantification (UQ) produces a measure of performance that users can apply as part of a credibility assessment for a given Digital Twin. VVUQ for Digital Twins should be a continual process that adapts to changes in the OME and its digital representation, data inputs, and decisions made [53]. The credibility assessment of Digital Twins may also include factors beyond VVUQ.

Digital Twin testing needs a test system comprising a set of tests for both the OME and its Digital Twin. The test system should also define what an acceptably valid Digital Twin should look like. Grieves proposed a virtual testing method for manufactured products, which can be adapted to Digital Twins [54]. For example, suppose the test system can run a set of tests, and the results of the Digital Twin can't be distinguished from those of the OME within a predefined probability threshold. In that case, the Digital Twin can be regarded as a reasonable representation of the OME. Trust in a Digital Twin also involves trust in the data collected from the OME, the model used in the Digital Twin, the data updating procedure, and the recommended decisions. All these aspects should have a measurable uncertainty, whose existence means that validation (comparison with reality) needs to be treated as a statistical process. Comparison of actual data with model results can be used to estimate the probability that the Digital Twin is a consistent representation of the OME.

Currently, there is no standard process for reporting VVUQ for digital twins. Developing robust VVUQ processes for digital twins remains a challenge. A potential new part of ISO 23247 on this topic could provide guidelines on and methodologies for how to measure uncertainty, how to perform VVUQ and testing for Digital Twins, how to select or construct a credibility assessment framework that supports these activities, and how to assess the credibility of the developed Digital Twins.

## 7.6 Digital Twins and the Metaverse

The metaverse can support monitoring the manufacturing system in real time, both visually and from a metric standpoint. It could also provide users with an immersive experience. This is now possible because of the maturity of technologies for virtual reality (VR), augmented reality (AR), and extended reality (XR), which can enhance users' visualization experience for manufacturing. For example, it has been demonstrated that AR technologies can be integrated with three-dimensional geometrical product specification and verification standards and practices [55].

A significant feature of the metaverse is the immersive visualization experience along with its human–machine interface. The hardware and software technologies developed for the metaverse can be used by the Digital Twin framework for manufacturing, especially in cases where there is human involvement. For example, the user domain (shown in Fig. 2) and user entity of the ISO 23247 standard can use human–machine interfaces provided by the metaverse. Alternatively, a metaverse may be a parallel virtual world that may subsume some of the Digital Twins of a manufacturing enterprise that the metaverse represents. A new part of the ISO 23247 series could include the metaverse concept, its definition, possible scenarios for integrating with manufacturing Digital Twins, including human Digital Twins, guidelines, and methodologies for such integration.

## 7.7 Extending the Framework to Specific Sectors

Based on the generic framework provided by the initial four parts of the ISO 23247 series, extensions can be developed as new parts of the standard for specific manufacturing sectors such as biomanufacturing, semiconductor manufacturing, and additive manufacturing. The new parts may include specialization of the Digital Twin framework by adding new functional entities or modifying existing functional entities to fit the new requirements. The new parts may also have the use cases implemented for those manufacturing sectors. These use-case implementations may, in turn, help identify new standardization requirements for that manufacturing sector.

The emerging biomanufacturing sector can use the generic framework to develop its Digital Twins, which may constitute a new part of the ISO 23247 series. Similarly, Digital Twins for additive manufacturing may have substantial potential to improve process control, and a new part in the ISO 23247 series can be dedicated to additive manufacturing. In semiconductor manufacturing, an extension of the standard can be developed to address challenges such as obtaining datasets for constructing Digital Twin models and cybersecurity associated with the Digital Twin. Other standards development organizations may adopt the current ISO 23247 series to create Digital Twins for their customer industries.

## 8 Summary

Digital Twins are becoming more prevalent in a wide variety of industries, including manufacturing. However, a standards-based ecosystem of Digital Twins has not yet been established. Developing and integrating Digital Twins presents significant challenges. Foundational work is needed to support an open marketplace for Digital Twin developers, users, and technology and service providers. This includes the development of standardized frameworks, reference models, interface specifications, and VVUQ methodologies to provide a solid foundation for ensuring the interoperability, validity, security, and trust of Digital Twins. This Chapter focuses on applying Digital Twins in manufacturing within a framework of standards, identifies current challenges, reviews relevant standardization efforts, and introduces the ISO Digital Twin framework standard for manufacturing, ISO 23247. This work also discusses some potential research directions and future standardized topics. A use case is presented to illustrate the Digital Twin development process by applying relevant standards. The example is obtained from the Digital Twin development efforts for a robot work cell in a Digital Twin Lab at NIST.

Future efforts include performing measurement science research to support the development and integration of Digital Twins in manufacturing, working with Industrial consortia and standards development organizations to prioritize the standardization topics (e.g., interoperability and VVUQ), formulating working groups and project teams to develop the new parts of the Digital Twin standards, and enhancing the Digital Twin Lab to serve as a digital twin testbed to support digital twin prototyping and standards development and testing.

**Disclaimer** Certain commercial products and systems are identified in this chapter to facilitate understanding. Such identification does not imply that these software systems are necessarily the best available for the purpose. No approval or endorsement of any commercial product by NIST is intended or implied.

## References


1. Allen D (2021) Digital twins and living models at NASA [Slides]. In: ASME digital twin summit. https://ntrs.nasa.gov/api/citations/20210023699/downloads/ASME%20Digital%20Twin%20Summit%20Keynote_final.pdf
2. Grieves M (2002) Completing the cycle: using PLM information in the sales and service functions [Slides]. In: SME management forum. Troy, MI
3. Khan A, Shahid F, Maple C, Ahmad A, Jeon G (2022) Toward smart manufacturing using spiral digital twin framework and twin chain. IEEE Trans Industr Inf 18(2):1359–1366
4. Negri E, Berardi S, Fumagalli L, Macchi M (2020) MES-integrated digital twin frameworks. J Manuf Syst 56:58–71
5. Jain S, Narayanan A (2023) Digital twin–enabled machine learning for smart manufacturing. Smart Sustain Manuf Syst 7(1):111–128
6. Galli E, Fani V, Bandinelli R, Lacroix S, Le Duigou J, Eynard B, Godart X (2023) Literature review and comparison of Digital Twin frameworks in manufacturing. Proc Eur Counc Model Simul 2023:428–434
7. Grieves M, Vickers J (2017) Digital Twin: Mitigating unpredictable, undesirable emergent behavior in complex systems. In: Kahlen F-J, Flumerfelt S, Alves A (A c. Di) Transdisciplinary perspectives on complex systems. Springer International Publishing, pp 85–113. https://doi.org/10.1007/978-3-319-38756-7_4
8. Lu Y, Liu C, Kevin I, Wang K, Huang H, Xu X (2020) Digital Twin-driven smart manufacturing: connotation, reference model, applications and research issues. Robot Comput-Integr Manuf 61:101837
9. Accenture (2021) Think thread first: surf the wave of product data. Retrieved 6 Oct 2023, from https://www.accenture.com/us-en/insights/industry-x/thread-first-thinking
10. Singh V, Willcox KE (2018) Engineering design with digital thread. AIAA J 56(11):4515–4528
11. Shao G, Jain S, Shin J (2014) Data analytics using simulation for smart manufacturing. In: Tolk A, Diallo SY, Ryzhov IO, Yilmaz L, Buckley S, Miller JA (eds) Proceeding of the 2014 winter simulation conference. IEEE, pp 2192–2203
12. ISO-23247-1 (2021) Automation systems and integration—Digital Twin framework for manufacturing—Part 1: overview and general principles. Retrieved on 10 Oct 2023, from https://www.iso.org/standard/75066.html
13. ISO-10303 (2021) Industrial automation systems and integration—product data representation and exchange—Part 1: Overview and fundamental principles, Retrieved on 21 Oct 2023, from https://www.iso.org/standard/72237.html
14. MTConnect (2022) MTConnect standardizes factory device data. https://www.mtconnect.org/
15. ISO (2020) ISO 23952-2020. Automation systems and integration—Quality information framework (QIF)—An integrated model for manufacturing quality information. Retrieved on 11 Oct 2023, from https://www.iso.org/standard/77461.html
16. ISO-23247-2 (2021) Automation systems and integration—Digital twin framework for manufacturing —Part 2: reference architecture. Retrieved on 10 Oct 2023, from https://www.iso.org/obp/ui/en/#iso:std:iso:23247:-2:ed-1:v1:en
17. STEP Tools (2020) Demonstration of Three ISO 23247. Retrieved on 30 Nov 2023 from https://www.youtube.com/watch?v=wbsC_qzB8us

18. IEC-62832 (2020) Industrial process measurement, control and automation—digital factory framework—Part1: General principles, Retrieved on 12 Oct 2023, from https://webstore.iec.ch/publication/65858
19. IEEE-P2806 (2019) System architecture of digital representation for physical objects in factory environments, Retrieved on 15 Oct 2023, from https://standards.ieee.org/project/2806.html
20. IEC 63278 -1. (2023) AAS (Asset Administration Shell) for industrial applications - Part 1: Asset Administration Shell structure. https://webstore.iec.ch/en/publication/65628
21. ISO/IEC-30141 (2018) Internet of Things (IoT)—reference architecture, Retrieved on 11 Oct 2023, from https://www.iso.org/standard/65695.html
22. ISO/IEC-21823-1 (2019) Internet of things (IoT)—interoperability for IoT systems—Part 1: framework, Retrieved on 11 Oct 2023, from https://www.iso.org/standard/71885.html
23. ISO/IEC/IEEE-15288 (2015)Systems and software engineering—system life cycle processes, Retrieved on 24 Oct 2023, from https://www.iso.org/standard/63711.html
24. DTDL(Digital Twin Definition Language) (2022) Digital twin definition language (DTDL) for models. Retrieved on 29 Sept 2023, from https://learn.microsoft.com/en-us/azure/digital-twins/concepts-models#digital-twin-definition-language-dtdl-for-models
25. HLA (High Level Architecture) (2010) IEEE standard for modeling and simulation (M&S) High level architecture (HLA)—federate interface specification. IEEE Comput Soc
26. OPC-UA (2017) OPC-UA services specification, Retrieved on 11 Oct 2023 from https://opcfoundation.org/developer-tools/specifications-unified-architecture#:~:text=OPC%20Unified%20Architecture%20Specification,more%20secure%20and%20scalable%20solution
27. MTConnect-OPC UA (2019) OPC-UA companion specification for MT connect. https://www.mtconnect.org/opc-ua-companion-specification
28. ISO/IEC-20922 (2016) Information technology—Message queuing telemetry transport (MQTT) v3.1.1, Retrieved on 26 Oct 2023, from https://www.iso.org/standard/69466.html
29. ISO/IEC-17826 (2016) Information technology—cloud data management interface (CDMI). Retrieved on 26 Oct 2023, from https://www.iso.org/standard/70226.html
30. ISO-13374 (2007) Condition monitoring and diagnostics of machines—Data processing, communication and presentation—Part 2: Data processing, Retrieved on 21 Oct 2023. From https://www.iso.org/standard/36645.html
31. ISO/IEC-30161 (2022) Internet of Things (IoT)—requirements of IoT data exchange platform for various IoT services, Retrieved on 23 Oct 2023, from https://www.iso.org/standard/53281.html
32. Automation ML (2021) Standard data exchange in the engineering process of production systems, Retrieved on 10 Oct 2023, from https://www.automationml.org/wp-content/uploads/2021/06/AutomationML-Brochure.pdf
33. CMSD (2010) Core manufacturing simulation data (CMSD) standard, https://www.sisostds.org/DesktopModules/Bring2mind/DMX/API/Entries/Download?Command=Core_Download&EntryId=31457&PortalId=0&TabId=105
34. ISO-15531 (2017) Industrial automation systems and integration—industrial manufacturing management data—Part 44: Information modelling for shop floor data acquisition, Retrieved on 26 Oct 2023, from https://www.iso.org/standard/71064.html
35. ISO-14649 (2011) Industrial automation systems and integration—physical device control—data model for computerized numerical controllers—Part 201: Machine tool data for cutting processes, Retrieved on 26 Oct 2023, from https://www.iso.org/standard/60042.html
36. ISO/IEC-21823-2 (2020) Internet of things (IoT)—interoperability for IoT systems—Part 2: Transport interoperability, Retrieved on 11 Oct 2023, from https://www.iso.org/standard/80986.html
37. ISO/IEC-21823-3 (2021) Internet of things (IoT) - interoperability for IoT systems—Part 3: semantic interoperability, https://www.iso.org/standard/83752.html
38. ISO/IEC-21823-4 (2022) Internet of things (IoT)—interoperability for IoT systems—Part 4: Syntactic interoperability, Retrieved on 11 Oct 2023, from https://www.iso.org/standard/84773.html

39. ASME (American Society of Mechanical Engineers) (1989) Digital representation for communication of product definition data. Retrieved on 10 Oct 2023, from https://standards.globalspec.com/std/437642/ASME%20Y14.26M
40. ISO-10303-105 (2014) Industrial automation systems and integration—product data representation and exchange—Part 105: Integrated application resource: Kinematics, Retrieved on 21 Oct 2023, from https://www.iso.org/standard/64294.html
41. ASME (American Society of Mechanical Engineers) (2009) Data specification for properties of machining and turning centers, ASME B5.59-2
42. ISO-13399 (2006) Cutting tool data representation and exchange—Part 1: overview, fundamental principles and general information model, Retrieved on 26 Oct 2023, from https://www.iso.org/standard/36757.html
43. ISO-16400 (2020) Automation systems and integration—equipment behavior catalogues for virtual production system—Part 1: overview, Retrieved on 26 Oct 2023, from https://www.iso.org/standard/73384.html
44. PMML (2018) The predictive model markup language (PMML) 4.3, Retrieved on 11 Oct 2023 from http://dmg.org/pmml/v4-3/GeneralStructure.html
45. Zhang L, Zhou L, Horn BK (2021) Building a right Digital Twin with model engineering. J Manuf Syst 59:151–164
46. Hua EY, Lazarova-Molnar S, Francis DP (2022) Validation of digital twins: challenges and opportunities. In: Feng B, Pedrielli G, Peng Y, Shashaani S, Song E, Corlu C G, Lee L H, Chew E P, Roeder T, and Lendermann P (eds) Proceedings of the 2022 winter simulation conference. Singapore, pp 2900–2911
47. Kibira D, Weiss BA (2022) Towards a digital twin of a robot workcell to support prognostics and health management. In: Feng B, Pedrielli G, Peng Y, Shashaani S, Song E, Corlu CG, Lee LH, Chew EP, Roeder T, Lendermann P (eds) Proceedings of the 2022 winter simulation conference. IEEE, pp 2968–2979
48. ASME V&V 10-2019 (2023) Standard for verification and validation in computational Solid Mechanics. Retrieved on 11 Oct 2023, from https://webstore.ansi.org/standards/asme/asme10 2019
49. ASME V&V 20-2009 (2023) Standard for verification and validation in computational Fluid Dynamics and Heat Transfer. Retrieved on 11 Oct 2023, from https://www.asme.org/codes-standards/find-codes-standards/v-v-20-standard-verification-validation-computational-fluid-dynamics-heat-transfer
50. ASME V&V 40-2018 (2023) Assessing credibility of computational modeling through verification and validation: Application to medical devices. Retrieved on 11 Oct 2023, from https://www.asme.org/codes-standards/find-codes-standards/v-v-40-assessing-credibility-computational-modeling-verification-validation-application-medical-devices
51. Shao G, Frechette S, Srinivasan V (2023) An analysis of the new ISO 23247 series of standards on digital twin framework for manufacturing. In: Proceedings of the 2023 international manufacturing science and engineering conference
52. Hussain M, Masoudi N, Mocko G, Paredis C (2022) Approaches for simulation model reuse in systems design—A review. SAE Int J Adv Curr Practices Mobility 4(2022-01-0355), pp 1457–1471
53. National Academies, Sciences, Engineering, Medicine (2023) Foundational research gaps and future directions for digital twins. National Academies Press, Washington, DC. http://nap.nationalacademies.org/26894
54. Grieves M (2006) Product lifecycle management: Driving the next generation of lean thinking. McGraw Hill, New York
55. Pérez L, Rodríguez-Jiménez S, Rodríguez N, Usamentiaga R, García DF (2020) Digital Twin and virtual reality-based methodology for multi-robot manufacturing cell commissioning. Appl Sci 10:3633

# NASA's Michoud Assembly Facility: Developing a Factory Digital Twin

**Greg Porter, John Vickers, Robert Savoie, and Marc Aubanel**

**Abstract** This chapter presents a brief introduction and history of NASA's Michoud Assembly Facility (MAF) and current ongoing initiatives to reduce cost, improve efficiency, and eliminate deficiencies/rework using Digital Twin (DT) technologies. NASA's mission to develop technologies and vehicles to literally reach for the stars is by its nature very costly when looked at on a per-unit basis. Commercial space flight companies, both in competition with and in partnership with NASA, are challenging NASA to explore ways to continue its unique mission while driving down costs, particularly as these complex programs like the Artemis rocket and its components transition from development to production. DT technologies have proven to be effective in many commercial applications, as well as programs for the Department of Defense, and hold substantial potential to help NASA achieve needed production enhancements and cost reductions. The Michoud Assembly Facility is serving as a prototype Factory DT for NASA as described in this chapter.

**Keywords** NASA Michoud Assembly Facility (MAF) · Digital Transformation (DT) in Manufacturing · Environmental Monitoring and Analysis · Simulation-driven Manufacturing · Augmented Reality/Virtual Reality (AR/VR) Applications · Real-time Simulation and Design · Unity and Unreal Engine Applications · Integration of CAD and Gaming Environments · Reality Capture Techniques · Efficient 3D Modeling · Real-time Tracking Sensors · Digital Thread Integration · Automated Data Provisioning · Cloud Computing for Digital Twins ·

G. Porter (✉)
Sev1Tech, Woodbridge, USA
e-mail: greg.porter@sev1tech.com

J. Vickers
NASA, Huntsville, USA
e-mail: john.h.vickers@nasa.gov

R. Savoie
Loyola University New Orleans, New Orleans, USA
e-mail: bobby.savoie@outlook.com

M. Aubanel
Louisiana State University, Baton Rouge, USA
e-mail: maubanel@lsu.edu

 
M. Grieves and E. Hua (eds.), *Digital Twins, Simulation, and the Metaverse*, Simulation Foundations, Methods and Applications, https://doi.org/10.1007/978-3-031-69107-2_8

Kubernetes Orchestration · Pixel Streaming for Interactive 3D Rendering · Predictive Maintenance in Industrial Internet of Things (IIoT) · Generative AI for Digital Twins · Automated Point Cloud Segmentation · Ontology-based Relationships

## 1 Introduction

Throughout its storied history, NASA has faced three major challenges that most organizations do not have. These challenges are: (a) the systems they create are very expensive; (b) they make very few of these systems; and (c) the systems they make have not been made before [1]. The difference confronting NASA, and particularly NASA's human space flight programs today, is the recent development of a commercial space industry and the successes achieved by some of the companies in that industry.

NASA is no longer competing only with other nations like Russia and China. NASA now finds itself both competing with and partnering with commercial space flight companies, particularly in Low Earth Orbit (LEO). SpaceX, for example, now makes regular flights to the International Space Station, which is located in LEO approximately 240 miles (400 km) above the earth. As such, it is no longer necessary for NASA to build a separate spacecraft capacity to service LEO and compete with these successful commercial space companies. In this case, the three issues delineated above are no longer applicable.

NASA is now focused on the next steps where the three challenges are still applicable—returning to the moon to establish a permanent lunar presence and eventually sending the first humans to mars. The moon is approximately 240,000 miles from the earth, which is a totally different challenge than traversing the 240 miles to LEO and mars is 230 million miles from earth, which is a massive step even from the moon.

These are the type of challenges that fall into NASA's mission space. The development of the Space Launch System (SLS), the Orion Capsule and the Exploration Upper Stage (EUS), all components of the Artemis rocket that have already circled the moon, fit within NASA's three key challenges. However, as SLS, Orion, EUS, and other elements of the Artemis program move from development to production, and as commercial space companies continue to develop more and more capabilities, often in partnership with NASA, there is growing pressure on NASA to bring down the costs of these complex systems to maintain its outreach to the moon, mars and beyond. The DT technologies described in this chapter are one set of effectual technologies that hold substantial promise for helping NASA to enhance production and reduce the costs associated with continuing its mission to reach for the stars.

## 2 History of NASA and Michoud

The NASA Michoud Assembly Facility (MAF) in New Orleans, Louisiana, has a long and storied history. It was originally built in 1940 by the Higgins Industries company as a production facility for cargo aircraft and tank engines during World War II. NASA chose MAF in 1961 as the site for the manufacturing and assembly of the first stages of the Saturn I and Saturn V launch vehicles (Fig. 1), which were used to send astronauts to the moon during the Apollo program. MAF also played a key role in the construction of the Space Shuttle, producing the external tank (Fig. 2) from the first launch in 1981 to the mission in 2011. NASA's space shuttle fleet flew 135 missions and was the delivery vehicle to support construction of the International Space Station.

In recent years, MAF has been involved in the development of new space exploration technologies, including the Orion spacecraft (Fig. 3) and the Space Launch System (SLS) (Fig. 4). MAF is one of the largest manufacturing facilities in the world, with over 800 acres of land and 40 acres of manufacturing space under one roof. Today, MAF is a vital asset to NASA's space exploration program and to the state of Louisiana. It is a symbol of American ingenuity and innovation, and it will continue to play a key role in the future of space exploration.

**Fig. 1** First stages of Saturn V rockets being assembled at the Michoud factory in the 1960s

**Fig. 2** Final external tank at Michoud Embarks on new mission of discovery

**Fig. 3** Readying Orion for Flight, The NASA team at the Michoud Assembly Facility in New Orleans has completed the final weld on the first space-bound Orion Capsule

**Fig. 4** NASA Attaches First of 4 RS-25 Engines to Artemis I Rocket Stage Engineers and technicians at NASA's Michoud Assembly Facility in New Orleans have structurally mated the first of four RS-25 engines to the core stage for NASA's Space Launch System rocket that will help power the first Artemis mission to the moon

New programs have allowed MAF to evolve and adopt new physical infrastructure and technologies where needed and in an incremental fashion to meet program needs such as the largest friction stir welding tool in the world, built for SLS and located in the MAF Vertical Assembly Center (Fig. 5). However, legacy facilities like MAF pose significant challenges to the adoption of widespread new technologies, hindering innovation and progress. Government and industry alike have failed to focus on the challenge of how to bring innovation to legacy sectors [2].

These legacy sectors make up the majority of the US economy. There are nearly 300,000 factories in the USA, of which 90% have fewer than 100 employees. Research indicates that it's likely less than 10% of small and medium-sized manufacturing enterprises in the USA are extensively adopting emerging technologies [3].

Legacy facilities are often designed to accommodate outdated technologies and processes, making it difficult to integrate new equipment or infrastructure. Physical limitations such as limited space, outdated electrical and computer systems, or inadequate support infrastructure, can render the implementation of new technologies impractical or extremely difficult.

Cultural inertia is often one of the biggest challenges. Established practices and routines within legacy facilities can create a culture resistant to change. Employees accustomed to traditional methods may be hesitant to embrace new technologies, even if they offer potential benefits. To overcome these challenges, organizations

**Fig. 5** 170-foot-tall, 78-foot-wide world-class welding tool that will be used to build the core stage of America's next great rocket, the Space Launch System

must adopt a strategic approach to technology adoption and foster a culture of innovation. To remain competitive and not become obsolete, organizations must invest in new technology and processes and allocate sufficient resources to acquire, implement, and integrate new technologies. Organizations can encourage employees to embrace new ideas, provide opportunities for experimentation, reward successful innovation efforts, and create a supportive environment. Finally, they might partner with consultants, or research institutions to gain expertise, access cutting-edge solutions, and navigate implementation challenges.

Today, these legacy challenges cause space systems development to take too long and cost too much, and this high cost of operating in the space industry limits new entrants and competitiveness. NASA and industry must deliver ever-more capable products at reasonable cost. Manufacturing technology advancements must play a critical role to improve US industry leadership in today's hypercompetitive global environment. The future of space manufacturing will be shaped by the digital transformation and particularly by a new concept of radical innovation in design and manufacturing called the DT.

While the terminology has changed over time, the basic concept of the DT model has remained fairly stable from its inception in 2002 [1]. The concept of the DT dates back to a presentation at an SME conference in 2002 by Dr. Michael Grieves [4]. The term DT was later coined by John Vickers around the year 2010 [5], and because of their working relationship was attached to Dr. Grieves concept.

The advancement of the DT approach has been tremendous. Today a Google Scholar search yields more than 80,000 results and a Google search produces more than 25,000,000 results. The DT approach has also evolved from its origins in manufacturing to vast applications in defense, business, transportation, science, medicine, and a DT of the earth's climate which is maybe the most complex system in existence. More than anyone else, Dr. Grieves continues to shape the landscape of DTs.

To propel MAF into the future NASA has awarded Louisiana State University (LSU) a grant to produce the first real-time virtual representation, a DT of MAF. The DT of the facility will mirror its real-life counterpart and replace the physical trial-and-error approaches with computational data-driven modeling to improve design, reduce cost, and improve quality.

The MAF DT will also replace the outdated model room shown in Fig. 6. Scale modeling or Replica Twins [6] has existed for millennia. Leonardo Da Vinci was famous for creating intricate scale models of catapults, paddleboats, and even mechanized robots during his life to present to the local rulers [7].

Today, the scale model for utilitarian purposes is likely obsolete. LSU will utilize the same technology used by the entertainment industry. The LSU College of Art & Design and the LSU Digital Media Arts & Engineering program will lead the construction of the DT that will eventually be housed at the LSU Campus. This involvement of the Art department comes from a lesson from the Disney Imagineers that don't allow the engineers to participate in the creation phase of a major attraction until the artists have produced the concept. The reason is that engineers would compulsorily favor physics over imagination.

**Fig. 6** Model room at the Michoud Assembly Facility

The MAF DT will be an incredibly valuable tool, both for space industry professionals, who will be able to test a design at our facility from anywhere in the world, as well as for educators, who will be able to bring the excitement of spacecraft engineering to their students.

## 3 Benefits of the Digital Twin

The benefits that may be realized from DT, digital thread, and Augmented Reality/Virtual Reality (AR/VR) technology depend on the use cases specific to particular factory applications, equipment, manufacturing processes/environment, challenges, and desired outcomes. For MAF, and particularly the production of the SLS and EUS, the initial challenges were due to the sheer size of the MAF manufacturing facility, primarily building 103, and the massive size of the SLS and EUS "finished products." The initial steps were to combine laser scanning of the facility, equipment, and tooling with the design of the SLS and EUS to ensure the "product" could be manufactured at MAF without impediments such as transferring parts or a completed tank through the various openings/doorways in the factory without hitting any obstacles that could halt production. This step has been underway for years and provided an excellent starting point for the work described later. While this work did not constitute a true DT, it certainly made clear the benefits and cost savings that could be achieved by generating a DT of MAF and the technologies associated therewith. The items listed below provide a high-level description of the potential benefits that can be realized at MAF.

- Environmental Interrogation—floor levelness, floor load, environmental sensors, floor shifting with the tide. The floor of MAF is concrete and covers 40 acres. Since it was built basically at sea level near the coastal marshes of Louisiana, the floor moves up and down slightly with the tide. Further, the floor is not uniform or completely level. Understanding the movement of the floor and scanning the floor to identify low spots where moisture or water might gather is invaluable when planning the layout of the production process and understanding the environmental effects on equipment and components being manufactured.
- Simulation—cranes, manufacturing processes—cranes are but one type of basic tool that is needed in the manufacturing process. The location, lift capacity, maintenance status, etc., are all critical to the process. Cranes must also be considered when planning the movement of large parts since they tend to be lower than the ceiling height and could pose an impediment.
- Factory optimization—tracking parts movement over time (discussed in detail later)
- AR for remote locations such as clean rooms—having a technician or other personnel use AR when performing work in a hard-to-reach location, such as a clean room, allows engineers and quality personnel to see what the technician sees without taking the time to enter the clean room or reach the remote location.
- AR/VR-based training and work instructions—Reducing non-conformances and rework by as much as 50%, as has been proven in other aerospace-related production facilities [8].
- Predictive/Preventative maintenance—For cranes, tooling, welders, etc.
- NASA Metaverse—Connecting all NASA sites, monitoring in real-time rockets or rocket sections being shipped from MAF to Kennedy or between any other NASA Centers. Supply Chain monitoring across all NASA sites.

## 4 Gamification of DTs

One of the largest shifts in how we interact with digital artifacts comes down to real-time interaction with real-world physics and data in fully realistic 3D environments. Automobile design, simulating crash tests, architecture, medical visualization, manufacturing processes, and training are all being revolutionized by real-time engines. This allows engineers and designers to manipulate and design via computers in real time.

Two of the popular video gaming development engines are Unreal and Unity. However, both game engines are used extensively outside of the video game development field where these engines were initially conceived. In 2022, the total revenue of the video game industry in the USA reached an impressive $56.6 Billion [9]. As per a report from the Entertainment Software Association unveiled on July 10th, 2023, 65% of the American population engages in playing video games, with a resounding 96% acknowledging the perceived benefits of these interactive experiences [10]. Beyond the realm of entertainment, video games have become influential in shaping designs for applications across various sectors. Platforms such as LinkedIn [11] and renowned news publications like The New York Times [12] bear the imprint of game concepts, fundamentally influencing human–computer interaction. This phenomenon contributes to narrowing the gap between the average LSU student and a NASA employee, both possessing a comparable level of computer literacy and sharing a common experiential understanding of how we interact with intricate systems.

Within the realm of DT development on gaming platforms, the capability to engage with the physical environment in real time allows us to iterate and conceptualize solutions virtually, offering a means to explore scenarios before any real-world implications. This shift in interaction has democratized tools that were once exclusive to skilled programmers, extending their utility to professionals such as NASA engineers without a background in software development. This allows us to leverage familiar user experiences, particularly those drawn from commonplace interactions with computers and video games, to lower the barrier of entry, and drive more engagement from users.

Numerous techniques become accessible in the virtual realm that are unattainable in the constraints of the physical world. The virtually limitless budget afforded by the negligible costs associated with creating a virtual rocket, in stark contrast to the expenses incurred in manufacturing the actual rocket component, empowers us to freely explore a multitude of ideas on a computer. This financial advantage enables iterative processes to unfold many times for a single design, enabling collaborative iterations with engineers and designers. Additionally, autonomous iterations facilitated by artificial intelligence (AI) enhance the speed of the iterative process, surpassing the capabilities achievable by humans alone. Consequently, with AI's help, we find solutions to complex problems on a computer that transcends the boundaries of human ingenuity.

A real-time 3D engine such as Unity or Unreal has another enormous advantage. It is multi-platform and multi-user. This allows us to support a large variety of devices interoperating together in the same simulation. For example, using the DT platform, an operator with an AR Headset like the Microsoft HoloLens 2 could interact with an engineer at another facility on their computer, tablet, or phone. This allows for an operator on the floor with the 3D glasses in the facility to interact with an engineer at another NASA location who not only sees what the operator can see but has access to the full DT as well. This technology allows them to communicate in a rich manner that could not be achieved by phone or messaging alone. It contextualizes the problems they are facing and again reduces the communication gap between personnel.

A major challenge we have faced is the enormity of MAF. Both Unity and Unreal support were skeptical about recreating a ~ 2 million square feet facility that would contain thousands of lights, as lighting is heavy to compute. Additionally, we needed to achieve a very high level of accuracy and details on the building and all tooling and equipment within it. To achieve this, we use 3D laser scanning (discussed in more detail in Sect. 5), which requires the processing of trillions of polygons in order to represent the entire facility. These challenges led us to select Unreal 5 game engine as the backend platform for the current iteration of the DT platform. There are a few features in Unreal 5 that we found beneficial in overcoming these challenges: Nanites and Datasmith imports.

Nanites replaces regular geometry within the gaming engine. Rendering using Nanites is a fundamental shift in how to make a level, or in our case a DT, performant. Traditionally, for 3D simulations we are "draw call bound" which results in a finite number of polygons that the GPU can keep in memory. The ability or inability to render a level in a game engine is the direct result of how many polygons are present. In large environments such as the recreation of MAF, software developers would have to use many tricks and complex techniques to lower the polygon count. However, since Nanites is screen resolution bound and not "draw call bound," we can have a much larger polygonal count.

For example, if we are 1000 feet away from a piece of equipment within the DT, the model may only consist of 10,000 polygons. But, as we approach the equipment, the polygons increase significantly to provide an extremely realistic representation of the equipment. A large piece of equipment can easily contain millions of polygons. Ultimately, this allows us to achieve high accuracy and model at the highest level of detail (LOD) possible. The engine will scale the model automatically to what is needed to represent it realistically on screen. This is a game changer for us as it allows us to generate a single model. With prior engines, we had to represent objects with multiple levels of detail, as the engine now takes care of the LOD.

Based on previous experiences, we knew one of the large challenges we would face on the project was working with engineers and architects who use CAD data to be able to view their work in a polygonal real-time environment. However, the compatibility between CAD and polygonal models is poorly supported. Additionally, the techniques to design a piece of manufacturing equipment and a game differ greatly. When designing manufacturing equipment, it is normal to create the entire model in great detail, as a single model. In game development, modeling is typically broken into small pieces and then placed together, analogous to building using Lego bricks to build a model versus having a single piece be the whole model.

The solution to this problem is Datasmith. It allows us to take models created in CAD and split them into smaller polygonal objects that we can then directly import into Unreal 5. This closes the gap between the DT platform's development workflow and the CAD modeler's workflow by increasing collaboration and reducing the amount of rework needed. Furthermore, by incorporating these tools and data objects into the digital thread (discussed in Sect. 6), we can fully understand and track the relational dependencies.

By using Unreal Engine 5, we also help solve another critical problem: cyber security. By making the DT platform widely available to many different devices such as computers, cell phones, tablets, and AR/VR gear, and people, there is an increased risk of a data breach. To alleviate these cyber security concerns, we are using pixel streaming. Pixel streaming allows us to run the DT platform from remote servers which prevents users from having to download the software. Instead, operators just connect to the DT platform with a web browser (this topic is discussed in greater detail in Sect. 7).

Another key factor for the project is to create a user interface that is easy to use, while allowing immersive interactions. To do this, the team is relying on sandbox-styled games like SimCity (Maxis). The goal is to create an extremely detailed environment with many base features to drive engagement by the operators.

Our reliance predominantly rests on sandbox-style games such as SimCity (Maxis) for our DT development. Our emphasis lies in fostering experimentation and facilitating changes within the operational facility. Despite representing intricate details, our goal is to encourage operators to experiment and craft scenarios that go beyond the scope envisioned by the DT's software designers. This approach not only ensures simplicity in controlling a complex system but also poses a primary challenge for our team: crafting a user interface that strikes a balance between ease of use and enabling robust interaction.

Through the creation of tools designed to address envisioned use cases, we anticipate diverse applications beyond the team's initial considerations. Drawing inspiration from the transformation witnessed in the video game industry, where developers released products before completion, allowing players to contribute to the final version of the game, our approach is similarly dynamic. The DT serves as a catalyst for this project team to significantly alter NASA's operational landscape, contributing to the overarching objectives of achieving lower costs, enhanced reliability, and ultimately, the development of an improved rocket for future mars missions.

## 5 Unlock Business Value Through Comprehensive Reality Capture in DTs

As mentioned in the history of NASA and Michoud section, Michoud is an older facility and has gone through lots of changes over the years to accommodate different space programs. As such, it was important for the project to begin with capturing the "as-built" state to form the foundation of the DT. To do so, we have employed several techniques which, when merged, we call reality capture. These techniques help us deliver the NASA DT as an "as-built" replica of the physical world in a digital context, ultimately delivering substantial business value and return on investment. The fusion of 3D laser scanning, photogrammetry, and advanced sensing technologies elevates the precision and utility of DTs, offering a wide array of applications across industries.

### 5.1 *Precision Through 3D Laser Scanning and Photogrammetry*

In the pursuit of precision, imperative when building rockets, our 3D laser scanning techniques to capture the "as-built" state of the factory, equipment, tooling, and flight hardware are highly refined and continually tested. The utilization of laser scanning provides a meticulous and highly accurate method for spatial data capture, ensuring modelers have a robust foundation to work with. The incorporation of target-based registration, which is the practice of using artificial targets, or points and planes extracted from scan points, to align multiple scans, further refines this process, increasing the alignment and accuracy of the captured data. This precise dataset serves as a crucial starting point for our 3D modelers, providing them with a detailed and reliable representation of the physical environment through the resulting point clouds.

In looking for ways to increase efficiency, our DT platform introduces a unique capability: the ability to display point clouds without the necessity of 3D models, also referred to as meshes. This feature proves invaluable in scenarios where time and cost considerations dictate the omission of a full 3D model. Point clouds within the DT retain all the functionality and richness of a traditional 3D model, providing a comprehensive and immersive experience. However, it is crucial to acknowledge a tradeoff: while point clouds offer a time-saving alternative for visualizing the "as-built" world, they require more computational power than their 3D model (mesh) counterparts. The increased computational requirements underscore the importance of balancing modeling intricacy with available compute resources.

Advancement in technology has paved the way for increased utilization and exploration of photogrammetry, defined as the process of deriving metric information about an object through measurements made on photographs of the object [13]. With high-definition photographs serving as the data source, it is applicable, at low costs, to capture 3D morphology, including photo-realistic textures in high resolution with large overlaps. Our modelers use this during modeling to implement realistic textures and color palettes to create Level of Detail (LOD) 400–500 meshes and textures. Ultimately, the goal of photogrammetry is automatic generation of 3D models from photographs, resulting in a significant leap in reality capture capabilities. However, current practices often require manual interactions such as geometric modeling, object identification, and monitoring [14]. Despite the rapid strides in this exciting subset of reality capture, which may one day replace most 3D modeling for as-built objects and environments, we believe that, for the precision required in most areas of our project, photogrammetry has not yet reached its full potential. We will keep experimenting with and exploring the technology.

## 5.2 *Efficiency in 3D Modeling*

The transition from point clouds to 3D models is where our efficiency begins to take center stage. Using Machine Learning (ML) models, discussed in detail within the "Artificial Intelligence, Machine Learning, and DTs" section, we can automatically extract objects from point clouds. This process greatly reduces the time 3D modelers must spend trying to manually separate objects from much larger point clouds. The smaller object-oriented point cloud segmentations allow the 3D modelers to produce highly accurate 3D models that accurately replicate the physical environment. The conversion process of smaller, object-oriented point cloud segmentations into 3D models is not only swift, but remarkably precise, enabling the creation of virtual replicas that mirror reality in astonishing detail.

These 3D models within the DT platform serve as a virtual representation of the physical environment, offering a wealth of insights for decision-making processes and coordination; NASA and its contractors can leverage to conduct spatial analysis, simulate various scenarios, and optimize designs. Additionally, the intricate details captured through laser scanning later translated into 3D models empower decision-makers with a holistic understanding of the environment, allowing for informed data-driven choices in various aspects of project planning, execution, and maintenance.

## 5.3 *Dynamic Spatial Awareness with Real-Time Tracking Sensors*

The integration of real-time location tracking sensors using ultra-wideband technology not only introduces a new dynamic dimension to the DT, but also paves the way for optimizing operations within Michoud. This reality capture approach goes beyond asset monitoring. It serves as a catalyst for comprehensive factory flow, adherence to safety protocols, and optimal resource management.

In the context of factory optimization, the continuous tracking of assets in real time becomes instrumental. Decision-makers, using the DT, can understand workflows and asset movements, gaining more transparency into the intricacies of operational processes. This historical tracking data provides a valuable foundation for optimizing factory layouts, identifying bottlenecks, and increasing overall operational efficiency. The DT, infused with this real-time location information, becomes a dynamic tool for not just monitoring, but actively shaping and refining the factory environment.

The real-time location data plays another important role in ensuring swift and accurate asset location within the expansive Michoud facility. The ability to quickly locate objects, equipment, and tooling enhances response times, minimizes downtime, and contributes to a more agile and responsive operational ecosystem.

Furthermore, the precision of location tracking using ultra-wideband enables a deeper understanding of how objects move within the real-world environment. This knowledge allows us to understand reality to accurately simulate movements within the DT. By precisely tracking the location and movement of assets, the DT can simulate and predict scenarios with a high degree of accuracy, offering a powerful resource for scenario planning, predictive maintenance, complex crane lifts, and operational optimization.

The relational linking of sensor tags to other data systems through the digital thread is a critical aspect of this integration. It creates a flow of information allowing the DT to tap into disparate systems containing details about the assets being tracked, prior/future work orders, active capacities, and much more.

## *5.4 Environmental Insights for Sustainability and Efficiency*

The integration of environmental sensors adds a layer of intelligence to DTs by continuously monitoring factors such as temperature, humidity, and volatile organic compounds. This real-time data is invaluable for optimizing environmental conditions, ensuring occupant comfort, and identifying opportunities for sustainability initiatives. The return on investment is manifested in energy savings, improved workplace conditions, and a positive impact on the organization's environmental footprint.

The integration of environmental sensors offers another layer of reality capture to increase the capabilities and realism of DTs. Within Michoud, we have implemented temperature, humidity, and volatile organic compound sensors. This real-time data further enhances the optimization of environmental conditions, uncovering avenues for sustainability initiatives. The return on investment materializes in tangible benefits such as energy savings, better workplace conditions, reduction in the organization's environmental footprint, and increased quality assurance in manufacturing processes.

In aerospace manufacturing, temperature and humidity play a significant role in the outcomes of diverse processes like welding and chemical applications. By tracking this data and utilizing it within ML algorithms, the DT platform becomes a catalyst for predicting manufacturing outcomes and prescribing ideal conditions for optimal results.

The integration of external weather data with internal sensor information further amplifies the power of the DT. The combination allows for a comprehensive understanding of how external conditions impact the internal environment. By leveraging external weather data, organizations gain insights into the broader environmental context, enabling proactive adjustments to internal conditions based on anticipated external changes. This interconnected approach fortifies the DT with a holistic view, ensuring that the manufacturing environment remains adaptive and responsive to the dynamic mix of internal and external factors.

As NASA and the rest of the world place greater emphasis on sustainability initiatives, the environmental data generated by these sensors can help identify areas for energy savings, enabling organizations to optimize resource utilization, reduce energy consumption, and minimize environmental impact. All of this can be displayed and interacted with through the DT.

In essence, the efficiency in 3D modeling achieved through laser scanning techniques and the use of point clouds within the DT, not only streamlines the modeling process, but when combined with endless data from the digital thread, unlocks invaluable insights. Reality capture positions NASA and its contractors at the forefront of precision-driven decision-making during the manufacturing and assembly processes, to advance innovation, precision, sustainability, and efficiency in the dynamic realm of space exploration and research.

## 6 Digital Threads: Catalysts for Transformative Data Management

MIT defines a digital thread as a data-driven architecture that links together information generated from all stages of the product lifecycle and is envisioned to be the primary or authoritative data and communication platform for a company's products at any instance of time [15].

At the practical level, implementing the digital thread augments existing enterprise integrations. This evolution is based on advancements within information technology, driven by well-known and used innovations such as cost-effective and nearly limitless computation and storage capabilities through cloud computing. Additionally, lighter, and more reliable data transmissions, inspired by optimizations realized in internet streaming, contribute to the seamless flow of information between systems. This contemporary landscape is further shaped by faster and stronger databases, new ways of controlling versions from disparate systems, affordable and dependable sensors, and the application of sophisticated techniques in physical object scanning and image processing.

The integration of these innovations not only strengthens the existing data infrastructure but also enriches the product development process measurably. The digital thread allows for comparison, including visualizations, of product data across diverse lifecycle stages. For example, it facilitates the comparative analysis of "as-built" digitalized tolerance measurements against the original design in a computer-aided design (CAD) file. This integration introduces closed feedback loops, contributing to a more comprehensive understanding and optimization of the product development lifecycle.

Value also arises from the automated provisioning of updated authoritative data sources, making the digital thread itself the authoritative source. It allows data to stay within its source system, exposing only necessary metadata until the actual data is needed. This empowers subject matter experts to retain control over their information. Unlike traditional data workflows requiring resource-intensive Extract, Transform, Load (ETL) workflows to replicate data into a centralized data warehouse, the digital thread renders these processes obsolete by allowing data to reside within its originating system. This streamlined approach conserves organizational resources and enhances agility and responsiveness.

Beyond data control, digital threads have established themselves as the cornerstone in DT architecture by facilitating the near real-time synchronization of digital system models and their corresponding DTs. Establishing connections between disparate systems fosters a relational understanding of the data, systematically breaking down data silos and providing organizations with a comprehensive view that leads to novel and insightful observations. The relationships between systems can be created through ontologies, which represent a shared, explicit specification of a conceptualization of the domain of knowledge [16]. It provides a formal specification of the vocabulary of concepts and the relationships among them, in a domain of interest [17].

In the diverse ecosystem of NASA, comprising vendors, contractors, and tenants, each entity operates with distinct software, processes, and scopes of work. To deploy an effective digital thread in such an ecosystem, it must be platform agnostic, allowing entities to continue their operations as usual. Complex systems can be interconnected through innovative tools such as the digital thread, utilizing modern-day Application Programming Interfaces (APIs). These APIs enable seamless interactions with the digital thread, facilitating native integration and data exchange.

## 7 DT Accessibility

Traditionally, operating an application with 3D graphics or models required the operator to run a system with a high-end Graphics Processing Unit (GPU). This has limited the number of users able to use and interact with the application. On this project, we sought to solve this problem, enabling us to deliver a massively complex 3D render involving point clouds, models, and high-fidelity textures from anywhere within a factory while sticking true to being platform agnostic. We have successfully done so using technologies such as multi-tiered cloud computing, Kubernetes, and pixel streaming.

### *7.1 Cloud Computing: Flexibility, Cost-Efficiency, Scalability, and GPU Acceleration*

Cloud computing has become a linchpin for businesses seeking flexibility, cost-efficiency, and scalability in their IT infrastructure. One pivotal advancement in cloud computing is the availability of GPU resources, significantly enhancing the performance of applications that require intensive graphical processing. This is particularly beneficial for workloads such as ML, simulations, and realistic rendering of DTs. By leveraging cloud services with GPU support, organizations can offload the burden of managing the procurement and deployment of specialized hardware, ensuring that resource-intensive tasks are executed efficiently. By leveraging cloud services, our project has been able to offload the burden of managing physical hardware, allowing us to focus on developing automated and highly scalable DT platforms, interconnected digital threads, and near real-time 3D model data catalogs.

The pay-as-you-go model of cloud computing ensures cost-effectiveness, since we only have to pay for the resources we consume. Moreover, the scalability of cloud resources enables us to effortlessly handle fluctuations in demand by using autoscaling and rightsizing best practices for modern-day architectures, ensuring optimal performance during peak times, and cost savings during periods of reduced activity. Cloud computing services have enabled us to innovate rapidly and remain agile in today's dynamic digital landscape. Furthermore, by using security groups, we implicitly deny all access and explicitly allow access, giving us granular control over infrastructure and applications.

## *7.2 Kubernetes: Streamlining Container Orchestration for Scalability and Reliability*

Kubernetes, an open-source container orchestration platform, has revolutionized the way applications are deployed, scaled, and managed. One of its primary benefits is its ability to automate the deployment, scaling, and operation of application containers, providing a consistent and resilient environment while being platform-agnostic and open-sourced. Kubernetes facilitates the efficient orchestration of containers, ensuring optimal resource utilization and seamless scaling to meet varying workloads. With features like automated load balancing and self-healing capabilities, Kubernetes enhances the reliability and availability of applications, making it a cornerstone technology for modern, cloud-native, hybrid, or on-premises architectures.

## *7.3 Pixel Streaming: Interactive 3D DT Anywhere, Anytime*

Unreal Engine Pixel Streaming introduces a paradigm shift in content delivery by enabling the streaming of interactive 3D content directly to a user's device. This technology allows resource-intensive 3D graphics to be rendered on a server running in the cloud or on-premises and streamed to end-users in real time. This provides many benefits such as eliminating the need for users to possess high-end hardware to run demanding applications, facilitating collaboration on complex 3D environments in real time, and presenting 3D applications in web-based applications. Pixel streaming has empowered us to deliver an immersive DT to a broader audience, irrespective of device specifications, fostering a more inclusive and accessible DT platform.

Unreal Engine's Pixel Streaming capabilities represent a revolutionary shift in content delivery, enabling the real-time streaming of interactive 3D content directly to users' devices. This technology allows resource-intensive 3D graphic applications to be processed on a server and dynamically streamed to end-users. Notably, Pixel Streaming provides many advantages, including but not limited to, eliminating the need for users to possess high-end hardware to run demanding 3D applications and facilitating real-time collaboration on complex 3D environments in real time.

Adding to its versatility, Pixel Streaming integrates with modern website design frameworks like React and Next.js. This integration provides an added layer of flexibility, allowing us to embed our interactive 3D DT into a modern web application. This not only enhances the user experience but also opens new possibilities for creating new engaging tie-ins with the DT platform like dynamically streaming data from the digital thread. The intersection of Pixel Streaming and web design methodologies represents a convergence of worlds, fostering a balanced blend of immersive 3D content and the latest trends in web development.

Beyond these capabilities, Pixel Streaming acts as a gateway to collaborative experiences within the DT through simulations and augmented reality (AR) based scenarios. It facilitates multiple users interacting with each other, whether connecting through AR devices, phones, or computers. This breakthrough allows users to share and jointly manipulate 3D content in real-time encouraging collaborative decision-making and enhancing the overall user engagement. Seamlessly integrating all these capabilities mixed in with federated identities, coupled with the digital thread as the single source of truth, allows us to ensure the delivery of an immersive environment, utilizing a security best practice known as least privilege.

## *7.4 Putting It All Together*

The convergence of Kubernetes, cloud computing, and Pixel Streaming lays a robust foundation for DTs. In this advancing technological landscape, the integration of these tools is proving to drive innovation across not only NASA and aerospace industries alike, but also education, manufacturing, energy, defense, real estate, marketing, and many others.

By containerizing Pixel Streaming and deploying it on Kubernetes within a cloud environment, we provide a consistent and portable DT that is buildable, testable, and deployable through continuous integration/continuous deployment (CI/CD) pipelines. Deploying on Kubernetes introduces unparalleled scalability, allowing us to efficiently handle varying workloads and fluctuations in user demand, even scaling down to zero resources when no one is using the DT platform. This dynamic scalability, coupled with the inherent accessibility of cloud resources, ensures optimal performance regardless of geographical location or personal hardware specifications.

Ultimately, this leads to cost reduction by optimizing resource utilization and eliminating unnecessary expenses. The orchestrated collaboration between Kubernetes, cloud computing, and Pixel Streaming reflects not only the current but also the future state of technology, underscoring a vision where interactive, visually compelling content is universally accessible, affordable, and responsive to the evolving demands of a global user base.

## 8 Maximizing Returns Through Artificial Intelligence, Machine Learning, and DTs

In the dynamic landscape of DTs, the strategic integration of AI and ML not only revolutionizes operational patterns but also serves as a jumping point for maximizing return on investment. As organizations increasingly harness the power of DTs, the infusion of intelligent technologies amplifies their transformative potential. This section delves into key applications within the MAF DT, focusing on our DT assistant, point cloud segmentation, and predictive maintenance, all underpinned by a central theme—the realization of substantial returns on investment. By exploring how AI and ML elevate decision-making, streamline workflows, and enhance predictive capabilities, we unveil the many ways in which NASA stands to gain significant value and efficiency from their investments in these cutting-edge technologies.

### *8.1 Digital Twin Assistant and the Digital Thread*

The DT assistant, intricately tied to the digital thread, represents a groundbreaking development. Leveraging large commercially built Large Language Models (LLMs) through transfer learning, the DT assistant gains a profound understanding of NASA data beyond its initial training data, which is a large corpus of publicly accessible internet data. The use of transfer learning, where we only retrain the last layer of the model, democratizes access to sophisticated language models, enabling the reuse of expansive knowledge for enhanced contextual understanding. This saves significant time and cost when developing a custom solution involving LLMs.

Conversational scoping is a crucial aspect, ensuring the DT assistant remains focused on the specific data we have trained it on. This precision in scoping allows for accurate and relevant responses, aligning with data privacy and security considerations. Additionally, the assistant demonstrates a unique ability to comprehend relationships between disparate data systems through the digital thread. This interconnected intelligence enhances the DT's capacity to provide nuanced information and insights, allowing for more informed decision-making across diverse domains.

### 8.2 Point Cloud Segmentation

Automated point cloud segmentation has emerged as a pivotal application within our DT project, specifically in expediting the 3D modeling workflow. By using state-of-the-art Learnable Region Growing for Class-Agnostic Point Cloud Segmentation ML model, we have been able to automate the segmentation of point clouds. By transforming extremely large, dense point cloud data into structured segmented point clouds we have significantly reduced the time and effort traditionally associated with manual segmentation.

### 8.3 Predictive Maintenance

The convergence of AI, ML, and DTs facilitates predictive maintenance strategies that revolutionize Industrial Internet of Things (IIoT) applications. Capturing IIoT data over time allows for a deep understanding of patterns within the data, enabling the prediction of potential issues before they occur. This capability extends beyond routine machinery maintenance, delving into complex scenarios such as predictive crane maintenance and assessing the effects of environmental factors like temperature and humidity on critical processes like welding and chemical applications.

In summary, the harmonious integration of AI and ML technologies within DTs transforms these virtual replicas into intelligent, dynamic entities. The DT Assistant, point cloud segmentation, and predictive maintenance represent key applications, showcasing the immense potential for enhanced decision-making, operational efficiency, and innovation across diverse industries. This synergy underscores the pivotal role of AI and ML in shaping the future trajectory of DT technologies.

## 9 Conclusion

In this chapter, we have covered the intricacies of the NASA Michoud Assembly Facility (MAF) DT project, a groundbreaking initiative at the forefront of precision engineering and exploration. As we conclude our exploration of the five key dimensions: gamification of DTs, comprehensive eality capture of DTs, digital threads, DT accessibility, AI, ML, and DTs of this comprehensive endeavor, it is evident that the integration of cutting-edge technologies is reshaping the landscape of space exploration and manufacturing.

The historical context provides a foundation for understanding the evolution of the Michoud facility and NASA's rich legacy. This historical perspective underscores the significance of adapting traditional facilities to the demands of contemporary space programs, setting the stage for the transformative journey into the digital realm.

Reality capture techniques, comprising 3D laser scanning, photogrammetry, and real-time tracking sensors, lay the groundwork for an unparalleled level of precision in creating the DT. The commitment to capturing the "as-built" state of the physical world reflects not only an attention to detail but a strategic investment in technologies that deliver substantial business value and return on investment.

Efficiency in 3D modeling, dynamic spatial awareness, and environmental insights contribute to the active and adaptive nature of the DT. Decision-makers are empowered with tools for spatial analysis, scenario simulation, and design optimization, fostering a holistic understanding of the environment. The introduction of real-time location tracking sensors, coupled with environmental sensors, amplifies operational efficiency, safety, and sustainability within the Michoud facility.

The concept of the digital thread emerges as a unifying force, seamlessly integrating information across the product lifecycle. Cloud computing, Kubernetes, and pixel streaming introduce a new era of accessibility, flexibility, and cost-efficiency. The democratization of 3D graphics ensures that the DT is not just a high-end GPU endeavor but a universally accessible and inclusive platform, aligned with the principles of modern technological landscapes.

In the final dimensions explored, we witness the strategic integration of AI and ML technologies, positioning the DT as a dynamic entity capable of substantial returns on investment. The DT Assistant, point cloud segmentation, and predictive maintenance showcase the transformative potential of these intelligent technologies, enhancing decision-making, operational efficiency, and innovation across diverse industries.

As we conclude this chapter, it is clear that the NASA Michoud Assembly Facility DT project is not merely a technological endeavor but a pioneering journey into the future of precision engineering and exploration. The intersection of historical legacy, eality capture, digital threads, and intelligent technologies propels NASA into a realm where informed decisions, efficient workflows, and sustainable practices converge.

## References

1. Grieves M, Vickers J (2017) DT: mitigating unpredictable, undesirable emergent behavior in complex systems. In: Kahlen F-J, Flumerfelt S, Alves A (eds) Trans-disciplinary perspectives on system complexity. Springer, Switzerland, pp 85–114
2. Bonvillian W, Singer PL (2017) Advanced manufacturing: the new American innovation policies. The MIT Press, Cambridge, Massachusetts, ix, 401 pages
3. Vickers J (2022) Advanced manufacturing transformation. In: Aerospace & Defense Manufacturing Industry Report 2022. SME, p 3
4. Grieves M (2002) Completing the cycle: using PLM information in the sales and service functions [Slides]. In: SME management forum, Troy, MI
5. Piascik R et al (2010) Technology area 12: materials, structures, mechanical systems, and manufacturing road map. NASA Office of Chief Technologist
6. Grieves M, Hua E (2024) Defining, exploring, and simulating the DT metaverses. In: Grieves M, Hua E (eds) DTs, simulation, and metaverse: driving efficiency and effectiveness in the physical world through simulation in the virtual worlds, Forthcoming. Springer, Heidelberg
7. Anonymous (2018) The history of plastic scale modeling [cited 2023 December 1, 2023]. Available from: https://www.megahobby.com/blog/the-history-of-plastic-scale-modeling/

8. Fink C (2019) Enterprise AR use cases. Available from: https://www.forbes.com/sites/charlie fink/2019/02/26/enterprise-ar-use-cases/amp/
9. Clement J (2023) Annual Revenue of the U.S. Video Game Industry 2016–2022. Available from: https://www.statista.com/statistics/249996/annual-revenue-of-the-us-video-game-industry-by-segment/
10. Anonymous (2019) Essential facts about the computer and video game industry. Available from: https://www.theesa.com/wp-content/uploads/2019/05/ESA_Essential_facts_2019_final.pdf
11. Lindemann T (2019) How LinkedIn uses gamification to boost engagement. Available from: http://thomas-lindemann.com/gamification-en/how-linkedin-uses-gamification-to-boost-engagement/
12. Tribbey C (2018) Viacom, NYT share how they gamify the data experience. Available from: https://www.mesaonline.org/2018/07/31/viacom-nyt-share-how-they-gamify-the-data-experience/
13. Mikhail EM, Bethel JS, McGlone JC (2001) Introduction to modern photogrammetry. John Wiley & Sons
14. Qin R, Gruen A (2021) The role of machine intelligence in photogrammetric 3D modelling—an overview and perspectives. Int J Digital Earth 14(1):15–31
15. Singh V, Wilcox KE (2018) Engineering design with digital thread. AIAA J 56(11):4515–4528
16. Gruber T (1993) A translation approach to portable ontology specifications. Knowl Acquis 5(2):199–220
17. Ganon M (2007) Ontology-based integration of data sources. In: 2007 10th international conference on information fusion, Quebec, QC, Canada, pp 1–8. https://doi.org/10.1109/ICIF.2007.4408086

# Distributed AI Modeling and Simulation for Smart Airport Digital Twin Applications

**Kostas Alexandridis, Soheil Sabri, Jeff Smith, Bob Logan, Katalin Bartfai-Walcott, and Doug Migliori**

**Abstract** This chapter overviews modeling and simulation methodology for smart airport Digital Twin metaverse applications, emphasizing the development of a distributed agent-based modeling implementation framework. It outlines current approaches and applications of modeling and simulation concerning optimization, sustainability, and digital transformation, along with theoretical, statistical, and empirical approximations to model validation and accuracy assessment. It also provides a rudimentary Agent-Based Modeling (ABM) categorization and classification of current models. We provide a key perspective to airport planning model development, including resource allocation, workflow design principles, and process optimization. An example reference and design of distributed and intelligent agent architecture for a smart airport is provided, along with the design elements for agent allocation, communication, and orchestration mechanisms for system architecture. Finally, we provide a series of inferences, insights, and principles to guide the design,

K. Alexandridis (✉)
Orange County Public Works, OC Survey Geospatial Services, Santa Ana, CA, USA
e-mail: kostas.alexandridis@ocpw.ocgov.com

S. Sabri
Urban Digital Twin Lab, School of Modeling, Simulation and Training,
University of Central Florida, Orlando, Florida, USA
e-mail: soheil.sabri@ucf.edu

J. Smith
Sierra Nevada Corporation, Nashua, NH, USA
e-mail: jeff.smith@sncorp.com

B. Logan
Rockport Software, Slough, England, UK
e-mail: bobl@rockportsoft.com

K. Bartfai-Walcott
Ambient Enterprises Inc., California, USA
e-mail: kati.walcott@ambiententerprises.com

D. Migliori
Event Driven Systems, Irvine California, USA
e-mail: doug@eventdriven.systems

 
M. Grieves and E. Hua (eds.), *Digital Twins, Simulation, and the Metaverse*, Simulation Foundations, Methods and Applications, https://doi.org/10.1007/978-3-031-69107-2_9

development, and implementation of distributed AI and agent-based modeling and simulation methodologies for smart airport and nested smart cities' methodological approaches.

## 1 Introduction

A growing body of practical and scientific work recognizes the importance of Agent-based Modeling (ABM) and simulation in enhancing transportation operations. For example, passenger flow modeling is one of the widest applications of ABM in terminal models, including railways and airports. This approach has been used for many use cases, such as improving non-aeronautical revenue, passenger satisfaction, hazard mitigation and evacuation, and energy efficiency [1, 2]. However, there are limited scientific reports on system architectures that incorporate airport planning and operations into a holistic digital environment. Furthermore, the current top-down management approach in many airports and other transportation hubs overlooks the human and social factors as essential dimensions in the operational planning and service allocation [3]. Given a progressive increase in the deployment of sensors, monitoring equipment, and high-performance computing technologies, the traditional modeling, simulation, and prediction approaches could be enriched with real-time and streamlined data.

This chapter concerns optimized ABMs calibrated with real-time data. The current state-of-the-art ABM supports a distributed AI-enabled metaverse, facilitating transparency and flexibility of management and operations in complex environments [4]. This method can also enable an adaptive system of systems (SoS) where stakeholders collaborate and understand the complex implications of their decisions and action scenarios.

We propose a distributed AI-based modeling and simulation architecture for smart airports. To do this, the chapter first explores the state-of-the-art applications of ABM in airports. We will explore the current applications of modeling and simulation and ABM in different dimensions, including human systems, physical and operational aspects, and sustainability. We pay specific attention to the Modeling and Simulation in transportation hubs in a subsection. Then, we investigate the validation approaches such as spatial, temporal, systemic, and behavioral. These will be translated into our proposed architectural principles. The next section describes an airport planning model, the definition of its components, rules, resource criteria, and usage. Using the content of this airport planning model, we propose a smart airport ABM design, and then we provide the visual metaverse aspects of an intelligent Digital Twin framework for smart Airports in the next section. Finally, the chapter provides the concluding remarks and sets an agenda for the development and implementation of a new generation of ABM integrated into the Digital Twin architecture to enhance Airport Planning.

## *1.1 Current Applications*

Given the complexity of airport operations, this chapter focuses on three interrelated dimensions: (a) the human dimension (passenger flow); (b) the physical and operational dimensions (terminals and airlines), and; (c) the sustainability dimension (energy efficiency). In this section, we review and synthesize the current state of the art in modeling, simulation, and virtual representation of airport operations, as well as social and behavioral theories and validation methods to identify the progress and limitations in the applied and scientific environments.

### 1.1.1 The Human Dimension

Recently, a group of studies focused on behavioral studies and incorporated humans' discretionary activities (e.g., shopping duty-free) into the passenger flow models [1, 5]. From a business perspective, these models improved airport terminal analytics to support understanding the non-aeronautical revenue within the aviation industry. For instance, these models can improve the airports' operational strategies that reduce passenger queue time or increase passenger free time, potentially enhancing airports' terminal performance through improved efficiency, increased revenue, and reduced cost [1]. Other benefits of such models are investigating the movement of a large number of people within confined areas and linking to the definition of impacts that can be seen in terms of time spent in each processing unit, preference of discretionary activity, and level of service (LOS) in the processing areas [6].

Furthermore, the discretionary attributes combined with social theories, such as social force, improve our understanding of passenger behavior during an emergency or hazard mitigation [7]. Specifically, after the calibration of these ABM models, the outputs can be used to improve the design elements of pedestrian facilities and egress routes [8]. It will provide an opportunity to explore different scenarios, such as using portable "obstacles" to stabilize flow patterns and make them more fluid in a real-time and 3D environment.

Another important factor related to the human dimensions when considering airport-related systems is that human behavior in routing passengers and passenger flows is *non-deterministic, often volatile, and highly uncertain.* Unlike purely physical systems or robots that can be logically routed and queued, human dispositions, behavior, attitudes, and satisfaction play a key role in routing decisions. Uncertainty, unpredictability, and latent patterns of collective behaviors result in challenges, problems, and decision-related systemic failures [8–10].

### 1.1.2 Physical and Operational Dimensions

Several studies indicated the importance of physical and operational strategies for passenger flow control. For example, increasing the service process capacity and social distancing during the COVID outbreak informed a better intervention and decision-making process in major transportation hubs [11]. Examining the inbound passenger flow lines through the average inbound time of passengers from each node (parking to the entrance and thereafter) and the average queue length for each kind of passenger service equipment facilitates gaining an insight into the terminal layouts [12]. The latter can also be modeled as the average service/processing time of each service, including passenger service equipment, while average queue length and wait times can be derived from combining service times and passenger volumes interactively [13].

Other studies looked into integrating pedestrian facilities planning and staff assignment for transfer capacity, transfer average time, and level of service [14]. The passenger flow line regulated the entrance and transition gates, such as bidirectional automatic fare gates (BAFGs), and avoided passenger congestion [15]. Given the time-varying passenger demands, this operational problem could be improved through real-time/near real-time data.

Undoubtedly, the airlines, as main stakeholders, play a crucial role in airport operations, and their operational model contributes to the smoothness of passenger flow. The service process capacity and staff assignments mentioned above could affect airlines' operational strategies. Airline's approach to passenger boarding has implications for passengers' satisfaction, the operational timing, and the energy used to air condition the boarding bridges [16]. As such, the physical and operational dimensions are interrelated to human activities and sustainability.

### 1.1.3 Sustainability

Airport terminals consume more energy than normal public buildings on average [17]. Studies indicated that three factors are correlated to the airport terminal's energy consumption: (a) passenger flows, (b) meteorological parameters, and (c) supply fan frequency (for the zonal airport terminal HVAC system) [18]. Given these factors, the role of ABMs in understanding the environmentally sustainable performance of airports is significant. We argue that the energy consumption of airport terminals poses a substantial challenge to the carbon neutrality policy due to the high energy demand and round-the-clock operation of these terminals. Integrating geospatial data and analytics into the ABM allows the development of the spatio-temporal passenger distribution model to describe the transient passenger distribution pattern of an airport terminal [19, 20]. The outputs of such models are used as input into the energy simulation tools to estimate the energy consumption of airport terminals in different scenarios [21].

Furthermore, passenger flow plays a vital role in the terminal's energy consumption and indoor environmental conditions [22]. A prediction model of passenger flow, developed by [22], confirmed the operation optimization and terminal's layout design to address energy efficiency. Alternatively, energy consumption for each service level can be aggregated based on overall service use [13].

### 1.1.4 Modeling and Simulation for Transportation Hubs

Recent research has highlighted the value of hybrid models in transportation hubs and, specifically, airport studies [23]. To improve the reliability of models, the combination of modeling and simulations and the integration of emerging technologies such as Machine Learning (ML) and Artificial Intelligence (AI) for optimization and predictive analysis are suggested.

For example, the passenger movements, multi-gate and multi-destination nature, passenger service processing nodes, and their operational patterns in airports justified the hybrid ABM and discrete event system (DES) models to develop a multi-function simulation tool to help inform decision/policymakers in different operational phases of planning, design, development, and implementations for addressing issues like public health emergencies and emergency evacuations [11].

Example application integrated agent-based simulations and machine learning methods, such as *random forests*, enable broader model optimization and estimation robustness for transportation hub's facility planning and staff assignment [24]. However, some challenges remained unresolved for the hybrid models due to the limited availability of simulation platforms, high demand for computational resources, and limitations in validation and optimization [11].

In macro-to-micro simulation models as a group of hybrid models, agent micro-behaviors and spatio-temporal movements are often combined to understand and model broader system dynamics. For example, in an airport model [15], passenger micro-behaviors and movements aid the understanding of system-wide efficiency patterns and parameters while enabling optimized configuration or efficient arrangement of system resources.

In addition, models of flow dynamics in terms of spatial and temporal movements and shifts are categorized as prediction models. Of particular importance is understanding optimization and flight arrangements to predict passenger flows in airport terminals or using the reverse problem formulae, using predictive passenger flows across space (terminal sections, gates, security points) and time (peak times, off-peak times) to optimize resource and system provisions or estimate carrying/critical capacity flows for the terminal system.

### 1.1.5 Validation Approaches

Often, model validation and accuracy issues fall within four related categories and groups. Specifically, these can be (a) spatial or geostatistical, (b) temporal, (c) systemic or engineering, and (d) behavioral or intelligence-related. Validation and reliability represent a cornerstone for the evolution of metaverse in the future mobility [25].

### Spatial/Geostatistical

Normal statistical model validation and accuracy assessment concern *what* and *to what degree* something occurs or emerges from the model behavior. Geostatistical model validation approaches not only must incorporate standard statistical methods but additionally must include *where* and *to which scale* such patterns and model behaviors occur [26]. Issues of spatial accuracy often involve spatial statistics about spatial autocorrelation, spatial heteroskedasticity, and scale variance (or invariance). For example, validation patterns that pass accuracy testing at 1:100,000 scales might fail at 1:10,000 scales and vice versa, rendering scaling up or down for model dynamics problematic or completely fallacious. Modeling and simulation patterns of behaviors and results must often be able to stand scaling-up and down approaches and thus require multi-resolution validation data or data sources that vary in spatial intensity or at least sampled across a varying degree of spatial resolution.

### Temporal

Assessing and validating temporal modeling and simulation dynamics requires a measurement of scale and uncertainty. Temporal scales range from historical, near-time, real-time, near-term, to long-term patterns. Each scale often requires a different regime or battery of model validation methods and tools. More often than not, temporal model dynamics are interwoven and closely coupled with related spatial dynamics in ways that are not easily distinguishable or cannot be readily decoupled. Spatiotemporal modeling and simulation models often require multi-dimensional model validation and accuracy assessment, spatio-temporal stratified sampling, and datasets (e.g., data cubes or multi-dimensional data samples). In other cases, temporal modeling dynamics exhibit deep uncertainty, are riddled with incomplete information, or lack historical data support for statistical validation. In such cases, model assessment and validation methods can include subjective and latent statistical methods of assessment or conditional/assumptional modeling assessment (e.g., finding critical conditions where model assumptions fail or lead to unexpected results or patterns).

### Systemic/Engineering

Examples of these validation and accuracy challenges include assessing system variable inclusion, issues of *equifinality* [27], such as issues of resource conflicts, and mutually exclusive and competitive system dynamics. Validation approaches addressing these challenges must not only have a solid foundation on rich and diverse data sources but also take into account the integrated nature of system behaviors, account for often amphi-directional forces and drivers of change, and whole-life-cycle engineering aspects from constructive to generative, to emergent mechanisms and dynamics.

Behavioral/Intelligence

In many cases, micro-finite data patterns on individual behaviors are often desired and needed to validate model estimations and results. For example, Scala et al. [28] used historical data and data analytics methods to validate their modeling and simulation. They generated a passenger "Activity-Travel-Diary" model from different sources, such as passenger's data (personal mobile phone, apps), airport data (airport WiFi, GPS, scanning facilities), and flight information (flight schedules and gate allocation) and successfully validated the simulation results. Creative uses of existing data and multi-source data pluralism can provide valuable insights into model behaviors and predictive patterns emerging from modeling and simulation approaches. On the other hand, behavioral and cognitive patterns and data cannot be used strictly to establish causal predictive behaviors and require additional heuristic, probabilistic, or propabilogic assessments and model validation methodologies. Agent-baMicro-finite particularly sensitive in terms of statistical validation methods, as their design and definitions encapsulate and embed a certain degree of dynamism, randomness, and uncertainty. Heuristic approximations and assessment of realities also require non-static methods of assessment and repeated measures such as statistical bootstrapping, Monte Carlo simulations, or simulation ensemble methods.

## *1.2 Architecture Principles*

### 1.2.1 Two Categories of ABM

This study focuses on two categories of agent modeling: (a) phenomena-based modeling; and (b) exploratory modeling [29]. In phenomena-based modeling, one uses agent rules to create known phenomena represented as a reference or aggregate pattern. An example of this with respect to an airport planning model is to change a traffic distribution pattern that may be derived from various sources including Airport Operator or Air Traffic Management.

In exploratory modeling, one creates a set of agents, using their behavior to explore emergent patterns. An example of this exploratory modeling with respect to an airport planning model is given in [30] where one simulates an airport with passenger agents passing through several airport zones, varying parameters such as boarding gates, shops, and arrival patterns, to observe the emerging behaviors.

We expect to use both methods, with the exploratory modeling providing the basic agent framework and parameters and the phenomena-based method used to replace the simplified distribution patterns used in the exploratory model.

As one refines a model with either modeling category, conformance to agent modeling standards such as the Foundation for Intelligent Physical Agents[1] (FIPA),

[1] www.fipa.org.

its Agent Communication Language[2] (ACL), (fipa.org/repository/aclspecs.html) and the burgeoning OMG Agent and Event Metamodel [31]) will connect phenomena or exploratory-based agent modeling exploration in a more formal Model Based System Engineering (MBSE) architecture.

### 1.2.2 Metaverse

Recent digital innovations and advancements, such as Digital Twins, metaverse, and parallel systems, provide an ecosystem for mirroring real-world entities, their dynamics, and interactions. These technologies have been proposed and customized for adoption in several domains, such as the gaming industry, art, communication, fashion, and education [32]. Recently, these technologies received significant attention among scholars in applied science and engineering sectors, such as manufacturing [33], energy [34], infrastructure planning [35, 36] and urban transportation [37, 38].

However, most of the current studies and technical reports remained at the conceptual level and provide general descriptions of the values of immersive and interactive environments for more sustainable, inclusive, and economically viable decision-making. Technologies such as extended reality, sensor-based object recognition, location tracking, and haptic augmented reality systems in the metaverse interactive environment are regarded as potential solutions for complex systems like airports, but more empirical studies are required to evaluate their feasibility [39]. Because of their complexity, airports will require a wide variety of Digital Twins from different sources and a wide variety of use cases. This will require interoperability of these Digital Twins to work together seamlessly. A Digital Twin metaverse that supports this capability will need to have certain defined features and capabilities that will need to be articulated [40].

### 1.2.3 Challenges

Most current approaches, however, have failed to provide a robust, reliable, and scalable solution to incorporate real-time data in the calibration, validation, and sensitivity of tools. The existing data collection methods often lead to biased data [41]. Most data used to calibrate the models are historical or experimental. Accordingly, agent-based models such as passenger flow models are highly simplified. The lack of data led to the introduction of many assumptions to the model, which decreased their reliability. In addition, in many cases, the airport terminals are considered individual entities, whereas these buildings are connected to broader and mass buildings/infrastructures and catchment areas in their precinct and city (e.g., parking, city transportation).

---

[2] fipa.org/repository/aclspecs.html.

The complexity of airport systems cannot be addressed by linear and *ad hoc* modeling and simulation. One of the issues for this fragmentation is the lack of data standardization and interoperability. This might be due to the lack of an ontological framework across different actors and stakeholders in airport operations.

As such, emerging technologies such as Metaverse and Digital Twins, which improve the understanding of human-environment interactions, play a crucial role in airport operations and decision-making.

### 1.2.4 Proposed Reference Architecture

We propose integrating complex-adaptive and multi-agent-based systems into the Digital Twin framework to focus on human-environment interactions. We also emphasize the significant role of geospatial intelligence, including data processing, advanced analytics, and visualization, in improving the accuracy and granularity of models.

### 1.2.5 Multi-Agent-Based Systems

Multi-agent architectures, in general, are ideal for linking Digital Twin use cases. The multi-agent network on the "digital" side can be arranged to resemble the "real" side structure. Each agent on the "digital" side can freely communicate with the sensors on the "real" counterpart. Parallels between the Digital Twin and agents are given in Table 1.

In our case, the phenomena-based modeling part approximates actual airport traffic with known patterns and varying initial parameters (e.g., in and out traffic arrival rates, number of agents, boarding gates, shops, check-in boxes, baggage belts, and station lag times). The exploratory-based modeling part tests the consequences of using different interaction protocols, varying the same parameters. Many commer-

**Table 1** Relationships between Digital Twin and agent entities

| Digital Twins (DT) | Agents |
|---|---|
| • An organized set of digital models representing a real-world entity designed to address specific issues/uses | • Evolved and planned behavior using agents with singular functions to address specific issues and uses |
| • Updated in relation to reality, with a frequency and precision adapted to issues and uses | • Updated in relation to reality, with a frequency and precision adapted to issues and uses |
| • Equipped with advanced operating tools, including the ability to understand, analyze, predict, and optimize | • Equipped with the ability to monitor, understand, analyze, predict, and optimize |

cial frameworks rapidly create agent applications using the Foundation for Intelligent Physical Agents (FIPA) and Agent Communication Language (ACL) standards, using tools, e.g., AgentSpeak, JACK, JADE, GAMA, Mason, Repast, and NetLogo. Among these applications, NetLogo [42] provides an environment to rapidly understand the value of ABM Digital Twin integration because it:

- Supports the FIPA and ACL standards through extensions;
- There is built-in support for connecting GIS data;
- Is being discussed in the Digital Twin Consortium, Mobility and Transportation Working Group, to implement an air traffic use case used to interact with different types of agents (e.g., positioner, evaluator, service provider, and user agents) and has been used for similar applications [30], and;
- Has human-like (Belief, Desire, and Intention) and communication extension code that could conform to the agent specification [43] by replacing existing NetLogo extensions.

NetLogo provides a conceptual and theoretical overview providing a powerful tool for the initial prototype. It is believed that the models should be simulator-independent. Connections to the metaverse virtual reality, evident in the example of the John Wayne Airport (JWA) terminal of the airport planning model that follows, will enable the usage of realistic airport maps and data. Integrating real and near-time data, along with the 3D reality capture imagery and the Digital Twin graph interface, will enable distributed multi-agent modeling and simulation architectures to propagate through the metaverse visualizations, modify existing visualizations, generate synthetic 3D data scenes, and offer theoretical and empirical metaverse versions of the predictive future simulation results. The goal of the agent simulation to the airport SoS is to provide a degree, magnitude, or measure of proposed types of agent satisfaction by performing parameter sensitivity analysis in a scalable, independent manner. Our agent approach will aid an eventual split of monolithic systems into multi-agent networks, simplifying the airport's overall architecture and improving controllability and reliability since control functions can become autonomous and failing agents can be restarted. The next section provides a guideline for applying such models in airport planning.

## 2 Airport Planning Model

This section provides the workflow of airport planning and operational models. These workflows will create a basis for the smart airport Agent-based Modeling design and implementation explained in Sect. 3.

**Table 2** Examples of environmental scenarios

| Environment | | |
|---|---|---|
| Range | Climate | Social |
| Gold | Severe Weather (below −10°) | Severe Disruption |
| Silver | Poor visibility (−10° to 0°) | Air Traffic Strike |
| Bronze | Medium visibility (0°–10°) | Train Strike, Olympics |
| Normal | Normal visibility (10°–27°) | Normal |

## 2.1 Context-Planning Capabilities

Many parties are involved in generating schedules for all activities across the airport. The multiple interdependencies between schedules mean that it may take weeks or months to develop a default schedule. The business rules that a schedule may need to take into account are not always written down, making it more difficult and time-consuming to generate the schedule and to verify it against operational scenarios.

Airports operate within an ecosystem, with multiple parties providing and consuming services within the ecosystem. Whether a hub or point-to-point, an airport acts as an intermodal transportation hub supporting a wide variety of transport modes, which will increasingly consist of unmanned vehicles subject to automated control. Interoperability in the SoS depends on the effective exchange of information based on a common understanding (human and machine) between the parties, including the airport, airlines, air traffic management, and a multitude of other participants in the ecosystem. Common ontologies and associated data standards are fundamental to meaningful information exchange.

Consequently, airport planning is not conducted in isolation. The airport operator provides information on demand and capacity to airlines. Control authorities such as air traffic management, and local aviation authorities impose constraints such as the minimum time between landing two aircraft of type F. Each organization has its own capabilities, including landing aircraft, refueling aircraft, immigration, check-in, baggage handling, airfield maintenance, etc. Other factors, such as climate conditions (visibility level) and social events (see example in Table 2), will increase the level of complexity in the operations.

The capabilities to do work require resources such as aircraft and surface transport vehicles, terminal facilities, airport equipment and facilities (runways, taxiways, stands, etc.), and ground handlers. These resources not only need to be deployed but they must also align with many other organizations. Planning activity occurs in many different places in many different organizations and the interdependencies must link for the operation to work effectively.

All work in an airport operation occurs in an environment, usually consisting of climatic conditions such as a working temperature range, and social conditions such as train driver strike. Plans are created for an environmental scenario category. Environment scenarios can change quickly.

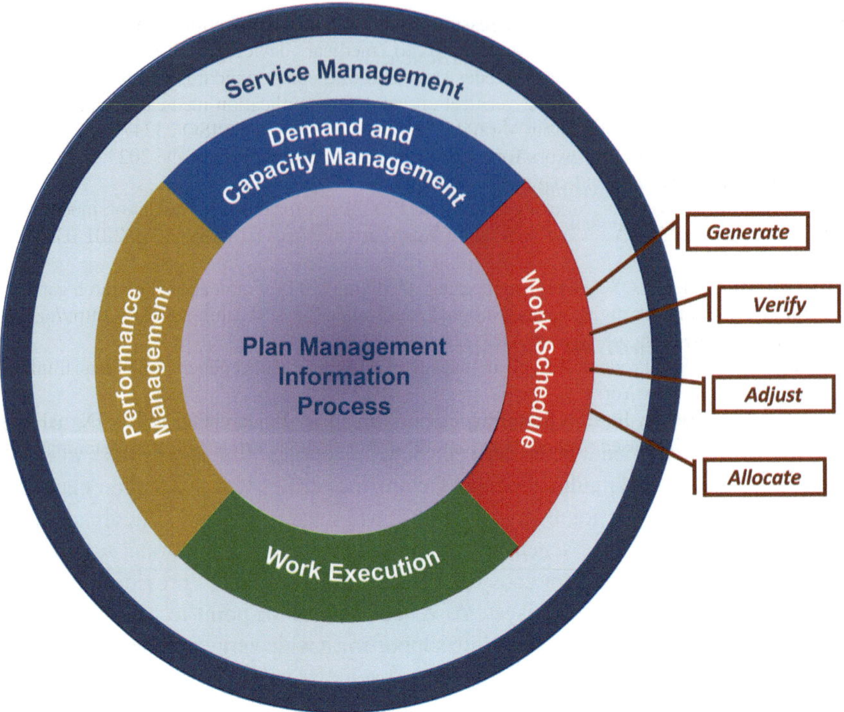


**Fig. 1** Plan management cycle and stages

## *2.2 Planning Cycle*

All organizations involved in an airport operation have their planning cycle typified by the following diagram in Fig. 1. Four stages of the plan management cycle are demand and capacity management, work scheduling, work execution, and performance management. The ideal situation is a smooth flow of information across these four stages.

This high-level perspective helps develop planning activities' processes and information requirements. Considering the power of Digital Twin for real-time data exchange and analytics and Metaverse for cross-collaboration in a virtual space, it is crucial to understand the business process of airport operation, which is explained here. Default schedules are created for all work activities based on a load-balanced capacity profile in a specified environmental scenario, taking account of resource usage constraints. The planner obtains the demand for airport resources from demand and capacity management, including the scenario it operates within, for example, a normal day. This constitutes the required capacity profile. Demand and capacity management also provide the resource usage constraints for the scenario.

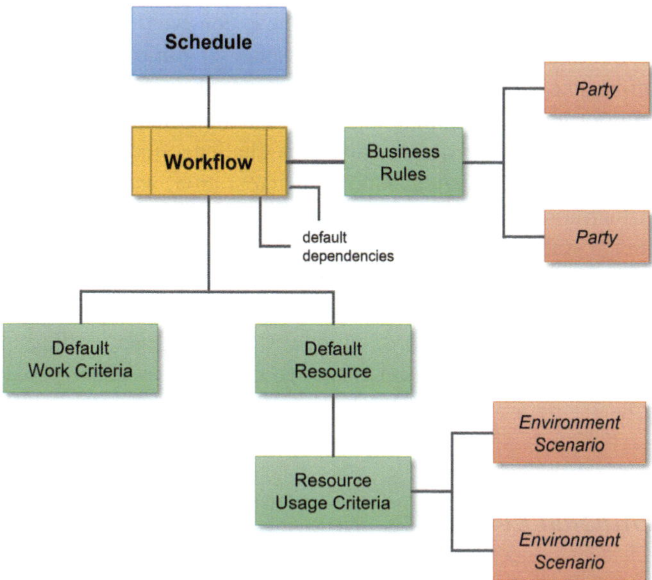


**Fig. 2** Workflow schedule standard model

The planner defines the business rules for the airport, airline, and government work units within the airport for the provided scenario, including the default work dependencies, the default work criteria, and the default resources to execute the work (for example, in Fig. 2).

All schedules are verified against forecast assumptions, usually via a manual process or automatically by process simulation or flow emulation. However, real-time data is lacking in most of the cases. The outcome is a set of work units that do or do not meet performance targets. Once default schedules have been verified and activated, they typically need to be adjusted on the day to take account of circumstances such as late arriving or departing flights, bad weather, and staff absence.

Given the lack of real-time data analytics and a single point of truth for stakeholder collaborations (Metaverse) changes to plans may take many hours, particularly if they have to be modified manually. Resources such as gates, stands, and baggage reclaims are unlikely to be allocated efficiently by time-consuming manual planning processes. This results in delays for flights, delays for passengers, and increased airport costs.

On the effective date and time of the schedule, party role, transport, equipment, and facility resources are allocated to the work. A system that automatically transfers and integrates on-time data to automate the plan update is critical. Automating the planning processes would reduce the time required to assess the impact of changes from hours to minutes. The corresponding impact would be reduced delays, reduced queues, and reduced costs.

Integrating data between applications can be expensive to develop and challenging to maintain. Standards for defining and exchanging information greatly reduce the time, effort, and cost involved. A plan's standard model improves the current exchange of information to support multiple stakeholder requirements and is extensible to incorporate future requirements. This challenge indicates the critical role of Digital Twins for on-time/real-time data integration, analysis, and scheduling updates. Furthermore, given the high complex and multi-stakeholder nature of airports' planning cycle, a multi-agent-based approach designed to handle the interaction of stakeholders (as agents) will enrich the Digital Twin's efficiency. To achieve this goal, leveraging standard data schema is crucial. The next section provides an overview of enriching The Aviation Community Recommendation Information Services (ACRIS) data and information exchange standards by agent-based modeling technologies.

## 2.3 *Operational Level Integration*

Planning closely relates to the operational and functional cycles at an airport. Such operational levels are tied to the airport's physical and virtual or computational infrastructure functions. Figure 3 provides an overview of the proposed ACRIS Semantic Model [44] integration with the coupled Digital Twin and agent-based modeling technologies. The three basic groups of category package modules of the semantic model (namely entities, moments, and motivations) are closely related to agent classes (as entities), to agent actions (as moments), and agent belief-desire-intention (BDI) goal architecture [45, 46] (as motivations). Therefore, according to Fig. 3, a Digital Twin for a Smart Airport can be enriched by the standard semantic models (ACRIS) coupled with agents' behavior to realistically reflect the airport's operation.

More specifically, the knowledge organization of the ACRIS Semantic Model integrates a number of operational modules, packages and data elements at the airport operational level [47] as can be seen in the following Fig. 4. Each of the three fundamental objects (entities, moments, motivations), is mapped to each of the element object library modules shown on the right side of Fig. 4. Each of these libraries in turn has a number of nested attributes, operational layers and data associated with them, and can be linked to the resources available to multi-agent BDI goals, strategies, decisions, and actions.

At the visual, 3D, and metaverse levels of reality capture, these operational elements of the DT-ABM-ACRIS integrative architecture can be visually, geospatially, and locationally oriented in virtual metaverse space. The following images in Fig. 5 demonstrate an example of the reality capture process of the John Wayne Airport (JWA) terminal in Orange County, California. The combination of stationary, mobile, and aerial LiDaR scanning technologies allows us to fully capture and closely associate centimeter-accuracy locational attributes of captured scenes, 3D objects and point-cloud coordinates with the developmental aspects of the agent-based modeling and simulation process.

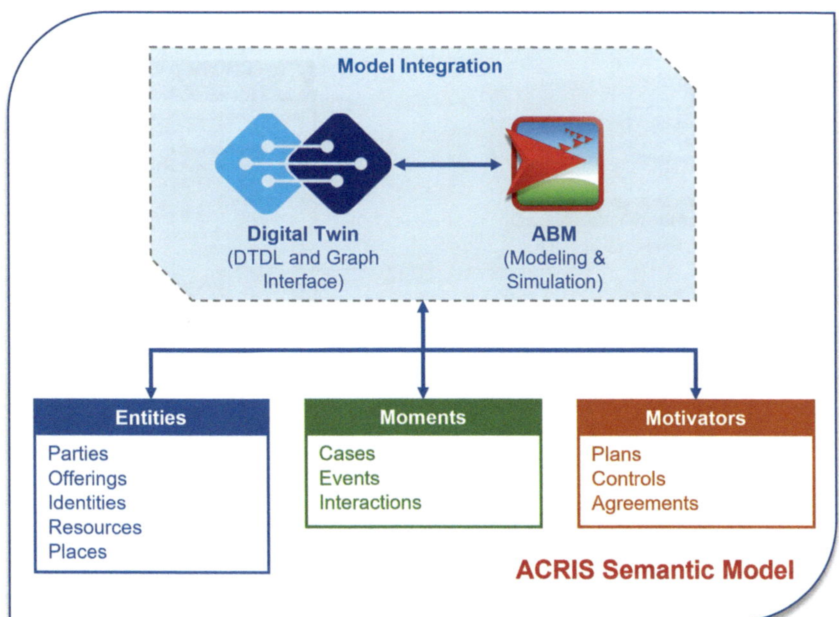


**Fig. 3** Digital Twin and agent-based model integration with basic module package organization of the ACRIS semantic model

## 2.4 *Definitions*

### 2.4.1 Allocation

Allocations are a manifestation of the schedule in the current time. Allocation refers to the assignment of resources to do work according to a published schedule. For example, a schedule of aircraft landing slots will identify the runway facility assigned to each particular landing slot. The runway facility will be allocated to the landing at the scheduled time and unallocated to the landing when it is complete.

### 2.4.2 Business Rule

Business rules are used to define the work to be done, the default work criteria, the default resource types that are required to do the work, and the dependencies between the work units. For example, the work unit "Land Aircraft" applies to all aircraft types within the International Air Transport Association (IATA) aircraft type codes and will require the default runway facility resource and also the participation of the air traffic authority party role resource. The work unit "Taxi Aircraft to Stand" depends on the

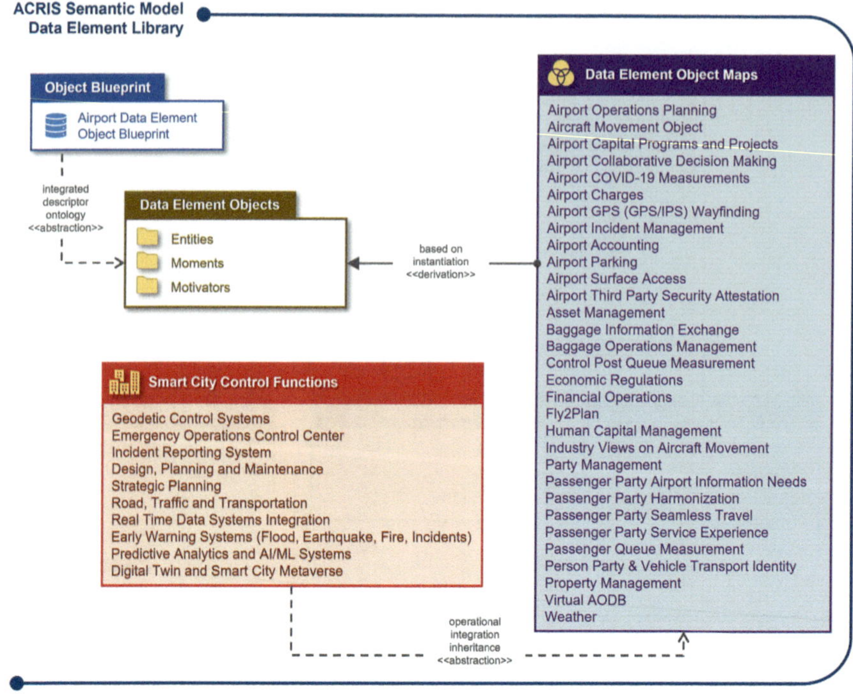


**Fig. 4** Package module and design organization of the modified ACRIS semantic model

"Land Aircraft" work unit having been successfully completed. This will apply to all aircraft types and will require a taxiway and air traffic authority as default resources (see example in Fig. 6).

### 2.4.3 Resource Criteria

Default Resource Criteria describe what a party wants to use resources for, such as equipment or facility resources. Resource criteria define what a particular facility or resource is required to be used for.

### 2.4.4 Resource Usage Constraint and Criteria

**Resource constraints (usage constraints)** specify the kind of work that a resource cannot be used, for example, where an explosive detection system cannot be used for Level 3 hold baggage screening. This is the opposite of resource usage criteria.

**Fig. 5** Infrastructure operational visualizations of the John Wayne Airport (JWA) in Orange County, California. Top Row: **a**, **b** LiDAR point-cloud and 360° photosphere reality capture; **c** geodetic monument examples at and around John Wayne Airport; **d** JWA monuments and high spatial accuracy geodetic control network; **e** Orange County GPS real-time network ties

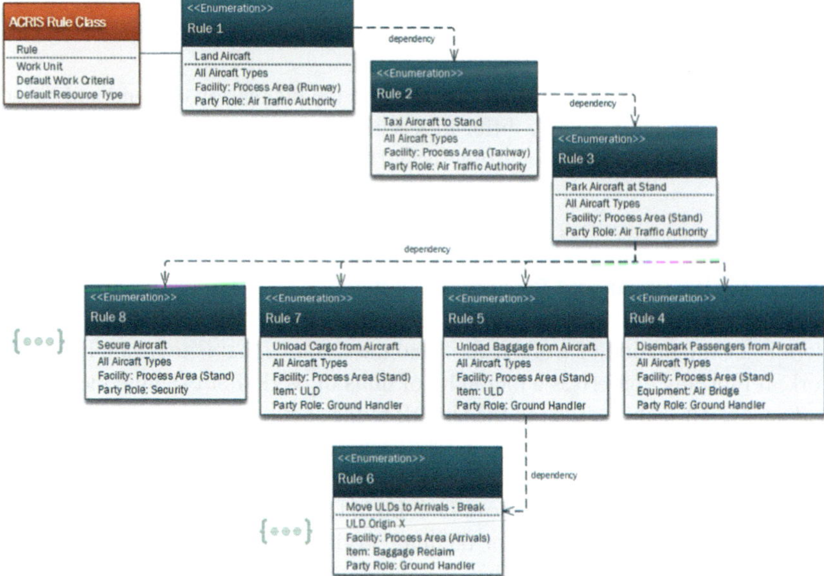


**Fig. 6** Example inheritance and dependency structure of airport operation business rules

**Table 3** Examples of integrated resource usage criteria

| Type[a] | Resource[b] | Usage[c] | Criteria[d] | Example | N[e] | B[e] | S[e] | G[e] |
|---|---|---|---|---|---|---|---|---|
| F | R-27L | A | F | A380 | 1.0 | 0.5 | 0.2 | 0.1 |
| F | R-27L | A | F | 747 | 2.0 | 1.0 | 0.5 | 0.2 |
| F | R-27L | A | D | | 3.0 | 2.0 | 1.0 | 0.5 |
| F | R-27L | A | C | | 4.0 | 3.0 | 2.0 | 1.0 |
| F | R-27L | A | B | | 4.0 | 3.0 | 2.0 | 1.0 |
| F | R-27L | A | A | | 5.0 | 4.0 | 3.0 | 1.0 |
| E | CD | B | SB | | 0.5 | 0.2 | 0.1 | 0.01 |
| E | CD | B | OOG | | 0.0 | 0.0 | 0.0 | 0.0 |
| E | OOGD | B | SB | | 0.1 | 0.1 | 0.1 | 0.01 |
| E | CBX | P | PS | | 6.0 | 3.0 | 2.0 | 1.0 |
| E | RCAR | B | SB | | 20.0 | 15.0 | 10.0 | 2.0 |

[a] F: Facility; E: Equipment
[b] R-#: Runway Number; CD: Check Desk; OOGD: OOG Desk; CBX: Cabin Bag X-Ray; RCAR: Reclaim Carousel
[c] A: Aircraft; B: Bag; P: Person
[d] A-F: Code; SB: Standard Bag; OOG: OOG Bag; PS: Passenger
[e] Operation cycles per minute: N-Normal, B-Bronze, S-Silver, G-Gold

**Resource Usage Criteria**, or just Usage Criteria, define what a particular facility or resource can be used for. This is the opposite of resource (usage) constraints. Resource usage criteria implicitly provide constraints on resources, for example standard check-in desks can deal with standard bags, but not Out Of Gauge (OOG) bags. An OOG desk can deal with standard bags as well as OOG bags, though at a reduced throughput in terms of bags per minute.

Usage criteria and capacities are provided to the planner by demand and capacity management and they may differ for each environmental scenario category. An example of a set of resource usage criteria and their characteristic properties is shown in Table 3.

### 2.4.5 Work Criteria and Progress Workflow

Work is a combination of work criteria and the resource that is to be applied to the work. It typically comprises a hierarchy of units of work. For example the work to land an aircraft includes taxiing the aircraft to the stand and parking it at the stand. Work may have dependencies, e.g., aircraft separation must be maintained to 5 min for code F, and may have a specified order in which it needs to be done, e.g., unload baggage from aircraft and then move Unit Load Devices (ULDs) to arrivals break.

Work Criteria is a standard for defining work to be done, including sufficient information so that the work can be:

- Scheduled with competent resources.
- Readily determined to have been completed or not.
- Objectively measured.

Work criteria identify the type of work to be done, e.g., (1) unload baggage from a British Airways 744 aircraft at 08:00 every Monday during the 2023 summer season, or (2) sort and load bags for business class passengers traveling on a British Airways flight to Boston. See Fig. 7 for further examples in the context of the required work.

Work progress is the quantifiable completion of work. The actual doing of the work is part of logistics and conformance. Monitoring the work progress by resources and assets is part of performance management.

In the plan model, a plan is a forward-looking work schedule. A work schedule is an association of facility, equipment, and party role resources to units of work in specified time periods. It takes business requirements into account, as provided in a load-balanced capacity profile for a defined scenario. It also incorporates the usage constraints of the resources that are to be included in the work schedule. Units of work should have a global unique identifier to enable accurate allocation and deallocation/reallocation of resources to the work.

# 3 Smart Airport ABM Design and Implementation

## *3.1 General Considerations*

As we indicated, the time-consuming and manual resource planning and allocations, such as gate stands and baggage reclaims, are highly challenging and often generate delays for flights, delays for passengers, and increased airport costs. The work schedule and sequence of activities can be designed, monitored, and controlled through agent-based modeling and simulation approaches implemented on a Digital Twin infrastructure. For instance, [14] examined the passengers' (as agents) satisfaction and the time-saving options through their movements toward different services such as passport controls, check-in counters of airline companies, boarding gates, and different types of shopping. This reflects the passengers' behavior, discretionary decisions (shopping), and the number of services (passport controls and boarding gates) to examine different scenarios.

The results of the 30-round simulation indicate that satisfaction decreases as service lines increase and user agents miss their flights. The increasing number of service lines can be due to delays in passport checks or limited airline check-in services.

Furthermore, the simulation results demonstrate that average time lengthens when a very high number of agents are included in the simulation. Reflecting the lack of optimized scheduling for 6216 incoming or outgoing passengers to/from the terminals.

Therefore, using two evaluation criteria (time savings and agent satisfaction) the potential benefits of using ABM are highlighted. However, it should be noted that this work is a proof of concept and NetLogo simplifies the process and reduces the number of agents to address the computing requirements. As such, the focus of

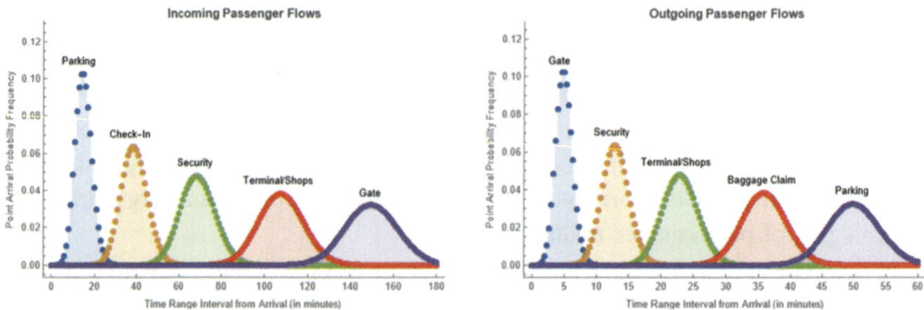


**Fig. 7** Arrival time probability distribution segmentation for incoming (left subgraph) and outgoing (right subgraph) passengers of a single flight in an airport terminal

NetLogo ABM is on a limited number of options. Therefore, further investigation and more robust ABM tools should be utilized to examine the ABM application in a Digital Twin.

Capturing passenger flow experiences related to travel times within, to, and from airports requires a firm understanding of segmentation across discrete and distinct spatiotemporal sections within any given airport terminal. Relevant literature has considered longitudinal methods of travel modes (e.g., [48]) in a smart city environment and statistically estimated wait and travel time distributions for citizens and travelers. Rather than modeling and simulating random passenger arrival times, we plan to use well-known Poisson probability distributions (see, for example, [49, 50] on how these distributions relate to human and socio-temporal dynamics). The following Fig. 7 showcases temporal distribution dynamics based on these parametric probabilistic inferences for both incoming and outgoing single flight passengers by airport experience segment.

While day-to-day operational reality in a single airport is by far more complex than the oversimplification shown in the figure (e.g., hundreds of inbound and outbound flights per day, many of them overlapping across the temporal continuum), a distributed agent-based simulation framework can combine these arrival probabilities based on daily airport schedule of flight arrival and departures and generate evidence-based, empirical mixture (e.g., multimodal) distributions to be used as Bayesian Priors for the BDI decision-making architecture.

## *3.2 Agent-Based Architecture*

### 3.2.1 Smart Airport Agent Typology

A number of conceptual elements of a smart airport agent-based architecture can be considered as fundamental elements of an agent typology. Employing a distributed modeling architecture requires the differentiation of agent archetypes. We can identify at least four fundamental types of agents, as illustrated in Fig. 8:

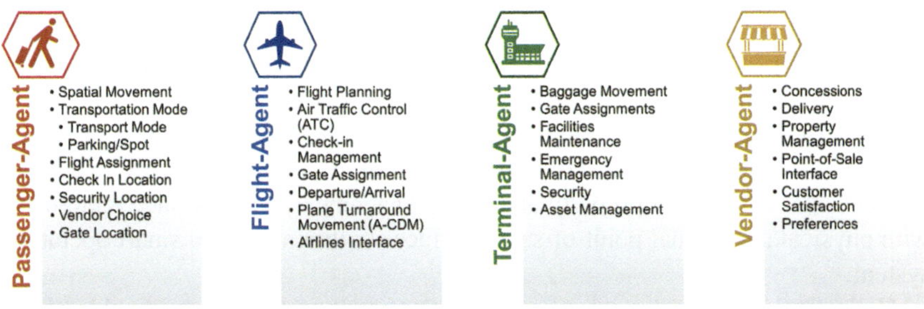


**Fig. 8** An example of smart airport agent typology considerations

- Passenger-type functional agents.
- Flight-related operational agents.
- Terminal-related operational agents.
- Vendor and concession-related functional agents.

Adopting such a distributed agent typology schema enables the introduction and use of an agent orchestration framework within a broader BDI agent architecture [45, 51]. Specifically,

**Passenger-Agent**: This archetypal agent typology captures and monitors basic passenger-agent spatial movements across different airport and terminal sections. It begins with the passenger transportation mode section by tracking the mode of transportation (e.g., public transport or service transportation) along with parking and parking spot allocation. Once passengers move into the main airport terminal (e.g., airline check-in section and location), it tracks flight assignments, thus associating each passenger agent with a specific flight module (consequently, linking the passenger-agent module with the flight-agent and terminal-agent modules). Further, along the passenger journey through the airport, the passenger-agents relate their location and assignment through the security check-in counters/locations, the choice of vendors, and vendor location for airport concessions (thus associating passenger-agent entities with vendor-agent entities). Finally, it tracks passenger movement through their departure/arrival gates and embarkation/disembarkation queues.

**Flight-Agent**: This typology provides a distributed agent-based framework for all flight-related components or modules related to smart airport operations. These include the check-in management process (from the airline and provider operational perspective), gate assignment for flights departing or arriving, time and process management for departures and arrivals, including delay and queue management, plane movement within an airport (e.g., from/to gates, etc.), plane turnaround and airport collaborative decision-making processes (A-CDM), as well a framework for modularity interfacing agent roles with airline operational systems.

**Terminal-Agent**: The third typology relates to smart airport terminal-based operations. These agents track and distribute baggage movements, gate management assignments (from an airport facility operational perspective), facility maintenance and integrated management (in terms of both temporal and spatial perspectives), airport security (physical and network), and smart asset management. These terminal-agent types interface directly with relevant system modules and provide distributed intelligent communication interfaces with physical system modules.

**Vendor-Agent**: This agent typology relates to distributed communication and action mechanisms related to vendor assignments, operations, product and service delivery mechanisms, concession property and title management, and interface (externally) with physical and virtual point-of-sale interfaces within an airport smart operations system.

Collectively, these examples of smart airport agent typologies are capable of providing the necessary role assignments and functions that enable a smart, robust, adaptive, and intelligent framework for agent actions, resource (management/allocation/distribution), time and space interactions with physical movement, behavioral and cognitive response mechanisms, and property and facility management functions. While not necessarily exhaustive (in fact, alternative and augmented additional agent types and typology groups can be designed and used), they form a cohesive and tightly coupled set of agent roles that, through their exchanges and interactions, can give rise to the emergence of complex and intelligent distributed system behaviors and operational mechanisms.

## 4 Implementation in a Metaverse Visual Framework

This section focuses on the visual metaverse aspects of an intelligent Digital Twin framework. It is mainly based on an award-winning TM Forum Moonshot Catalyst project (URN M23.0.567) [13]. The project features innovative and active public-private partnerships across a range of industry consortia, public organizations, and airports. The remaining section content provides a short and concise outline of the reference architecture and design as well as the structural and operational intelligence modeling and simulation elements used in enhancing a Digital Twin coupled with metaverse experience, supporting the implementation of the agent-based intelligent ontology semantic planning framework described in the previous sections.

Figure 9 provides an overview of the key elements of the interoperable airport ecosystem for a metaverse approach. The elements range from those of high abstraction/fundamental composability (e.g., public transport, airports, airlines, passengers, agencies), to more specific and metaverse domain-explicit concepts (retail, telecom, AECO, traffic management, etc.). The intent-based service composition and orchestration, along with information exchange and cross-domain interoperability, are governed by a *common and shared ontology* framework, such as the ACRIS semantic framework described in Sect. 3.

The relevant reference implementation outline example is shown in Fig. 10. As the *back-end* metaverse framework interacts with the entities, moments, and motivators (e.g., person recognition and hazard recognition service layers), the *mid-end* layers provide a design and reference orchestration for services like application communication, scenario modeling and simulation, data transformation and interoperability,

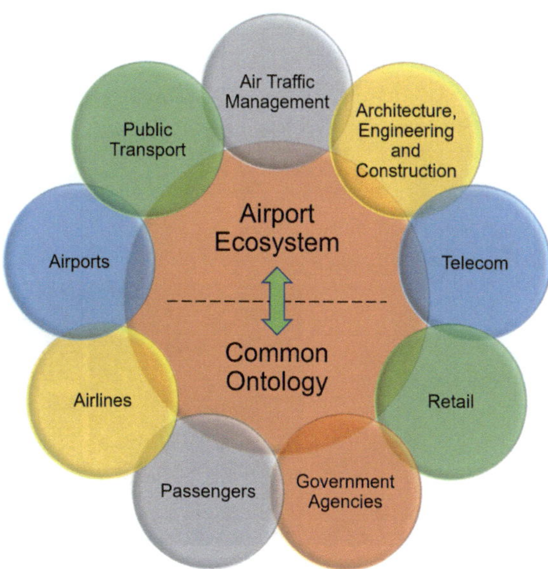


**Fig. 9** Elements of the interoperable airport ecosystem metaverse approach. The scalable architecture includes and encapsulates a common semantic ontology for cross-domain interoperability

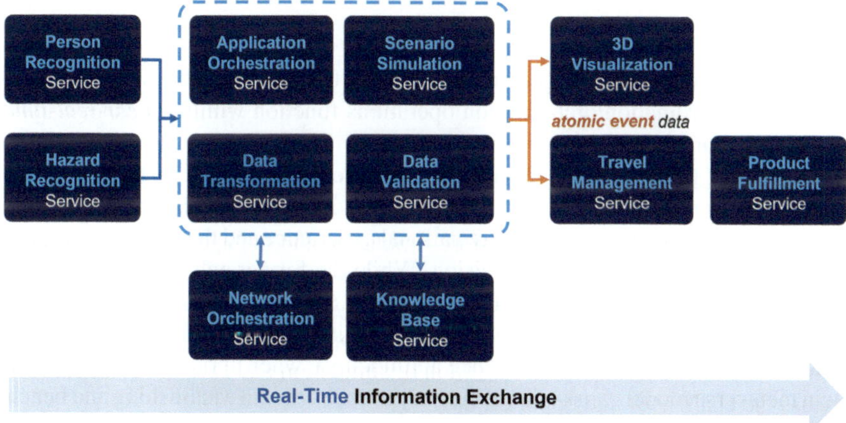


**Fig. 10** Real-time information exchange components and layers for metaverse reference implementation for the Digital Twin smart airport model

and validation and assessment services. Both network-centric and knowledge-centric orchestration and coordination service layers operate across a range of coordinated layers to achieve integrative delivery and visualization services. The latter (3D visualization services) allow for metaverse immersive and interactive experiences while enabling and facilitating the provisioning of other end-user (*front-end*) virtualization

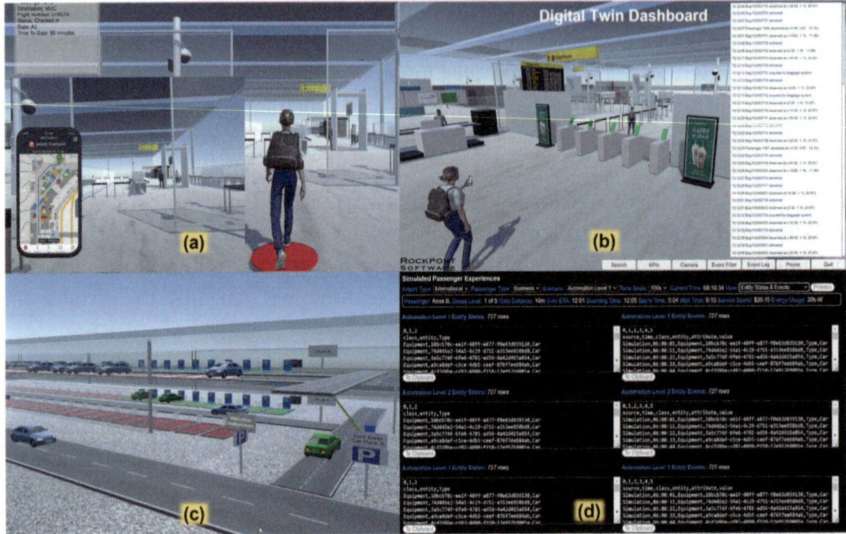


**Fig. 11** Metaverse visualization and real-time Digital Twin dashboard operations: **a** a visual display of navigational agent spatial movements; **b** visual check-point operations and service message exchange real-time operations; **c** showcase of interactive airport parking allocation and navigation; **d** passenger-based dashboard for real-time event-driven scenario simulations

services, such as travel management and product fulfillment-all the service orchestration and coordination/orchestration operations function within a *near-real-time, high-fidelity* functional metaverse environment.

In terms of the metaverse visualization experience (see Fig. 11), the nature and characteristics of Digital Twin 3D immersion allow for both enhancing the visual intelligence and locational/spatial navigational experience and the near real-time service, data, and communication provision. While the former enhances and supports *visual intelligence* operations, the latter promotes and encapsulates implementation based on high accuracy, fidelity, and interoperability. Taken together, these two fundamental principles (visual intelligence and fidelity), when applied within a Digital Twin metaverse model, work complementary to each other in establishing and benchmarking a conceptual framework for *intelligent Digital Twins*.

The overview of the overall smart airport and intelligent Digital Twin metaverse implementation is provided in Fig. 12. The approach focuses on predictable, accurate, and visually immersed outcomes stemming from automation, intelligent decision-making, and inter-agent distributed intelligence. Both internally (within events, service provisioning, resource allocation components, and orchestration mechanisms) and externally (inputs and outputs such as data and services), it is essential to strive to achieve a high degree of cross-standards interoperability and composability.

Some of the benefits of integrative metaverse design, control, and operations include the ability of airport management and operational planning capable to simulate and experimenting with alternative scenarios before implementation, reducing

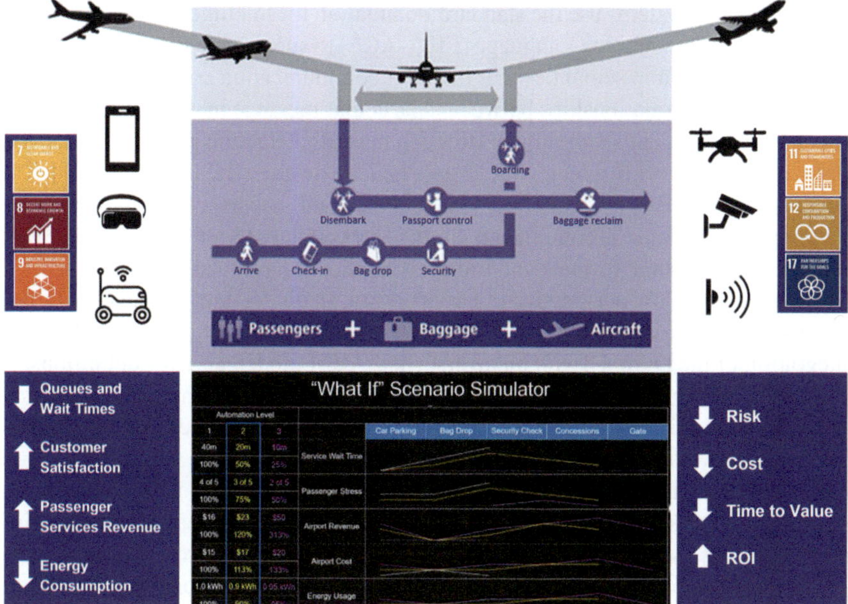


**Fig. 12** Visual conceptualization of the intelligence airport metaverse framework

cost (design and implementation) while managing risk and risk expectations (including risk aversion and risk propensity attitudinal dispositions). As a result, an intelligent metaverse Digital Twin framework streamlines time-to-value operations while, in the near- and long-term, promoting higher return-on-investment ratios. At the same time, the Digital Twin framework enables the development and establishment of broader sustainability and robustness characteristics at the airport level. These might include emergency and incident response or reporting operations, addressing sustainability goals and criteria, and forming a more innovation-minded workforce.

## 5 Conclusions and Future Work

In this chapter, we proposed a new generation of ABM that leverages the metaverse and Digital Twin capabilities. Also, we highlighted the significant role of real-time data for a more realistic representation of Airport operational activities and planning processes. The aim is to optimize and enable airport planning with predictive analysis to understand and adapt to unpredictable issues. We argued that the complexity of airport systems cannot be addressed by linear and ad hoc modeling and simulation methods. To address this challenge, we described two methods of categorizing ABMs to set up both the use of phenomena and exploratory-based models in the airport planning domain and introduce agent and communication standards. It was shown that many agent systems use the standard Foundation for Intelligent Physical Agents and Agent Communication Language (FIPA-ACL) to support the sharing of message content with common semantics that includes communication context. This context explicitly defined relationships between a particular message in relation to a context such as a sender agent's current workflow, an agreed interaction protocol, goals, and plans, or with respect to a receiving agent's beliefs, desires, and intentions [53]. The OMG Agent and Event Metamodel [31] standard will supersede FIPA-ACL but is planning to include ACL compliance. Future work will trade-off ACL working with the message exchange formats and transport/application protocols.

We provided a key perspective to airport planning model development, including an ecosystem for context planning, a planning cycle with a workflow standard model, integration of the Digital Twin, metaverse, and ABM model, and key definitions.

The provided conceptual model has practical and theoretical implications. From a practical perspective, different operational activities can benefit the role of Metaverse and Digital Twins to test the human-environment interactions in various scenarios. This capability can be used for the interaction of different stakeholders, such as airlines, third-party services, and safety and security teams. The agent-based modeling and simulation, calibrated with real-time data, is a new line of research that needs further exploration to offer a reliable and trustworthy output for decision-makers in airport operations. In addition, an ontological framework is critical to address the interoperability and harmonization of complex data and analytics in an airport ecosystem.

Finally, this document provided an agenda for developing a proof of concept and pilot study to be implemented in an actual case and an examination of the idea of a new generation of ABM integrated into the Digital Twin architecture and detailed information and guidelines for Airport Planning, which can be used for ABM development.

**Acknowledgements** The work presented in this chapter draws from a fertile collaborative environment for discussions, discourse, and collaboration of a number of collective organizations. Digital Twin Consortium (DTC) working groups are one of these organizations. S. Sabri and K. Alexandridis serve as co-chairs of the DTC Academia and Research Working Group, while D. Migliori serves as one of the co-chairs of the DTC Mobility and Transportation Working Group. Many of these ideas and concepts emerged through its collaborative environment and the author's interactions within it. The parent industry standards development organization Object Management Group (OMG) is another key organization we wish to acknowledge. Jeff Smith serves as co-chair of the OMG Analysis and Design Task Force and co-submitter of the AgEnt specification. TM Forum and the Airports Council International (ACI) provided the necessary background for the development of ideas, projects, and resources. Specifically, the metaverse implementation related to passenger experiences, provided in Sect. 4, is an integral part of an award-winning Moonshot Catalyst program of TM Forum and represents an active collaboration of many industry and public organizations. D. Migliori, B. Logan, and K. Alexandridis serve as key organizational and industry champion members of the project. Furthermore, the integration of the ACRIS Semantic Model described in Sect. 3 owes its roots and fundamental concepts to the ACI's Aviation Community Recommended Information Services (ACRIS) model approach. B. Logan is one of the key architects and custodians of the ACRIS semantic model organization and its resources. Finally, the authors wish to acknowledge a number of individuals and organizational representatives for the exchange of ideas and organizational support: Dan Isaacs, general manager of OMG and CTO of DTC for his support of DTC working groups and organizational facilitation; Members of various DTC working groups for feedback and review of this work and ideas over various stages; Member-organizations of the TM Forum Moonshot Catalyst Project No. M23.0.567 for supporting and recognizing the transformative concepts.

## References

1. Mekić A, Mohammadi Ziabari SS, Sharpanskykh A (2021) Systemic agent-based modeling and analysis of passenger discretionary activities in airport terminals. In: Aerospace, vol 8, p 162, June 2021. 3 citations (Crossref) [2023-12-04] Number: 6 Publisher: Multidisciplinary Digital Publishing Institute
2. Cheng L, Reddy V, Fookes C, Yarlagadda P (2014) Agent-based modelling simulation case study: assessment of airport check-in and evacuation process by considering group travel behaviour of air passengers, vol 568–570, pp 1859–1864, Trans Tech Publications Ltd, 2014. 4 citations (Crossref) [2023-09-25]
3. Zhao C, Dai X, Lv Y, Niu J, Lin Y (2023) Decentralized autonomous operations and organizations in transVerse: federated intelligence for smart mobility. IEEE Trans Syst Man Cybern Syst 53(4):2062–2072
4. El Jaouhari A, Arif J, Samadhiya A, Kumar A, Jain V, Agrawal R (2023) Are metaverse applications in quality 4.0 enablers of manufacturing resiliency? An exploratory review under disruption impressions and future research. TQM J
5. Kleinschmidt T, Guo X, Ma W, Yarlagadda PK (2011) Including airport duty-free shopping in arriving passenger simulation and the opportunities this presents. In: Proceedings of the 2011 Winter Simulation Conference (WSC), (Phoenix, AZ, USA), pp 210–221, IEEE, Dec 2011. 2 citations (Crossref) [2023-09-25]
6. Tesoriere G, Campisi T, Canale A, Severino A, Arena F (2018) Modelling and simulation of passenger flow distribution at terminal of Catania airport. In Proceedings of the international conference of computational methods in sciences and engineering 2018 (ICCMSE 2018), (Thessaloniki, Greece), p 140006. 17 citations (Crossref) [2023-09-25]
7. Ma W, Fookes C, Kleinschmidt T, Yarlagadda P (2012) Modelling passengers flow at airport terminals—individual agent decision model for stochastic passenger behaviour. In: Pina N, Kacprzyk J (eds) Proceedings of the 2nd international conference on simulation and modeling methodologies, technologies and applications (SIMULTECH). SciTePress Digital Library, pp 109–113
8. Helbing D, Buzna L, Johansson A, Werner T (2005) Self-organized pedestrian crowd dynamics: experiments, simulations, and design solutions. Transp Sci 39:1–24. 1019 citations (Crossref) [2023-09-25]
9. Doostmohammadi M, Fragniere E, Holdsworth R (2020) Crowd behaviour modelling developments through mixed integer programming: the case of Airport queue management. In:2020 9th International Conference on Industrial Technology and Management (ICITM), pp 32–36
10. Bandarian F, Kordani A (2022) Evaluation of passenger behavior in the baggage claim area of the Airport passenger terminal. In: Mathematical problems in engineering, vol 2022. 0 citations (Crossref) [2023-09-25]. Hindawi Limited
11. Zhu H, Liu S, Li X, Zhang W, Osgood N, Jia P (2023) Using a hybrid simulation model to assess the impacts of combined COVID-19 containment measures in a high-speed train station. J Simul. 2 citations (Crossref) [2023-09-25] Taylor and Francis Ltd
12. Liu X, Fu H (2023) Simulation and optimization of inbound passenger flow line in large-scale railway station. In: Kong X, Falcone F (eds) Proceedings of SPIE—the international society for optical engineering, vol 12708, SPIE. 0 citations (Crossref) [2023-09-25]

13. TM Forum (2023) Transforming passenger experiences with continuous decision intelligence (Moonshot Catalyst URN M23.0.567)
14. Zhang H, He B, Lu G, Zhu Y (2022) A simulation and machine learning based optimization method for integrated pedestrian facilities planning and staff assignment problem in the multi-mode rail transit transfer station. In: Simulation modelling practice and theory, vol 115. 9 citations (Crossref) [2023-09-25] Elsevier B.V
15. Peng J, Wei Z, Wang S, Qiu S (2023) Toward dynamic regulation of bidirectional automatic fare gates: a macro-to-micro simulation approach. In: Simulation modelling practice and theory, vol 124. 1 citations (Crossref) [2023-09-25] Elsevier B.V
16. Delcea C, Cotfas LA, Milne R, Xie N, Mierzwiak R (2022) Grey clustering of the variations in the back-to-front airplane boarding method considering COVID-19 flying restrictions. In: Grey systems, vol 12, no 1, pp 25–59. 8 citations (Crossref) [2023-09-25] Emerald Group Holdings Ltd
17. Zellers SJH, Bradford WC, Snowden CH, Knoedler RJ, Alexander JE, Coalson MB, Sonkin J, Foley C, Stinson W, Tansey P, Smith JF, Quinn J, Williams K, Whitfield-Smith L (2024) Airport energy resiliency roadmap. Transportation Research Board, Washington, DC
18. Xianliang G, Jingchao X, Zhiwen L, Jiaping L (2021) Analysis to energy consumption characteristics and influencing factors of terminal building based on airport operating data. In: Sustainable energy technologies and assessments, vol 44. 4 citations (Crossref) [2023-09-25] Elsevier Ltd
19. Gu X, Xie J, Huang C, Ma K, Liu J (2022) Prediction of the spatiotemporal passenger distribution of a large airport terminal and its impact on energy simulation. In: Sustainable cities and society, vol 78, p 103619. 5 citations (Crossref) [2023-09-25]
20. Gu X, Xie J, Huang C, Liu J (2022) A spatiotemporal passenger distribution model for airport terminal energy simulation. In: Indoor and built environment, vol 31, pp 1834–1857. 1 citations (Crossref) [2023-09-25]
21. Li Z, Zhang J, Mu S (2023) Passenger spatiotemporal distribution prediction in airport terminals based on insect intelligent building architecture and its contribution to fresh air energy saving. In: Sustainable cities and society, vol 97, p 104772. 0 citations (Crossref) [2023-09-25]
22. Lin L, Liu X, Liu X, Zhang T, Cao Y (2023) A prediction model to forecast passenger flow based on flight arrangement in airport terminals. In: Energy and built environment, vol 4, no 6, pp 680–688. 1 citations (Crossref) [2023-09-25] KeAi Communications Co
23. Janssen S, Sharpanskykh A, Curran R (2019) Agent-based modelling and analysis of security and efficiency in airport terminals. In: Transportation research part C: emerging technologies, vol 100, pp 142–160. 18 citations (Crossref) [2023-12-04]
24. Zhang X, Chen G, Shi J (203) Simulation of guided crowd evacuation scheme of high-speed train carriage. Int J Simul Modell 22(1):110–120. 0 citations (Crossref) [2023-09-25] DAAAM International Vienna
25. Lahoti N (2022) How the metaverse will change transportation and future mobility
26. Darvishi M, Ahmadi G (2014) Validation techniques of agent based modelling for geospatial simulations. Int Arch Photogrammetry Remote Sens Spatial Inf Sci XL-2-W3:91–95. 8 citations (Crossref) [2023-12-04] Conference Name: WG II/1, WG II/4, ICWG II/IV, WG IV/7, The 1st ISPRS International Conference on Geospatial Information Research (Volume XL-2/W3) - 15–17 November 2014, Tehran, Iran Publisher: Copernicus GmbH
27. Smaldino P (2023) Modeling social behavior: mathematical and agent-based models of social dynamics and cultural evolution. Princeton University Press. Google-Books-ID: aOS3EAAAQBAJ
28. Scala P, Mota M, Blasco-Puyuelo J, Garcia-Cantu O, Blasco C (2022) A novel validation approach for validating the simulation model of a passengers' airport terminal: case study Palma de Mallorca airport. In: European modeling and simulation symposium, EMSS. 0 citations (Crossref) [2023-09-25]
29. Wilensky U, Rand W (2015) An introduction to agent-based modeling: modeling natural, social, and engineered complex systems with NetLogo. MIT Press, Google-Books-ID, LQrhBwAAQBAJ

30. Carbo J, Sanchez-Pi N, Molina JM (2018) Agent-based simulation with NetLogo to evaluate ambient intelligence scenarios. J Simul 12:42–52. 9 citations (Crossref) [2023-09-25] Publisher: Taylor & Francis _eprint: https://doi.org/10.1057/jos.2016.10 Northwestern: https://ccl.northwestern.edu/2016/carbo.pdf
31. OMG (2022) Agent and Event Metamodel (AgEnt) joint revised submission. "AgEnt" Team 0.4-01, Object Management Group
32. Vlăduţescu, Stănescu G (2023) Environmental sustainability of metaverse: perspectives from Romanian developers. Sustainability (Switzerland) 15(15)
33. Khalaj O, Jamshidi M, Hassas P, Mašek B, Stadler C, Svoboda J (2023) Digital twinning of a magnetic forging holder to enhance productivity for industry 4.0 and metaverse. Processes 11(6)
34. Ma Z (2023) Energy metaverse: the conceptual framework with a review of the state-of-the-art methods and technologies. Energy Inf 6(1)
35. Negi P, Singh R, Gehlot A, Kathuria S, Thakur A, Gupta L, Abbas M (2023) Specific soft computing strategies for the digitalization of infrastructure and its sustainability: a comprehensive analysis. Arch Comput Methods Eng
36. Naderi H, Shojaei A (2023) Digital twinning of civil infrastructures: current state of model architectures, interoperability solutions, and future prospects. Autom Constr 149
37. Pamucar D, Deveci M, Gokasar I, Tavana M, Köppen M (2022) A metaverse assessment model for sustainable transportation using ordinal priority approach and Aczel-Alsina norms. In: Technological forecasting and social change, vol 182
38. Deveci M, Pamucar D,Gokasar I, Martinez L, Köppen M, Pedrycz W (2024) Accelerating the integration of the metaverse into urban transportation using fuzzy trigonometric based decision making. In: Engineering applications of artificial intelligence, vol 127
39. Grupac M, Negoianu A-E (2023) Immersive extended reality and sensor-based object recognition technologies, socially-oriented location tracking and simulation modeling tools, and artificial vision and haptic augmented reality systems in the metaverse interactive environment. Rev Contemporary Philos 22:226–243
40. Grieves M, Hua E (2024) Defining, exploring, and simulating the Digital Twin metaverses. In: Digital Twins, simulation, and metaverse: driving efficiency and effectiveness in the physical world through simulation in the virtual worlds, Forthcoming. Springer, Heidelberg
41. Xu Z, Bai Q, Shao Y, Hu A, Dong Z (2022) A review on passenger emergency evacuation from multimodal transportation hubs. J Traffic Transp Eng (English Edition) 9:591–607. 2 citations (Crossref) [2023-09-25]
42. Wooldridge M, Jennings NR (1995) Intelligent agents: theory and practice. In: The knowledge engineering review, vol 10, pp 115–152. 3452 citations (Crossref) [2023-09-25] Cambridge University Press
43. Benfield SS, Hendrickson J, Galanti D (2006) Making a strong business case for multiagent technology. In: Proceedings of the fifth international joint conference on autonomous agents and multiagent systems, (Hakodate Japan). ACM, pp 10–15. 21 citations (Crossref) [2023-09-25]
44. Airports Council International (2019) ACI-ACRIS semantic model of SOA, tech rep, Airports Council International (ACI)
45. Grimm V, Berger U, Bastiansen F, Eliassen S, Ginot V, Giske J, Goss-Custard J, Grand T, Heinz SK, Huse G, Huth A, Jepsen JU, Jorgensen C, Mooij WM, Muller B, Pe'er G, Piou C, Railsback SF, Robbins AM, Robbins MM, Rossmanith E, Ruger N, Strand E, Souissi S, Stillman RA, Vabo R, Visser U, DeAngelis DL (2006) A standard protocol for describing individual-based and agent-based models. Ecol Modelling 198(1–2):115–126. 1896 citations (Crossref) [2023-09-25]
46. Alexandridis KT (2006) Exploring complex dynamics in multi agent-based intelligent systems: theoretical and experimental approaches using the Multi Agent-based Behavioral Economic Landscape (MABEL) model. Ph.D. Dissertation, Purdue University, West Lafayette, Indiana, May 2006. ISBN: 9780542864230 Publication Title: Forestry & Natural Resources Volume: Doctor of Philosophy 3232142

47. Alayande S, Airports Council International (2017) ACRIS semantic model guide to the knowledge layer. Technical Report 1.1, Airports Council International (ACI)
48. Tenkanen H, Toivonen T (2020) Longitudinal spatial dataset on travel times and distances by different travel modes in Helsinki Region. Sci Data 7:77. 16 citations (Crossref) [2023-09-25] Number: 1. Nature Publishing Group
49. Agresti A (2003) Introduction: distributions and inference for categorical data. In: Categorical data analysis, Wiley Series in Probability and Statistics. John Wiley & Sons - Wiley InterScience, 2nd edn, pp 1–35, Section: 1
50. Crane R, Sornette D (2008) Robust dynamic classes revealed by measuring the response function of a social system. In: Proceedings of the national academy of sciences, vol 105, pp 15649–15653. 535 citations (Crossref) [2023-09-25]
51. Wai Sk, WaiShiang C, Khairuddin MAB, Bujang YRB, Hidayat R, Paschal CH (2021) Autonomous agents in 3D crowd simulation through BDI architecture. JOIV : Int J Inf Visualization 5:1–7. 2 citations (Crossref) [2023-12-04] Number: 1
52. Palanca J, Rincon JA, Carrascosa C, Julian VJ, Terrasa A (2023) Flexible agent architecture: mixing reactive and deliberative behaviors in SPADE. Electronics 12:659. 1 citations (Crossref) [2023-12-04] Number: 3. Multidisciplinary Digital Publishing Institute
53. Poslad S (2007) Specifying protocols for multi-agent systems interaction. ACM Trans Autonomous Adaptive Syst 2:15–es. 94 citations (Crossref) [2023-12-04]

# The First Real-Time Digital Twin of a Nuclear Reactor

**John Darrington, Ben-oni E. Vainqueur, and Christopher Ritter**

**Abstract** In this chapter we examine the first fully operation digital twin of a fissile nuclear reactor. In 2023, Idaho National Laboratory (INL) partnered with Idaho State University (ISU) to create and run a digital twin of their AGN-201 fissile nuclear reactor. This reactor can produce up to 5 watts and has proven to be the perfect testbed for discovering how best to implement a digital twin of a larger reactor system. INL created an integrated, cloud-based digital twin capable of measuring data from the reactor in near real-time and built the foundation for running operations such as machine learning and analytics in near real-time in an offsite location. This experiment is a key step in building a digital twin of a larger reactor system and has helped highlight many potential pitfalls and problems that such an endeavor might face. This experiment has also shown the great promise that a cloud-first approach has when creating digital twins.

**Keywords** Red vs. blue · Surrogate models · Isolation forest · Nuclear proliferation · Augmented reality · Mixed reality · Virtual reality · Real-time capture · Data analysis · Azure cloud · Microsoft azure · Grownet · Rust · Digital twin security · Aqueous seperations · Anomaly detection system · Physics accessibility · Microsoft hololens

## 1 A Brief History of INL's Nuclear and Digital Twin Work

The Idaho National Laboratory (INL), established post-World War II, has played a significant role in the development of nuclear technology, and now Digital Twins. Initially named the National Reactor Testing Station, it was the site where, in 1951, the Experimental Breeder Reactor I (EBR-I) first generated electricity from nuclear power.

J. Darrington (✉) · B. E. Vainqueur · C. Ritter
Digital Engineering Battelle Energy Alliance, National Laboratory USA, Idaho Falls, Idaho, USA
e-mail: john.darrington@inl.gov



M. Grieves and E. Hua (eds.), *Digital Twins, Simulation, and the Metaverse*, Simulation Foundations, Methods and Applications, https://doi.org/10.1007/978-3-031-69107-2_10

During the 1960s and 1970s, INL focused on researching advanced reactor designs and improving nuclear safety. By the 1980s, the laboratory shifted its attention to environmental management, particularly the remediation of nuclear waste.

In recent decades, INL has continued to work on nuclear reactor technology, emphasizing extending reactor lifespans, advancing next-generation reactors, and contributing to small modular reactor (SMR) research. It also engages in cyber-security for nuclear facilities and supports nuclear nonproliferation and security initiatives.

Today, INL is involved in developing nuclear energy technologies with an increased emphasis on addressing climate change. Throughout its history, INL has had a substantial influence on the evolution of nuclear energy policy and infrastructure in the United States.

For the past 3–5 years, the Idaho National Laboratory (INL) has been championing the use and creation of DTs in the nuclear industry. This consists of developing virtual DTs to evaluate potential reactor misuse, and for diversion scenarios for sodium fast reactors and high-temperature Pebble Bed Reactors [1].

According to the United States Department of Defense, Digital Engineering is a comprehensive approach that relies on authoritative sources of system data and models, seamlessly spanning multiple disciplines to support activities throughout a product's entire lifecycle, from conceptualization to disposal [2]. Unlike traditional static documents, digital engineering emphasizes the use of interconnected and dynamic models to describe a product. These models are integrated across diverse platforms to facilitate the development of a design product that can be effectively supported throughout its operational life [3, 4].

Digital Twins leverage the principles of digital engineering to create a model that represents either an existing physical product or the design of a future physical product [4]. This concept allows for the creation of two distinct types of Digital Twins: a virtual Digital Twin, which simulates the behavior of a physical product through modeling and simulation, and a Digital Twin of a physical system, which combines a physical asset with computational models to provide a virtual representation of the asset.

## 2 The Impact of Digital Twins in the Nuclear Sector

The adoption of Digital Twins has begun to revolutionize various aspects of nuclear plant operations, maintenance, and training [1, 5, 6]. Plant reference simulators serve as a rudimentary form of Digital Twins, providing operators with a training tool. However, these simulators are not directly connected to the actual power plant. Furthermore, Digital Twins are increasingly being applied in the realm of condition monitoring. Online performance and condition data can be collected through sensors, which are then processed by various artificial intelligence and machine learning algorithms to provide valuable insights.

Training of operators has seen improvements using plant reference simulators, which, despite acting as a form of Digital Twin, have their limitations due to the lack of a real-time connection with the actual nuclear power plants. However, these simulators are critical for providing operators with a realistic, yet controlled environment to hone their skills without impacting plant operations.

Additionally, the integration of DTs is becoming increasingly prevalent within the current nuclear fleet, marking a significant advance in enhancing operational efficiency, maintenance practices, and training methodologies [1, 5, 6]. Digital Twins serve as dynamic, virtual models of the nuclear plants and can significantly aid in simulating plant processes, facilitating a proactive approach to plant management.

Furthermore, the application of DTs extends into the realm of condition monitoring. By employing an array of sensors, continuous streams of performance and condition data are captured in real-time. This influx of data is not merely collected; it is analyzed and leveraged by advanced AI/ML algorithms. The result is a highly informed, data-driven approach to monitoring the plant's health, which can predict potential issues, optimize performance, and reduce downtime. By tapping into the capabilities of AI and ML, the data harnessed through DTs is transformed into actionable insights, allowing for smarter, more efficient plant management. This integration points to a future where digital and operational technologies converge to create a more reliable and safe nuclear power infrastructure.

### *2.1 Aqueous Separations*

Another notable application of Digital Twins in development pertains to aqueous metal separations. The Idaho National Laboratory (INL) is actively involved in the creation of a Digital Twin that emulates an aqueous separation process utilizing centrifugal contactors. This Digital Twin is expected to play a vital role in monitoring, implementing models, and ensuring the security of separations equipment and processes. The preliminary development of the Digital Twin is taking place at INL's Solvent Extraction Laboratory, where two teams are concurrently engaged in its development. One team is integrating conventional sensors, such as optical sensors, while the other team is incorporating nontraditional sensors, such as acoustic sensors. To achieve the ultimate objective of a fully functional Digital Twin, various components are being integrated into a unified system through the creation of adapters. These adapters connect a central data repository, DeepLynx, to various project nodes, including LabView data streams, machine learning models, chemical models, international safeguards support, and visualizations.

The fusion of Digital Engineering and Digital Twins (DTs) is revolutionizing the nuclear sector, transforming product conceptualization and design and leading to an era marked by increased efficiency, precision, and innovation. These initiatives not only deepen our grasp of sensor integration and analytical methods but also yield modular components that are adaptable for a variety of uses. Additionally, they establish a template for data-driven evaluations that are instrumental in the creation and oversight of emerging aqueous metal separations technologies.

## 2.2 Design

DTs offer transformative potential for nuclear reactor design, particularly during the basic and final stages. These sophisticated virtual models can be instrumental in evaluating both the reactor's nuclear components for safeguards and the Nuclear Power Plant's (NPP) overall security. A pivotal role of a DT at this stage is to ensure the design incorporates the necessary infrastructure for security and safeguards monitoring, including strategic placement of sensors within the reactor core and facility to maintain operational knowledge.

## 2.3 Security

Moreover, DTs can facilitate the early identification of security vulnerabilities, enabling the International Atomic Energy Agency to address these issues proactively, thus mitigating the expense of subsequent modifications. As a design progresses, a specialized DT can be crafted to serve a dual role, bridging operational processes with domestic and international safeguard and security protocols.

Operational DTs enhance the efficacy of both international and domestic inspections by pinpointing potential discrepancies for on-site verification and help maintain regulatory compliance. They enable real-time monitoring, promptly notifying authorities of abnormal events. Crucially, DTs inform security strategies, differentiating between deceptive tactics and actual threats, thus guiding responsive measures—a consideration increasingly relevant for remote security management of small autonomous reactors. The use of DTs allows access to information that is critical for accurately predicting the diversion or misuse of a significant quantity of fuel, through the collection of control rod data and a selection of assembly power over the span of a year for sodium fast reactors [5]. For PBR reactors, sufficient statistical analysis of diversion scenarios can be determined using gamma analysis [6].

In addition to advancements in nuclear safeguards, researchers have been exploring the potential of advanced sensors and their ability to transmit data directly to both a monitoring agency and the reactor itself [7]. Research in the field of industrial control systems has the potential to enable a framework in which both the reactor operator and the IAEA can receive independent streams of information. While this research may not be directly related to safeguards, it plays a crucial role in elucidating the intricacies of sensor-based monitoring for Small Modular Reactors (SMR) and Medium Reactors (MR) systems.

Previous collaborations with the IAEA inform optimal sensor placements, benefiting both operational safety and compliance monitoring. Thus, a DT embodies an integrated approach to the nuclear sector's safety, security, and safeguards, commonly referred to as the '3S by design' concept, enhancing oversight and ensuring continuous improvement.

## 3 Physical Description of the AGN-201

The AGN-201 is a compact nuclear reactor with a power output of 5.0 watts, situated at Idaho State University. This reactor has been designed for both research purposes and practical training [8]. Its structure comprises two primary components: the core region and the ex-core elements.

The core region encompasses nine fuel disks that consist of a homogeneous mixture of polyethylene and uranium dioxide fuel, enriched uranium. The core's approximate dimensions are 24 cm in height and 25.6 cm in diameter [68]. The core is divided into two sections connected by a thermal fuse made of polystyrene, which contains twice the fuel loading of the fuel disks. The thermal fuse serves as a safety feature, it will melt and separate the core into two pieces, thereby terminating the chain reaction if the temperature rises above a specified threshold.

To sustain criticality, four fuel control rods are inserted into the core. Two safety rods remain fully always inserted, while a coarse control rod (CCR) and a fine control rod (FCR) are employed to adjust, reduce, or maintain the reactor's power output. Additionally, there is a central irradiation facility passing through the core's center, allowing for experiments to be directly inserted into the core.

The core is enclosed by a graphite reflector contained within an aluminum core tank. Surrounding this tank is an additional layer of graphite reflector, followed by a lead shield, and finally, a water shield. Above the core, there is a removable graphite thermal column that can be taken out for experimental purposes, and below the core are the control rod drive mechanisms. The complete configuration of the AGN-201 reactor is illustrated in Fig. 1.

## 4 Risks and Hurdles, Challenges in Implementing Digital Twins

There are numerous advantages of using DT technologies. They contain the ability to encompass many of the necessary international safeguards and security of reactors, but there are also challenges that arise when implementing a DT. Using any digital product such as a Digital Twin, artificial intelligence, and machine learning, risk and uncertainty are introduced into the system, and can possibly become a high risk. Some of the challenges of using these products are:

1. Absence of regulatory guidance and requirements for compliance and acceptance of use.

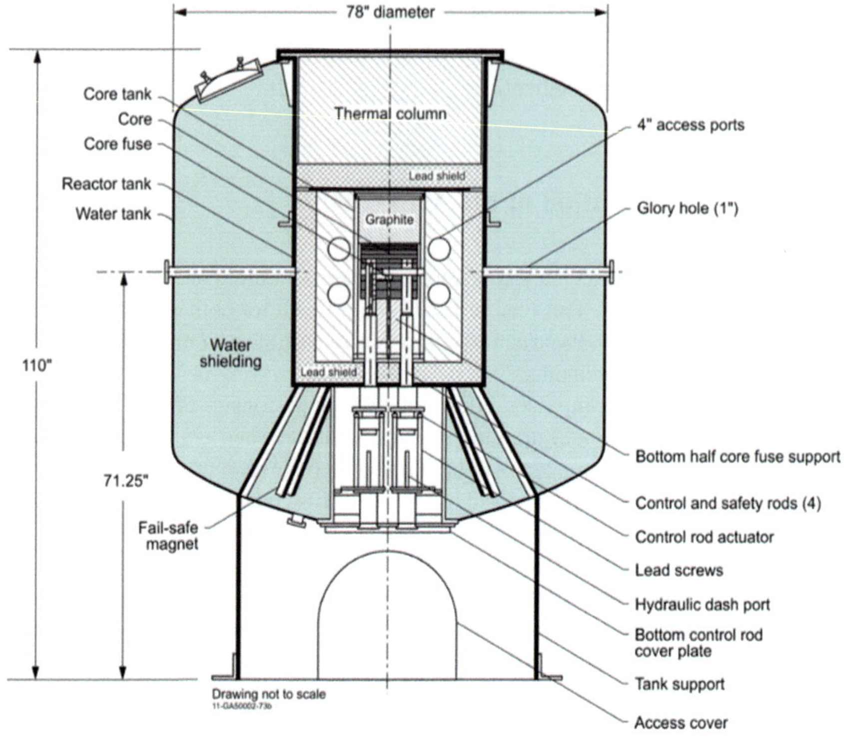


**Fig. 1** AGN-201 reactor, reflector, and shielding core

2. Absence of any meaningful acceptance of international standards for design (including cybersecurity) of DTs, AI, and ML.
3. Ensuring the workforce has the essential skills to design, evaluate, implement, sustain, and update technologies as needed.
4. Performing the proper verification and validation of its performance over the facility lifecycle.
5. Assuring user trust through proof, explanations and recurrent performance testing/evaluation.

## 5 Technical Implementation

Digital Engineering and the development of digital twins necessitate the use of highly sophisticated tools and software [26]. DTS allow the early detection, deterrence, and response to potential threats by leveraging AI/ML algorithms for real-time monitoring of reactor data streams. Idaho State University and the Idaho National Lab partnered up to create a digital twin centered around the AGN-201 reactor.

The digital twin is composed of several components. First, data is acquired through the Data Acquisition System (DAS) and uploaded into the open-source INL-developed data lake DeepLynx[1] through the INL-developed command line tool called Jester. Once data has been ingested into the data lake, various machine learning and AI operations occur on the data and feed their results back into the data lake. These results are available on the data lake's web interface or through a beta operator windows program also developed by INL. Additionally, this data fuels user interaction through augmented reality and a real-time model of the reactor and the data through INL-developed augmented reality programs.

Each section of this chapter will highlight a specific part of the diagram below, as well as diving into the reasons behind certain technical decisions.

## 5.1 *Jester*

Jesteris an open-source tool developed by Idaho National Laboratory that transfers data from sensor systems to the DeepLynx data lake. This software was developed and open-sourced as part of the ISU digital twin efforts but was designed with modularity and extendibility in mind. Jester has an extensive plugin system which allows it to work on a myriad of different systems and with different file types.

Both Jester and an ISU specific plugin were developed as part of this effort. This plugin is responsible for working with Jester to inform how the AGN-201's data acquisition system outputs the sensor readings and how to take those readings and ingest them into DeepLynx. This plugin is written in Rust and is designed to tail various CSV files output by LabVIEW and send them on an interval to the DeepLynx data lake.

### Rust

Rust is a multi-paradigm, general-purpose programming language that emphasizes performance, type safety, and concurrency. It enforces memory safety, ensuring that all references point to valid memory, without requiring the use of a garbage collector or reference counting present in other memory safe languages. Rust's rich type system and ownership model guarantee memory safety and thread-safety—enabling you to eliminate many classes of bugs at compile-time.

Rust was the preferred language due to having a smaller learning curve compared to C++ , and a faster processing time than Python. Having a fast runtime and memory safety are extremely important since the AGN-201 reactor's DAS machine is small and has limited memory. Choosing the Rust programming language allowed ISU and INL the ability to safely and securely work with a nuclear reactor to power the digital twin.

[1] https://github.com/idaholab/Deep-Lynx.

## 5.2 DeepLynx

DeepLynx is an open-source data lake focused on enabling complex projects to embrace digital engineering. Unlike other data warehouses, DeepLynx users can use an ontology to custom-define how their data will be represented. DeepLynx enables users to store their data in a graph-like format, ensuring that connections between data can be easily seen and understood. This data lake also allows users to store tabular, or time-series, data such as the data coming in off of sensors and IOT.

DeepLynx is hosted on Microsoft Azure for Government cloud platform leveraging the Azure Kubernetes system (AKS) to deploy and manage infrastructure. The Microsoft Azure cloud platform grants DeepLynx the ability to connect all other processes and software developed for this digital twin (Fig. 2). It is important to note that while fast, this process is not considered "real-time" but "near real-time" due to the network latency between each process communicating over the web (Fig. 3).

**PostgreSQL**

PostgreSQL a free and open-source object-relational database management system (RDBMS) that can store and load data over time through associated pairs of times and values. PostgreSQL possesses an incredible number of features to enhance performance, security, programming extensions, and configuration. PostgreSQL powers

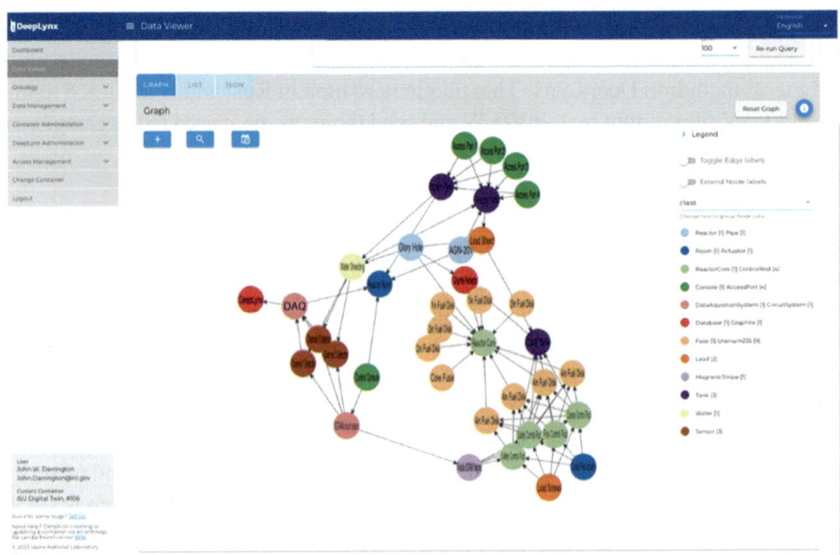


**Fig. 2** Data flows into the data acquisition system from the reactor. Jester is used to push data from the DAS into a DeepLynx cloud instance. The data is analyzed using machine learning algorithms implemented through Juypter notebooks managed through Papermill, and pushes those results back into DeepLynx, which then can be viewed by users through DeepLynx's UI or the operator Window's program

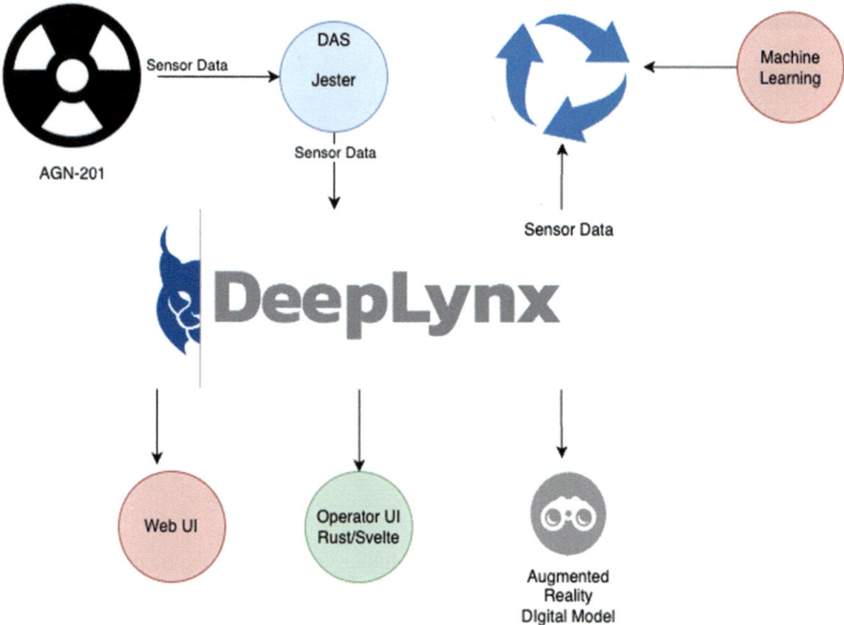


**Fig. 3** DeepLynx representation of the AGN-201 in graph format. This representation of the data allows a human readable view of the time-series data collected from the AGN-201 reactor

data mining to obtain patterns or models from the gathered data from all kinds of environments [25]. It has the power that relational database management systems have with the bonus of object-oriented features such as defining complex data types, overloading functions to work with different argument data types, and defining inheritance relationships between tables.

**Timeseries Data**

DeepLynx is designed to not only store a graph representation of the AGN-201 reactor, but also serves as the storage for the data read from the sensors themselves, as previously stated. Currently, DeepLynx leverages TimescaleDB for its storage and querying of time-series or tabular data. This is another open-source technology which was adopted by the DeepLynx team.

Data is processed only in near real-time due to network connectivity via HTTPS. The different components of Fig. 1 are connected via TCP sockets using the HTTPS format which is not necessarily the best for high throughput data. This stops the system from producing real-time results. Currently the system runs in about thirty second loops, including the machine learning and AI portions of the flow (Figs. 4 and 5).

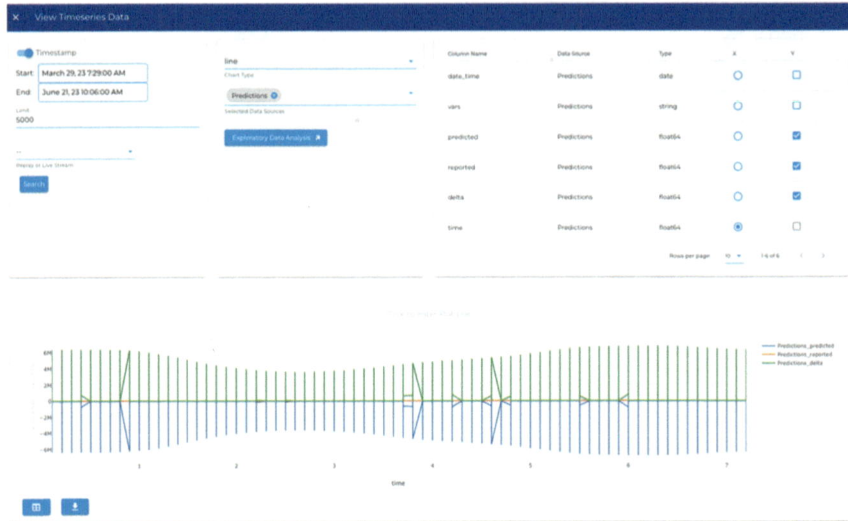


**Fig. 4** DeepLynx representation of data from the Multilayer perceptron algorithms

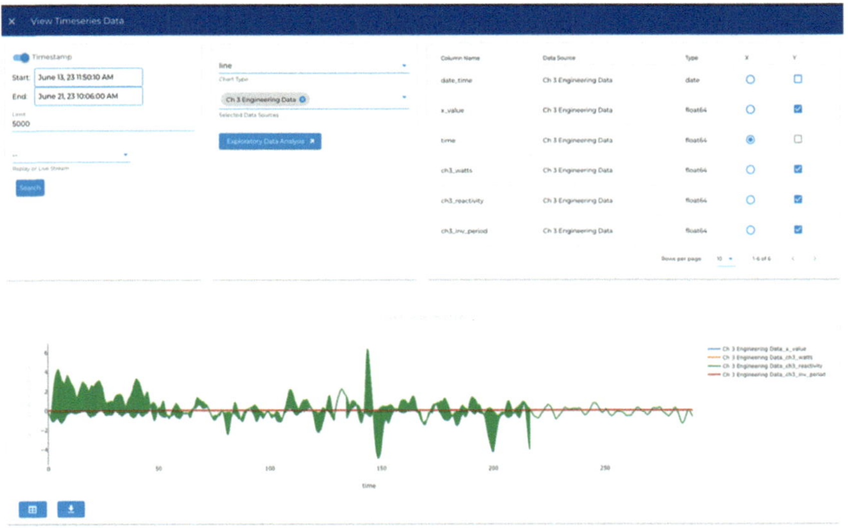


**Fig. 5** DeepLynx representation of AGN-201 Engineering sensor data

## 5.3 *Visualization Technology*

We have made significant progress in realm of digital twin visualization for the AGN-201 project. We have developed programs using various tools to enable operators and monitors to easily identify anomalies and issues in reactor operation.

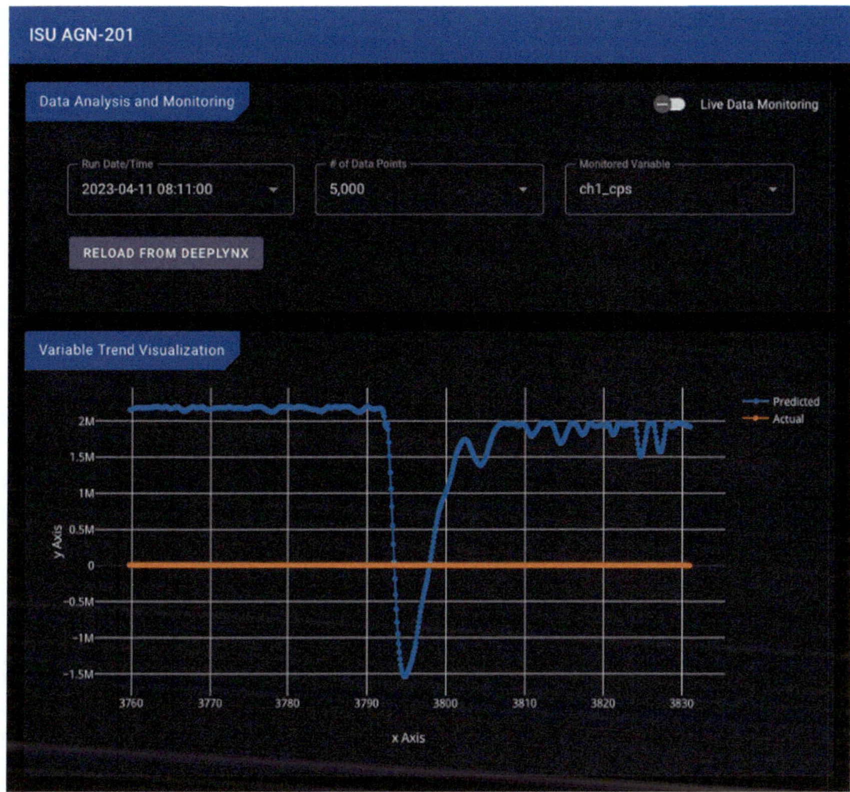


**Fig. 6** The operator UI allows operators to view trends and monitor the reactor in a human readable format. This is a screenshot of the Windows desktop application

## Operator UI

The operator UI is a beta desktop application designed to give insights into reactor function to a reactor operator and monitor. This system incorporates the machine learning data ingested by DeepLynx from the program listed previously and displays it to the end user in an easy to digest format. This desktop program will allow operators to digest the data from the prediction models and use that information to understand risks and problems in the reactor proactively. The operator UI was created by using a combination of Tauri,[2] an open-source tool that builds optimized, secure, and frontend-independent application for multi-platform deployment, and Svelte,[3] an open-source front end component framework/language (Figs. 2 and 6).

[2] https://tauri.app/.

[3] https://svelte.dev/.

**Augmented Reality Using Unity and HoloLens**

Microsoft HoloLens is an augmented reality (AR)/mixed reality (MR) headset with positional tracking technology developed and manufactured by Microsoft. In combination with Unity, a cross-platform game engine, an AR interactive view of the AGN-201 reactor can be used by operators to see data that flows in. The AR systems pulls in the predictions for the raw data from Deep Lynx and displays it through the HoloLens using unity.

## *5.4 Machine Learning*

A key aspect of the AGN-201 digital twin is the machine learning and artificial intelligence detection programs. These programs take the live data from the reactor, pulling it directly from DeepLynx, and performs various tasks to determine different things. We will discuss the actual models and mathematics later on this chapter, for now we will focus only on the technology involved in the operations.

**Papermill**

Papermill[4] is an open-source tool that allows you to execute Jupyter Notebooks an parameterize them. This allows the execution of workflows without manually having to combine different notebooks. INL-developed a tool in Rust to manage Papermill's execution of various Jupyter notebooks (Fig. 1, Sect. 2.1.1). This allows the system to use the memory efficiency/security of Rust, while leveraging the power/flexibility of Python for machine learning. This allows us to maintain a highly available, highly resilient system while at the same time letting data scientists work in a familiar system and paradigm. This system is contained in a soon to be open-sourced Docker[5] image.

**DuckDB**

DuckDB[6] is an in-memory process/in process SQL OLAP database management system. It leverages columnar storage which makes it ideal for analytics and allows a simpler data format that data scientists can read. We utilize DuckDB as a temporary store for the data from DeepLynx, putting it into an easily queried and read paradigm that allows data scientists to focus on their algorithms instead of the data fetching and network concerns of communicating with DeepLynx.

---

[4] https://papermill.readthedocs.io/.

[5] https://www.docker.com/.

[6] https://duckdb.org/

### 5.5 Conclusion

INL and ISU was able to build a DT with near real-time system for anomaly prediction and general analysis of the AGN-201 reactor. There were several lessons learned from this process that will continue to advance the understanding of how DTs are implemented. First, the DT is near real-time due to the limitations of data transmission through https and security firewalls. This constrained the DT to leverage batch processing instead of streaming the data from the AGN-201 reactor. Second, rust is a newer language and needed a lot of debugging to establish a dependable interop between rust and python to effectively manage and analyze reactor data being processed by Deep Lynx.

## 6 Physics Model Design "Physics-Based Surrogate Models"

**Citation Key**: [2] = ~ \cite{scikit-learn}.

The creation of a surrogate model (SM) for simulating a nuclear reactor involves two main components: a mathematical model for reactor physics and a simplified model for the reactor's kinetic equations (PKE-SM). This model gets its information from extensive simulations that mimic both normal and abnormal reactor operations.

Using a high-fidelity physics modeling software called Serpent, these simulations consider factors like water temperature and control rod positions to determine the reactor's status—whether it's operating normally (critical), shut down (subcritical), or in an unsafe condition (supercritical). The results from these simulations are fed into algorithms to create the SM, which can predict the reactor's behavior based on those inputs.

During the reactor's operation, the SM continuously checks conditions like water temperature and control rod positions to ensure the reactor is functioning as expected. If these parameters stay within a safe range, it signifies that the reactor is in a stable, critical state (Fig. 7).

### 6.1 Model Framework

The model framework consists of two primary components: (1) Model Development and (2) Model Implementation and the models attempt to predict the following: Ch2 (Watts), Ch3 (Watts), FCR (cm), and CCR (cm).

Our methodology in machine learning (ML) serves as an automated system for quality assurance and control (QAQC), as well as a framework for probing different methods. The core of our strategy is the formulation of models that predict outcomes for critical components such as Coarse Control Rods (CCR), Fine Control Rod Heights (FCR), and the power outputs of Channel 2 and Channel 3 (measured in

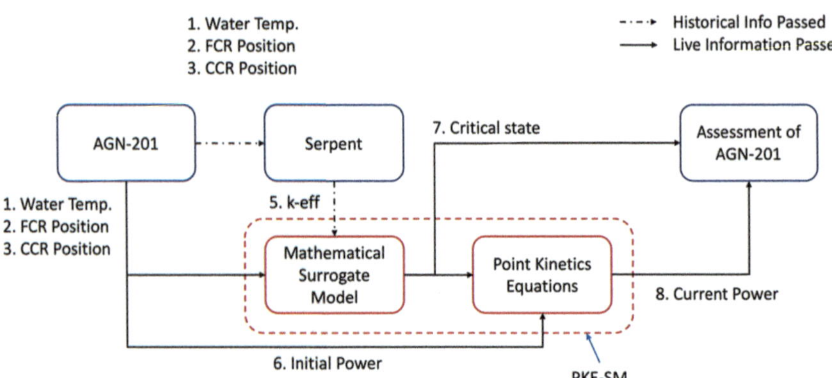


**Fig. 7** Flow of data from the AGN-201 through the mathematical SM and PKE-SM

Watts). We achieve this by employing all available variables, excluding the target variable, to serve as predictors within our models. The predictor variables, detailed in Table 1, include Channel 1 count rate (Ch1, CPS), power readings from Channel 2 (Ch2, W) and Channel 3 (Ch3, W), positional measurements of Fine Control Rods (FCR, cm) and Coarse Control Rods (CCR, cm), Temperature (Temp, °C), and Inverse Period (Inv-P, sec).

In this research, we have adopted three distinct model architectures, namely Multilayer Perceptron (MLP) Regression, GrowNet, and TabNet, each bringing a unique perspective to the analysis of tabular data.

Multilayer Perceptron Regression, drawing upon classical neural network paradigms, serves as a foundational model to set a benchmark for performance comparison [9]. On the other hand, GrowNet is a novel gradient-boosted approach that employs a composite of simpler neural networks, which are iteratively combined to form an intricate and robust model adept at optimizing a variety of loss functions [10]. A notable feature of GrowNet is its comprehensive corrective phase, which fine-tunes the parameters across all contributing networks to enhance the model's predictive capacity.

TabNet, employing a sequential attention mechanism within a deep learning context, stands out by sequentially parsing through variables to identify and prioritize the most significant ones in relation to the overall dataset features [11]. This focus on salient variables allows TabNet to make more informed predictions by effectively allocating computational attention where it's most impactful.

Together, these three methodologies represent the spectrum from tried-and-true to state-of-the-art, encapsulating a broad range of capabilities in tabular data analysis and providing a robust suite of tools for developing alongside digital twins.

## 7 ML Model Development and Framework Design

The model development portion of the framework consists of the following steps:

1. **Training Data Preprocessing**: The training data set is cleaned using a rule-based quality assurance/quality control (QAQC) along with a standard normalization. First, the data is scrubbed by reviewing the data for errors that violate the reactor's known data thresholds, specifically identifying and fixing values that exceed power maximum, and fall underneath reactor power minimums. This also consists of populating any missing and incorrectly formatted missing data (NaN values). Each variable is then adjusted using a standard normalization function. Finally, the mean values and standard deviation values are stored for subsequent access.
2. **Model Training**: The training data is split into training, testing, and validation datasets. This function is serialized so that a unique model is developed for each variable.
3. **Meta Data Storage**: This step allows the model implementation step to access these values for consistent data normalization.
4. **Trained Model Export**: Each trained model is exported. The twin can accommodate any file format that is appropriate for a specific model type.

### *7.1 Model Implementation*

The near real-time model implementation portion of the framework consists of the following steps:

1. **Data Ingestion**: The framework incrementally ingests data via SQL queries, formats it to match the original training data, and stores the information in memory.
2. **Data Preprocessing**: The framework isolates each data time step individually and performs an automated QAQC and standard normalization using the mean and standard deviation values it retrieves from the meta data produced during model development.
3. **Prediction**: The trained model files are accessed and used to predict the values of each variable (FCR, CCR, water temperature, Ch1 power, Ch2 power, Ch3 power) for the isolated and preprocessed data set.
4. **Denormalization**: The model predictions are returned to their original units by undoing the normalization function.
5. **Result Storage**: The model results include the reported and predicted values for each variable along with how those values differ. Under normal conditions while the reactor is critical, the reported and predicted values should be nearly identical. Deviation between those values is indicative that either (1) additional model training is required, (2) sensor failure, or (3) the reactor is not operating

under expected conditions. The model does not attempt to identify which of those three scenarios is taking place, but the discrepancy is stored in the results so that it can be investigated further.

## 7.2 Computation and Using High Performance Computing (HPC)

Creating a nuclear safeguards digital twin leverages the combined power of cloud computing and high-performance computing (HPC). While the bulk of the digital twin safeguards reside on the cloud, they collaborate with the INL HPC for optimization algorithms and physics computations. This setup endows the digital twin with the robust computational strength of HPCs and the flexibility and adaptability of cloud storage.

Simultaneously, a groundbreaking digital twin has been developed, tested, and validated for the autonomous control of live assets within the Deep Lynx ecosystem. This advancement, funded through the INL Laboratory Directed Research and Development (LDRD) initiative, incorporated tools like Deep Lynx, the ML adapter, and physics simulations grounded in the Multiphysics Object-Oriented Simulation Environment (MOOSE). The INL's Microreactor Agile Non-Nuclear Experimental Testbed (MAGNET) serves as a testing platform for microreactor technologies. Notably, on March 30, 2022, the LDRD digital twin was employed in a MAGNET test involving a single heat pipe. The test showcased the digital twin's prowess in forecasting temperature benchmarks and adjusting MAGNET's temperature set point autonomously, enabled by the seamless two-way communication between Deep Lynx and the MAGNET data acquisition system.

The digital twin is slated for hosting on a blend of the Microsoft Azure cloud and HPC infrastructure. Given the rigorous computational requirements of the Serpent simulations, an HPC environment is best suited for rapid simulation processing. While the Serpent simulation is contained within the INL HPC cluster, other aspects of the digital twin will be housed in the Azure cloud. This arrangement offers several advantages, such as scalability, cost-efficiency, enhanced security, and deeper insights into hardware and network dynamics. For seamless data exchange between the cloud and HPC, specific components and a data-flow framework are imperative. For instance, Deep Lynx in the cloud can initiate an HPC processing request. An HPC adapter fetches required files from Deep Lynx and secures them in the Azure storage. Within the HPC, a service node constantly scans the Azure storage for fresh requests and related files. Post-processing, results get transferred back to the cloud storage. This HPC adapter consistently monitors the storage, relaying results back to Deep Lynx once detected.

# 8 Visualization

Extended Reality (XR) encompasses a suite of technologies including Virtual Reality (VR), Augmented Reality (AR), and Mixed Reality (MR) that merge the virtual and physical worlds. VR immerses users in a completely virtual environment, whereas AR enhances reality with virtual overlays, and MR combines the two to facilitate intuitive interactions between human, computer, and environmental elements (Milgram and Kishino, 1994). Each of these technologies offers distinct user experiences. VR immerses users in a completely virtual environment, disconnecting them from the physical world. Idaho National Laboratory (INL) leverages these XR capabilities in projects that span Nuclear, Integrated Energy, and National and Homeland Security missions (Idaho National Laboratory, n.d.).

At the forefront of Digital Engineering, INL integrates visualized data streams, visualizations, and physics-based models to create comprehensive digital twins. These digital twins enable semi-autonomous design, autonomous operation, and real-time anomaly detection, thus transforming complex data into accessible and actionable insights.

## 8.1 The Advantages of Incorporating XR for Data Analysis

Extended Reality, often abbreviated as XR, is an umbrella term encompassing various immersive technologies, such as Virtual Reality (VR), Augmented Reality (AR), and Mixed Reality (MR).

The primary advantage of integrating XR into data analysis processes lies in its ability to facilitate the interpretation and representation of complex data. Traditional data analysis techniques often rely on subject matter experts to manually collect, examine, and report anomalous data. By providing operators with a spatial view of captured data, XR simplifies the process of identifying and plotting data points. This enhancement makes data and anomalies more engaging, enabling operators to readily identify unknown signals in real-time or near real-time.

## 8.2 QR Code Scanning for Precision

QR code scanning plays a significant role in our workflow by enabling precise positioning of virtual assets relative to the physical world. This process allows engineers to conduct design reviews, test ergonomics, fitment, escape routes, and address practical challenges, ultimately enhancing the design and testing phases (Idaho National Laboratory, 2023).

Our interaction libraries, integrated via SDKs, enable us to manipulate objects, create interactive menus, toggle layers, and utilize QR code scanning for precise model positioning within physical spaces. This spatial modeling approach facilitates design reviews, ergonomic assessments, and fitment evaluations. It also addresses unforeseen challenges, such as navigating tight spaces, thereby enhancing the overall design process. This integration allows the visualization of dynamic states such as flow rates, temperature heat maps, and fuel rod positions, thus eliminating the need for intermediate data interpretation (Smith et al., 2021).

## 8.3 Unity and OpenXR

Utilizing the game creation engine Unity (Unity Technologies n.d.), INL has developed a robust environment for visualizing digital twins. This platform was chosen for its versatility in building projects across various platforms and its supportive community.

INL begins by converting Computer-Aided Design (CAD) models into a mesh that is easier to render using PIXYZ, a tool designed to optimize models by simplifying geometry without compromising on detail (Unity Technologies n.d.-a). The primary advantage of integrating XR with data analysis lies in its capacity to facilitate the interpretation and representation of intricate datasets. Conventional approaches often rely on subject matter experts to manually collect, examine, and report anomalous data. Our approach, however, offers spatial visualization capabilities, simplifying the process of data identification and plotting. This real-time or near real-time engagement enables operators to swiftly detect and understand unknown signals and anomalies, elevating our decision-making capabilities.

While HoloLens 2 remains our primary choice, the industry's direction emphasizes cross-platform compatibility. OpenXR platform enables headset makers to define hardware specifications while utilizing standardized interaction methods, ensuring that applications can be developed for multiple headsets with ease (Idaho National Laboratory 2023). Following the conversion process, these assets are imported into Unity, where INL engineers craft interactive experiences tailored for XR headsets like the HoloLens 2 (Microsoft n.d.), known for its Mixed Reality Toolkit (MRTK) that facilitates the development of immersive applications with features such as hand tracking, hand poses, speech recognition, eye tracking, gaze tracking, speech recognition, and other interaction methods (Microsoft n.d.-a). Alternatively, the industry is moving toward the OpenXR standard, which promises cross-platform compatibility for XR applications (The Khronos Group n.d.).

### 8.4 Connecting Real-Time Data with Digital Twins

To achieve a comprehensive digital twin, we interconnect real-time data from existing assets, a process successfully applied to MAGNET and the ISU Twin. Utilizing data acquisition (DAQ) systems and importing data into DeepLynx is our preferred method. DeepLynx seamlessly integrates data, providing real-time insights into system states. XR headsets eliminate the need for data interpretation, allowing users to visualize the system's status intuitively. This visualization includes flow rates, temperature or concentration heat maps, fuel rod positions, conflict areas, and dynamic animations of construction processes. Thanks to DeepLynx, these visualizations can be dynamically updated as new data flows in, accessible across various platforms.

Unity's "GameObjects" and their associated components and scripts, written in C#, facilitate the creation of interactive digital twins. These objects can represent various 3D objects, text, or abstract constructs within the scene hierarchy (Unity Technologies n.d.-b).

### 8.5 Collaborative XR Experience with Serval Multiplayer

An innovative feature utilized by INL is the Serval Multiplayer server, which enables collaborative sessions by synchronizing state data among multiple users, even though they operate in individual sessions. By allowing users to scan the same QR code, they can view models in synchronized positions while maintaining individual sessions, ensuring privacy. This enhances the collaborative process, allowing simultaneous interaction with and visualization of the digital twin (Serval Project n.d.).

In conclusion, INL's integration of 3D visualization and XR technologies into digital twins represents a significant advancement in the field, offering enhanced data interpretation, real-time operational insights, and improved collaborative capabilities.

## 9 Case Study of the Impact the AGN-201 Digital Twin: Red Versus Blue Test and Conclusion

**Necessary Terminology**

CCR Coarse control rod.
DT Digital twin.
FCR Fine control rod.
IFML Isolation forest machine learning.
INL Idaho National Laboratory.

ISU Idaho State University.
ML Machine learning.
PKE-SM Point kinetics equations surrogate model.

## 9.1 Overview

In the past year, the reactor operators at Idaho State University (ISU) operated the AGN-201 reactor under various conditions without informing the digital twin (DT) team at Idaho National Laboratory (INL) to perform both offline and online detection of hypothetical proliferation attempts. These data were transmitted to DeepLynx during the operation and stored for analysis by the DT team. During this time, knowledge of the actual operations was kept from the DT team to allow for an offline analysis and estimate of events. Three separate analyses (reactor physics assessment, machine learning predictions, and anomaly detection) were performed by the DT team to examine the DT's ability to detect and quantify off-normal operations. This provided first-of-a-kind online and offline monitoring of a nuclear reactor using a DT, proves the impact, and necessity of the utilization of digital twin's in the realm of security, and validation of critical systems.

## 9.2 AGN-201 Operations

The red/blue team test involved the ISU reactor supervisor (red team) modifying the operational parameters of the AGN-201 reactor, within regulatory limits but in ways the DT had not previously seen, to simulate a potential bad actor nefariously gaining access to the reactor to produce a dangerous isotope or other illicit use. The DT analysts (blue team) then analyzed the resulting outputs from the reactor and determined if/when anomalous actions were taken.

The test consisted of three separate alterations to the reactor with different amounts of reactivity change. From highest to lowest reactivity change, the list of alterations to the reactor is:

- Inserting a polyethylene rod into the central irradiation facility.
- Inserting a cadmium foil cover into the central irradiation facility.
- Removing two graphite blocks from the thermal column.

All the alterations were performed while the reactor was operating, and the operator and reactor supervisor tried to make these changes as hidden as possible.

After a sequence of first determining if the reactor was critical using a fine control rod (FCR) and a coarse control rod (CCR), a "nefarious" was emulated by placing cadmium foil cover into the central irradiation facility to act as a poison. The cadmium was inserted into the reactor approximately 20 min after the polyethylene rod was removed, resulting in the FCR.

## 9.3 *Online Detection of Reactor Operations*

To enhance the Digital Twin (DT) of the AGN-201 reactor with an effective anomaly detection system, the team implemented the Isolation Forest machine learning algorithm during the Red versus Blue Test. This algorithm is particularly good at spotting unusual patterns in data without the need for previously identified examples of such anomalies.

The team first standardized the data, then fine-tuned the Isolation Forest model to the specific characteristics of the DT's data. They also set a 'contamination parameter' to estimate the expected ratio of anomalies in the data.

As the reactor operates, the algorithm continuously analyzes incoming data, deciding if each piece is normal or not. If it detects what seems to be an anomaly, it marks and logs the occurrence for later review. This system is key in promptly identifying and responding to any irregularities that might arise.

The initial operation was used to assess the acceptability of the reactor physics model and anomaly detection model for predicting off-normal conditions (i.e., simulated proliferation attempts). Given the success of both models, two additional reactor operations were performed to assess the online capabilities of the DT system. This ensured that data was being transferred from the AGN-201 digital acquisition system to DeepLynx, where it could be analyzed in near real-time. This process allowed the reactor operators at ISU to operate for approximately an hour, and then within the next 20 min, the data could be assessed and discussed between the two teams. The ability to provide a near real-time assessment of reactor operations was a first for a DT in nuclear reactor operations.

## 9.4 *Takeaways from Red Versus Blue Test*

Overall, the initial red/blue team exercise was a success in showing that a Digital Twin framework could effectively transfer data from AGN-201 via DeepLynx and analyze it for off-normal operations. This was demonstrated for both offline and online operations to provide confidence that the methodologies present would be transferable to the near real-time detection of anomalies.

Through the creation of a digital twin, the reactor physics surrogate and anomaly detection models were able to capture the major anomalous behaviors that the reactor operations team performed. The integrated anomaly detection algorithm within the twin displayed a remarkable aptitude in capturing many of the off-normal operations, homing in on sensor fluctuations to identify divergences from the norm. This achievement is particularly noteworthy given the complexity of the reactor's operational dynamics and the nuanced variations that were successfully flagged, showcasing the algorithm's precise tuning and responsive nature to dynamic operational changes, and the impact of using DT's in this realm.

Future work will entail wrapping in the PKE-SM to enhance the capabilities of the reactor physics surrogate model. This will allow us to capture changes in power as a function of time rather than just examining $k_{eff}$. For the reactor physics models, an additional area of focus will be on determining what levels of perturbation are quantifiable, meaning how small of a perturbation can be detected before it falls within the range of uncertainty in the model, sensor data, etc. For both the reactor physics surrogate and anomaly detection models, we plan to focus on detecting startups, power changes, and shutdowns to prevent flagging these as anomalies.

Through further testing, we were able to provide a near real-time assessment of reactor operations and were able to detect nearly all the anomalous behaviors of the reactor.

## *9.5 Conclusion*

The Digital Twin of the Idaho State University's AGN-201 was a resounding success, despite the difficulties and hurdles in both operation in implementation. The Idaho National Laboratory team was able to successfully create a near real-time twin of a functioning nuclear reactor, a first in the industry. Not only did they create a way for monitoring the twin using readily available cloud services but worked toward creating a framework for monitoring the correct usage of said reactor to stop potential bad actors.

This work will lay the foundation for future digital twins in the Idaho National Laboratory's sphere of influence and has proven that they have the capability to handle larger projects.

## References

1. Ritter C, Hays R, Browning J, Stewart R, Bays S, Reyes G, Schanfein M, Pluth A, Sabharwall P, Kunz R, Shields A, Koudelka J, Zohner P (2022) Digital twin to detect nuclear proliferation: a case study. J Energy Resour Technol 144(10):03, 102108. [Online]. Available: https://doi.org/10.1115/1.4053979
2. "Digital engineering strategy," U.S Department of Defense, Tech. Rep. [Online]. Available: https://man.fas.org/eprint/digeng-2018.pdf
3. Yadav V et al (2021) The state of technology of application of digital twins. U.S. Nuclear Regulatory Commission, Tech. Rep. TLR/RES-DE-REB-2021-01, 2021.[Online]. Available: https://www.nrc.gov/docs/ML2116/ML21160A074.pdf
4. Yadav V et al (2021) Technical challenges and gaps in digital-twin-enabling technologies for nuclear reactor applications. U.S. Nuclear Regulatory Commission Tech Rep TRL/RES-DE-REB-2021-17. [Online]. Available: https://www.nrc.gov/docs/ML2136/ML21361A261.pdf
5. Stewart R, Shields A, Gleicher F, Bays S, Schanfein M, Reyes G, Ritter C (2023) Utilizing a digital twin to analyze and monitor proliferation scenarios in a sodium-cooled fast reactor. J Sci Global Secur (Submitted)

6. Wen S, Stewart R, Shields F, Gleicher A, Bays S, Reyes G, Schanfein M, Ritter C (2023) Utilizing advanced statistics to determine anomalistic conditions in pebble-bed reactors. J Nucl Mater Manage (Submitted)
7. Browning J, Slaughter A, Kunz R, Hansel J, Rolston B, Wilsdon K, Pluth A, McCardell D (2022) Foundations for a fission battery digital twin. Nucl Technol 208(7):1089–1101. [Online]. Available: https://doi.org/10.1080/00295450.2021.2011574
8. Pope C, Pheonix W (2021) Idaho State University AGN-201 low power teaching reactor—An overlooked gem, ser. nuclear reactors spacecraft propulsion, research reactors, and reactor analysis topics
9. Safety analysis report of the idaho state university agn-201m research reactor (2003). Tech. Rep. License No. R-110, Docket No. 20–284
10. Sercan A, Pfister T (2021) TabNet: attentive interpretable tabular learning. Proc AAAI Conf Artif Intell 35(8):6679–6687. https://doi.org/10.1609/aaai.v35i8.16826
11. Murtagh F (1991) Multilayer perceptrons for classification and regression. Neurocomputing 2(5–6):183–197

# Evolution of Simulation and Digital Twin in Health Care: From Discovery to Design and Integration

**Yue Dong, Amos Lal, Alexander S. Niven, and Xiang Zhong**

**Abstract** Over the past two decades, simulation and Digital Twin (DT) technologies have become increasingly prevalent in health care. These technologies have significant potential to advance modern medicine, enhance clinical decision-making and team performance, improve healthcare delivery, reduce cost, and improve patient outcomes. This chapter provides an overview of these technologies and their emerging applications in the field of health care, including opportunities to accelerate discovery in basic science, delivery of more realistic training opportunities that advance clinician competence and interprofessional teamwork, and an efficient and cost-effective approach to analyze, improve, and monitor clinical workflows, healthcare delivery systems, and their performance. This emerging field fosters multidisciplinary research among healthcare professionals, information technology experts, engineers, and data scientists, all working together to better serve our society. Widespread adoption of these technologies will require solutions to address several technical, ethical, and regulatory challenges. These solutions will require a close collaboration between industry, academic centers, and government to develop a thoughtful approach that aligns implementation and integration of simulation and DT technologies into the healthcare workplace with measurement of meaningful outcomes to ensure broad-based access benefit from these important tools.

Y. Dong (✉) · A. Lal · A. S. Niven
Mayo Clinic, 200 First Street SW, Rochester, MN 55905, USA
e-mail: dong.yue@mayo.edu

A. Lal
e-mail: lal.amos@mayo.edu

A. S. Niven
e-mail: niven.alexander@mayo.edu

X. Zhong
Univesity of Florida, Gainesville, FL 32603, USA
e-mail: xiang.zhong@ise.ufl.edu



M. Grieves and E. Hua (eds.), *Digital Twins, Simulation, and the Metaverse*, Simulation Foundations, Methods and Applications, https://doi.org/10.1007/978-3-031-69107-2_11

# 1 Introduction

Modern health care is a complex adaptive system that requires highly skilled interprofessional team members to work together using advanced equipment and standard processes to care for patients with a wide variety of diseases with steadily increasing acuity over time. While the rapid pace of advancement in medical science offers exciting possibilities and new hope to patients and their family within our global community, it also demands constant changes in an already complicated and strained healthcare delivery system. The practice of medicine and current healthcare outcomes fall far below the level of quality and safety achieved by high-reliability industries, such as aerospace and manufacturing. As a result, institutions have become increasingly focused on the "value equation" of the care they deliver, generally defined as the quality of care divided by the total cost of patient care over time [1].

Over the past two decades, simulation and Digital Twin (DT) technology and their applications have become increasingly prevalent in health care. These tools can provide increasingly accurate representations of the pathobiology behind patient conditions, opportunities to monitor patient health and optimize disease diagnosis and management, enhance and evaluate the performance of clinicians and their healthcare teams, and analyze and improve healthcare delivery and systems. Using a systematic implementation strategy that includes key stakeholder collaboration, these technologies offer the opportunity to accelerate medical research discovery and improve healthcare system performance and patient outcomes with greater efficiency and at a reduced cost. This chapter provides an overview of these technologies and their emerging applications in health care.

## *1.1 Healthcare Evolution from Individual Treatment to Health Delivery Integration: A Systems Perspective*

The twentieth century witnessed a wide range of groundbreaking scientific discoveries in medicine, including antibiotics, anesthesia, blood transfusions, vaccines, Computed Tomography (CT) scans, magnetic resonance imaging (MRI), and extracorporeal therapies, including dialysis and extracorporeal Membrane Oxygenation (ECMO), and more [2]. The twenty-first century promises an even faster pace of discovery, with rapid advances in genomics offering a customized approach to common and complex diseases using precision medicine, and the digital revolution challenging our traditional approach to care delivery.

These advancements have had a profound impact on human health and have significantly shaped both the current practice of modern medicine and the benefits it provides to society. Healthcare delivery systems have evolved into complex ecosystems of clinical practice environments that offer a wide variety of services to scale. However, these healthcare improvements come with their own set of challenges. Population aging and growth, the increasing prevalence of chronic conditions, and advanced diagnostic and treatment modalities are both driving cost and reducing access to and equity of care. As the complexity of disease management has increased, there is a growing demand for a collaborative, multidisciplinary, and team-based approach that includes physicians, nurses, pharmacists, and respiratory therapists and technicians. These teams require a clear understanding of each other's roles and responsibilities and a high level of communication and teamwork to ensure timely and coordinated care delivery.

Given its complexity, health care today has evolved into a "system of systems" of healthcare delivery focusing on not only just safe and timely medical care, but also better management of resources to meet demand at scale across geographic locations and time. The Systems Engineering Initiative for Patient Safety (SEIPS) 3.0 framework could serve as a guiding model in this complex landscape [3]. It offers a holistic approach to integrating various components such as people, tasks, tools, and organization, thereby providing a comprehensive view of healthcare delivery systems. These "systems" are also often stratified using different levels, including biology and disease, patients and clinicians, and care delivery processes at the bedside, within the hospital, and throughout healthcare systems as a whole [4]. A System of Systems (SoS) in health care refers to integrating and interacting various independent systems within the healthcare sector to achieve a more comprehensive, efficient, and effective healthcare delivery.

In complex healthcare systems, the overall system performance is directly linked to the functionality of each subsystem and the interconnections between them. Individual components, such as patient care, administrative processes, and healthcare technology, must not only function effectively in isolation but also integrate seamlessly. The synergy between these subsystems is crucial for the overall system performance. Medical errors and gaps in care quality have been well documented. However, fixing them is not easy due to the complex and unpredictable work environment common in medical practice [5, 6]. The recent global COVID-19 pandemic and its impact further exposed these persistent vulnerabilities in healthcare systems worldwide [7–9]. The pandemic has served as a catalyst for change, underscoring the importance of improving current healthcare delivery mechanisms. It is imperative to enhance the efficiency of healthcare delivery while maintaining the highest standards of quality care and use innovative solutions and technologies to improve healthcare access and distribution on a global scale. This systems view of healthcare delivery also provides a framework that we can use to better understand how simulation and DTs can be used to offer important solutions to these complex problems.

## *1.2 Technology Advancement in Health Care*

Over the past 30 years, we have witnessed a technological revolution that has significantly impacted all aspects of our lives, including health care. Personal computers, mobile devices, the internet, cloud computing, and artificial intelligence/machine learning (AI/ML) have led to an explosion of data related to biology discovery, disease diagnosis, management, and healthcare delivery systems optimization. These data-driven decision support tools can amplify and augment human capacity to improve the efficiency, quality, and safety of health care during the whole journey of patient experience (prevention, management, recovery, and wellness).

Simulation is the imitation or representation of one act or system by another. The Society for Simulation in Healthcare describes four main purposes of simulation—education, assessment, research, and health system integration to facilitate patient safety [10]. Simulation-based training started in the discipline of anesthesiology and has grown rapidly in other areas of medical education, with the development of a growing variety of computer-controlled manikins and task trainers. These simulators can be programmed to recreate a variety of standardized clinical scenarios, allowing students to practice their clinical reasoning, technical (intubation, central line insertion, etc.), and teamwork skills (communication) in a safe and controlled environment. Simulation-based education has been shown to be an effective way to improve student knowledge and skills and translate to better patient outcomes [11].

Computer simulation involves creating a computer-based model or virtual representation of a system or process to study and analyze its behavior.

As an extension to traditional computer simulation, DT is a virtual representation of real-world entities and processes synchronized at a specified frequency and fidelity. DTs' systems transform business by accelerating holistic understanding, optimal decision-making, and effective action. DTs use real-time and historical data to represent the past and present and simulate predicted futures [12]. It is a digital replica of a physical system that can be used to monitor, predict, and optimize the system's behavior. DT is a virtual representation of a physical asset replicated virtually through a data connection, making it possible to link the system with its virtual counterparts in a bi-directional way. The "bi-directional" exchange of information, which synchronizes the virtual system response to match the physical system, distinguishes DT from traditional digital simulation, which is often considered independently operated. Specifically, a digital model has no interaction with the physical system. A digital shadow is updated with data from the physical system but does not inform its control. A DT arises when the physical system data is used to update the digital model, and the resulting simulation is used to control the physical system [13]. Very few works in literature have realized the control loop and are considered as true DTs in use, and thus, our review also includes simulations that are digital models or digital shadows [14].

## *1.3 Healthcare Applications*

DTs have various applications in health care, such as training, diagnosis, management, and care delivery for patients, providers, and healthcare organizations. One way to categorize DTs is based on Siemens' framework, which has three levels of twinning [15].

First, product twining provides a virtual-physical connection to analyze how a product performs under various conditions and adjust in the virtual world to ensure that the physical product will perform exactly as planned in the field. For product twinning, one example is the DTs of medical devices (e.g., digital radiological devices) [16, 17]. These efforts focus on the physical device/product twinning (similar to the concept of DTs for management in manufacturing) and are mostly used to monitor its status, diagnose issues, and test solutions remotely, ultimately optimizing its performance and reducing the risk of malfunctions. From digital devices to digital/virtual patients, precision medicine is one area that exemplifies the application of both product twinning and process twinning. Digitally replicating the human body (from cell to organ biological/physiological systems) allows for in silico clinical trials to examine the prevention, early detection, and targeted treatments of many diseases. These DT human body/organ systems have been used for drug development and treatment recommendations [18]. Another example of virtual patients is optimizing health care at the individual level, i.e., "personalized health monitoring," with the goal of healthcare management and promoting healthy behavior. It can be used to create personalized models of patients that can be used to monitor their health status, predict the course of disease, and optimize treatment plans [19]. This application is still focused on modeling a virtual patient instead of the healthcare system at large and is mainly focused on healthy people's daily life and their living environment instead of a specific healthcare facility. The third level is system twinning, which includes using DTs to improve hospital operation processes and workflows by allowing managers to tweak inputs and see how outputs are affected without the risk of upending existing workflow [20]. The system performance can be captured, analyzed, and acted on operational data, providing insights for informed decisions to maintain effective interactions among the components of the system at the system level.

### *1.4 Outline of the Chapter*

DTs have risen at the intersection of Industry 4.0 and the Internet of Things (IoT), relying on converging technologies in big data analytics, pervasive sensing, and cloud computing infrastructure. In this new era, simulation and DTs promise to enhance healthcare delivery by improving efficiency, flexibility, and patient-centered care, all of which are crucial in a post-pandemic world. This book chapter highlights how technological advancements have enriched health care with valuable data and digital assets, making it imperative to utilize these resources to optimize the entire spectrum of healthcare services. We will delve into the applications, benefits, challenges, and future prospects of DTs and simulation technologies in health care, particularly in the context of a post-COVID-19 world acute care setting. We will explore how these technologies can serve as invaluable tools for healthcare professionals, clinicians, researchers, IT professionals, engineers, and data scientists, all working together to serve society. In Sect. 2, we will explore simulation and DT applications in biology and disease management. In Sect. 3, we will discuss the latest professional training applications using simulation and DTs' technology. In Sect. 4, we will present applications that support decision-making for hospital operations and management. Then, in Sects. 5 and 6, we will provide a vision of simulation and DTs in health care and discuss the challenges and opportunities.

## 2 Simulation and Digital Twin Applications at the Biology and Disease Management

Modeling and simulation have emerged as a powerful tool for understanding complex biological systems and processes in addition to the traditional theory and experiment [21]. Utilizing mathematical models, simulation allows researchers and clinicians to predict disease outcomes, test treatment interventions, and optimize healthcare delivery. There are two common approaches for physiology and disease modeling: rule-based and data-driven modeling. Each of these approaches offers distinct advantages and challenges [22]. The rule-based approach relies on predefined rules and knowledge about biological systems to model disease processes. These rules are typically derived from clinical guidelines, expert opinions, or well-established physiological pathways and experienced-based learning in clinical settings. It has the benefits of transparency (by incorporation of directed acyclic graphs in the creation and execution of rules), easy control of model behavior, and stability for output, which make it easy to accept for clinicians [23, 24]. At the same time, it also has stability and flexibility limitations because it requires expert knowledge and comes with a possibility of bias. The mechanistic model is designed to represent the underlying mechanisms of a system—the processes and interactions that lead to observed phenomena. They are often contrasted with empirical models, which are purely based on observed data without assumptions about the underlying physiology processes. The data-driven model uses large datasets to investigate inter-relationships using different statistical and data mining methods. It is based on systems' biology and systems' immunology and has expanded rapidly in the last decade because of more accessible data from EMR and more computer power [25, 26].

### *2.1 Understanding Physiology and Treatment*

The simulation technologies have revolutionized medical science by understanding physiology. Various virtual training platforms now offer immersive experiences that closely mimic real-world clinical scenarios based on physiology [11]. These platforms are designed to be interactive, allowing medical students and healthcare professionals to engage in problem-solving, diagnosis, and treatment planning in a risk-free environment. The virtual training modules often include a wide range of case studies covering various diseases, conditions, and patient demographics, providing a comprehensive learning experience. The lab and vital signs will change based on the intervention by user input. The virtual patient offers a practical understanding of disease mechanisms and treatment protocols, thereby bridging the gap between academic learning and clinical practice. This hands-on approach enhances the learner's ability to make informed clinical decisions in a safe environment that will not harm patients, ultimately leading to better patient outcomes.

Archimedes Diabetes Model was designed to simulate individual patients and how they would respond to various treatments for diabetes. It incorporated a wide range of variables, from glucose metabolism to treatment protocols, to generate realistic patient outcomes [27, 28]. This model serves as one of the early successes in utilizing computer-based physiology to understand and manage diabetes. This was also an early implementation of using directed acyclic graphs to explain the complex pathophysiology and rules based on the clinical environment. The model underwent rigorous validation, and its predictions were found to be highly consistent with the results of randomized controlled trials. This level of accuracy made the Archimedes Diabetes Model a valuable tool for clinicians, researchers, and policymakers alike. It not only helped in understanding the complex physiology of diabetes but also played a crucial role in optimizing treatment plans and healthcare policies related to diabetes management.

Agent-based models can simulate the interactions between various biological agents, such as cells and molecules, in each biological system [29, 30]. These agent-based models are particularly useful for understanding complex, multiscale biological systems where traditional modeling approaches may fall short. By providing a more nuanced understanding of biological systems, agent-based models have paved the way for more targeted and effective treatments, bridging the gap between basic biological research and clinical applications. This approach is offering insights that are directly applicable to patient care and treatment optimization. The research group led by Vodovotz explores the use of omics data and mathematical modeling to understand and explore various critical illness management strategies by integrating data-driven and knowledge-based modeling approaches [22]. This approach aims to integrate large-scale data with computational modeling to improve research and clinical applications in diseases involving severe inflammation and immune responses.

BioGears is an open-source, comprehensive human physiology engine that has been instrumental in driving medical education, research, and training technologies [31]. BioGears aims to provide accurate and consistent physiology simulation across the medical community. It can be used as a standalone application or integrated with other simulators and sensor interfaces making it a valuable asset in the healthcare simulation community. Similar to BioGears, the Pulse Physiology Engine offers robust physics-based circuit and transport solvers [32]. It includes a common data model for standard models and data definitions, a software interface for engine control, and a verification and validation suite. Pulse's architecture is designed to reduce model development time and increase usability, making it a go-to solution for many in the healthcare simulation community.

Emerging roles of virtual patients in the age of AI have been explored, and rule-based systems have shown promise in healthcare modeling. These systems can simulate patient responses based on a set of predefined expert rules, making them useful for training and diagnostic purposes. Dr. Lal and team have developed and verified a DTs model of critically ill patients using the causal AI approach to predict the response to specific treatment during the first 24 h of sepsis [23]. This expert rule-based DTs focus on creating individualized patient models for better medical education and clinical decision support [33]. This work was previously developed specifically for sepsis in critically ill patients but has now evolved into multiple commonly seen clinical scenarios in the intensive care setting. The primary goal is to provide a safe testing bed for the learners and the clinicians to evaluate a proposed intervention (useful or otherwise) in an in silico environment, before actually performing those interventions on the patient at the bedside. The envisioned result is to avoid any preventable harm to the patient by pre-testing the interventions that carry any uncertain risks. The utility of such DTs in health care can serve multiple purposes, including advances in medical education delivery, in silico research, and adjunct clinical decision support at the bedside.

## 2.2 *In Silico Clinical Trials*

In silico clinical trials have emerged as a groundbreaking approach in medical research, offering many opportunities and advantages over traditional clinical trials. Utilizing computational models and simulations, in silico trials allow for the testing of medical interventions or devices in a virtual environment, thereby revolutionizing the way clinical research is conducted [34, 35].

One of the most compelling advantages of in silico trials is the significant reduction in costs. Traditional clinical trials often require extensive resources, including patient recruitment, site management, and long-term follow-ups, all of which contribute to high operational costs. In contrast, in silico trials can be conducted with minimal overhead, as they rely on computational power rather than physical infrastructure and human resources. This cost-effectiveness makes it feasible to explore a broader range of research questions and to conduct multiple trials simultaneously, thereby accelerating the pace of medical innovation. Time is another critical factor that holds a distinct advantage for in silico trials. Traditional trials can take several years to complete, from initial planning to final data analysis. In silico trials, however, can be executed in a fraction of the time. The ability to quickly simulate different scenarios and interventions enables researchers to arrive at conclusions more rapidly, thereby speeding up the "time to market" for new treatments and medical devices. Perhaps, the most significant benefit of in silico trials is the elimination of risks to human subjects. Despite rigorous ethical standards, traditional clinical trials always carry some risk to the participants. In silico trials remove this concern entirely, as they are conducted in a virtual environment. This not only ensures the safety of potential patients but also allows for the testing of treatments that might be considered too risky for traditional trials. By offering a safer, faster, and more cost-effective alternative to traditional methods, in silico clinical trials are poised to become a cornerstone in the future of medical research. They offer a promising pathway for the development of new treatments and medical technologies, with the potential to significantly improve patient outcomes and healthcare systems globally. Important components of biomedical ethics, such as informed consent, algorithm fairness and biases, intellectual property law, data privacy, safety, and transparency, should be considered alongside the regulatory issues of DTs [36]. Computer modeling and simulation can be adopted from other industries to aid in various stages of medical device development, testing, clinical evaluations, and failure analysis, leading to cost reduction [37].

Recently, the FDA has published reports to showcase how modeling and simulation can be used for scientific research and regulatory decision-making [38]. FDA also published guidance providing recommendations to assess the Credibility of Computational Modeling and Simulation in Medical Device Submissions [39]. Virtual clinical trials represent a significant shift in how clinical research is conducted. By leveraging DT platforms, researchers can gather large data more quickly and from a broader population base. Simulations and virtual environments offer a unique opportunity for policymakers to test the implications of healthcare policies in a controlled, risk-free setting [40]. By modeling the outcomes of proposed policies, stakeholders can anticipate their effects and refine them before implementation, ensuring that new regulations are both effective and efficient.

## 3 Health Professional Training Through Simulation and Digital Twin Technology

Over the past two decades, there has been a significant expansion in clinical training for healthcare professionals through various simulation technologies. These technologies range from task trainers, mannequins to more advanced Virtual Reality (VR) and Augmented Reality (AR) systems. The learners include the multidisciplinary clinical team (physicians, nurses, pharmacists, and respiratory therapists). While traditional task trainers have been around for many years, modern technology has allowed for the expansion and sophistication of simulation training programs. These programs now cover a wide array of skills, from procedural skills to non-technical skills like communication, compassion, and empathy. Numerous studies have shown that simulation-based training is associated with improved care processes and better patient outcomes [41].

### 3.1 Computer-Driven Mannequins and Task Trainers in Medical Training and Assessment

Modern mannequins used in medical training have evolved to become highly sophisticated, computer-driven devices. These mannequins can be controlled remotely or in-room, offering a range of clinical scenarios for trainees to practice clinical skills and teamwork [42, 43]. Those mannequins provide invaluable hands-on experience for clinicians in a controlled, risk-free environment for various specialties and disciplines [41, 44, 45]. The COVID-19 pandemic has forced the healthcare industry to adopt distance simulation using various technologies to find alternative approaches for traditional onsite clinical training [46, 47]. Simulation has been used for skill assessment and evaluation in many disciplines [48–50]. The use of VR and AR in healthcare training is growing rapidly to offer immersive learning experiences for learners that traditional methods cannot match [51–53]. Game-based training modules have also gained popularity, providing an engaging way for healthcare providers to hone their skills [54]. Although the cost of hardware can be a limiting factor, the benefits, such as improved skill retention and real-world applicability, often outweigh the initial investment.

### 3.2 Research Potential: Investigating System Vulnerabilities

Simulation technologies not only provide clinical skill training but also serve as research tools for human factor analysis. They offer a reproducible clinical environment where the interaction of system factors (clinician, technology, and workflow) can be investigated independently of patient factors [55, 56]. Researchers can use these technologies to investigate system vulnerabilities, test new workflows, new technology innovations, and even simulate the impact of potential policy changes before clinical implementation [57–60]. This simulation-based research extends the utility of simulation technologies as integral tools for overall healthcare system improvement [61].

## 4 Hospital Operations and Management

### 4.1 Leverage Data Insights from EMR, IoT, AI

The past 15 years have seen a data revolution in health care fueled by electronic medical records' (EMRs) implementation and adoption. This shift from paper to digital documentation has unlocked a wealth of patient data, improving team communication and clinical decision support, informing disease treatment plans, and tracking outcomes. Further, the integration of EMRs with the IoT and Radio

Frequency Identification (RFID) technologies creates a connected ecosystem for real-time patient flow monitoring and streamlined inventory management benefits from this data influx [62–64].

By collecting and analyzing this data, healthcare professionals can leverage predictive modeling to anticipate patient needs and prescriptive analytics to design personalize treatment plans, ultimately resulting in more informed decisions, improved patient care, and efficient healthcare operations [65]. Recent advancements in AI, computer vision, and Large Language Model (LLM) are opening up a new era in the healthcare innovation using simulation and DT [66, 67]. These technologies extend the capabilities of traditional data collection and analysis, offering a more holistic view of healthcare delivery from outpatient settings to inpatient care, and even post-discharge home care.

## *4.2 Operation Management and Resource Allocation*

Healthcare operations' management aims to improve the efficiency of business processes within healthcare facilities to reduce overcrowding, waiting times, and delays. One of the main challenges in healthcare operations is managing the unpredictability caused by variations in patient demand, staffing capacity, and resource availability. For instance, changes in the number of patients or types of cases can affect staff schedules, patient flow, and patient room utilization. According to a rapid literature review of papers published in 2002–2022, the use of DTs for healthcare systems management is an emerging topic [68]. With availability of real-time patient and hospital operation data, healthcare managers can make informed decisions about staff deployment based on patient needs and workflow efficiencies. Similarly, data analytics can be used to optimize inventory levels, ensuring that medical supplies are ordered and utilized most effectively, thereby reducing waste and costs. The key advantage of DTs in health care is the ability to leverage real-time data to model complex systems and processes that involve many interdependent variables [69]. This allows for dynamic, evidence-based decision-making by integrating a large amount of heterogeneous data and real-time data queries to achieve better resource allocation at the hospital level [70].

Trauma centers, emergency departments, and ICUs are systems whose processes are subject to large variability and are very time-sensitive, and thus, have attracted substantial attention for process improvement and patient safety assurance using DTs. For instance, a trauma DT is used to digitize and support the process of severe trauma management, considering it as a physical asset that is mirrored by two DTs [71]. In Augusto et al. 2018, a DT of an emergency unit was developed to optimize the pathway of patient care in the unit. The system accounts for various arrival processes to account for massive arrivals in case of a crisis and determine the best available leverages to optimize the operations of the system [72]. DTs of ICU processes are used to identify inefficiencies in patient flows, optimize patient care by clinical staff at the enterprise level, and use remote monitoring to detect process faults and anomalies [68, 73]. Healthcare leaders can shift from reactive decisions to proactive optimization based on data-driven insights from DTs. Hospital-level model by using predictive decision support model that employs real-time service data is drawn from the systems and devices [74]. Their model enables assessing the efficiency of existing healthcare delivery systems and evaluating the impact of changes in services without disrupting the daily activities of the hospital. Along the same line, Karakra et al. 2020 developed discrete event simulation and DTs through a system called HospiT'Win that allows for tracking the pathways of patients inside the healthcare organization to manage growing demand and decrease waiting times [69]. Rodriguez-Aguilar et al. 2020 proposed a digital healthcare system initiative through multi-paradigm simulation [75]. Computer modeling has also been used to simulate infectious disease transmission dynamics, optimize the vaccination strategy, and test public policies before clinical implementation [76–78].

## 5 Challenges and Opportunities: Ushering a New Era in Healthcare Simulation and Digital Twins

The adoption of simulation and DT technologies in health care is not without its challenges [79, 80]. One of the most pressing issues is the need for accurate and complete data. Incomplete or erroneous data can significantly impact the effectiveness of these technologies in both training and real-world applications. Also, modeling complex biological systems involves numerous variables and nonlinear interactions, making it challenging. Regulatory hurdles, such as compliance with healthcare standards and data protection laws, further complicate the adoption process. Ethical considerations are paramount when using patient data in the DTs, especially when it comes to patient and clinician data privacy and confidentiality. The collection and use of patient data must adhere to strict ethical guidelines to ensure that individual privacy is respected. Moreover, there is a risk of bias in data collection and analysis, which could inadvertently lead to unequal healthcare delivery. Addressing these ethical and bias concerns is crucial for the responsible deployment of these technologies. Data sharing and interoperability present another set of challenges. Different healthcare systems often use different data formats and standards, making integration a complex task. The lack of interoperability can hinder the seamless exchange of information, thereby limiting the effectiveness of simulation and DTs technologies in a multi-system environment [81].

For simulation and DT tools to be effectively integrated into health care, there needs to be a set of clearly defined performance metrics for systems evaluation. These metrics should measure impacts of simulation and DTs across various domains, including disease outcomes, patient outcomes, system outcomes, and return on investment. The development of such comprehensive evaluation metrics will enable stakeholders to assess the effectiveness of DT technologies objectively.

Despite these challenges, there are numerous opportunities for innovation. For instance, multiscale simulation is a powerful tool for understanding complex biological systems and healthcare processes. By simulating the behavior of individual molecules, cells, and tissues, multiscale simulations can provide insights into how diseases develop and how drugs work. Multiscale simulations are being used to understand the molecular basis of diseases, such as Alzheimer's disease and drug development [82, 83]. Meanwhile, DTs can be used to track the progress of patients, monitor their vital signs, and predict when they may need medical attention. DTs have been used to monitor cardiac and cancer patients for personalized treatment planning [84, 85]. Additionally, AI algorithms can be used to analyze the data generated by simulations and DTs. This data can be used to train AI models to make predictions and identify patterns. AI models can be used to improve the accuracy of diagnoses, recommend treatments, and develop new drugs. The synergistic relationship between multiscale simulation, DTs, and AI offers unprecedented capabilities in health care, from real-time analytics to predictive and prescriptive modeling for disease prevention, diagnosis, and management. The integration of multiscale simulation, DTs, and AI is creating a new era of precision medicine and personalized health care. Data-driven decision-making is revolutionizing disease management and patient care. By leveraging predictive analytics, healthcare providers can anticipate disease progression and intervene earlier. This proactive approach improves patient outcomes and reduces the burden on healthcare systems. Simulation and DTs will empower clinicians and operational managers to use real-time data to streamline workflows and enhance the quality of care. By analyzing patterns in healthcare delivery, they can identify bottlenecks and inefficiencies, leading to policies that align with the goals of all stakeholders and provide safe, effective, efficient, and equitable patient care.

The multidisciplinary nature of this field allows for the convergence of healthcare professionals, engineers, data scientists, IT experts, AI specialists, etc. Such collaboration can lead to innovative solutions that address the existing challenges. Moreover, as these technologies become more integrated into health care, there will be a growing need for workforce training and upskilling to ensure that healthcare professionals can effectively leverage these advanced tools.

While technological advances in health care are promising, their successful adoption hinges on robust change management, implementation science strategies, and workforce upskilling. Crucially, comprehensive training is needed to equip healthcare professionals with the skills to leverage new digital capabilities effectively. This includes technical competencies, data literacy, human–AI collaboration, cybersecurity, and ethics. Investing in change management, implementation science, and strategic workforce development will be critical for healthcare organizations to capitalize on transformative technological advances. Several major societies and groups are active in related fields. The Society for Simulation in Healthcare (SSH) is a global community of medical educators who use various technologies, such as manikins and task trainers, to deliver educational interventions for skills and teamwork training [86]. The Winter Simulation Conference includes many engineers working on modeling and simulation in various industries [87]. The Interagency

Modeling and Analysis Group (IMAG) is a government group of program officials from multiple federal government agencies supporting research funding for modeling and analysis of biomedical, biological, and behavioral systems. IMAG is focused on research across the biological continuum across different scales or levels of resolution with modelers from multidisciplinary research communities [88]. The Medical Device Innovation Consortium (MDIC) primarily focuses on device innovation, using Computational Modeling and Simulation (CM&S) to reduce product development costs, speed up time to market, and better serve patients with safe and effective medical devices [37]. The Virtual Physiological Human Institute for Integrative Biomedical Research, commonly known as the VPH Institute, is another significant player in this field. Its mission is to ensure that the Virtual Physiological Human is fully realized, universally adopted, and effectively used in research and clinical settings [89]. The Europe Digital Twin in Healthcare (EDITH) project aims to capitalize on the growing trend of interest in Digital Twins [90]. EDITH creates a roadmap for future development, allowing stakeholders to exchange best practices, analyze ecosystems and data flows, and identify vulnerabilities. In addition, groups are working to prepare data for modeling. Mobilizing Computable Biomedical Knowledge is an international community focused on ensuring that biomedical knowledge in computable form is findable, accessible, interoperable, and reusable. A recent report from the National Academies of Sciences, Engineering, and Medicine presented an intergraded research agenda to advance the field. DTs can be a critical tool for decision-making base on the synergistic combination of models and data [91].

## 6 Conclusion

The emergence of simulation and DT technologies marks the beginning of a new era in health care. These technologies are practical tools that can improve disease management, care planning, and resource allocation. While there are undoubtedly challenges, such as technical complexities, ethical and regulatory hurdles, overcoming these obstacles will enable us to fully integrate simulation and DT technologies to support healthcare system digital transformation that serves all patients with safety, effectiveness, patient-centeredness, timeliness, efficiency, and equity.

## References

1. Yong PL., Olsen L, Jm M (2010) Value in health care: accounting for cost, quality, safety, outcomes, and innovation: workshop summary
2. Brandt AM (2012) A reader's guide to 200 years of the New England journal of medicine. N Engl J Med 366(1):1–7
3. Carayon P et al (2020) SEIPS 3.0: human-centered design of the patient journey for patient safety. Appl Ergon 84:103033

4. Wickramasinghe N et al (2007) Healthcare system of systems. IEEE Int Conf Syst Syst Eng 2007:1–6
5. Institute of Medicine Committee on Quality of Health Care (2000) In A., To Err is human: building a safer health system, In: Kohn LT, Corrigan JM, Donaldson MS (eds) To Err is Human: building a safer health system. National Academies Press, National Academy of Sciences, Washington (DC)
6. Makary MA, Daniel M (2016) Medical error—the third leading cause of death in the US. Br Med J 353
7. Blumenthal D et al (2020) Covid-19—Implications for the health care system. N Engl J Med 383(15):1483–1488
8. Kaye AD et al (2020) Economic impact of COVID-19 pandemic on healthcare facilities and systems: international perspectives. Best Pract Res Clin Anaesthesiol 35:293–306
9. Domecq JP et al (2021) Outcomes of patients with coronavirus disease 2019 receiving organ support therapies: the international viral infection and respiratory illness universal study registry. Crit Care Med 49(3)
10. Society for simuation in healthacre: what is simulation? (2024) 22 Apirl 2013. Available from: http://ssih.org/about-simulation
11. Cook DA, Erwin PJ, Triola MM (2010) Computerized virtual patients in health professions education: a systematic review and meta-analysis. Acad Med 85(10):1589–1602
12. Digital Twin Consortium (2024) Available from: https://www.digitaltwinconsortium.org/faq/
13. Kritzinger W et al (2018) Digital Twin in manufacturing: a categorical literature review and classification. IFAC-PapersOnLine 51(11):1016–1022
14. Cockrell C et al (2022) Generating synthetic data with a mechanism-based critical illness digital twin: demonstration for post traumatic acute respiratory distress syndrome. bioRxiv, p 517524, 22 Nov 2022
15. SIEMENS (2024) What is a digital twin? 2024; Available from: https://www.sw.siemens.com/en-US/technology/digital-twin/
16. Pesapane F et al (2022) Digital twins in radiology. J Clin Med 11(21)
17. Bjelland Ø et al (2022) Toward a digital twin for arthroscopic knee surgery: a systematic review. IEEE Access 10:45029–45052
18. Hernandez-Boussard T et al (2021) Digital twins for predictive oncology will be a paradigm shift for precision cancer care. Nat Med 27(12):2065–2066
19. Rivera LF et al (2019) Towards continuous monitoring in personalized healthcare through digital twins. In: Conference of the centre for advanced studies on collaborative research
20. Vallée A (2023) Digital twin for healthcare systems. Frontiers Digital Health 5
21. Pool R (1992) The third branch of science debuts. Science 256:44–47
22. Vodovotz Y (2023) Towards systems immunology of critical illness at scale: from single cell 'omics to digital twins. Trends Immunol 44(5):345–355
23. Lal A et al (2020) Development and verification of a digital twin patient model to predict specific treatment response during the first 24 hours of sepsis. Crit Care Explor 2(11):e0249
24. Lal A et al (2020) Artificial intelligence and computer simulation models in critical illness. World J Crit Care Med 9(2):13–19
25. Alquraishi M, Sorger PK (2021) Differentiable biology: using deep learning for biophysics-based and data-driven modeling of molecular mechanisms. Nat Methods 18:1169–1180
26. Yue R, Dutta A (2022) Computational systems biology in disease modeling and control, review and perspectives. NPJ Syst Biol Appl 8
27. Eddy DM, Schlessinger L (2003) Validation of the archimedes diabetes model. Diabetes Care 26(11):3102–3110
28. Schlessinger L, Eddy DM (2002) Archimedes: a new model for simulating health care systems–the mathematical formulation. J Biomed Inform 35(1):37–50
29. An G (2022) Specialty grand challenge: what it will take to cross the valley of death: translational systems biology, "True" precision medicine, medical digital twins, artificial intelligence and in silico clinical trials. Frontiers Syst Biol 2

30. An G et al (2009) Agent-based models in translational systems biology. Wiley Interdiscip Rev Syst Biol Med 1(2):159–171
31. McDaniel M et al (2019) A whole-body mathematical model of sepsis progression and treatment designed in the BioGears physiology engine. Front Physiol 10
32. Bray A et al (2019) Pulse physiology engine: an open-source software platform for computational modeling of human medical simulation. SN Compr Clin Med 1:362–377
33. Rovati L et al (2024) Development and usability testing of a patient digital twin for critical care education: a mixed methods study. Front Med 10
34. Pappalardo F et al (2019) In silico clinical trials: concepts and early adoptions. Briefings Bioinf
35. Viceconti M et al (2021) In silico trials: verification, validation and uncertainty quantification of predictive models used in the regulatory evaluation of biomedical products. Methods 185:120–127
36. Lal A et al (2022) Regulatory oversight and ethical concerns surrounding software as medical device (SaMD) and digital twin technology in healthcare. Ann Transl Med 10(18):950
37. Landscape report and industry survey on the use of computational modeling and simulation in medical device development (2023). Available from: https://mdic.org/resource/cmslandscape report/#download_form
38. Successes and opportunities in modeling and simulation for FDA (2022)
39. Guidance document (2023) Assessing the credibility of computational modeling and simulation in medical device submissions, U.S.F.D. Administration, Editor
40. Romero-Brufau S et al (2021) Public health impact of delaying second dose of BNT162b2 or mRNA-1273 covid-19 vaccine: simulation agent based modeling study. BMJ 373:n1087
41. Cook DA et al (2011) Technology-enhanced simulation for health professions education: a systematic review and meta-analysis. JAMA 306(9):978–988
42. Cooper JB, Taqueti VR (2004) A brief history of the development of mannequin simulators for clinical education and training. Qual Saf Health Care 13:i11–i18
43. Gaba DM (2004) The future vision of simulation in health care. Qual Saf Health Care 13:i10–i12
44. Schmidt E et al (2013) Simulation exercises as a patient safety strategy a systematic review. Ann Intern Med 158(5_Part_2):426–432
45. McGaghie WC et al (2014) A critical review of simulation-based mastery learning with translational outcomes. Med Educ 48:375–385
46. Buléon C et al (2022) The state of distance healthcare simulation during the COVID-19 pandemic: results of an international survey. Adv Simul 7(1):10
47. Flurin L et al (2022) International virtual simulation education in critical care during COVID-19 pandemic: preliminary description of the virtual checklist for early recognition and treatment of acute illness and iNjury program. Simul Healthc 17(3):205–207
48. Ryall T, Judd BK, Gordon CJ (2016) Simulation-based assessments in health professional education: a systematic review. J Mult Healthc 9:69–82
49. Boulet JR, Murray DJ (2010) Simulation-based assessment in anesthesiology: requirements for practical implementation. Anesthesiology 112(4):1041–1052
50. Hayden JK et al (2014) The NCSBN national simulation study: a longitudinal, randomized, controlled study replacing clinical hours with simulation in Prelicensure nursing education. J Nurs Regul 5(2, Supplement):S3–S40
51. Kononowicz AA et al (2019) Virtual patient simulations in health professions education: systematic review and meta-analysis by the digital health education collaboration. J Med Int Res 21
52. Foronda CL et al (2024) A comparison of virtual reality to traditional simulation in health professions education: a systematic review. Simul Healthc 19(1s):S90-s97
53. Woodall WJ et al (2024) Does extended reality simulation improve surgical/procedural learning and patient outcomes when compared with standard training methods?: A systematic review. Simul Healthc J Soc Simul Healthc 19(1S):S98–S111
54. Gentry SV et al (2019) Serious gaming and gamification education in health professions: systematic review. J Med Internet Res 21(3):e12994

55. Small SD (2007) Simulation applications for human factors and systems evaluation. Anesthesiol Clin 25(2):237–259
56. Deutsch ES et al (2016) Leveraging health care simulation technology for human factors research: closing the gap between lab and bedside. Hum Factors 58(7):1082–1095
57. Ahmed A et al (2011) The effect of two different electronic health record user interfaces on intensive care provider task load, errors of cognition, and performance. Crit Care Med 39(7):1626–1634
58. Arriaga AF et al (2013) Simulation-based trial of surgical-crisis checklists. N Engl J Med 368(3):246–253
59. Patterson MD et al (2013) In situ simulation: detection of safety threats and teamwork training in a high risk emergency department. BMJ Qual Saf 22(6):468–477
60. Yeh VJ et al (2024) Using mobile and remote simulation as a research methodology for health care delivery research. Simul Healthc
61. Goldshtein D et al (2020) In situ simulation and its effects on patient outcomes: a systematic review. BMJ Simul Technol Enhanc Learn 6(1):3–9
62. Asamoah DA et al (2016) RFID-based information visibility for hospital operations: exploring its positive effects using discrete event simulation. Health Care Manag Sci 21:305–316
63. Haddara M, Staaby A (2018) RFID applications and adoptions in healthcare: a review on patient safety. Procedia Comput Sci 138:80–88
64. Abkari SE, Jilbab A, Mhamdi JE (2020) RFID medication management system in hospitals. Int J Online Biomed Eng 16:155–168
65. Sutton RT et al (2020) An overview of clinical decision support systems: benefits, risks, and strategies for success. NPJ Digital Med 3
66. Cristofoletti M et al (2023) Towards a computer vision-based approach for digital twin implementation. Int Conf Intell Metaverse Technol Appl (iMETA) 2023:1–6
67. Wang Y et al (2024) TWIN-GPT: digital twins for clinical trials via large language model
68. Elkefi S, Asan O (2022) Digital twins for managing health care systems: rapid literature review. J Med Internet Res 24(8):e37641
69. Karakra A et al (2019) HospiT'Win: a predictive simulation-based digital twin for patients pathways in hospital. IEEE EMBS Int Conf Biomed Health Informatics (BHI) 2019:1–4
70. Peng Y et al (2020) Digital twin hospital buildings: an exemplary case study through continuous lifecycle integration. Adv Civil Eng 2020:1–13
71. Croatti A et al (2020) On the integration of agents and digital twins in healthcare. J Med Syst 44(9):161
72. Augusto V, Murgier M, Viallon A (2018) A modelling and simulation framework for intelligent control of emergency units in the case of major crisis. Winter Simul Conf (WSC) 2018:2495–2506
73. Zhong X et al (2022) A multidisciplinary approach to the development of digital twin models of critical care delivery in intensive care units. Int J Prod Res 60(13):4197–4213
74. Karakra A et al (2018) Pervasive computing integrated discrete event simulation for a hospital digital twin. In: 2018 IEEE/ACS 15th international conference on computer systems and applications (AICCSA), p 1–6
75. Rodríguez-Aguilar R, Marmolejo-Saucedo JA (2020) Conceptual framework of digital health public emergency system: digital twins and multiparadigm simulation. EAI Endorsed Trans Pervasive Health Technol 6:e3
76. Currie CSM et al (2020) How simulation modelling can help reduce the impact of COVID-19. J Simul 14:83–97
77. Romero-Brufau S et al (2021) Public health impact of delaying second dose of BNT162b2 or mRNA-1273 covid-19 vaccine: simulation agent based modeling study. The BMJ 373
78. Zhang T et al (2023) Data-driven modeling and analysis for COVID-19 pandemic hospital beds planning. IEEE Trans Autom Sci Eng 20:1551–1564
79. Fackler J, Hankin J, Young T (2012) Why healthcare professionals are slow to adopt modeling and simulation. In: Proceedings of the 2012 winter simulation conference (WSC)

80. Combs CD, Combs PF (2019) Emerging roles of virtual patients in the age of AI. AMA J Ethics 21(2):E153-159
81. Jacoby M, Usländer T (2020) Digital twin and internet of things—Current standards landscape. Appl Sci
82. Stefanovski L et al (2021) Scales in Alzheimer's disease: biological framework for brain simulation with the virtual brain. Frontiers Neuroinformatics 15
83. Joshi A et al (2020) Opportunities for multiscale computational modelling of serotonergic drug effects in Alzheimer's disease. Neuropharmacology 174:108118
84. Coorey G et al (2022) The health digital twin to tackle cardiovascular disease-a review of an emerging interdisciplinary field. NPJ Digit Med 5(1):126
85. Stahlberg EA et al (2022) Exploring approaches for predictive cancer patient digital twins: Opportunities for collaboration and innovation. Front Digit Health 4:1007784
86. Society for Simulation in Healthcare (SSH) Council for accreditation of healthcare simulation programs accreditation standards
87. 2024 Winter Simulation Conference (2024). Available from: https://meetings.informs.org/wordpress/wsc2024/
88. IMAG: Interagency Modeling and Analysis Group (2024)
89. Virtual Physiological Human Institute (2024). Available from: https://www.vph-institute.org/
90. Europe Digital Twin in Healthcare (2024). Available from: https://www.digitaleurope.org/projects/ecosystem-digital-twins-in-healthcare-edith/
91. National Academies of Sciences, Engineering, and Medicine. Foundational Research Gaps and Future Directions for Digital Twins (2024). Available from: https://www.nationalacademies.org/our-work/foundational-research-gaps-and-future-directions-for-digital-twins

# Digital Twins for Sustainable Semiconductor Manufacturing

**Ala Moradian**

**Abstract** This chapter explores the applications of Digital Twins in smart semiconductor manufacturing, highlighting their potential to drive a more sustainable industry. Digital Twins enable advanced monitoring of chemical and energy consumption, as well as other environmental impacts within semiconductor fabrication processes. The chapter reviews various initiatives and applications that utilize Digital Twins for smart monitoring and provides guidance on extending these efforts to further optimize semiconductor manufacturing. This includes enhancing device performance, yield, and cost of ownership, while minimizing environmental impact. The chapter concludes that achieving truly smart manufacturing requires a holistic approach focused on optimizing specific equipment design and processes (such as deposition, etch, lithography, chemical mechanical planarization) and fostering cohesive collaboration among stakeholders from academia and industry. Addressing the gaps in understanding physics, chemistry, sensor technology, software infrastructure, data security, and establishing universal standards and protocols for data sharing and integration are essential for realizing the full potential of Digital Twins in the semiconductor industry.

**Keywords** Digital twins · Semiconductor manufacturing · Sustainability · Smart manufacturing · Simulation · Modeling

## 1 Introduction

With the arrival of digital transformation innovations such as, Generative AI[1] and LLMs,[2] Industry 4.0, autonomous vehicles, and Digital Twins the demand for semiconductor chip manufacturing continues to grow. As a result, sustainability measures

[1] Artificial Intelligence.

[2] Large Language Models.

A. Moradian (✉)
Applied Materials Inc, Santa Clara, USA
e-mail: Ala_moradian@amat.com

 
M. Grieves and E. Hua (eds.), *Digital Twins, Simulation, and the Metaverse*, Simulation Foundations, Methods and Applications, https://doi.org/10.1007/978-3-031-69107-2_12

and reducing the environmental impact will become a key area of focus for manufacturers as well as governments and local energy regulators. To address this sustainability challenge, manufacturers along the entire value chain need to work both individually as organizations and collectively as an industry to define goals and agree on actions. Major semiconductor manufacturers have begun advocating for the industry to enable a smart and digital future while simultaneously reducing environmental impacts.

In situ monitoring of energy, water and gas consumption of semiconductor manufacturing plants will be the first critical step. As the famous Lord Kelvin once mentioned, what you cannot measure you cannot improve. By synchronizing data from multiple sensors, and leveraging domain knowledge and models, detailed on-demand reports of carbon footprint and related consumption parameters from both the fab and sub-fab can be generated. In addition to the monitoring the consumptions, Digital Twins of manufacturing equipment can enable what-if scenarios with virtual wafer processing and estimating their environmental impact. Ultimately, manufacturing equipment in the fab and the auxiliary equipment in the subfab would need to be enhanced with built-in intelligence to optimize process and operational conditions for not only on-wafer process outcome but also the environmental impact of the whole manufacturing flow of future semiconductor devices. This sustainability-driven goal would require calibrated processes and hardware Digital Twins that can realistically predict impact of any changes to the system, not only at the environmental impact level, but also for the integration and performance of the semiconductor devices.

## *1.1 Digital Twins and Promises for Smart Manufacturing*

Smart manufacturing represents a transformative shift in the industrial landscape, harnessing the power of digital tools to optimize production processes. At its core, this paradigm leverages the systematic collection of data streams from an array of sensors, notably IoT[3] devices, which are strategically integrated into manufacturing environments. The linchpin of this revolution is the deployment of Digital Twins, software products or models created baed on physics or data-driven algorithms. These twins encompass a diverse array of technologies, from machine learning to hybrid and heuristic models. They play a pivotal role in the analysis of real-time data, enabling manufacturers to gain unprecedented insights for monitoring, diagnostics, process enhancement, and asset control. The importance of this digital ecosystem cannot be overstated, as it empowers businesses to enhance efficiency, reduce downtime, and ultimately compete more effectively in an increasingly dynamic global market.

With the foundational understanding of smart manufacturing and the pivotal role that Digital Twins play in optimizing industrial processes, it becomes imperative to delve into the broader landscape of Digital Twins across diverse industries, with an emphasis on manufacturing. By examining real-world use cases and exploring

[3] Internet of Things.

the evolving nature of Digital Twins, we will gain a comprehensive understanding of their significance in manufacturing and the broader industrial context. In a later section we will dive into the use of Digital Twins in semiconductor manufacturing and the important use cases in addressing the carbon footprint, energy, water, and chemical consumptions of the industry.

### 1.1.1 Definition of Digital Twin

The concept of Digital Twins was first proposed by Michael Grieves in 2002 in the context of product lifecycle management (PLM) [1]. Since then, there has been multiple definitions suggested among which the definition by NASA and Grieves are more broadly accepted. NASA described the Digital Twin as follows: "A Digital Twin is an integrated multiphysics, multiscale, probabilistic simulation of an as-built vehicle or system that uses the best available physical models, sensor updates, fleet history, etc., to mirror the life of its corresponding flying twin" [2]. Grieves proposed that the basic Digital Twin model has three components, viz. physical entity in real space, virtual entity in virtual space, and the networks of information data that tie the physical and virtual entities or space together [3]. Digital Twin (DT) constructs a virtual replica of a physical scenario to monitor the behavior of the ongoing process, with condition monitoring, detection of anomalies, and prediction of future trends within the process. Based on the literature, the definitions of the Digital Twin is (see Fig. 1):

- the digital representation of a physical entity with possible data and models. The term "data" refers to the data from all the processes obtained during the run time and the system's development phase [4];
- it always incorporates synchronization with its related physical entities [5];
- it is a simulation model for the working of the related physical entity or process [6].

The Digital Twin Consortium (DTC) which promotes awareness, adoption, interoperability, and development of Digital Twin technology, defined Digital Twin as a virtual representation of real-world entities and processes, synchronized at a specified frequency and fidelity [7].

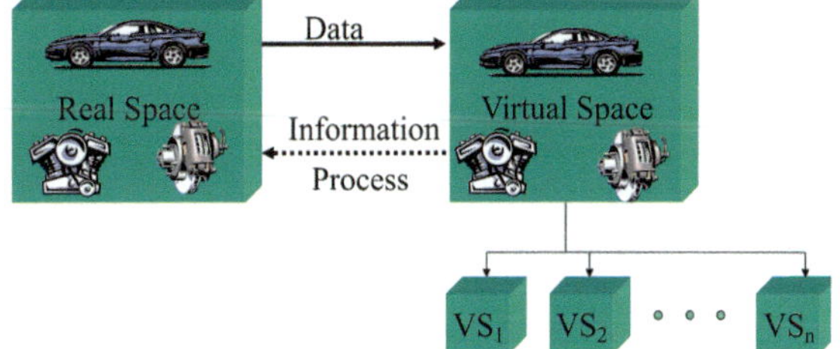


**Fig. 1** Concept of Digital Twin was coined in SME management forum completing the cycle: using PLM information in the sales and service function [1]

Another definition is based on ISO 23247 [8], which described a Digital Twin as a "fit-for-purpose digital representation of an observable manufacturing element (OME) with synchronization between the OME and its digital representation". OMEs include personnel, equipment, materials, processes, facilities, environment, products. Figure 2 shows a diagram of Digital Twin in manufacturing based on this definition. According to a report by National Academy of Sciences focused on the foundational research gaps and future directions for Digital Twins, the definition was proposed as [14]: a set of virtual information constructs that mimics the structure, context, and behavior of a natural, engineered, or social system (or system-of-systems), is dynamically updated with data from its physical twin, has a predictive capability, and informs decisions that realize value. The bidirectional interaction between the virtual and the physical is central to the digital twin.

Feng et al. 2023 demonstrated an implementation of the ISO23247 for data requirements for Digital Twins in additive manufacturing [8].

DTC introduced the Digital Twin Capabilities Periodic Table (CPT) which is an architecture and technology agnostic requirements definition framework, see Fig. 3. It is aimed at organizations who want to design, develop, deploy, and operate Digital Twins based on use case capability requirements versus the features of technology solutions [9].

Several systematic bibliometric analyses have been conducted in the literature targeting Digital Twin framework in the context of smart manufacturing. As an example, Warke et al. 2021 surveyed evolution, background, and implementation of critical enabling technologies such as data-driven decision-making, machine learning, artificial intelligence, and deep learning. In their study, Warke et al. 2021 leveraged Scopus and Web of Science databases from 2016–21 for their analysis [10]. Dalibor et al. 2022 presented a systematic mapping of 356 Digital Twins publications to characterize the applications in different domains. 20 different application domains were presented, but only 70% of the studied work corresponded to the manufacturing industry. Those applications were related to monitoring and controlling either before production, to improve the design process, or during product lifetime [11].

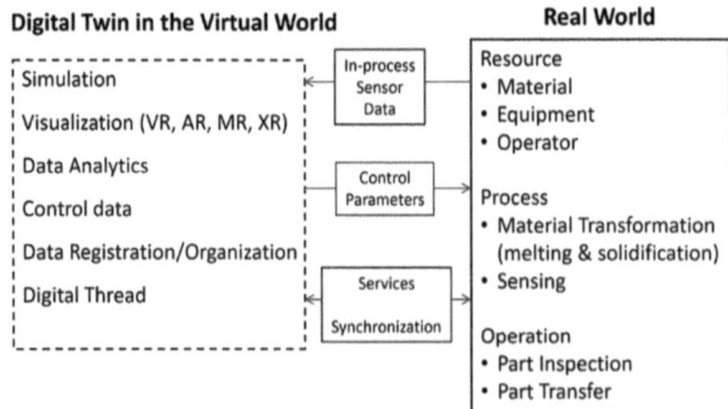


**Fig. 2** Diagram of Digital Twin implementation based on ISO23247 [8]

| 1<br>Data Acquisition & Ingestion | 9<br>Synthetic Data Generation | 17<br>Enterprise System Integration | 23<br>Edge AI & Intelligence | 29<br>Prediction | | 39<br>Basic Visualization | 45<br>Dashboards |
|---|---|---|---|---|---|---|---|
| 2<br>Data Streaming | 10<br>Ontology Management | 18<br>Eng. System Integration | 24<br>Command & Control | 30<br>Machine Learning ML | | 40<br>Advanced Visualization | 46<br>Continuous Intelligence |
| 3<br>Data Transformation | 11<br>Digital Twin (DT) Model Repository | 19<br>OT/IoT System Integration | 25<br>Orchestration | 31<br>Artificial Intelligence AI | 35<br>Prescriptive Recommendations | 41<br>Real-time Monitoring | 47<br>Business Intelligence |
| 4<br>Data Contextualization | 12<br>DT Instance Repository | 20<br>Digital Twin Integration | 26<br>Alerts & Notifications | 32<br>Federated Learning | 36<br>Business Rules | 42<br>Entity Relationship Visualization | 48<br>BPM & Workflow |
| 5<br>Batch Processing | 13<br>Temporal Data Store | 21<br>Collab Platform Integration | 27<br>Reporting | 33<br>Simulation | 37<br>Distributed Ledger & Smart Contracts | 43<br>Augmented Reality AR | 49<br>Gaming Engine Visualization |
| 6<br>Real-time Processing | 14<br>Data Storage & Archive Services | 22<br>API Services | 28<br>Data Analysis & Analytics | 34<br>Mathematical Analytics | 38<br>Composition | 44<br>Virtual Reality VR | 50<br>3D Rendering |
| 7<br>Data PubSub Push | 15<br>Simulation Model Repository | 52<br>Device Management | 54<br>Event Logging | 56<br>Data Encryption | 58<br>Security | 60<br>Safety | 51<br>Gamification |
| 8<br>Data Aggregation | 16<br>AI Model Repository | 53<br>System Monitoring | 54<br>Data Governance | 57<br>Device Security | 59<br>Privacy | 61<br>Reliability | 62<br>Resilience |

Data Services Integration Intelligence UX Management Trustworthiness

**Fig. 3** Digital Twin capabilities periodic table proposed by DTC [9]

Figure 4 shows network analysis of co-occurrence of keywords used in Warke et al. 2021 study from Scopus database.

Several national funding initiatives for Digital Twin development and proliferation have emerged to address technology requirements, data structure, cyber-security provisions, and capability development in diverse industrial sectors. One of such examples is *Change2Twin*, a forward-looking initiative funded by EU since 2020 on a variety of industrial applications. This EU-funded project helps manufacturing

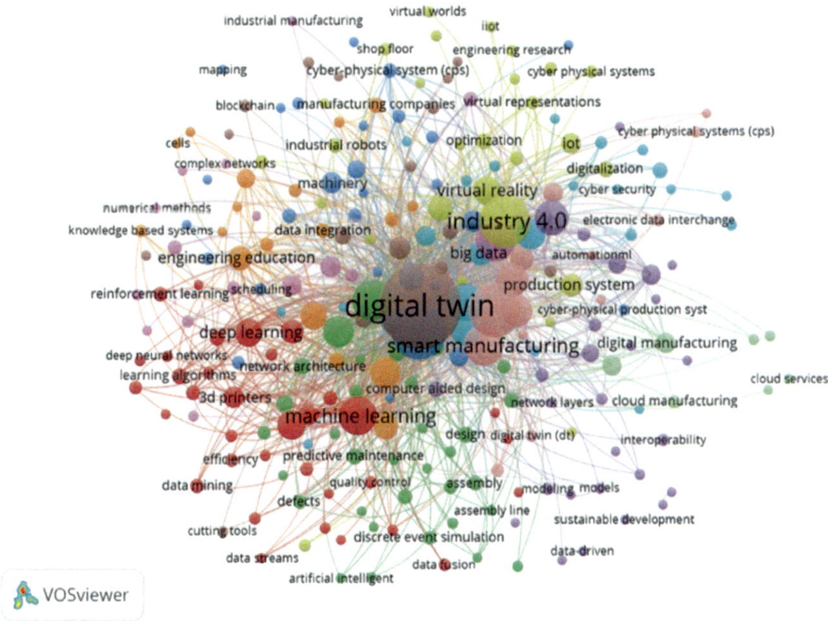


**Fig. 4** Network analysis of co-occurrence of keywords in Scopus [10]

subject matter experts (SMEs) in their digitization efforts to deploy Digital Twins. They are a consortium of 18 partners from all over Europe and promote the idea that the concept of Digital Twin is one of the big game-changers in manufacturing allowing companies to significantly increase their global competitiveness [12].

With a value of $12.9 billion in 2022, the Digital Twin market is expected to grow at a CAGR of 35% to 40% through 2030, according to industry analysts. This market growth is primarily driven by the increasing adoption of enabling technologies such as AI, enterprise internet of things (IoT) platforms, augmented reality (AR), and virtual reality (VR) [13]. In a comprehensive report by National Academies (funded by the Department of Defense—Air Force Office of Scientific Research and Defense Advanced Research Projects Agency, the Department of Energy, the National Institutes of Health, and the National Science Foundation) the opportunities enabled by Digital Twin were defined and the foundational research and resources needed to support the development of Digital Twin technologies were identified [14]. The report presented critical future research priorities and an interdisciplinary research agenda for the field, including how federal agencies and researchers across domains can best collaborate.

## *1.2 Digital Twins for Environmental Impact and Sustainability*

Digital Twins play a pivotal role in enhancing sustainability efforts across various industries. The ability to simulate and analyze operations in real-time allows businesses to identify and implement sustainable practices. As industries increasingly prioritize environmental responsibility, Digital Twins emerge as powerful tools to monitor, analyze, and minimize carbon footprints across diverse sectors. In manufacturing, these twins optimize energy consumption and production efficiency, reducing carbon footprints. In agriculture, Digital Twins aid precision farming, minimizing resource wastage. Smart buildings employ Digital Twins to optimize energy usage, contributing to eco-friendly urban development. Below is a few examples.

Data from two enterprises was used as a case study in an industrial DevOps research project called *Titan* [15], which was focused on methods and tools for integrating and analyzing big data from IoT devices in industrial manufacturing and used to estimate power consumption. The authors proposed measures that can be implemented in an industrial DevOps analytics platform, the *Titan Control Center* [16]. Henning et al. 2021 discussed two industrial pilot cases, where analyzing power consumption data can serve the goals of reporting, optimization, fault detection, and predictive maintenance. In a pilot implementation of a power consumption analytics platform, they showed how measuring real-time data processing, multilevel monitoring, temporal aggregation, correlation, anomaly detection, forecasting, visualization can be implemented with a microservice-based architecture (see Figure 5), stream processing techniques, and the fog computing paradigm, see Fig. 6 [16].

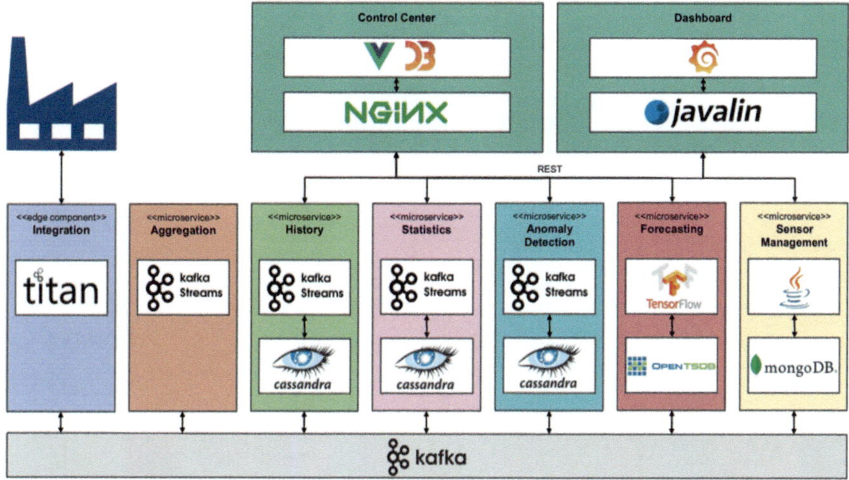


**Fig. 5** Microservice-based pilot architecture of the Titan Control center for analyzing electrical power consumption [16]

Seegrun et al. 2023 conducted a systematic literature review (SLR) through Scopus and Web of Science. Their query retrieved 523 records on Scopus and 281 on Web of Science (data extraction in October 2022). After the elimination of 249 duplicates, the dataset contained 555 records. These publications were categorized according to their industry focus.

From the search results, application of Digital Twins were allocated throughout a product's lifecycle and assessed in terms of their technological maturity and sustainability scope. To analyze the state of the art of research on Digital Twins in the context of sustainability and to frame directions for future research, a SLR was conducted to address the following research questions (RQs):

- **RQ1**: What are the application scenarios of Digital Twins in the context of sustainability in the manufacturing industry?
- **RQ2**: In which product lifecycle (PLC) phases are Digital Twins applied in the context of sustainability?
- **RQ3**: What is the technology readiness level (TRL) of the Digital Twin applications?
- **RQ4**: For the attainment of which sustainability objectives are Digital Twins currently applied?
- **RQ5**: What are the key directions for future research in the manufacturing industry for sustainability-focused applications of Digital Twins?

The authors divided a product lifecycle into three stages: Beginning-of-life (BoL), Middle-of-life (MoL) as well as End-of-Life (EoL) which are in turn subdivided into the following lifecycle phases:

- BoL: Planning (BoL1), product development (BoL2), production (BoL3)

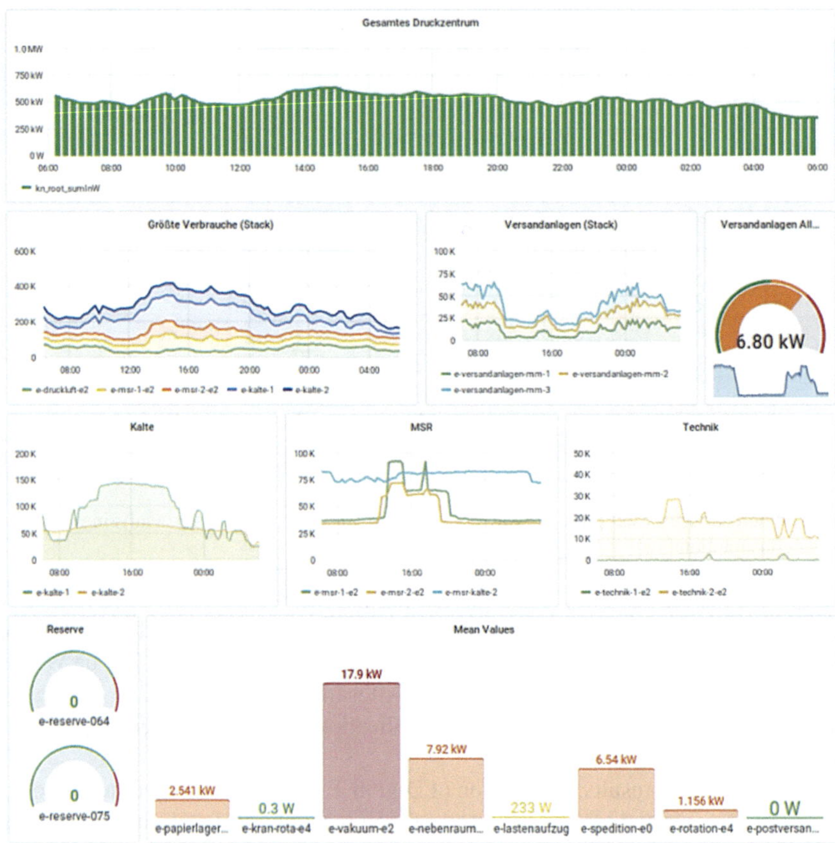


**Fig. 6** Titan dashboard implemented in Grafana [16]

- MoL: Distribution (MoL1), usage (MoL2), service (MoL3)
- EoL: Reuse/remanufacturing (EoL1), recycling (EoL2), disposal (EoL3).

In terms of technological maturity of the implementations of Digital Twin identified in the research by Seegrun et al. 2023, five levels of technology readiness levels (TRL) were defined, see Fig. 7:

- TRL 1: Basic principles observed (e.g., SLR, survey)
- TRL 2: Technology concept formulated (e.g., conceptual framework)
- TRL 3: Experimental proof of concept (e.g., validated conceptual framework)
- TRL 4: Technology validated in academic environment.
- TRL 5: Technology demonstrated in relevant environment [17].

Figure 8 shows the share of the addressed lifecycle phases within the literature body. Most Digital Twin approaches proposed are primarily located in the BoL stage or span across multiple lifecycle phases.

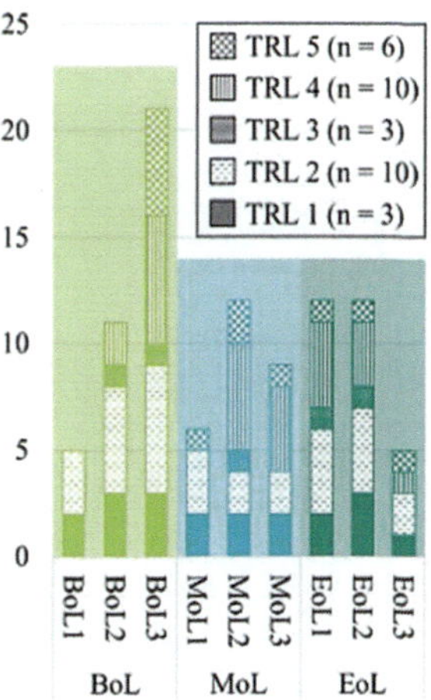


**Fig. 7** Distribution of literature body according to product lifecycle and technology maturity level [17]

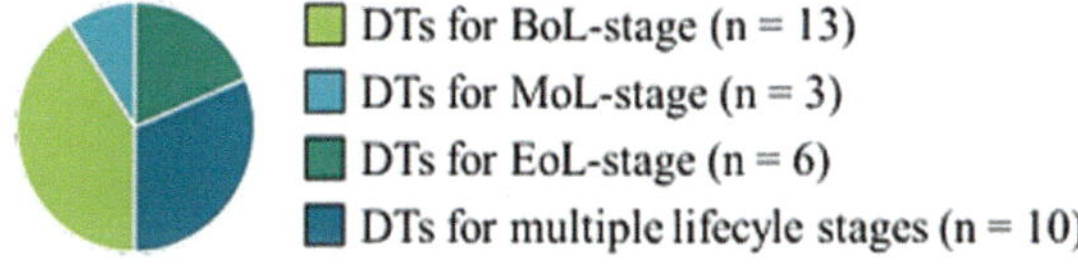


**Fig. 8** Application potentials of Digital Twins within the literature body based on the stage in product lifecycles [17]

XMPro inc provides a generic Digital Twin platform that can track real-time data, provide decision support From multiple data sources [18]. In one example, the platform was used to provide a virtual representation of the wind farm (see Figure 9) synchronized at a high frequency and fidelity, which enables real -time monitoring, analysis, and optimization. By integrating data from sensors and historical performance, the Digital Twin can predict the wind farm's behavior and identify potential issues before they occur, enabling proactive measures to be taken. The Digital Twin remote operations center provides a platform for remote decision-making, allowing operators to monitor and predict changes without disrupting operations, resulting in improved performance, reduced maintenance costs, and enhanced safety.

Abdune et al. 2023 proposed a data-driven methodology for integrating the energy consumption model into Digital Twins using techniques such as segmentation and regression. It relies on power absorption measurement of industrial equipment to generate energy consumption related parameters to be fed into the DT model to monitor the current operating condition of the physical system. A case study on an industrial robot was used to validate and assess the performance of the approach in a laboratory environment [20].

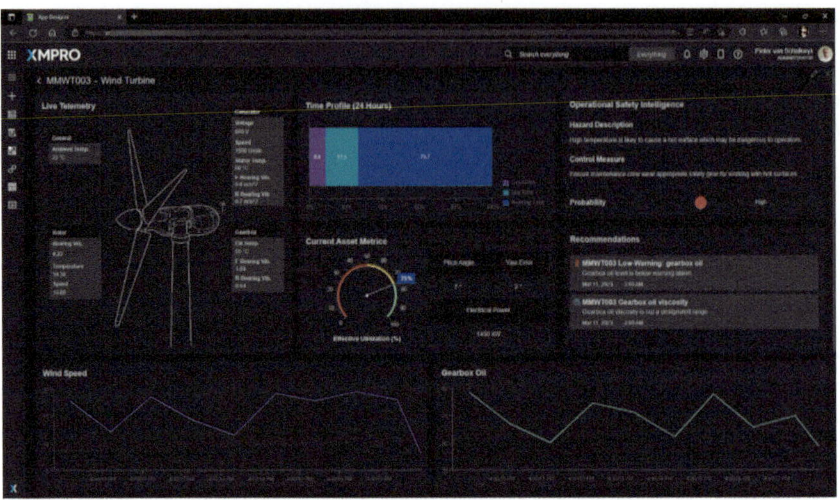


**Fig. 9** Wind farm Digital Twin dashboard developed in XMPro [19]

### 1.2.1 Categories of Digital Twins

Physics-based category

Physics-based modeling describes the behavior of a system using physics equations and first principles. For example, Asrai et al. (2018) provided a modeling approach that can trace the energy flow to compute the energy consumed in milling processes based on active mechanisms of energy conversion within a machine tool [21]. Lv et al. (2017) analyzed the spindle acceleration energy consumption of machine tool (CNC) lathes based on the computation of the moment of inertia for the spindle drive system [22]. For many situations physics-based modeling methods are challenging to implement in the real-world applications because these models require the foundational understanding of the governing physics and accurate physical properties for materials, and domain knowledge which isdifficult to quantify or estimate. In one example, a virtual machine tool to estimate the energy consumption during machining condition was also presented [23]. A genetic algorithm was then used to reduce energy consumption. Seow et al. (2013) proposed an approach to estimate the impact of direct and indirect factors of manufacturing on energy consumption through what-if scenario simulations in order to reduce energy [24].

Data-Based Category

Data-based models are also referred to as data models, empirical models, or artificial intelligence methods. This category of Digital Twins mostly depend on time-series statistical analyses and machine learning techniques. It can be considered as statistical modeling based on systematic evaluation of experimental results. As an example for energy consumption monitoring and forecast, Karanjkar et al. (2018) proposed an IoT-driven Digital Twin for energy optimization of an automated printed circuit board assembly line with various sensors, as well as an open-source software platform for data fusion and the energy Digital Twin [25]. Data-driven techniques do not involve explicitly modeling the underlying physical system. Artificial Neural Network (ANN) is the most used algorithm; Support Vector Regression and Random Forest are other much used models [26]. Some researchers used unsupervised deep learning to select and extract features of energy consumption combined with supervised deep learning to predict energy of machine tools [27]. Another approach proposed was based on sensors to monitor the energy to distinguish between cases in which a large amount of machine data is available allowing machine learning methods and those in which only basic process information are accessible [28].

Intel published a white paper on optimizing factory performance using Digital Twins and listed some of their Automated Factory Solutions (Intel® AFS). Two of the examples from these solutions include Intel® Factory Recon which enables manufacturers to use game-like, immersive graphics capabilities to instantly visualize their operations better than ever before—how they're running now, how they ran in the past, and how they might run in the future. This tool is expected to help with reducing Mean Time to Repair (MTTR) for factory incidents. Another example is Factory Optimizer which is an AI-based control layer for simulation optimization and offline analysis. It's a tool that allows engineers to quickly and easily test changes to factory assets or conditions [29]. TSMC has implemented data models and machine learning for quality management, such as auto defect classification (ADC) using Human-in-the-loop (HITL) approach which leverages both human and machine intelligence to create machine learning models [30]. Similarly, Micron uses AI to improve yield and quality of memory device manufacturing [31].

#### Heuristic and Semi-empirical

Considering the complexity of many manufacturing processes, in particular in multi-physics, multiscale (length and time), it is imperative to fill the gaps in the theory and fundamental physics, chemistry, material science, etc. with empirical insights. Heuristic models are generally experience-based methods which are used to reduce the need for calculations pertaining to some less studied characteristics of the problem, such as equipment size, performance, or process/operating conditions, nucleation, crystal growth faceting, etching in large aspect-ratios, etc. Such models are usually used to describe a phenomenological process where not all the details are understood from the scientific perspective. As an example, Chen et al. 2023 used a heuristic approach to study an automated smart manufacturing framework based on Digital Twins and Blockchain for IIoT.[4] The data used in the Digital Twins are all from the cluster generated after blockchain authentication [32]. Rehman et al. 2023

[4] Industrial Internet of Things.

proposes a heuristic approach for cognitive Digital Twin technology. The author proposed the heuristic approach as a feature selection tool to enhance the cognitive capabilities of a Digital Twin throughout the product design phase of production. The proposed approach was validated using the use-case of Power Transfer Unit (PTU) production [33].

### Hybrid-Based Category

Hybrid modeling refers to the situation where part of a model can be formulated on the basis of first principles and part of the model has to be inferred from data because of a lack of understanding of the mechanistic details [34]. Huang et al. 2022 proposed a modeling framework for hybrid learning-based Digital Twin for manufacturing process with the corresponding trial implementation. The authors constructed a data processing procedure to contextualize metadata sources across the process chain, and a modeling pipeline for the integration of production domain knowledge and AI techniques [35]. Langlotz et al. 2022 published a concept for Digital Twins that are modeled by physics-based and data-driven models. Furthermore, this concept is validated with the help of a use case that controls an energy management system of a model-scale factory by finding the optimal time for charging and using the battery [36].

Reduced order models (ROM) or surrogate models are critical to enable real-time Digital Twins and expedite the run-time of models. These methods are in particular critical for high-fidelity physics or hybrid models where solving massive set of equations resulted from discretization of computational domain into thousands or millions of grid point or mesh cells. Some of such methods are reviewed by Es-haghi et al. 2024 [37].

Another important area of research for enabling predictive Digital Twins, in particular for applications where the data is scarce (or sparse) are AI-assisted physics models, sometimes referred to as physics informed machine learning (PIML), physics-informed neural networks (PINN). Raissi et al. 2019 are among the pioneers of the PINN method [38]. In addition, Graph Neural Net (GNN) in another technique with increased popularity. Based on inherent advantages of graph-based structures for representing unstructured data, graph neural network has been receiving more and more attention in predicting flow fields defined on unstructured computing nodes. Some of these methods are based on Scarselli et al. 2009 [39] and Sanchez-Gonzalez et al. 2020 [40].

### State-Based Category

State-based models are related to the segmentation of the system into different functions and sections, each of which is typically represented by an average behavior or characteristics. For example, in the context of energy consumption and monitoring, at the machine level, the energy consumption of the various operating modes—switch-off, start-up, waiting, idle, ready for processing, and processing—are typically examined here. An average energy usage is calculated for each operating mode and processing step. The energy usage is then predicted by merging the average consumption for each operating mode as well as the transition from one state to another. Wang et al. (2019) used this Digital Twin approach to create an event-driven online machine state decision for an energy-efficient manufacturing system [41]. Jia et al. (2017) suggested a finite state machine to model energy consumption transient state during machining using the Pareto principal to identify the important ones and then constructing a state transition chart to measure their duration [42]. Others such as Dietmair and Verl 2009 proposed modeling machine energy consumption based on statistical discrete states [43].

### 1.2.2 Examples of Digital Twins Used for Sustainability Applications

The *interTwin project* is funded by the European Union Horizon Europe Program (2023-25) aimed at designing and building a prototype of an interdisciplinary Digital Twin engine (DTE), based on a co-designed Blueprint Architecture. The mission of the Digital Twin engine is to simplify & accelerate the development of complex application-specific Digital Twins that benefits researchers, business, and civil society. The effort attempts to extend the technical capabilities of the European Open Science Cloud with modeling & simulation tools, demonstrate data fusion with complex modeling & prediction technologies among other things. The program has described several deliverables, one of which is to support the climate change use cases and the implementation of the related impact decision support tools. Some of the use cases of the program include studying tropical storms change in response to climate change—The goal is to provide notebooks for scientists and policy makers for running analysis on tropical cyclones on future projection data. Other example of use cases are wildfire risk assessment in response to climate change, and flood early warning in coastal and inland regions [44].

Another European research project called *Trustworthy virtual experiments and Digital Twins (ViDiT)* was started, funded by the European Union through the European Partnership on Metrology, co-financed from the EU's Horizon Europe Research and Innovation Program and by the Participating States.

ViDiT focuses on virtual experiments and Digital Twins as key enabling technologies to achieve and realize European strategic policies devoted to sustainability and digitalization within the complex framework of Industry 4.0 and the European Green Deal. The consortium of this three-year project consists of 21 institutions: 8 National metrology institutes, 2 research centers close to industry, 5 universities and 6 companies. The project is coordinated by Physikalisch-Technische Bundesanstalt (PTB), based in Braunschweig, Germany [45].

Green material optimal selection (GMOS) in product design is a key issue for realizing sustainable manufacturing. To improve the accuracy and efficiency for

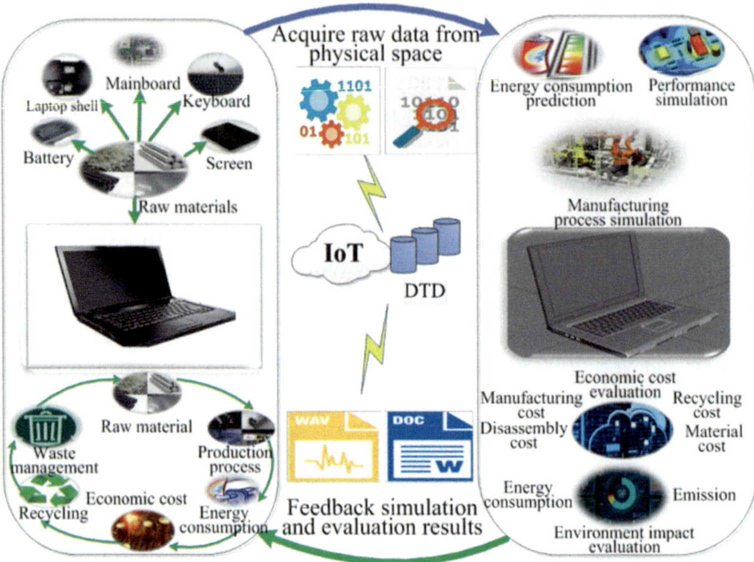


**Fig. 10** Optimal selection of green material for manufacturing based on Digital Twins [46]

green material optimal selection in product, a new method driven by Digital Twin was proposed by Xiang et al. 2019 [46]. In their method, first a high-fidelity Digital Twin of the physical product was developed. Then the actual data (physical) and the cyber (modeling) data was combined, and the performance of the green material selection was evaluated. The team used a laptop design example to showcase the methodology, see Fig. 10.

Tao et al. 2018 presented a five-dimensional model for Digital Twins. In Tao's model M_DT = (PE, VE, DD, Ss, CN), where PE refers to physical entity, VE is virtual entity, DD is Digital Twin data, Ss is service for PE and VE, and CN is connection among all PE, VE, and Ss. Xiang et al. 2019 proposed a variation of this methods for GMOS where the 5D variation for Digital Twin model (5D-EDTM) is divided into two parts M = (PE, VE, SE) + (DE, CE); one part consists of PE which refers to evolutionary model for physical space, VE refers to evolutionary model for virtual space, SE is the evolutionary model in service, DE stands for data evolutionary model, CE refers to connection model among all dimensions. Therefore, 5D-EDTM for GMOS can be divided into five models [6]. Implementation of the 5D-EDTM driven GMOS method applied to GMOS is depicted in Fig. 11.

As described earlier, Henning et al. [16] used the Titan architecture to monitor and analyze the power consumption for two pilot cases. Davila et al. 2023 performed a similar analysis included other types of energy different from electricity (i.e., gas, heat, compressed air) and relevant resources for sustainability, such as water and raw materials [47]. Their study also considered the privacy concerns of the data from different sources or stakeholders. Basically, they proposed a pre-processing data to

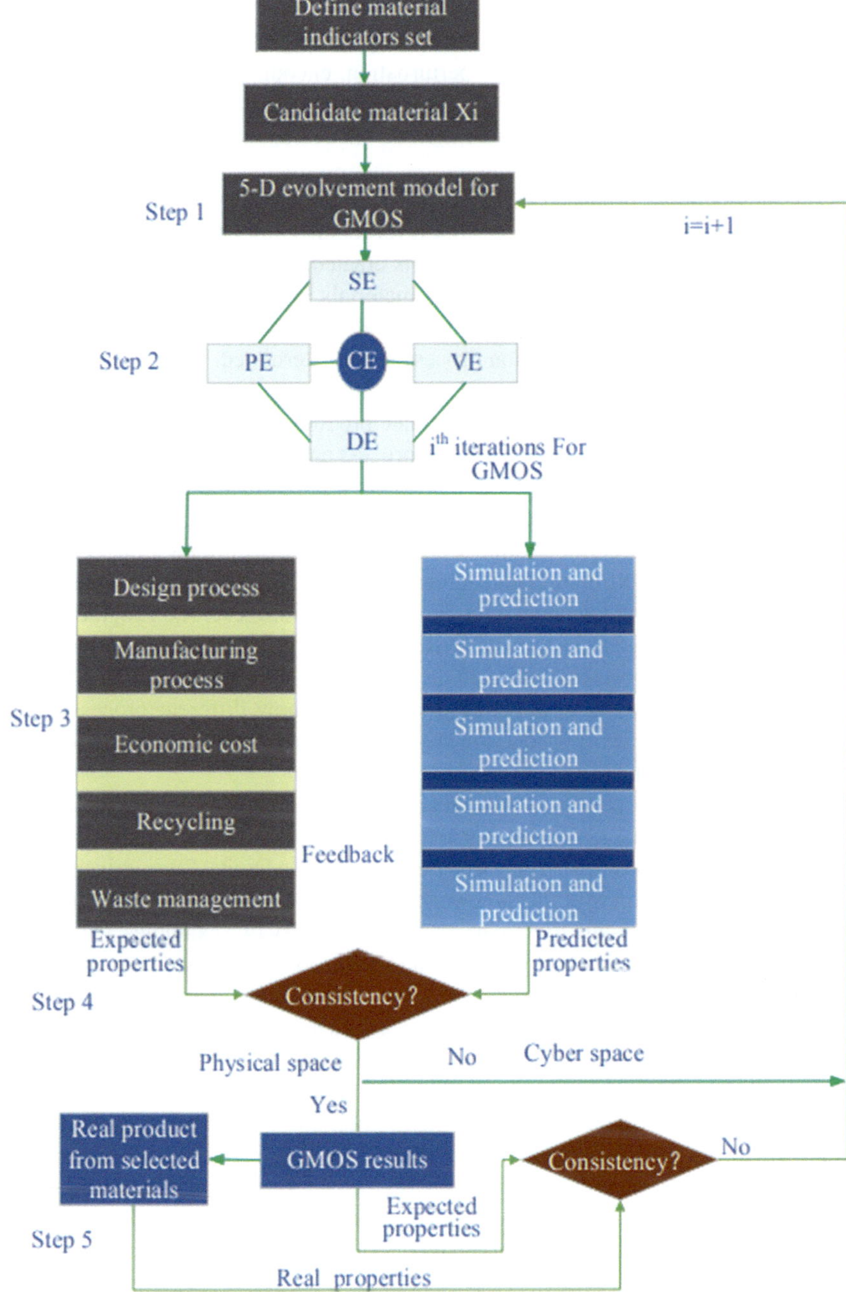


**Fig. 11** Digital Twin-based method for green material selection methodology, 5D-EDTM [46]

obfuscate the protected information about the production process or the machines used. There are many approaches to achieve privacy-preserving data sharing, for example encryption, anonymization, perturbation, encoding or more complex data synthesis (creating data synthetically). Choosing the appropriate technique depends mainly on how the data is going to be used downstream, and what level of privacy, or disclosure risk is desired.

The methodology proposed by Davila et al. 2023 consists of three levels for the development of the Digital Twin. In first level (connection and communication), a process model of the manufacturing steps is created. The first step is defining the system boundaries, starting with the raw materials, and ending with the finished product. Afterward, all the required inputs and outputs of each of the machines involved in the different production processes are determined. The different inputs can be electricity, heat, gas, compressed air, or materials, within others. The output are typically the processed materials, intermediate products and finally the product. All the relevant measurement points within the process are determined, to install sensors required for the measurement of the inputs. As a result of the process modeling, the flow of resources into and out of the system boundaries can be measured and monitored in real-time. The second level is concerned with data collection from all the machines and different data sources (and types), such as sensors, machines, production, supplier, and weather data. Finally, in the third level (i.e., cognitive, and analytical level), the data is used to perform energy efficiency and sustainability assessments and the costs and estimate the durations per product, see Fig. 12.

In an example, Davila et al. 2023 demonstrated how different processes that involve a set of machines or production steps can be considered for an overall energy efficiency and sustainability analysis.

Ma et al. 2022 proposed a sustainable smart manufacturing strategy based on information management systems for energy-intensive industries (EIIs) from the product lifecycle perspective. Their study encompasses a Digital Twin-driven operation mechanism, and an overall framework of big data cleansing and integration are designed to describe sustainable smart manufacturing. By two examples, the authors showed that the unit energy consumption and energy cost of production as well as the costs of environmental protection can be decreased by implementing their proposed strategy based on the 'cradle-to-gate' lifecycle data analysis [48].

Li et al. 2022 developed a data-driven hybrid petri-net (DDHPN) [49, 95] inspired by both the state-based energy modeling and machine learning for establishing the energy behavior meta-model. Gaussian kernel extreme learning machine was proposed to fit the instantaneous firing speed of energy consumption continuous transitions in DDHPN. DDHPN-based energy behavior model was driven by physical data under real-time working conditions, operating parameters, and production load for generating a virtual data space of energy management, See Figs. 13, 14, 15, and 16.

In the context of sustainable manufacturing, the Triple Bottom Line TBL supports the concept of sustainability in industries by clarifying three dimensions: economic, environmental, and social [50]. The concept of triple bottom line has been introduced over a decade ago and is being discussed more recently in the context of Digital Twins

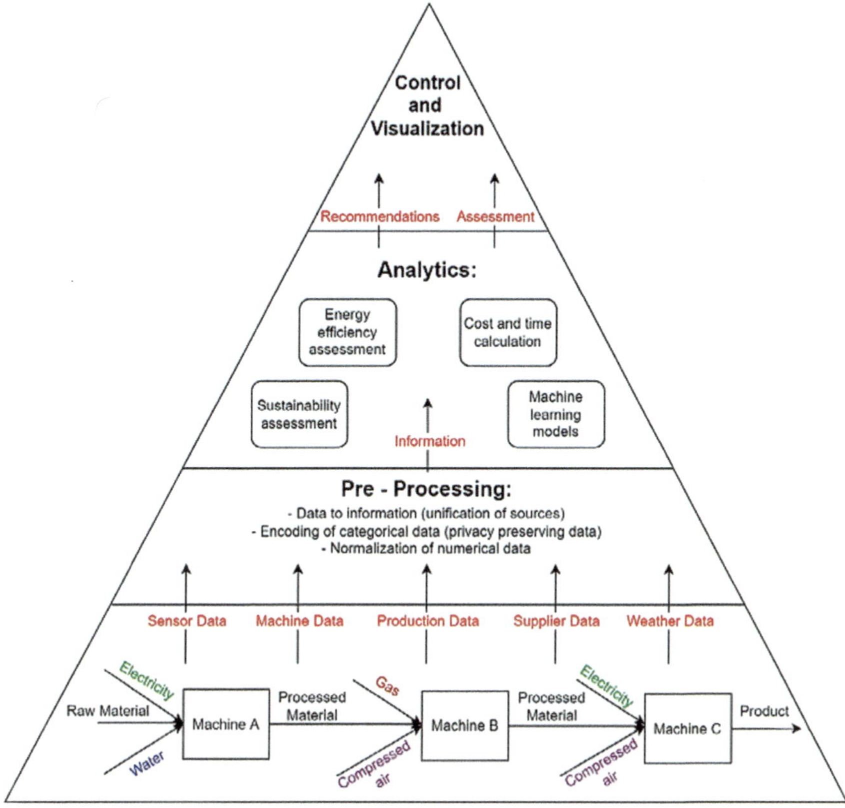


**Fig. 12** Sustainability Digital Twin architecture proposed by Davila et al. 2023 [47]

for sustainability. For example, Miehe et al. 2021 attempted to stimulate interdisciplinary scientific discourse by discussing the interpretation of sustainable production and the role of Digital Twins [51]. They concluded that although sustainability represents a key factor of future production, it is not conclusively defined in order to be technically applicable. Existing (bottom-up) approaches assessing the contribution of DTs to sustainable production are therefore not considered comprehensively from a contemporary perspective. In order to navigate the technical development of DTs in an appropriate direction in the sense of sustainable production, a further (top-down) perspective is necessary.

Chavez et al. 2022 identified suitable key performance indicators (KPIs) for a Discrete Event Simulation (DES) model and evaluated the impact in a drone factory in four scenarios that test final assembly processes. Based on this project (called *TWINGOALS* and funded by European Institute of Innovation and Technology (EIT) [5294]), the team integrated sustainability indicators in a simulation model of a drone assembly cell at the Stena Industry Innovation Lab (SII-Lab), based at Chalmers University of Technology in Gothenburg, Sweden. This study attempted to answer

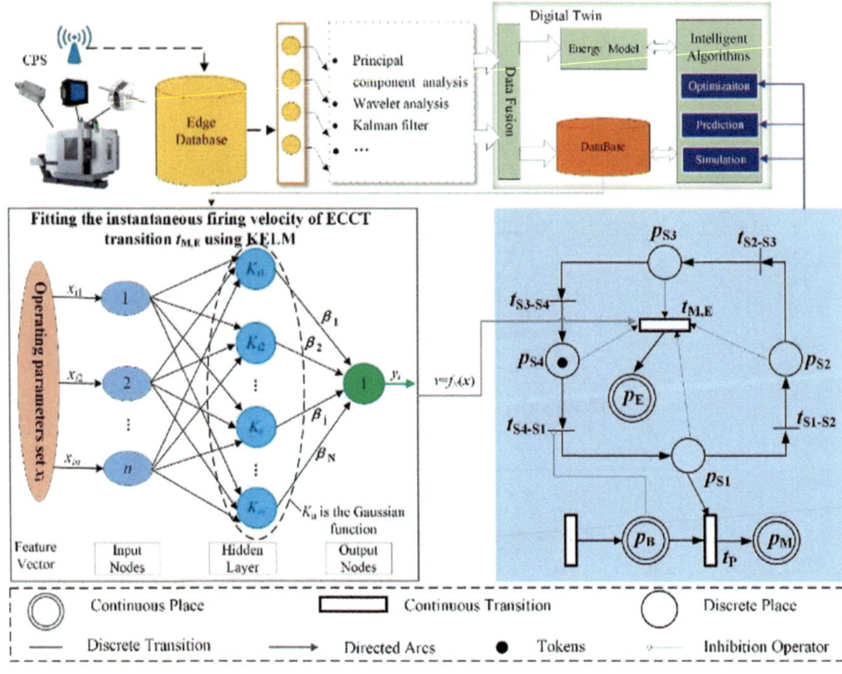


**Fig. 13** Energy behavior modeling based on a data-driven hybrid petri-net by Li et al. 2022 [49]

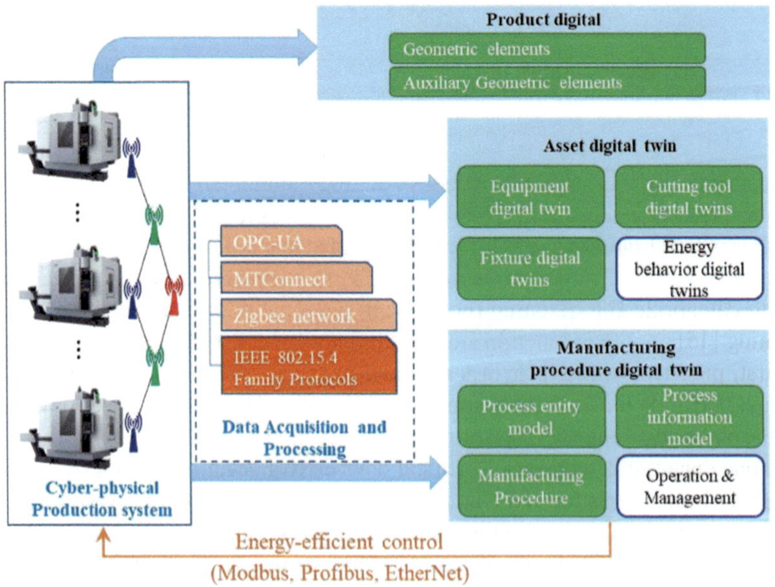


**Fig. 14** Energy management framework proposed by Li et al. 2022 for manufacturing [49]

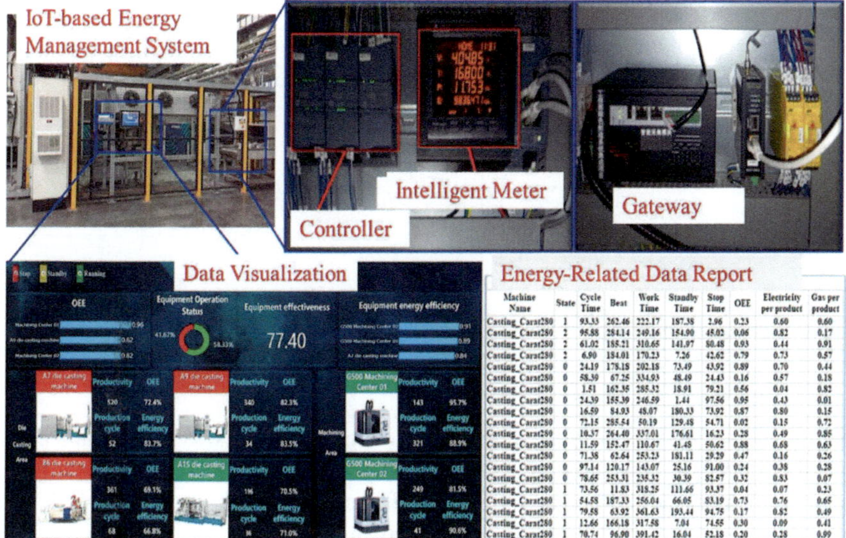


**Fig. 15** IoT-based data collection for energy management system [49]

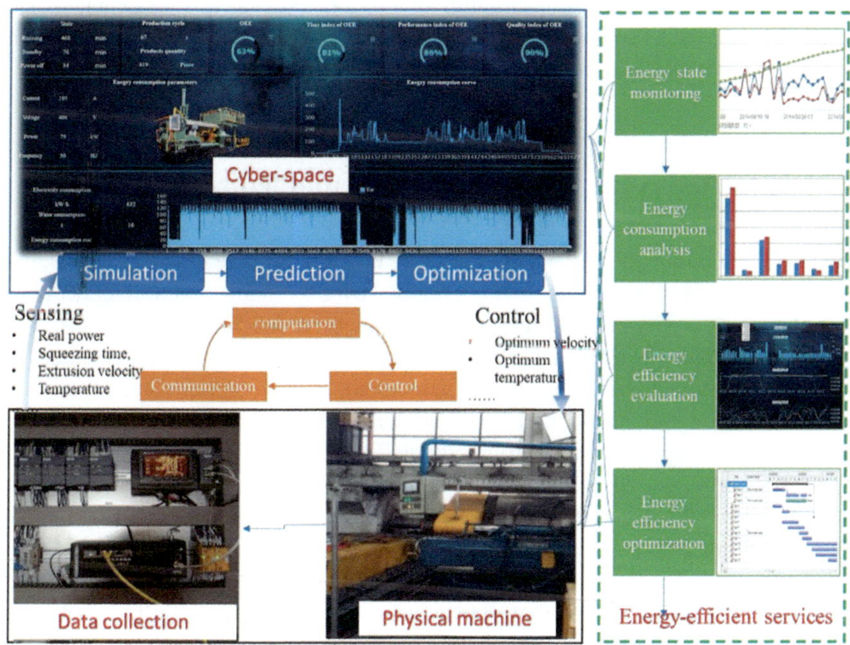


**Fig. 16** Prototype of a Digital-Twin-based energy management system [49]

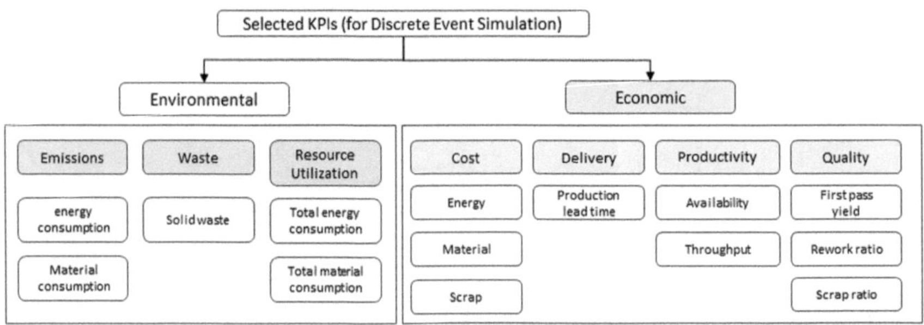


**Fig. 17** Incorporation of KPIs related to the triple bottom line principle in the product lifecycle [53]

two questions first identifying the most suitable KPIs to assess the sustainability of production systems through a DES (or Digital Twin), and second, identifying improvement sustainability opportunities of production systems. See Fig. 17 for an example of the selected KPIs to be used in a Digital Twin model with TBL considerations.

In a study performed by Banerji et al. 2021, a real-time agent-based optimization motion planning for energy consumption of a robotic cellular was investigated. The research proposed a framework a qualitative analysis, and a quantitative comparison based on energy consumption in a robotic experiment case [54].

## 2 Digital Twins for Semiconductor Manufacturing

Since the first publication by Michael Grieves in 2005 [1] there has been remarkable interests in developing and deploying Digital Twins to enable and accelerate smart manufacturing across many industries. Similarly, there has been extensive progress and developments by academia [55]. It is not a surprise that there has been an ever-growing excitement also by the semiconductor industry to partake in adopting the technology; however, semiconductor manufacturing is still in the very early innings of creating and capturing value from Digital Twins with proven efficacy.

In 2022, the US government implemented an industrial strategy to revitalize domestic manufacturing, strengthen American supply chains, and accelerate the industries of the future. The CHIPS and Science Act of 2022 [56], which buildsd on this progress, making historic investments that would potentially poise U.S. workers, local communities, and American businesses to win the race for the twenty-first century. This act promises to strengthen American manufacturing, supply chains, and national security, and invest in research and development, science, and technology in the industries of tomorrow, including nanotechnology, clean energy, quantum computing, and artificial intelligence.

The Industrial Advisory Committee (IAC) advises the Secretary of Commerce on the science and technology needs of the nation's domestic microelectronics industry, the national strategy on microelectronics research, the CHIPS research and development programs and other advanced microelectronics activities authorized under Section 9906 of the CHIPS Act, and opportunities for new public–private partnerships to advance the domestic microelectronics industry [90, 91]. The charter of this working group has been to look at the long-term research needs of the semiconductor industry. The working group will then need to understand where the gaps are, and then suggest priorities to the IAC as to where the focus areas should be for CHIPS funding and the National Semiconductor Technology Center (NSTC) that provide the best opportunities to sustain US leadership in semiconductor innovation. NSTC is a key component of the research and development program established by President Biden's CHIPS and Science Act. One of the recommendations from the IAC on the R&D Gaps Working Group has been the development and use of Digital Twins in semiconductor R&D and manufacturing. On May 6th, 2024, US Department of Commerce announced $285 million Funding Opportunity for a Digital Twin and Semiconductor CHIPS Manufacturing USA Institute [57]. CHIPS Digital Twins Manufacturing USA Institute The objectives of the Institute include: Convene stakeholders across the semiconductor production ecosystem; Improve the state of the art in manufacturing-relevant digital twins; Significantly reduce cost for U.S. chip development and manufacturing; Improve development cycle times of semiconductor product innovation; Advance digital twin-enabled curricula for training a domestic semiconductor workforce; and create a digital twin marketplace for industry to access digital models.

Industries are shaping themselves toward the development of customized and cost-effective processes to satisfy customer needs with the aid of a Digital Twins. Such Digital Twin frameworks are expected to enable the user to monitor, simulate, control, optimize, and identify defects and trends within ongoing processes, and reduces the chances of human prone errors. Digital Twins represent tremendous opportunities to propel semiconductor manufacturing by accelerating prototyping, testing, concept creating, process emulation, and knowledge sharing for talent development. To help manage the increase in complexity, we must use every tool in our toolbox. Recent advances in algorithms, modeling, and artificial intelligence, as well as the integration of sensors and metrology, offer great promise to help overcome these challenges. Predictive maintenance and diagnostics, chamber matching, accelerated process transfer, and equipment performance and productivity improvement are some of the benefits of Digital Twins for equipment and processes. Following up from a workshop by SEMI dedicated to Digital Twins for semiconductor manufacturing [58], The SEMI Smart Manufacturing Initiative published a summary whitepaper with industry members presenting a comprehensive view of Digital Twin technology in semiconductor manufacturing for achieving AI-driven autonomous factories, covering industry definitions, taxonomy descriptions, and challenges in development and deployment.

## 2.1 Framework for Semiconductor Capital Equipment Manufacturing

A successful Digital Twin framework for semiconductor manufacturing would conceptually provide a flexible platform where different dimensions of the complex ecosystem spanning from fleet-level down to on-wafer feature-level, and even materials engineering and atomistic-level, are captured and the interconnectedness among the different levels are considered. In an ideal future implementation of such a framework, the interconnectedness of processing reactors knobs and the quality and performance of the device fabrication will be a two-way relationship where the virtual fab, built upon a series of synchronized and high-fidelity Digital Twins, enables actionable insights (or even autonomous action) for optimized process development in terms of performance, sustainability, cost, time, or other optimization targets. In addition, that future ideal implementation, would provide insights for managerial implications regarding how the process flow, tool, processing reactors can be selected or re-arranged for overall optimum cost of ownership and revenue perspective.

In the realm of Digital Twins, frameworks and platforms specially tailored to the intricate and demanding landscape of semiconductor manufacturing have been a missing link. Recognizing this gap, Applied Materials proposed AppliedTwin™, envisioned to address the unique challenges of the semiconductor industry. AppliedTwin represents the entire integration flow of various processes in the semiconductor manufacturing chain. This holistic approach covers the entire spectrum of Digital Twins application from fab-level models (FabTwin™) down to device (DeviceTwin™), even extending its reach to encompass materials engineering. This comprehensive platform can make significant strides in bridging the gap between the Digital Twin concept and the complex needs of semiconductor manufacturing.

Per AppliedTwin™ framework, Digital Twin for semiconductor manufacturing is a virtual representation of equipment and process built from the best available models, along with sensors and metrology data, guided by domain knowledge, and synchronized at a specified frequency with the physical equipment and process. AppliedTwin predicts performance of the equipment and process with reasonable fidelity and provides actionable insights.

Furthermore, Digital Twins for semiconductor equipment and processes are built on foundations where critical behavior of an equipment and process is modeled and synchronized with the physical/actual system through sensors and metrology data. An essential element of the twin, however, is the domain knowledge which captures process and hardware know-how, see Fig. 18.

Through its multiple classes of twins - such as ChamberTwin™, ProcessTwin™, Tool Twin™, FleetTwin™ - AppliedTwin will provide a digital replica of the actual system (hardware and process) and can enable virtual experimentation and creating actionable insights, such as predicting the impact of reactor design or process twinning knobs on feature-level topology and device performance on wafer, see Fig. 19.

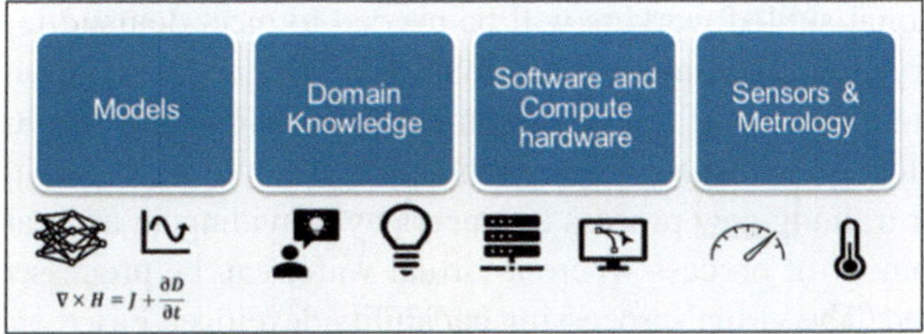


**Fig. 18** Foundations of Digital Twins for semiconductor manufacturing [59]

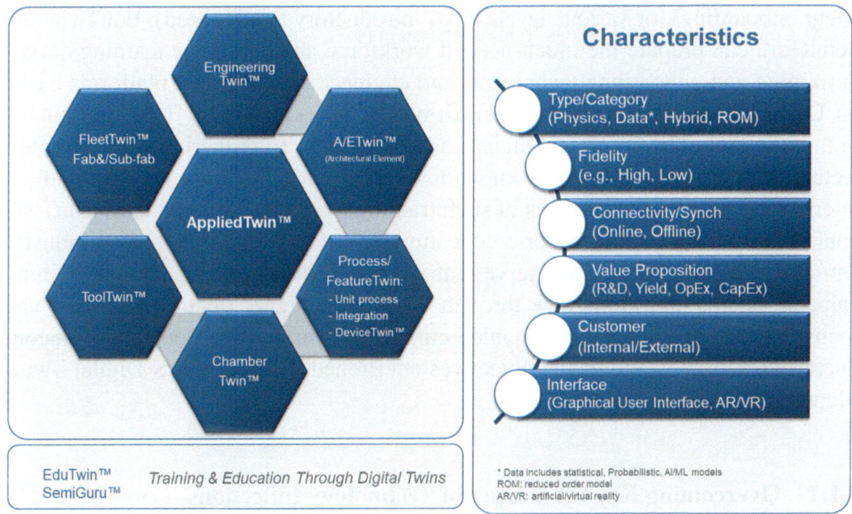


**Fig. 19** Classifications and characteristics of Digital Twins for semiconductor manufacturing [60]

Some of the characteristics of a twin in the context of semiconductor manufacturing are connectivity, user segment, and user interface where the user can interact with twin. the user interaction could of course consist of a graphical user intergace (GUI), augmented/virtual/mixed reality (AR, VR, MR), or a textual and/or real-time natural language communication enables by artificial intelligence. ChamberTwin, for example can be either always in live communication with the physical chamber or synchronized at a certain frequency. The twin could be also tailored to support the design modification of the equipment or tuning a process recipe at the R&D stage, ramp-up, or high-volume manufacturing. Such design modifications can be enabled with the fast developing generative design algorithms and surrogate/ROM model training and deployment methods. An example of ProcessTwin is described by Sathiyanarayanan et al. [61].

It is projected that there will be a tremendous need for STEM and skilled workers needed in near future in order to remain competitive in the semiconductor industry [62–64]. For example, a study by Deloitte suggests that by 2030, more than one million additional skilled workers will be needed to meet demand in the semiconductor industry [65]. To capture the domain knowledge (i.e., codifying the know-how) and sustainable workforce development, EduTwin™ and SemiGuru™ represent all the educational aspects of the twins, for example virtual unit or integrated processing for training new process engineers by providing an accurate replica of a specific equipment or process where a virtual wafer can be processed, and results can be analyzed. The virtual-processing capability developed based on the chamber, processes, or fleet of twins would support democratization of the semiconductor technology knowledge. In multiple scopes, focused on different segmentsof the workforce (,e.g., STEM students in high-schools, community colleges, universities, or talent onboarding), or various levels (e.g., introductory to advanced), EduTwin and SemiGuru can prepare the much-needed workforce and "bend the learning curve" in training and onboarding technicians and engineers. Educational platforms based on Digital Twins can prepare the much-needed workforce and "bend the curve" in training and onboarding technicians and engineers. To properly safeguard intellectual property and proprietary know-how, organizations should develop both an internal and an external versions of such training platforms. For instance, EduTwin could function as a more generic educational platform for training new industry entrants, while SemiGuru could serve as an advanced educational platform containing valuable details and know-how through twins with higher fidelity, accuracy, and focus. Advancing digital twin-enabled curricula for training a domestic semiconductor workforce is one of the objectives envisioned for the CHIPS Digital Twins Manufacturing USA Institute.

### 2.1.1 Overcoming Key Challenges of Technology Inflections

Digital Twins and virtual fabrication clearly offer great promise, but there are challenges that need to be overcome to tap their full potential in the semiconductor manufacturing environment. Semiconductor equipment and processes are highly multiphysics and multiscale environments, with a diverse set of length and times scales. It is extremely challenging to integrate these disparate length- and time-scale models. In many cases, a complete understanding of the underlying physics and chemistry for processes used in the semiconductor manufacturing is lacking. Therefore, a hybrid-model, that not only rely on the physics but also incorporate avaiable test data, are more desired. Another fundamental hurdle is having adequate sensors providing real-time data that can be placed within the harsh environments of the process chambers. The absence of appropriate sensors makes it difficult to build accurate predictive models and algorithms that account for all the nuanced interdependencies and nonlinear effects. Finally, there is the challenge of gathering adequate amount of useful data to train data models and capturing adequate level of physics to matching the models accurately with actual on-wafer parameters. This is exactly why Digital Twins are built based on the best models and will need to be revised and improved whenever more accurate data becomes available or there is a major change in our fundamental understanding of the processes.

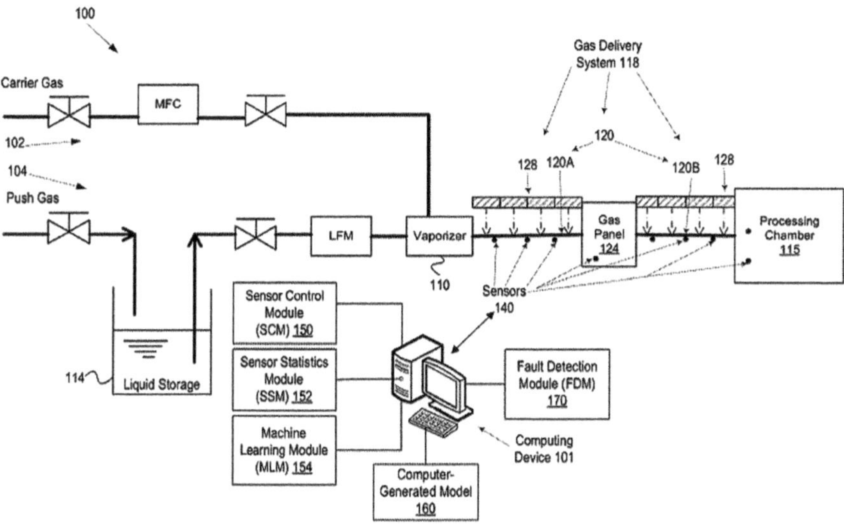


**Fig. 20** Schematic of Digital Twin for fab (FleetTwin) for heater jacker health and proper installation [66]

In one example of Digital Twins application in semiconductor manufacturing, Moradian et al. 2023 proposed using a digital representation of gas and chemical delivery to processing chambers for fault detection and diagnostics for heater jackets health and proper installation after preventative maintenance, see Fig. 20. This application is categorized as a FleetTwin which covers gas delivery for a fleet of processing chambers and can be incorporated in conjunction with other preventative maintenance, diagnostics, or productivity Digital Twins within the manufacturing execution system (MES) software infrastructure of a fabrication facility [66]. Another example is EcoTwin™, which accuratly captures energy and chemical consumption as well as carbon footprint of semiconductor manufacturing [92, 93(described in more details later in this chapter). EcoTwin can be deployed as a chamber-level twin (with inclusion of sub-fab consumptions related to the chamber), or extended all the way to the fleet-level, where all the chambers and other fab and sub-fab equipment are taken into account.

## 3 Sustainability in Semiconductor Manufacturing

The semiconductor industry is poised for significant growth, with expectations of reaching a staggering one trillion dollars in revenue by the year 2030 [67]. However, in light of the current energy consumption and carbon footprint associated with semiconductor manufacturing, it becomes imperative for the industry to make substantial

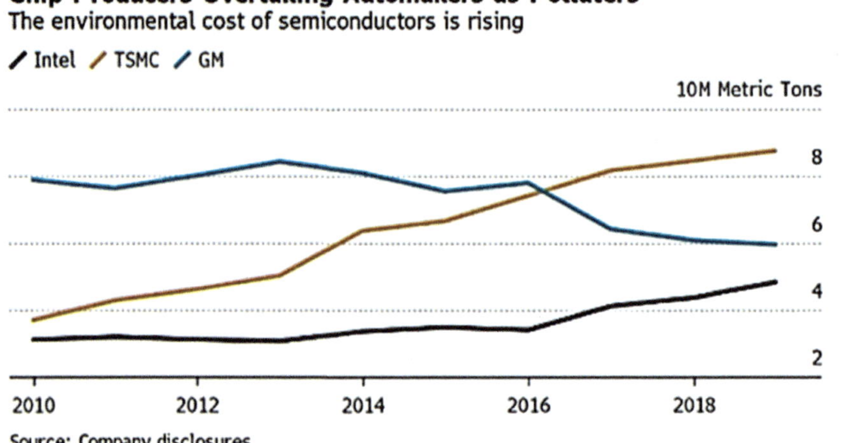


**Fig. 21** Carbon footprint of semiconductor manufacturing exceeding automotive industry [69]

advancements in resource efficiency and environmental sustainability. This transformation is essential not only to seize the forthcoming business opportunities in the next decade but also to ensure a greener and more sustainable future. Figure 21 compared carbon footprint of one of the major semiconductor manufacturers versus a representative of automotive industry and suggests that the semiconductor manufacturing has already exceeded the automotive industry in terms of environmental impact.

The current emissions from computing are almost 4% of the world total. This is already more than emissions from the airline industry and are projected to rise steeply over the next two decades. By 2040 emissions from computing alone will account for more than half of the emissions budget to keep global warming below 1.5 °C. Consequently, this growth in computing emissions is unsustainable. The emissions from production of computing devices exceed the emissions from operating them, so even if devices are more energy efficient producing more of them will make the emissions problem worse [68]. The recent surge in the use of AI and large language models, like ChatGPT, has highlighted the challenge of increased energy consumption. ChatGPT consumes over half a million kilowatts of electricity each day, an amount staggering enough to service about two hundred million requests. ChatGPT's daily power usage is nearly equal to 180,000 U.S. households, each using about twenty-nine kilowatts.

## 3.1 Energy Consumption and Carbon Footprint Challenges

Researchers led by Udit Gupta of Harvard University used publicly available sustainability reports from companies including TSMC, Intel Corp. and Apple Inc. to show that as computing becomes increasingly ubiquitous, "so does its environmental impact."

Information and computing technology is expected to account for as much as 20% of global energy demand by 2030, with hardware responsible for more of that footprint than the operation of a system. "Chip manufacturing, as opposed to hardware use and energy consumption, accounts for most of the carbon output,".

Power use is more dramatic; TSMC's annual electricity consumption is estimated by Greenpeace at 4.8% of Taiwan's entire usage, and more than that of the capital, Taipei. Greenpeace says that will rise to 7.2% once commercial production comes online of TSMC's newest fabs that will shrink the process further from the current leading-edge of 5 nm, or billions of a meter, to 3 nm chips [69].

It is also expected to consume 237 TWh of electricity globally in 2030, close to Australia's 2021 electricity consumption [70].

The Greenhouse Gas Protocol (GHG Protocol) offers companies different levels of commitments to advance their sustainability journey. Emissions from semiconductor device makers fall into different categories [71]:

**Scope 1** emissions arise directly from fabs, primarily from process gases with high global warming potential (GWP) that are used during wafer etching, chamber cleaning, and other tasks; they can also come from high-GWP heat-transfer fluids that may leak into the atmosphere when fabs use them in chillers. For example, $NF_3$ has a global warming potential that is 17.000 times more potent than $CO_2$ [72]. A fraction of these potent gases are released to the atmosphere and directly contribute to global warming.

**Scope 2** emissions arise directly from purchased electricity, steam, heating, and cooling equipment; the major sources include production tools and facilities/utilities.

**Scope 3** emissions include all other indirect emissions in a company's value chain; upstream emissions are those generated by suppliers or their products, while downstream emissions are related to the usage of products containing semiconductors. The number and quantity of materials used for IC manufacturing is steadily growing with every technology node (e.g., Si wafers, bulk gases, gas precursors, minerals, chemicals). The upstream production of these materials also leads to greenhouse gas emissions that must be factored into the emission of the IC chip manufacturing [72].

Majority of the semiconductor fabrication emissions are from scope 1 process gases and scope 2 electricity consumption of tools, see Fig. 22.

Semiconductor companies recognize that many of their customers have set aggressive net-zero targets for their supply chain, see Fig. 23. To address these concerns, some large semiconductor companies have begun to set ambitious sustainability commitments.

Figure 24 shows the comparison of total carbon footprint (or emission) in $kgCO_2$/Wafer for the different technology nodes from N28 to N5. The carbon footprint for

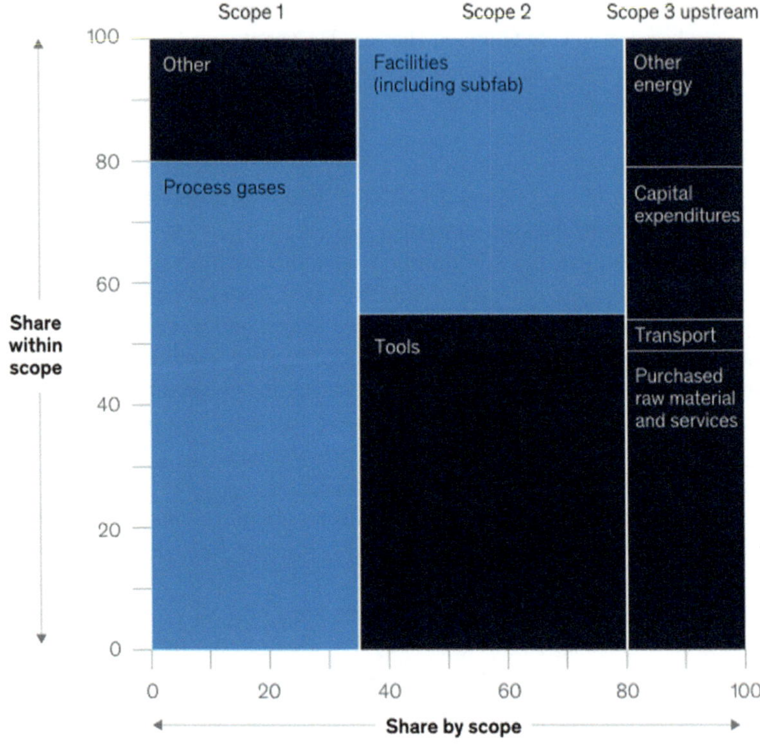


**Fig. 22** Majority of the semiconductor fab emissions are from scope 1 process gases and scope 2 electricity consumption of tools [71, 96]

full flow fabrication of an N5 device is almost two times larger than that of an N28 device technology [73].

Semiconductor companies are also committing to carbon reductions.

- Applied Materials is actively working with customers to look at how to reduce the energy used in the fab during the manufacturing of the chips and aims to achieve net-zero global emissions for scopes 1, 2 and 3 by 2040 (Applied Materials inc., 2023, [87]).
- Lam Research has committed to achieve 100% renewable electricity in 2030, net-zero operations (Scope 1 and 2) in 2040 and net-zero Scope 3 emissions by 2050 (Lam Research inc., 2023, [88]).

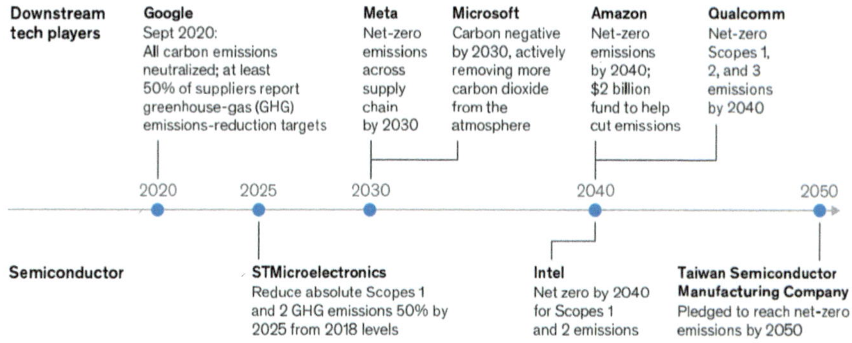


**Fig. 23** Many semiconductor companies and their customers have made aggressive net-zero commitments [71]

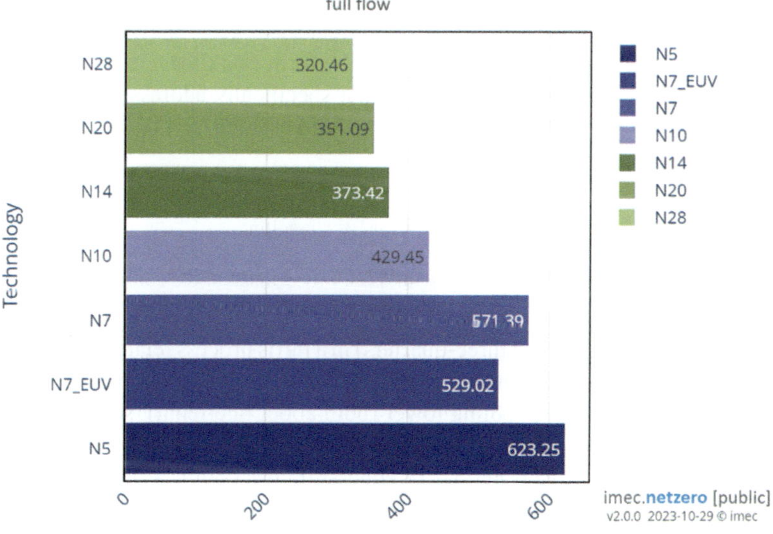


**Fig. 24** Evolution of total carbon footprint for different technology nodes [73]

- ASM International expects to achieve net-zero emissions for scopes 1, 2 and 3 by 2035 (ASMI, 2021, [89])
- ASML expects to achieve net zero GHG emissions from product use at customers by 2040 and emissions in operations (scope 1 and 2) by 2025 [74].

## *3.2 Standards Relevant to the Digital Twins and Sustainability*

SEMI S23 is the standard [75] that addresses concepts related to energy, utilities, and materials measurement and use efficiency of semiconductor manufacturing equipment . Semi S23 is intended to be a tool that can be used to analyze energy, utilities, and materials usage. Other relevant standards include ISO 50001, which provides guidance for energy management systems—Requirements with guidance for use, as well as a framework for continues monitoring and improvement (i.e., Plan, Do, Check, Act) [76]. Another relevant standard is ISO 20140 which provides directions for automation systems and integration and evaluating energy efficiency as well as other factors of manufacturing systems that influence the environment [77]. The German Standards Institute (DIN) requires the Asset Administration Shell (AAS)—a reference framework for Digital Twins [78]—to be suitable for containing sustainability data and to provide it at the end of a product's lifecycle for efficient disposal or recycling. DIN provides a similar concept in the form of a lifecycle record for technical plants [51]. Further standardizations that specifically address isolated aspects of sustainability (e.g., energy efficiency) in the field of Industry 4.0 and Digital Twins is IEC 62,832–3 [79, 80].

## *3.3 Water Consumptions in a Semiconductor Fab*

The largest use of water (about three-quarters) in a fab is process related, with much of that being converted to ultra-pure water (UPW) needed for production itself, followed by the facility scrubber and cooling tower (both about one-tenth), see Fig. 25. Fabs typically have separate circuits for ultrapure water (UPW), which can be hot and cold, and lower purity (LP) water. UPW generation is a complex, multistep process that also consumes significant amounts of power. Most fabs have some level of UPW reclamation, although rates vary widely among fabs and processes within a fab. Many processes associated with water usage also contribute to the industry's carbon footprint through their consumption of energy. The latest update to S23, SEMI's guidance for energy use in the fab, assigns a value of 9 kWh/m$^3$ to generate cold UPW, and 92 kWh/m$^3$ for hot UPW. S23 does not declare a value for wastewater management.

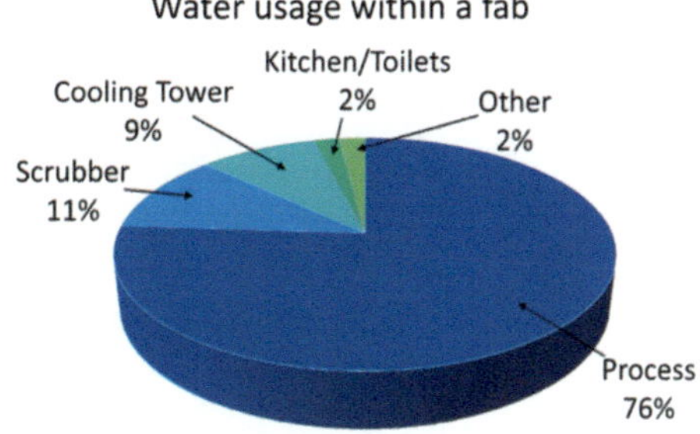


**Fig. 25** An estimate of water usage split in a semiconductor fab [82]

Frost and Hua published a study where they tried to quantify spatio-temporal impacts of the interaction of water scarcity and water use by the global semiconductor manufacturing industry. Their analysis is useful as a benchmark for global semiconductor industry water withdrawals and may assist OEMs in decisions about supply chain sourcing. This could also guide semiconductor manufacturers in prioritizing locations and time periods to implement water-saving technologies or employ less water intensive electricity sources. One highlight from this study was that in the case of semiconductor manufacturing, the most efficient way to reduce overall manufacturing water withdrawals is through reduction in fab electricity use. Reductions in electricity water use can also be achieved by using less water intensive sources of electricity, such as solar PV and wind, which is especially important during seasons of higher water scarcity [81].

The total water drawn from the network for different technology nodes is shown in Fig. 26.

## 4 Digital Twins for Monitoring and Optimizing Environmental Impacts in Semiconductor Manufacturing

With the arrival of digital transformation innovations such as generative AI, industry 4.0, autonomous vehicles, and Digital Twins the demand for semiconductor chip manufacturing continues to grow. As a result, sustainability measures and reducing the environmental impact will become a key area of focus for manufacturers as well as governments and local energy regulators. To address this sustainability challenge, companies need to work individually as organizations and collectively as an industry to define goals and agree on actions.

Efforts to address eco-efficiency in semiconductor manufacturing has been ongoing for the past decade, for instance Higgs et al. 2012 reviewed historic trends in energy use and environmental impacts of the finished products and identified some of the key challenges for reducing these impacts going forward [83]. By creating digital replicas of physical systems, Digital Twins can perform real-time monitoring,

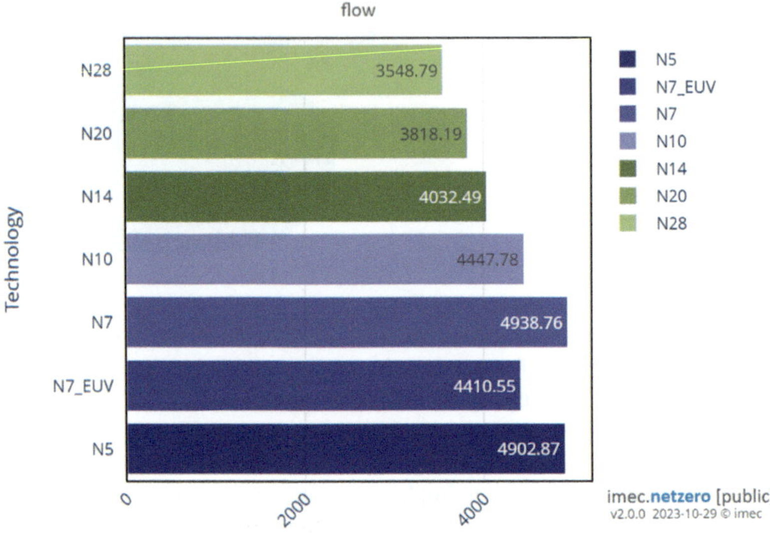


**Fig. 26** The total water drawn from the network for different technology nodes [73]

analytics, and simulation of energy systems, infrastructure, and ecosystems. This can aid in resource management, which is key to mitigating climate change [84].

Using Digital Twins for quantifying carbon footprint, energy, and chemical consumptions of semiconductor manufacturing equipment was first implemented in EcoTwin™ [85]. EcoTwin™ is a Digital Twin developed by Applied Materials that enables on-demand energy and chemical consumption of semiconductor manufacturing equipment. By synchronizing data from multiple sensors, and leveraging domain knowledge and models, detailed reports of carbon footprint and related consumption parameters from both the fab and sub-fab are automatically generated. EcoTwin Monitor provides transparency and actionable insights to identify opportunities for sustainable solutions in product and process development [86]. Figure 27 illustrates convergence and aggregation of multiple data streams from subfab and fab tools to portray a complete picture of consumptions and carbon footprint of the manufacturing site. Such a digital thread can be extended across multiple fabs and within as organization and across diverse geographical locations.

The implementation of EcoTwin is envisioned over a couple of distinct phases. In its Monitor phase, EcoTwin provides transparency and actionable insights to identify opportunities for sustainable solutions in product and process development, see Fig. 28. In the Explore phase, EcoTwin provides a myriad of tools aimed at consumption predictions, what-if scenarios, and beyond. Ultimately, the manufacturers may adopt and implement opportunities for autonomous decision-making and action-taking steps triggered by this system, at the chamber, tool, or fleet-level. This

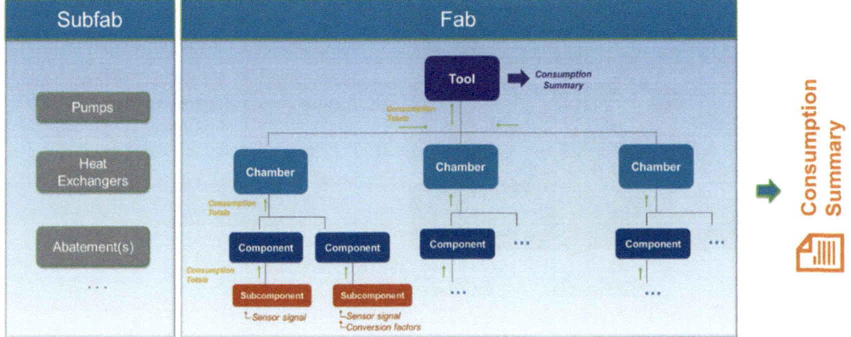


**Fig. 27** Schematic of data streams aggregated to capture environment impact of manufacturing process encompassing fab and subfab [85]

sustainability-driven goal would require calibrated processes and hardware Digital Twins that can realistically predict impact of any changes to the system, not only at the environmental impact level, but also for on-wafer process results. On-wafer process results, such as deposition/etch uniformity, film quality, resistivity, critical characteristic dimensions (CDs), etc., will remain as the primary priority for the industry. These key performance indicators in turn determine device performance, yield, and cost of ownership. Therefore, the calibrated and validated digital twins would need to evaluate and identify opportunities aimed at sustainability within the acceptable process windows. In other words, despite the importance of sustainability goals, the technology node enablement remains as the highest priority. Fabs will rely on digital twins as enablers for making trade-offs to ensure that the on-wafer process performance is not affected while sustainability targets are met in high-volume manufacturing (HVM) operations. The complexity of such Digital Twins, along with their necessity for all fabs and throughout the entire semiconductor manufacturing value chain requires the development of underlying technologies. This involves addressing gaps in the fundamental understanding of physics, chemistry, material science, and alternative chemicals, and their environmental impact. This topic would be a highly suitable research area for the CHIPS Digital Twin Manufacturing USA Institute.

Figure 29 shows a user-friendly illustration of consumptions for the system (tool-level shown) with the sub-systems, e.g., chambers and subfab auxiliary/supporting equipment. In Fig. 30 detailed consumption information for a chamber with its subcomponents (e.g., heaters, plasma source) are shown, as an example.

All the processing recipes ran during a selected date range can be evaluated and ranked based on environmental impact and their amounts of chemicals consumption. In addition, user can select specific recipes (e.g., best known-method process recipes for a desired technology application) and compared them in detail from environmental impact perspective; see Fig. 31.

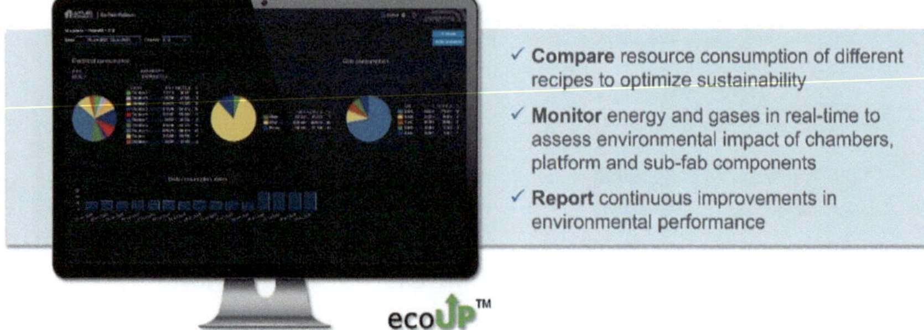


**Fig. 28** Overall capability snapshot of EcoTwin Monitor for semiconductor manufacturing carbon footprint and consumption monitoring platform [86]

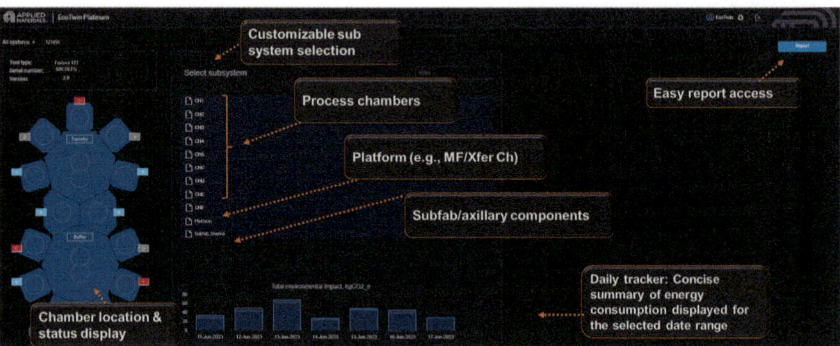


**Fig. 29** EcoTwin Tool-level dashboard – user can monitor overall impact as well as diving into each of the subcomponents in fab or subfab

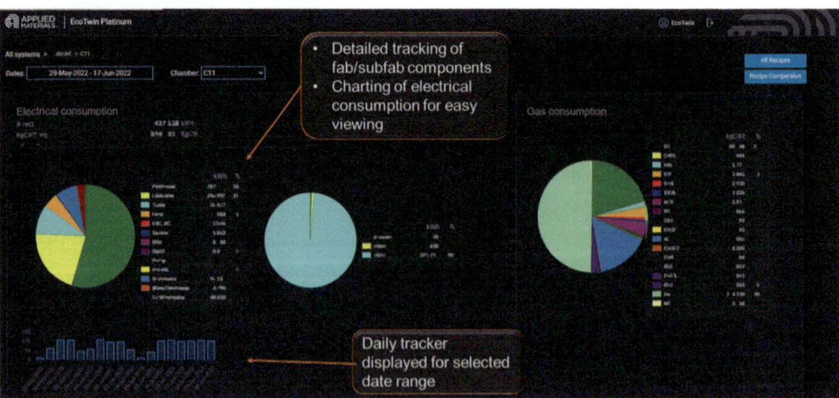


**Fig. 30** Chamber-level dashboard with detailed consumptions and carbon footprint contributions

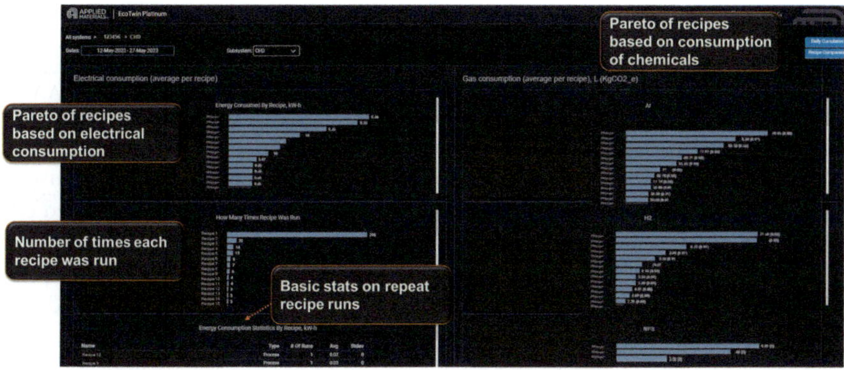


**Fig. 31** Process recipe consumptions, direct electrical, carbon footprints, and volumetric consumption of chemicals

# 5 Conclusion

The role of Digital Twin technology in semiconductor manufacturing is rapidly evolving, offering a powerful framework for tackling the complexities and challenges inherent in the industry. As Digital Twins with diverse fidelity levels and complexities, informed by both physics-based models and sensor data, gain traction across various sectors, their adoption in semiconductor manufacturing stands out as particularly promising.

Achieving truly smart manufacturing for semiconductor industry requires a holistic approach focused on optimizing specific equipment design and processes (such as deposition, etch, lithography, chemical mechanical planarization) and fostering cohesive collaboration among stakeholders from both academia and industry. Addressing the gaps in a) understanding the science, b) developing more accurate and new sensor technology, c) creating scalable software infrastructure with considerations for intellectual property and (IP) and data security, and d) establishing universal standards and protocols for data sharing and integration, are essential for realizing the full potential of Digital Twins in the semiconductor industry. In an era marked by the proliferation of AI-enabled applications, the emergence of large-language models, and the relentless pursuit of accelerated computing methods, Digital Twins present a compelling solution to address the multifaceted demands of research and development, ramp-up, and high-volume manufacturing in the semiconductor realm.

Beyond these overarching advantages, Digital Twins offer a multitude of specific benefits tailored to the semiconductor manufacturing process. From process recipe optimization and reactor design co-optimization for advanced technology nodes to enhancing productivity in process engineering and ensuring chamber-to-chamber matching, the potential applications of Digital Twins are vast and varied.

Moreover, Digital Twins pave the way for workforce development initiatives centered around virtual representations of manufacturing equipment, integration processes, and device design and verification. By leveraging Digital Twins of virtual fabs, personnel can engage in immersive training experiences and gain invaluable insights into the intricacies of semiconductor manufacturing.

In essence, the integration of Digital Twin technology holds the promise of revolutionizing semiconductor manufacturing, offering a pathway to enhanced efficiency, innovation, and competitiveness in an increasingly dynamic landscape. As the industry continues to evolve, Digital Twins are poised to play an indispensable role in driving progress and unlocking new possibilities on the semiconductor manufacturing frontier.

## References

1. Grieves M (2005) Product lifecycle management: the new paradigm for enterprises. Int J Prod Dev 2(1–2):71–84
2. Shafto M, Conroy M, Doyle R, Glaessgen E, Kemp C, LeMoigne J, Wang L (2010) Modeling, simulation, information technology and processing roadmap, NASA
3. Grieves M (2015) Digital Twin: manufacturing excellence through virtual factory replication
4. Negri E, Fumagalli L, Macchi M (2017) A Review of the roles of Digital Twin in CPS-based production systems. Proc Manuf 11:939–948
5. Tao F, Zhang H, Liu A, Nee AYC (2019) Digital Twin in industry: state-of-the-art. IEEE Trans Ind Inf 15(4):2405–2415
6. Tao F, Zhang M, Liu Y, Nee A (2018) Digital Twin driven prognostics and health management for complex equipment. CIRP Ann 67(1):169–172
7. DTC, DIgital Twin Consortium (2020) [Online]. Available: https://www.digitaltwinconsortium.org/hot-topics/the-definition-of-a-digital-twin/
8. Feng S, Jones A, Shao G (2023) ISO 23247–1 Automation systems and integration—Digital twin framework for manufacturing—Overview and general principles, International organization for standardization. In: Data requirements for digital twins in additive manufacturing, ASME manufacturing science and engineering conference, New Brunswick, NJ, US
9. Digital Twin Consortium (DTC), Capabilities Periodic Table (CPT) (2022) [Online]. Available: https://www.digitaltwinconsortium.org/initiatives/capabilities-periodic-table/
10. Warke V, Kumar S, Bongale A, Kotecha K (2021) Sustainable development of smart manufacturing driven by the Digital Twin framework: a statistical analysis. Sustainability 13:10139
11. Dalibor M, Jansen N, Rumpe B, Schmalzing D, Wachtmeister L, Wimmer M, Wortmann A (2022) A Cross-domain systematic mapping study on software engineering for Digital Twins. J Syst Softw 193:111361
12. Change2Twin Project, Change2Twin—Digital Twin for every manufacturing SME! (2020) [Online]. Available: https://www.change2twin.eu/
13. Land I, Niatas J, Serughetti M (2023) What are Digital Twins? A primer on virtual models (2023) [Online]. Available: https://www.synopsys.com/blogs/chip-design/digital-twins-semiconductor-industry.html
14. National Academies of Sciences, Engineering, and Medicine (2024) Foundational Research Gaps and Future Directions for Digital Twins. Washington, DC: The National Academies Press. https://doi.org/10.17226/26894.
15. Henning S, Hasselbring W (2012) The Titan Control center for industrial devops analytics research. Softw Impact 7:100050

16. Henning S, Hasselbring W, Burmester H, Möbius A, Wojcieszak M (2021) Goals and measures for analyzing power consumption data in manufacturing enterprises. J Data Inf Manag 3(1):65–82
17. Seegrün A, Kruschke T, Mügge J, Hardinghaus L, Knauf T, Riedelsheimer T, Lindow K (2023) Sustainable product lifecycle management with Digital Twins: a systematic literature review. Proc CIRP 119:776–781
18. XMPro inc (2023) [Online]. Available: https://xmpro.com/platform/. Accessed 11 Nov 2023
19. Green G (2023) XMPRO-Digital Twin consortium 2023: a digital twin-based remote operations center for wind farms. [Online]. Available: https://www.brighttalk.com/webcast/18347/592103
20. Abdoune F, Ragazzini L, Nouiri M, Negri E, Cardin O (2023) Toward Digital Twin for sustainable manufacturing: a data-driven approach for energy consumption behavior model generation. Comput Ind 150:103949
21. Asrai RI, Newman ST, Nassehi A (2018) A mechanistic model of energy consumption in milling. Int J Prod Res 56(1–2):642–659
22. Lv J, Tang R, Tang W, Liu Y, Zhang Y, Jia S (2017) An investigation into reducing the spindle acceleration energy consumption of machine tools. J Clean Prod 143:794–803
23. Lee W, Kim SH, Park J, Min BK (2017) Simulation-based machining condition optimization for machine tool energy consumption reduction. J Clean Prod 150:352–360
24. Seow Y, Rahimifard S (2011) A framework for modelling energy consumption within manufacturing systems. CIRP J Manuf Sci Technol 4(1):258–264
25. Karanjkar N, Joglekar A, Mohanty S, Prabhu V, Raghunath D, Sundaresan R (2018) Digital Twin for energy optimization in an SMT-PCB assembly line. In: 2018 international conference on internet of things and intelligence system (IOTAIS)
26. Walther J, Weigold M (2021) A systematic review on predicting and forecasting the electrical energy consumption in the manufacturing industry. Energies 14(4):968
27. He Y, Wu P, Li Y, Wang Y, Tao F, Wang Y (2020) A generic energy prediction model of machine tools using deep learning algorithms. Appl Energy 275:115402
28. Sossenheimer J, Walther J, Fleddermann J, Abele E (2019) A sensor reduced machine learning approach for condition-based energy monitoring for machine tools. Proc CIRP 81:570–575
29. Lee D, Meyer JS, Touloukian P, Story J, Schneider P, Sartini J (2023) White paper: how Intel® automated factory solutions delivers tremendous benefits to semiconductor manufacturing environments, Intel
30. TSMC—Machine Learning for Quality Management (2023) [Online]. Available: https://www.tsmc.com/english/dedicatedFoundry/services/apm_intelligent_packaging_fab/intelligentFab_automation
31. Smart sight: how Micron uses AI to enhance yield and quality (2023) [Online]. Available: https://www.micron.com/insight/smart-sight-how-micron-uses-ai-to-enhance-yield-and-quality
32. Chen H, Jeremiah S, Lee C, Park J (2023) A Digital Twin-based heuristic multi-cooperation scheduling framework for smart manufacturing in IIoT environment. Appl Sci 13(3):1440
33. Rehman A, Ahmed M, Begum S (2023) Cognitive Digital Twin in manufacturing: a heuristic optimization approach. In: Artificial intelligence applications and innovations; AIAI 2023. IFIP advances in information and communication technology, vol 676
34. Bhalode P, Metta N, Chen Y, Ierapetritou M (2020) Efficient data-based methodology for model enhancement and flowsheet analyses for continuous pharmaceutical manufacturing. Comput Aided Chem Eng 48(2020):127–132
35. Huang Z, Fey M, Liu C, Beysel E, Xu X, Brecher C (2023) Hybrid learning-based digital twin for manufacturing process: modeling framework and implementation. Robot Comput-Integr Manuf 82:102545
36. Langlotz P, Klar M, Yi L, Hussong M, Sousa FJ, Aurich JC (2022) Concept of hybrid modeled digital twins and its application for an energy management of manufacturing systems. Proc CIRP 112:549–554
37. Es-haghi MS, Anitescu C, Rabczuk T (2024) Methods for enabling real-time analysis in digital twins: a literature review. Comput Struct 297:107342

38. Raissi M, Perdikaris P, Karniadakis G (2019) Physics-informed neural networks: a deep learning framework for solving forward and inverse problems involving nonlinear partial differential equations. J Comput Phys 378:686–707
39. Scarselli F, Gori M, Tsoi AC, Hagenbuchner M, Monfardini G (2009) The graph neural network model. IEEE Trans Neural Netw 20(1):61–80
40. Sanchez-Gonzalez A, Godwin J, Pfaff T, Ying R, Leskovec J, Battaglia PW (2020) Learning to simulate complex physics with graph networks. In: Proceedings of the 37th international conference on machine
41. Wang J, Huang Y, Chang Q, Li S (2019) Event-driven online machine state decision for energy-efficient manufacturing system based on digital twin using max-plus Algebra. Sustainability 11(18):5036
42. Jia S, Tang R, Lv J, Yuan Q, Peng T (2017) Energy consumption modeling of machining transient states based on finite state machine. Int J Adv Manuf Technol 88:2305–2320
43. Dietmair A, Verl A (2009) A generic energy consumption model for decision making and energy efficiency optimisation in manufacturing. Int J Sustain Eng 2(2):123–133
44. interTwin Project, "interTwin," (2023) [Online]. Available: https://www.intertwin.eu/about-intertwin/
45. Schmelter S (2023) Trustworthy virtual experiments and digital twins (ViDiT)—Funded [Online]. Available: https://www.vidit.ptb.de/home
46. Xiang F, Zhang Z, Zuo Y, Tao F (2019) Digital Twin driven green material optimal-selection towards sustainable manufacturing. Proc CIRP 81:1290–1294
47. Davila R, Maria F, Schwark F, Dawel L, Pehlken A (2023) Sustainability Digital Twin: a tool for the manufacturing industry. Proc CIRP 116:143–148
48. Ma S, Ding W, Liu Y, Ren S, Yang H (2022) Digital twin and big data-driven sustainable smart manufacturing based on information management systems for energy-intensive industries. Appl Energy 326:119986
49. Li H, Yang D, Cao H, Ge W, Chen E, Wen X, Li C (2022) Data-driven hybrid petri-net based energy consumption behaviour modelling for digital twin of energy-efficient manufacturing system. Energy 239:122178
50. Almström P, Andersson C, Öberg A, Hammersberg P, Kurdve M, Landström A, Shahbazi S, Wiktorsson M, Windmark C (2017) Sustainable and resource efficient business performance measurement systems—The handbook, Mölndal, Mölndal: Billes Tryckeri
51. Miehe R, Waltersmann L, Sauer A, Bauernhansl T (2021) Sustainable production and the role of Digital Twins–basic reflections and perspectives. J Adv Manuf Process 3(2):e10078
52. TwinGoals (2020) EIT manufacturing and 'digital twins' improving tomorrow's manufacturing. [Online]. Available: https://eit.europa.eu/news-events/news/eit-manufacturing-and-digital-twins-improving-tomorrows-manufacturing
53. Chávez CAG, Bärring M, Frantzén M, Annepavar A, Gopalakrishnan D, Johansson B (2022) Achieving sustainable manufacturing by embedding sustainability KPIs in Digital Twins. In: Winter simulation conference, Singapore
54. Barenji AV, Liu X, Guo H, Li Z (2021) A Digital Twin-driven approach towards smart manufacturing: reduced energy consumption for a robotic cell. Int J Comput Integr Manuf 34(7–8):844–859
55. Kapteyn M, Pretorius J, Willcox K (2021) A probabilistic graphical model foundation for enabling predictive digital twins at scale. Nat Comput Sci 1(5):337–347
56. The White House—Statements (2022) Fact sheet: chips and science act will lower costs, create jobs, strengthen supply chains, and counter China. [Online]. Available: https://www.whitehouse.gov/briefing-room/statements-releases/2022/08/09/fact-sheet-chips-and-science-act-will-lower-costs-create-jobs-strengthen-supply-chains-and-counter-china/
57. U.S. Department of Commerce (2024) CHIPS for America Announces $285 million funding opportunity for a digital twin and semiconductor chips manufacturing usa institute, U.S. Department Of Commerce, [Online]. Available: https://www.commerce.gov/news/press-releases/2024/05/chips-america-announces-285-million-funding-opportunity-digital-twin

58. da Silva M, Somani K (2024) Digital twins in semiconductor manufacturing—semi smart manufacturing initiative, SEMI, San Jose, CA
59. Kelkar U (2023) Semiconductor equipment and processes need Digital Twins. [Online]. Available: https://www.appliedmaterials.com/us/en/blog/blog-posts/semiconductor-equipment-and-processes-need-digital-twins.html
60. Moradian A, Kelkar U (2023) SEMI semiconductor Digital Twin workshop, Milpitas, CA: SEMI
61. Rajesh S, Sadanandam N, Kailash M, Gowdra Thippeswamy S, Abhra R, Phillip S, Yi X, Yu L, Wei L (2024) Coupling reactor-scale and feature-scale simulations: ProcessTwin™ for unit processes. In: 2024 8th IEEE electron devices technology & manufacturing conference (EDTM), Bangalore, India
62. Johnson M (2023) USA today. [Online]. Available: https://www.usatoday.com/story/opinion/2023/09/13/stem-workforce-shortage-semiconductor-manufacturing-race/70721967007/
63. Patel P (2023) U.S. Universities are building a new semiconductor workforce. [Online]. Available: https://spectrum.ieee.org/chips-act-workforce-development
64. Rosso D (2023) America faces significant shortage of tech workers in semiconductor industry and throughout U.S. economy. [Online]. Available: https://www.semiconductors.org/america-faces-significant-shortage-of-tech-workers-in-semiconductor-industry-and-throughout-u-s-economy/
65. Weisz K, Stewart D, Simons C, Lewis T, Kulik B (2023) The global semiconductor talent shortage
66. Moradian A, L'Heureux JO, Sheng S, Mahakali R, Ramanathan K, Zhang L, Kelkar UM, Prabhu GB, Yuan Z, Oh J (2020) Parameter sensing and computer modeling for gas delivery health monitoring. United States Patent US11768984B2
67. Dickerson G (2020) Applied materials CEO Keynote—SEMICON Wes: make possible a better future
68. Vanderbauwhede W (2022) Frugal computing—On the need for low-carbon and sustainable computing and the path towards zero-carbon computing. IAB workshop on environmental impact of internet applications and systems
69. www.supplychainbrain.com, The chip industry has a problem with its giant carbon footprint (2021) [Online]. Available: https://www.supplychainbrain.com/articles/32910-the-chip-industry-has-a-problem-with-its-giant-carbon-footprint
70. Pelé A-F (2023) Semiconductor manufacturing on the way to net zero, [Online]. Available: https://www.eetimes.com/semiconductor-manufacturing-on-the-way-to-net-zero/
71. Göke S, Issler M, Liu D, Patel M, Spiller P (2022) Keeping the semiconductor industry on the path to net zero, McKinsey & Company. [Online]. Available: https://www.mckinsey.com/industries/semiconductors/our-insights/keeping-the-semiconductor-industry-on-the-path-to-net-zero
72. Ragnarsson L-Å, Rolin C, Shamuilia S, Parton E (2022) The green transition of the IC industry, imec—SSTS Program, Leuven
73. imec-NetZero, "imec.netzero" (2023). [Online]. Available: https://netzero.imec-int.com/. Accessed 10 Nov 2023]
74. ASML-Sustainability, "Our sustainability strategy" (2022) [Online]. Available: https://www.asml.com/en/company/sustainability
75. SEMI Standards (2023) SEMI S23—Guide for conservation of energy, utilities and materials used by semiconductor manufacturing equipment, vol Safety Guidelines, SEMI
76. ISO-50001 (2018) ISO 50001 Energy management systems—Requirements with guidance for use
77. Ghita M, Siham B, Hicham M, Griguer H (2021) Digital Twins based LCA and ISO 20140 for smart and sustainable manufacturing systems. In: Sustainable intelligent systems. advances in sustainability science and technology, Singapore, Springer
78. Boss B et al (2020) Digital Twin and asset administration shell concepts and application in the industrial internet and industrie 4.0. Plattform Ind 4:13–14

79. IEC-62832–3 (2020) BS EN IEC 62832–3 industrial-process measurement, control and automation. Digital factory framework. Institution British Standards
80. ISO-20140 (2018) ISO 20140 Automation systems and integration—Evaluating energy efficiency and other factors of manufacturing systems that influence the environment
81. Frost K, Hua I (2019) Quantifying spatiotemporal impacts of the interaction of water scarcity and water use by the global semiconductor manufacturing industry. Water Resour Ind 22:100115
82. Davis S (2022) Water supply challenges for the semiconductor industry. [Online]. Available: https://www.semiconductor-digest.com/water-supply-challenges-for-the-semiconductor-industry/
83. Higgs T, Brady T, Yao M (2012) Progress and challenges in EcoDesign of semiconductor products. In: Design for innovative value towards a sustainable society, Dordrecht
84. Verdict.co.uk: Intelligence, GlobalData Thematic (2023) Digital twins: key to addressing climate change. [Online]. Available: https://www.verdict.co.uk/digital-twins-combat-climate-change/
85. Moradian A (2023) SEMICON west, smart manufacturing: EcoTwin—an integrated solution for sustainability in semiconductor manufacturing, San Fransisco, CA
86. Applied Materials inc. (2023) "EcoTwin™ Eco-efficiency software. [Online]. Available: https://www.appliedmaterials.com/us/en/semiconductor/solutions-and-software/ai-x/ecotwin.html
87. Applied Materials inc. (2023) Driving a net zero 2040 playbook powered by collaboration, clean energy and innovation. [Online]. Available: https://www.appliedmaterials.com/us/en/corporate-responsibility/planet/net-zero.html
88. Lam Research inc. (2023) 2050 net zero strategy. [Online]. Available: https://www.lamresearch.com/company/environmental-social-and-governance/2050-net-zero-strategy/
89. ASMI (2021) ASM international aims to achieve net zero by 2035. [Online]. Available: https://www.asm.com/press-releases/2304062
90. NIST-IAC, (2023) Industrial advisory committee (IAC)—R&D gaps working group. [Online]. Available: https://www.nist.gov/system/files/documents/2023/02/08/Feb%207%20IAC%20Meeting%20R%26D%20Gaps%20WG%20Materials%20Final.pdf
91. NIST (2023) NIST seeks nominations for the industrial advisory committee. [Online]. Available: https://www.nist.gov/news-events/news/2023/05/nist-seeks-nominations-industrial-advisory-committee
92. Moradian A, Neville E, Kelkar UM, Denome MR, Kothnur P, Ramanathan K, Shah K, Trejo O, Meirovich S (2021) United States Patent US20220334569A1
93. Moradian A, Kelkar UM, Neville E, Trejo O, Meirovich S, Shah KB, Kher SS (2021) Uinted States Patent US20230185268A1
94. Annepannavar A, Gopalakrishnan D (2021) Digital Twins for sustainable production: modelling and simulation of a production system towards a Digital Twin. Chalmers University of Technology, Gothenburg
95. Li L, Mao C, Sun H, Yuan Y, Lei B (2020) Digital Twin driven green performance evaluation methodology of intelligent manufacturing: hybrid model based on fuzzy rough-sets AHP, II. Complex Spec Issue: Complex Econ Bus 4:2020
96. Burkacky O, Göke S, Nikolka M, Patel M, Spiller P (2022) Sustainability in semiconductor operations: toward net-zero production, McKinsey & Company. [Online]. Available: https://www.mckinsey.com/industries/semiconductors/our-insights/sustainability-in-semiconductor-operations-toward-net-zero-production

# Digital Twins for Robot Systems in Manufacturing

**Ali Ahmad Malik, Guodong Shao, and Jane Tarakhovsky**

**Abstract** The increasing need for industrial automation is driving the adoption of robotics, with new developments such as human–robot collaboration through autonomous mobile robots and collaborative robotic arms. While automation improves product quality and working conditions and lowers manufacturing costs, it can also limit manufacturing adaptability. Therefore, when integrating robots in manufacturing, there is a pressing need to simplify the methods to develop, install, reconfigure, and operate robot systems to achieve greater adaptability. This is where the concept of Digital Twin, which replicates the behavior of a complex system in a virtual environment for analysis and optimization, comes into play. In robotics, the Digital Twin technology is expected to address the challenges associated with designing, testing, commissioning, and reconfigurations. This requires a Digital Twin to accurately represent various dimensions of a robotic system under production variables. This study characterizes the components of a robot system that need to be modeled in a Digital Twin to create a trustworthy virtual replica of a physical robot system. A Digital Twin of this kind can be utilized throughout the lifecycle of the physical robot installation across various use cases.

**Keywords** Robotics · Digital Twin · Flexible automation · Manufacturing systems · Simulation · System design

A. A. Malik (✉) · J. Tarakhovsky
Department of Industrial and Systems Engineering, Oakland University, Rochester, MI 48309, USA
e-mail: aliahmadmalik@oakland.edu

J. Tarakhovsky
e-mail: janetarakhovsky@oakland.edu

G. Shao
Engineering Laboratory, National Institute of Standards and Technology, Gaithersburg, MD 20899, USA
e-mail: guodong.shao@nist.gov

 
M. Grieves and E. Hua (eds.), *Digital Twins, Simulation, and the Metaverse*, Simulation Foundations, Methods and Applications, https://doi.org/10.1007/978-3-031-69107-2_13

## 1 Introduction

Modern manufacturing systems are complex, with facilities evolving into large networks of data-connected mechatronic components [1]. The complexity of these systems stems from the quantity of information that spans various lifecycle phases, including design, development, commissioning, operations, and end-of-life [2, 3]. The elevated complexity and interconnectivity make the lifecycle management of modern-day manufacturing systems more challenging. Conversely, demand is increasing for manufacturing systems to possess resilience and adaptability [4]. Addressing these challenges in complex scenarios requires enabling smart manufacturing through digitalization, data connectivity, and the integration of machine learning [5]. Smart manufacturing, besides resilience, can bring cost reduction, enhance workers' well-being, and result in a better return on investment (ROI).

Industry 4.0, or the fourth industrial revolution, is the net sociotechnological impact of infusing emerging technologies such as additive manufacturing, machine learning, robotics, simulations, and the Internet of Things in products and their manufacturing systems. Advanced robotic automation stands out as one of the enablers for Industry 4.0 [6]. Modern installations strive to make robotic automation more flexible, adaptable, safe, and cost-effective than traditional robotics implementations. However, flexible approaches are lacking in developing plug-and-play hardware, programming the robots, control program generation, task scheduling, layout planning, safety assessment, and alignment with production plans.

Advancements in virtualization, sensing technologies, and computing power facilitate the realization of Digital Twins (DTs), which enable the testing and validation throughout the design, development, and control phases of a complex system in a virtual space. Different scientific domains increasingly recognize the potential value of DTs for managing complexity in areas such as manufacturing, transportation, aircraft, and space missions [7]. Manufacturing customization and reconfigurability are vital domains to manage through DTs [7].

Computer models provide a means to shorten the time needed to design, redesign, and deploy robot systems. Computer-based virtual models of physical systems can be beneficial for testing and validating the production before implementation [8]. While this method is consistent with traditional virtual modeling, the emerging "lifecycle" approach and real-time communication between physical and virtual systems are pivotal concepts of DTs [9].

Many studies have documented the potential advantages and relevance of employing DTs for robot systems [10–12]. It has also been observed that developing a trustworthy virtual replica of a robot system is time-consuming and demands advanced engineering skills and investment in different engineering software. Creating and deploying a DT of a robot system should be structured, simplified, and standardized to realize the needed ROI. It requires identifying the components of a robot system that are relevant to the purpose of its DT. Moreover, the flexibility of the DT itself is critical to ensure that the DT can effortlessly be adapted to evolving circumstances.

This chapter presents the importance of DTs in robotic installations within manufacturing systems. The components outlined in a DT of a robot system can assist researchers and practitioners in developing cost-effective, modular, and flexible DTs, thereby improving the resilience of robot installations. This is an essential step toward achieving adaptable manufacturing systems.

The key contributions of this chapter are to:

1. Present the components of a DT of a robot system for flexibility
2. Examine the lifecycle phases of a robot system that a DT can support
3. Apply DTs in robot systems in manufacturing settings
4. Present use cases that demonstrate the utilization of DTs in robot systems.

## 2 Challenges and Opportunities in Contemporary Manufacturing

The continuous drive to shorten product life cycles emerges as a significant transformation in the contemporary business landscape [13]. Emerging sociotechnological trends require shorter product development and launch timelines. In this setting, manufacturing companies leverage emerging hardware and software technologies and their potential opportunities [14, 15].

Aside from the rapid pace of changes, manufacturers face a shortage of skilled workers. The recent global exposure to the COVID-19 pandemic also displayed widespread disruptions in supply chains [16] partly because of a shortage of workers due to social distancing measures. Research studies identified that future factories could better address such challenges by adopting modular, flexible, and human-friendly automation solutions [17].

A way to develop human-friendly automation solutions is through flexible, collaborative robots. Technologies that facilitate the swift validation of new manufacturing strategies are also needed. Therefore, future manufacturing systems must not only be repurposable but also be designed, developed, commissioned, and reconfigured at a significantly faster pace [18].

DTs can be utilized to address the resilience requirements within a manufacturing system. For example, DTs can help reduce the time required to validate new manufacturing strategies, generate automation programs, and provide maintenance support. Additionally, DTs can harness real-time and historical data to offer insights for process optimization. Such assistance can potentially enhance the level of resilience that a manufacturing system can provide.

## 3 Robotic Automation in Manufacturing

Automation describes assigning physical and cognitive tasks to machines and software to boost production and decrease human effort [19] In manufacturing, adopting automation brings advantages such as enhancing workplace safety, efficiency, quality, and cost-effectiveness [20, 21]. However, this often comes at the cost of reduced production flexibility. At the heart of industrial automation lies industrial robots. The subsequent sections elaborate on the diverse types of robots employed in manufacturing facilities. Figure 1 shows various industrial robot types, including spherical, SCARA (Selective Compliance Assembly Robot Arm), delta, Cartesian, and humanoid robots [22]. The robot selection for a specific task is based on the nature of the tasks to be automated, available space, financial constraints, and process-related considerations. While these robots enhance manufacturing efficiency, their applications are limited in certain operations, such as assembly, which only constitutes 7.3% of robotic use [23].

### 3.1 *Traditional Industrial Robots*

Robots are the predominant force driving the industrial automation of physical tasks [22, 24]. These robots, characterized by fixed positioning, operation within enclosures, and time-consuming reconfiguration processes [25], fall into the category of fixed automation solutions. They can help achieve high production volumes but must strictly be separated from human interaction. They also demonstrate limited flexibility [26]. Industrial robots have proven successful in various sectors, including automotive, medicine, food, and electronics manufacturing [27]. The primary reason for the unsuitability of robots in assembly is human safety and the challenges of their reconfiguration [28].

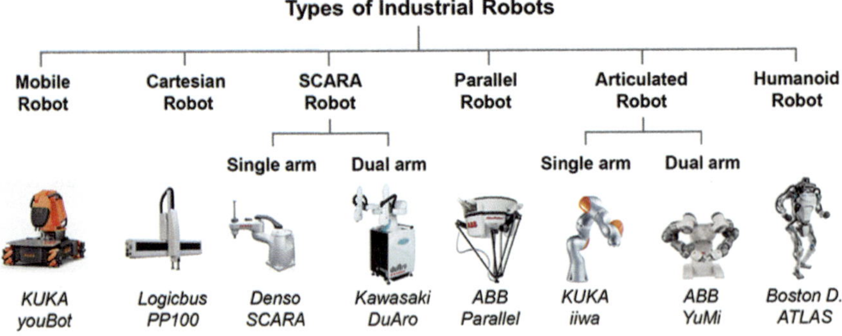


**Fig. 1** Various types of industrial robots

**Fig. 2** A human and robot coexisting in manufacturing [36]

## *3.2 Collaborative Robots*

Modern industrial robots are lighter, portable, easy to program, and safe. This change can meet the need for flexibility in terms of mobility, capability, and capacity [29]. These robots, designed for collaboration and coexistence with humans (Fig. 2), are commonly known as "cobots" or collaborative robots [30]. A collaborative robot can be defined as a mechanical device intended for direct physical interaction with humans, a concept first introduced by Colgate [31] and further developed by Kruger [32]. Cobots allow humans and robots to work together to harness the strengths of humans and machines. This concept, often called lean automation, exists at the convergence of human flexibility and machine efficiency [33].

Literature showcases diverse applications of cobots, spanning pick-and-place operations, assembly tasks, welding, inspection processes, and packing [34]. Moreover, cobots have been explored as a viable solution for rapidly repurposing factories in emergencies [17]. The predominant use of cobots has been in manufacturing small components such as those assembled into electronics, appliances, and electronic actuators [28, 35]. With the advancement of sensing and safety technologies, cobots are being considered to automate large and heavy components.

## *3.3 Autonomous Mobile Robots*

Autonomous mobile robots (AMRs) represent a distinctive category within collaborative robots [37]. They have proven highly effective in material handling applications in manufacturing settings [40]. Their adaptability and versatility make them well-suited for tasks requiring interactions with humans. Beyond manufacturing, AMRs have found applications in warehouses, military operations, healthcare, search and rescue missions, security, and home environments [38]. This versatility underscores the potential of mobile robots to automate operations in diverse fields.

Different robot types have standard features such as mechanical multijointed reprogrammable actuators, end-of-arm tooling, machine vision, positioning technology, and control programs. Adaptability is recognized as necessary for most modern-day robots. The following section presents a typical physical architecture of a robot installation.

## 4 Architecture of a Robot System in Manufacturing

Robotic systems in manufacturing settings are available in various designs, layouts, and configurations, influenced by specific use cases. A typical robot cell comprises multiple hardware and software components (Fig. 3). Articulated robot arms, with one or more reprogrammable mechanical joints, represent most robot installations [39]. One or more tools (end effectors) are attached to a robot manipulator's tool post to perform various functions. A robot controller oversees the operations of the robot system, connecting all external hardware through the input/output (I/O) interfaces of the controller. Various sensors are embedded in the robot body and integrated externally to monitor its performance and respond to emerging situations.

Robotic arms designed for human–robot collaboration (HRC) typically have power and force-limiting bodies, speed and separation monitoring, hand guiding, and emergency stop as stipulated in the ISO15066 safety standard for HRC [21]. These attributes ensure safe interaction between robotic arms and humans [40]. AMRs can

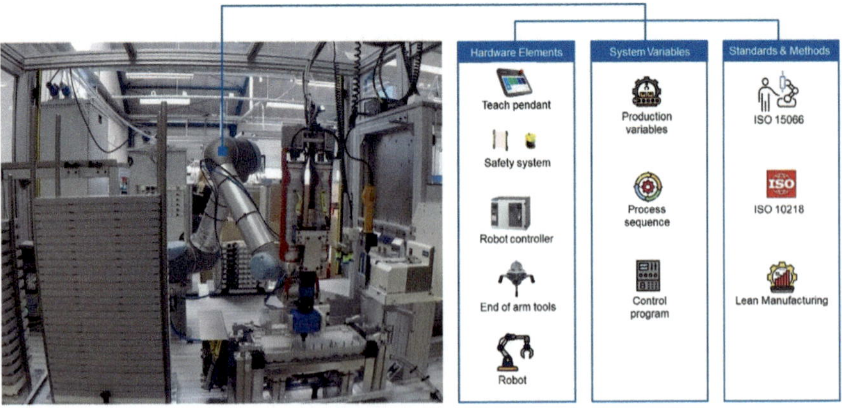


**Fig. 3** Typical components of a robot system in a manufacturing setting

also facilitate the mobility of a robotic arm within the robot system. Furthermore, collaborative robots must comply with the ISO19649 standard [41], which establishes terms related to mobile robots operating on solid surfaces and engaging in industrial robot applications. To effectively employ the concept of DTs, it is vital to model most, if not all, of these features of robot systems in their digital models.

## 5 DTs for Robots in Manufacturing

A DT is a virtual representation of the components and dynamics of an observable physical system [44]. DTs can mirror real-time operating conditions and predict the future behavior of a physical system [45]. The core concept of a DT involves creating a digital model of a physical system and linking each component of the digital model to its corresponding physical assets. In return, the virtual model must act as a front-runner of the physical system to predict or estimate its future behavior.

The present understanding of the DT concept originates from the idea of a "Conceptual Ideal for PLM" (Product Lifecycle Management) [42]. It proposes that every system is a subset of two other systems: the physical system in the physical world and a virtual system existing in virtual space, containing all necessary information about the physical system. The bidirectional relationship between the physical and digital systems can enhance product design, manufacturing, and service throughout the system's life cycle [43].

The methods of using an informational virtual model to represent the complexity of a physical system have evolved. In earlier times, the virtual model existed as a mental image [1], limited in its capacity to address questions about the system's performance. In the mid-twentieth century, creating virtual models became possible, starting with 2D CAD (computer-aided design) objects and advancing to 3D models and dynamic simulations. These virtual models are typically developed early in the system's lifecycle, i.e., during design. These models often become useless when the system transitions to the operation phase.

The linkage of digital models to their physical counterparts and the integrated intelligence throughout their lifecycle are now achievable (Fig. 4). This enables them to understand operational behavior and assist in addressing day-to-day production constraints. In this context, DTs can be categorized into DT prototypes and DT instances [44]. A DT prototype is used to refine system design, presenting optimal static and dynamic information to achieve desired outcomes. Meanwhile, a DT instance integrates monitoring, service, sensing, and behavioral information about the physical twin during its operations. DT instances exhibit predictive and interrogative behaviors, which prove beneficial during the operational and maintenance phases.

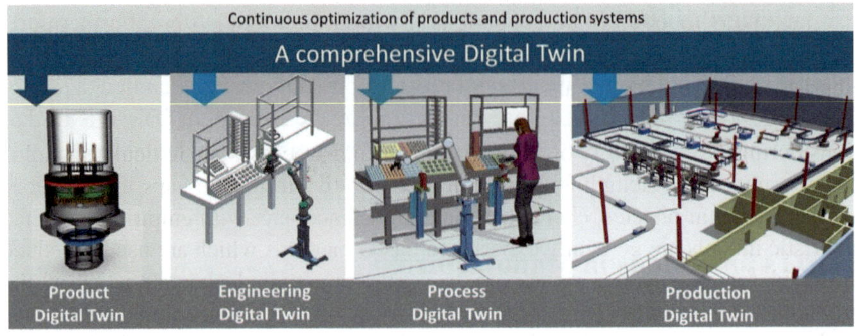


**Fig. 4** Scope of DTs in manufacturing systems

## 6 Lifecycle Phases of Robot Systems

A robotic system undergoes a comprehensive life cycle, commencing with its design and concluding at its end-of-life stage. Correspondingly, its DT follows a parallel life cycle, adapting to various scenarios and system evolution throughout the lifecycle (Fig. 5). The subsequent section describes the functions of a DT across multiple stages in a robotic system's life cycle.

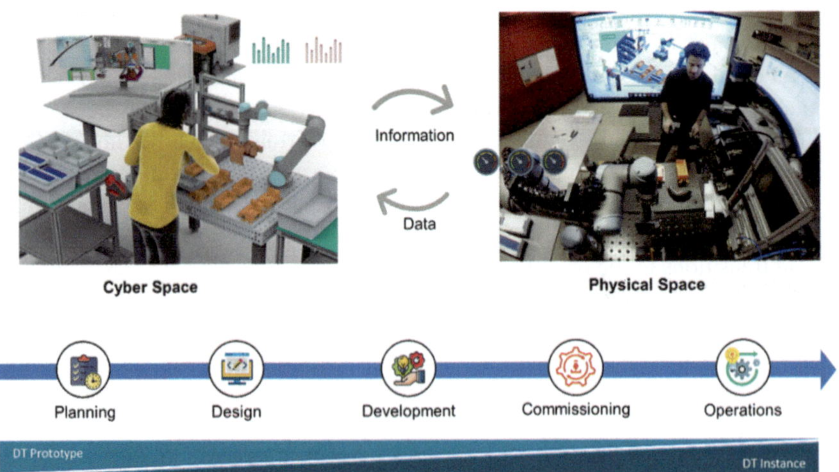


**Fig. 5** Concept of a Digital Twin system in human–robot collaboration

## 6.1 *Design of the Robot System*

In developing a new robot system, it is customary to construct virtual models before the actual physical counterpart is built. This virtual representation, which can also be referred to as a DT prototype, is conceived to conceptualize and finalize the system's appearance, specifications, the selection of off-the-shelf components, and the overall layout. Despite the absence of the corresponding physical counterpart during the design phase, the DT represents the intended physical twin. It enables the exploration of various what-if scenarios, facilitating swift, secure, and reliable design outcomes. The choice of robot manipulators, workstation design, layout, fixtures, and financial assessments are critical questions that must be addressed at the design stage.

## 6.2 *Commissioning the Robot System*

The results derived from the design phase provide information for developing the components of the physical system. The developed system then moves to the commissioning stage. In the case of a robot system, this stage may entail the creation of workstations, fixtures, feeding devices, and other hardware elements. The Bill of Materials (BOM) and Bill of Processes (BOP) can be generated, guiding the development of the physical system. Throughout this phase, the connection between the physical systems and their DTs can be established by linking the DT to an actual controller or programmable logic controller (PLC) to identify potential errors. This methodology is analogous to virtual commissioning (VC). VC, or hardware-in-the-loop simulations, reduces development time by facilitating virtual testing and integration well before actual commissioning. The physical robot can be live connected with its DT, allowing it to execute tasks as designed in the DT.

## 6.3 *Scheduled and Preventive Maintenance*

Maintenance is an essential component of most production systems. Emerging technologies such as augmented reality (AR) or chatbots can be integrated with a DT to optimize maintenance procedures, which can better assist maintenance personnel with enhanced visualization tools for fault detection and training tasks. Virtual reality (VR) is another visualization technology that can be integrated with the DT, particularly for training.

Maintenance can benefit from the DT technology in ways such as:

- Real-time Monitoring: Data capturing operating parameters, energy consumption, and system health.
- Predictive Maintenance: Use of machine learning within DT to predict potential failures or maintenance needs based on performance trends.

- Condition Monitoring: IoT sensors provide data on system health indicators, which can be integrated with DT for continuous condition monitoring.
- Asset Tracking: The usage and lifecycle of robotics assets, such as operating hours, replacement history, etc., can be tracked and help with proactive maintenance scheduling.
- Remote Diagnostics and Troubleshooting: Identifying issues remotely can help reduce downtime and improve overall efficiency.

### *6.4 Operations and Changeovers*

The most compelling application of DT technology lies in its application throughout the operational life of a robot system. Over time, a robot system may need changeovers, safety assessments, production analyses, and modifications. Assessing the overall equipment effectiveness (OEE) is another practical facet when evaluating a production system for continuous performance optimization. A DT plays a pivotal role in elevating the quality of these processes, thereby enhancing the performance and reliability of the robot system. This helps justify the investment in the creation and maintenance of its DT.

The DT developed during the design phase is extended to facilitate real-time communication with the physical system during operation, enabling behavioral analysis and performance optimization. At this stage, the system synchronizes the real-world data with the DT, enabling automated assessment cycles. This cyber-physical system integrates production planning and control databases to support scheduling production orders and changeovers. The DT is valuable in simplifying the reconfiguration or repurposing of the robot system in response to demand fluctuations.

## 7 Components of a DT for a Robot System

This section presents the fundamental components or modules comprising a robotic system's DT, as illustrated in Fig. 6. Traditionally, various tools are required to simulate each of these components. Connectivity protocols can enable near real-time communication between these components, streamlining the development of an accurate DT for a robotic system.

### *7.1 Static CAD Modeling*

The initial step in constructing a virtual manufacturing system is to create a 3D visualization using CAD software. There is a multitude of tools available for this

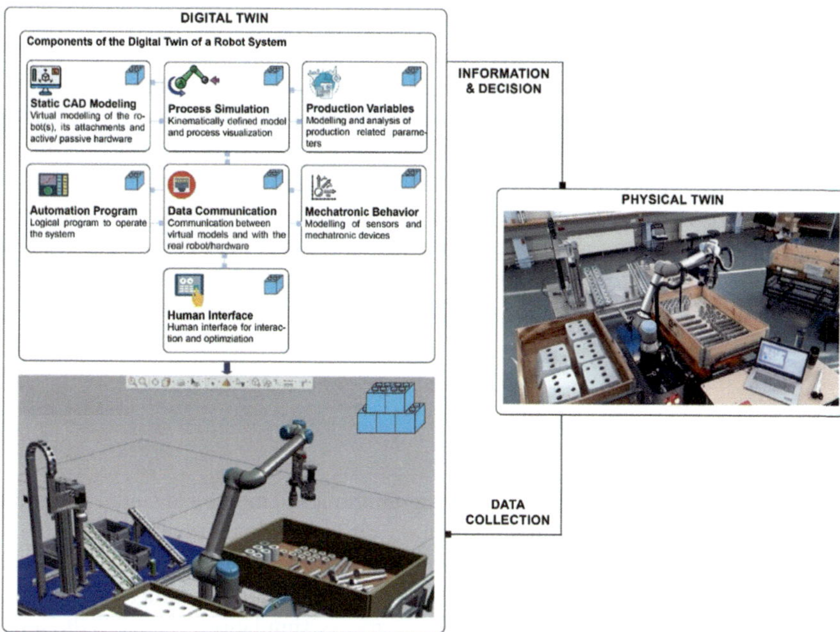


**Fig. 6** Components of DT of robot systems

purpose. This CAD data can be obtained directly from the equipment manufacturer, often in standard exchangeable formats such as STEP (Standard for the Exchange of Product Data) [45]. Robot manufacturers offer CAD models of their robots, and a similar practice is followed by manufacturers of related equipment, such as grippers, fixtures, feeders, and tables. Furthermore, many simulation tools feature a built-in library of proprietary and generic factory resource CAD models.

A critical step in preparing the CAD data is creating an assembly file and consolidating the individual CAD models of various devices and equipment. The virtual assembly model must represent the complete physical robot system being investigated. Each component can be assigned material properties and visualization to aid in subsequent analyses. This file can be exported to various exchangeable formats, with STEP being the most common standard format.

## 7.2 *Process Simulation*

The CAD data can be imported into a continuous simulation environment. Creating a dynamic simulation starts with defining the kinematics of each active resource within the system. It involves specifying position and location constraints, joint types, joint limits, and velocity limits. For example, a gripper may need to be defined for its motion kinematic joint types, limits, and action poses. The visualization/simulation of a DT is achieved through three steps: (1) creating the simulation model of a robot system along with its operation sequences, akin to a Gantt chart, (2) an event-driven continuous simulation that runs for a pre-determined time and controlled by an internal logic engine, and (3) the simulation is controlled through signals from a virtual PLC and other emulators. This simulation becomes the primary component of the DT for visualization, experimentation, and analysis. After the simulation, it can perform analyses (e.g., collision detection, layout assessment, cycle time estimates) and optimizations. Numerous proprietary tools are available to create this type of simulation, while open-source engines can also be utilized.

## 7.3 Automation Program

PLCs serve as industrial computers for programming and monitoring industrial robot systems. A critical step in commissioning a robot-based manufacturing system is creating and validating the automation program. Usually, this program is created later in the development stages. Developing, testing, and validating the automation program in a virtual space, along with process simulation, enhances the reliability of the system's performance. Each PLC has its programming tool, and open-source program development tools are also available. To ensure interoperability, PLC programs follow the IEC 61131 standard [46]. The IEC 61131-3, developed by the International Electrotechnical Commission (IEC), sets the standard for PLCs' syntax, semantics, and interoperability. The developed programs are downloaded onto a virtual PLC and interfaced with the simulation.

## 7.4 Mechatronic Behavior

A robot system includes sensors, actuators, feeders, fixtures, and other mechatronic elements (Fig. 3). Behavioral modeling of these devices enables an accurate virtual model of the entire system. The Functional Mock-up Unit (FMU) is a tool-independent, free standard crafted for dynamic model exchange and co-simulation. FMUs define a container and an interface for sharing dynamic simulation models through a combination of Extensible Markup Language (XML) files, binaries, and C-code. Both commercial and open-source tools are accessible for simulating the behavior of each device and interfacing it with the process simulation.

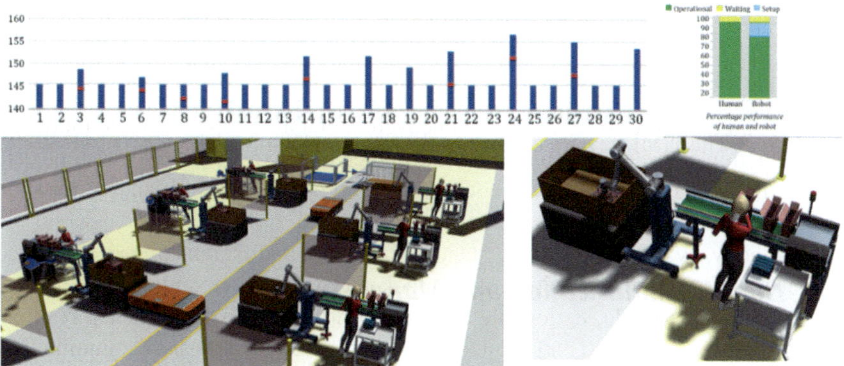


**Fig. 7** Production assessments through simulations

## *7.5 Production Variables*

The evaluation of production variables based on historical statistical data is not always required in the process modeling of a robot system. However, it is a component that can be added to a DT of a robot system for production-related analysis. In this phase, production-related parameters and throughput specifications are defined. Since these analyses are done in a stochastic simulation, a different tool is often required to perform stochastic modeling and interface it with the existing continuous process simulation model. This model simulates cycle time, startup time, setup time, potential failure scenarios, repair and maintenance requirements, shifts, worker allocation, and other relevant factors with statistical probabilities. An example of such a simulation is shown in Fig. 7, where six human–robot packaging stations are shown, and the simulation presents the operational time and waiting time for each resource. Discrete simulations can run thousands of trials based on probabilistic distributions derived from historical data. This analysis enables the prediction of the throughput of a robot system under the design variables.

## *7.6 Human–machine Interface*

In industrial settings, human–machine interfaces (HMIs) enable workers to interact with manufacturing systems or robots. This interaction involves conveying instructions such as start/stop commands, speed adjustments, and troubleshooting. Touchscreen HMIs are commonly employed in industrial settings for this purpose. An HMI is essential for facilitating continuous user interaction with the robot system in most robot work cells. Depending on the process and system design, modeling and integrating an HMI with the process simulation may be needed in a DT. A simulated HMI can communicate with PLCs and the process simulation. Using a virtual HMI allows end users to interact with the robot system in a way that is similar to its actual application. HMIs are designed following the ISA 101 standard, and validating them may require a DT model for thorough validation. A virtual HMI may be accessed on a computer screen or a handheld computing device communicating with the process simulation. The HMI developed during this step is downloadable to real HMIs for practical field applications.

## *7.7 Data Communication and Management*

In robot system DTs, real-world data are integrated into the simulation. This integration enables continuous assessments under variable conditions, enhancing the accuracy and responsiveness of the DT. System performance data, including logs and alerts, can be stored in a data repository for ongoing evaluations. Various sensors, tailored to specific requirements, can be employed in robotic systems to log parameters such as robot joint positions, machine vision data, collision events, task completion status, safety breaches, and cycle time. As exemplified by [1], a robot assembly system is connected to a cloud data repository through an internet-based router for data logging. It is an HRC assembly cell using a UR-5 robot. The performance-related logs from the assembly cell are stored and used in the simulation for design optimization (Fig. 8). These performance factors include the idle time for both the operator and the robot, human operator safety (collision occurrences), and completed cycle counts. The recorded data is then used for layout and robot path optimization in the simulation.

Data management organizes data into meaningful information, focusing on creating "Golden datasets" that undergo cleaning, transformation, validation, integration, and standardization. The data management lifecycle encompasses key stages:

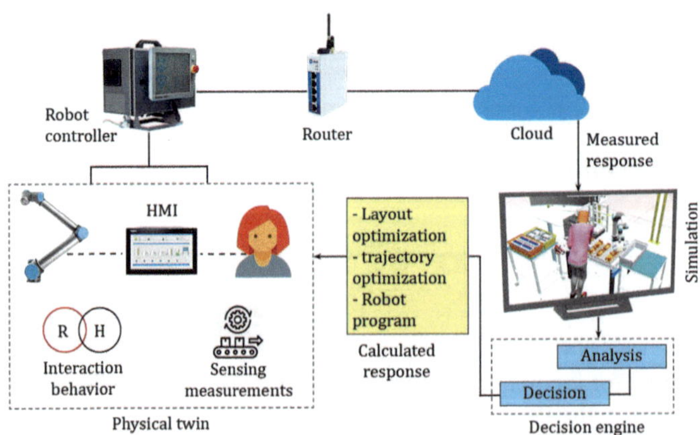


**Fig. 8** Communication of data logs from robot system [47]

data collection, transmission, storage, processing, fusion, and visualization. During data collection, hardware, software, and network resources are integrated. Data transmission involves various wireless and wired technologies. Data storage, required for processing and analysis, utilizes technologies such as cloud storage, NewSQL databases, NoSQL databases, and distributed file storage (DFS). After processing data, data visualization can be made available in formats required by AR or VR [48].

## 8 Applications of DTs in Robots

Deploying and utilizing DTs alongside robot systems in manufacturing facilities offers a range of benefits. This section presents various advantages of robot systems contributing to robot installations' flexibility, reliability, and safety.

### *8.1 Safety Assessment and Validation*

Manufacturing-related robot installations comply with safety standards, notably ISO 10218 [49] and ISO 13849 [50], which outline safety requirements for industrial robots. Moreover, collaborative robots align with safety specifications outlined in ISO 15066 [51], which is dedicated to collaborative robotic devices. Various safety measures, including emergency stops, human movement monitoring, and active collision avoidance, are integrated into robot cells to safeguard coexisting humans. The assessment and validation of robot systems for compliance with safety requirements are essential.

The DT model should accurately replicate the robotic system's physical characteristics, behaviors, and interactions (movements, sensors, and environment) to be valid. Under ISO 15066, a safety risk assessment is required following any physical alterations to the system. The risk assessment process can be streamlined using DTs within a controlled and risk-free environment. Risk identification can be conducted in the DT, maintaining a live connection with the physical robot. This approach identifies potential collisions in the virtual environment before the physical system is populated with hardware resources.

Evaluating potential accidents or injuries is essential in the context of HRC. Dynamic simulations within DT can evaluate safety performance under different operating scenarios (emergency, failure modes, and normal operations). Collision detection algorithms within the DT can identify potential failures and develop strategies for avoidance or mitigation to enhance safety.

Continuous monitoring using collision-related data logs can enable the assessment of the frequency of collisions between humans and robots over time, facilitating the optimization of robot paths to avoid such incidents. Fault tolerance mechanisms and redundancy in a DT can enhance safety and reliability. The DT can also offer training and awareness programs for operators, maintenance personnel, and other stakeholders. The DT models can also simulate the safety training scenarios within a virtual environment to enhance learning and preparedness.

### 8.2 Reduced Development Time

Studies have found that robotic automation projects frequently exceed the initially projected timeline [52]. This duration can be further extended, particularly with more complex tasks such as assembly and battery pack manufacturing. Challenges that emerge unexpectedly in the planning stage, impacting the project timeline, include issues related to process balancing, task scheduling, feeding methods, fixtures required, the necessity for multiple grippers, and safety complications. This prolonged integration timeframe is attributed to the nature of robot operations. Robots must coordinate their movements with other hardware (machine tools, equipment, end-of-arm tooling), peripherals (vision systems, force sensors), and humans, contributing to increased integration and operational complexity.

Various methods and frameworks have been documented for developing robot systems [53–55]. However, there is a growing need for novel approaches that focus on minimizing the time and effort required for integration, validation, and reconfiguration. DT prototypes can offer a high-fidelity and trustworthy digital model of the real system to reduce the chances of any errors that may arise at a later stage.

### 8.3 Robot Programming

Robot programming involves defining the paths, actions, and logical procedures for robots to perform the assigned tasks. Industrial robotic applications often require significant expertise and effort. Despite the promise of more accessible programming in the latest robots, manually programming complex robot paths remains time-consuming. With the continuous desire for customization and changing market dynamics, robot programming is not a one-time activity in their operational life. The needed flexibility and adaptability require easy ways of programming the robots.

To tackle this challenge, a DT can facilitate the intuitive development of robot programs offline within a graphical environment (Fig. 9). The robot program can be seamlessly downloaded from the DT directly to the connected robot by virtually testing the desired operation, complete with defined robot trajectories and logic. This approach streamlines the programming process, leading to a significant enhancement in efficiency during robot deployment.

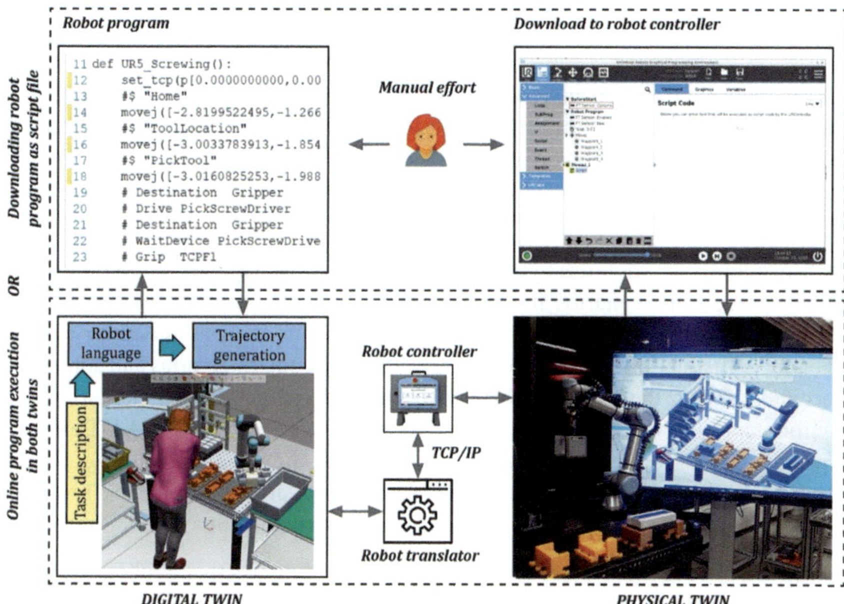


**Fig. 9** Generating robot program and controlling robot arm through its DT [47]

## 8.4 *Simplified Reconfigurations*

The demand for variety in manufacturing is rising, and assembly is recognized as a prime area with significant potential for introducing product diversity [56]. As we navigate the transition phase of the fourth industrial revolution, the notion of batch size one, focusing on personalized manufacturing, is increasingly gaining relevance. However, a challenge is that current robots often lack the flexibility necessary for rapid reconfiguration.

To realize the vision of batch size one, manufacturers require robots with features, hardware, and software solutions that facilitate effortless adaptation to changes and variations in product design. Swift reconfiguration of robots is indispensable for staying in sync with the dynamic demands of modern manufacturing, where rapid shifts in product specifications and design variations are commonplace. DT models emerge as a valuable tool in expediting these reconfigurations by automatically evaluating the automation potential for specific tasks. This includes generating a process sequence, automation programs, robot codes, and worker training opportunities. DT models contribute to streamlining the adaptation process, ensuring efficiency and agility in response to evolving manufacturing requirements.

## 8.5 Layout Planning and Optimization

In a robot installation, achieving an optimal arrangement of robots, fixtures, assembly parts, and associated hardware is critical to minimize footprints and shorten cycle times. In the case of collaborative robots, additional safety layers need attention. To enhance the design of the workstation layout, the following experiments can play a pivotal role:

**Collision analysis**: Identifying and mitigating collisions become paramount in the robotic process in scenarios involving confined spaces with multiple robots and associated hardware. Conducting a thorough collision analysis serves to identify potential collisions, allowing for the optimization of robot trajectories and equipment placement. Given the frequent changeovers and reconfigurations inherent in such environments, this collision analysis may need to be a routine activity. A DT emerges as a valuable tool for generating safe robot paths that circumvent potential collisions (see Fig. 10). By using DTs, robot paths can be proactively optimized. The result is a more efficient and collision-free operation, even in dynamically changing scenarios.

**Reach test**: The reach test is instrumental in determining whether a robot can access all desired locations within its workspace. This test aids in defining the most optimal locations for placing robots and relevant production equipment. Specific location points for robots are established, evaluating whether the robot can safely reach all desired locations. A grasp envelope is created for a human reach test to demonstrate the human arms' reachability without adopting unsafe body postures.

**Placement test**: The placement test seeks to identify optimal locations for robots, humans, and production equipment. The goal is to minimize cycle times, prevent

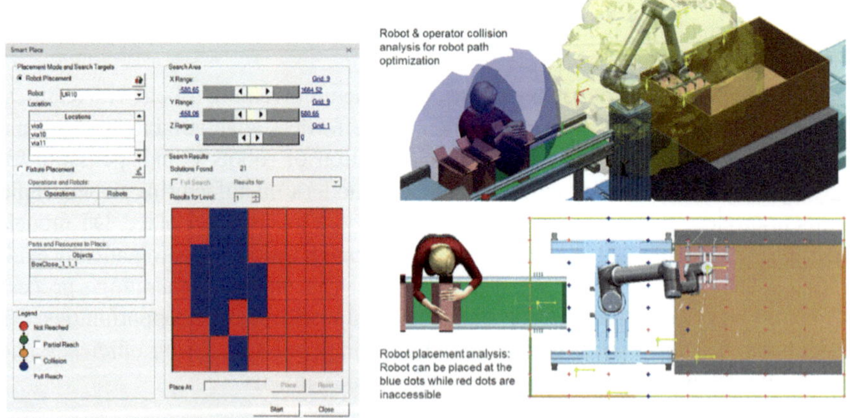


**Fig. 10** Layout planning of a robot cell in simulation

collisions, and ensure a safe working environment. The test can define a range of points from which a robot can reach a selected set of locations.

Other considerations may include designing the layout with flexibility and scalability to accommodate future changes and expansions. Modular design approaches and flexible workspace configurations enable easier reconfiguration and adaptation to evolving production needs. Another important consideration is ensuring the seamless integration of robotic installations with existing manufacturing systems, including conveyor systems and automated guided vehicles. Coordinating the layout design with other production processes (inventory management systems) is essential to optimize workflow integration and synchronization.

## 8.6 *Process Visualization*

A comprehensive simulation enables high-accuracy visualization of the detailed process. Even without making any computer-based analysis of collisions, robot path, or layout visualization, visualization allows visual assessment of the process. Potential errors are identified, and the required modifications are determined. As shown in Fig. 11, a simulation of a robotic process presents the assembly of medical ventilators.

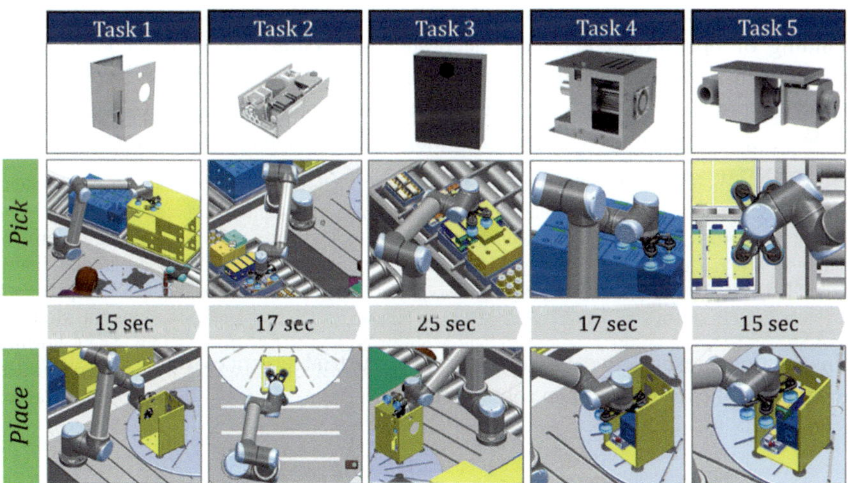


**Fig. 11** Sketch shows the assembly process sequence in a robotic assembly process

### 8.7 Control Program Generation

Robot systems with multiple robotic arms, sensors, actuators, feeders, and hardware devices typically operate based on a logical control program. Traditionally, the development and verification of this logical program are carried out during the commissioning phase. The validation and error-proofing procedures can be time-consuming and may uncover errors that could set the project back to earlier stages. However, leveraging a DT allows for a more streamlined approach. The control program can be generated directly from the simulation model and verified several times during the system development. This generated program is then downloaded to a PLC. The advantage of this approach is that, with each change in the system design, a new control program can be rapidly created, validated within the DT simulation, and then efficiently downloaded to the PLC. This dynamic adaptability enhances the efficiency and agility of the robot system, reducing the time and effort traditionally associated with the commissioning process.

### 8.8 Assessment of Return on Investment

ROI analysis is crucial for decision-making when implementing a robot system to automate a specific process. Without a detailed visualization, analysis, and understanding of operational behavior, accurately assessing the financial gains from investing in a robotic system becomes challenging. A comprehensive simulation-based DT offers an in-depth perspective on various phases of a robot system's lifecycle. It provides a more accurate and thorough assessment of ROI throughout its entire operational lifespan.

## 9 Practical Use Cases

This section presents two use cases from industrial practices that exemplify the practical application of DTs in robot systems.

### 9.1 DT-Based Development of a Robot Assistant in Wind Turbine Manufacturing

This case exemplifies the creation and utilization of a DT for a collaborative mobile robot aiming to automate assembly tasks in the manufacturing of wind turbines. Applying a mobile robot assistant for hybrid automation of wind turbine manufacturing is expected to give benefits such as reduced production costs, enhanced product

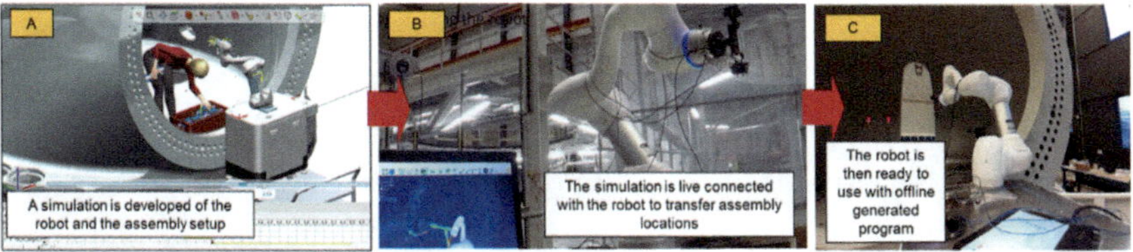


**Fig. 12** DT-supported deployment of a robot assistant in wind turbine manufacturing

quality, and improved working conditions. The DT technology was employed for the commissioning and reconfiguring the mobile robot, contributing to expedited design and validation.

The growing demand for higher rating capacity in wind turbines has driven the production of larger generators, extended blades, and taller towers. The assembly of these turbines entails the manipulation of large-sized components. Traditional automation is impractical due to frequent design alterations and diverse tasks, rendering it labor-intensive. However, human capabilities fall short in managing the assembly of these substantial components, resulting in extended lead times and increased overall costs for sustainable energy. To address these automation challenges, the case presents a solution consisting of a mobile robot equipped with a robotic arm that uses laser beams to assist operators in precisely positioning assembly components. This approach streamlined the assembly process and reduced the assembly time.

The simulated robot was connected to the physical robot in real time using an Ethernet connection (see Fig. 12). The robot's IP address was utilized in the simulation to establish this connection. A post-processor for Doosan Robotics was used. A robot post-processor defines how robot programs are generated for a specific robot controller.

During the design phase, a comprehensive workspace with the robot system was modeled in simulation-based DT. The DT model served as a design validation tool for each reconfiguration. During the development phase, the DT model was a reference for programming the robot. Assembly locations were extracted from the DT model and encoded in the robot program. Vision tests and safety risk assessments were performed using the DT. The DT was operational with the robot system to verify the changes and robot programming.

## 9.2 *DT of a Human-Robot Collaborative Assembly Cell*

This section illustrates the application of a DT in developing and operating a collaborative assembly cell (Fig. 13), emphasizing its significance in designing, developing, and operating human–robot production systems. In the design phase, the DT was

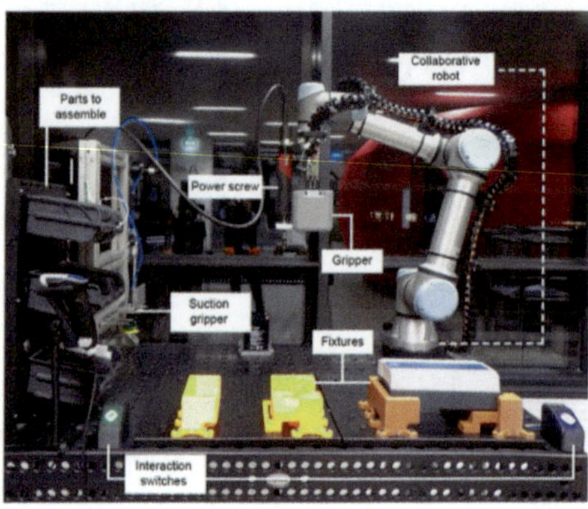


**Fig. 13** Humans and robots working collaboratively [47]

used to create and select the elements of the assembly system according to production requirements and its integration with the overall system. Dynamic simulation facilitated a quantitative assessment and a business value proposition for the proposed solution. The Task Simulation Builder in TPS was used to model human tasks for ergonomic assessments.

During the operational phase, using data loggers, the DT transitioned from manual data syncing to automated and real-time data syncing, enhancing its utility for error identifications. The DT proved valuable in dynamic task distribution based on task complexity rating and event-driven simulation. Other benefits included intuitive robot programming to reduce manual efforts, human safety assessments, the generation of data logs for critical actions through sensor integration, and the incorporation of artificial intelligence to enable the system to self-learn and make decisions based on past experiences.

## 10 Future Research Directions

Interoperability stands out as a critical challenge in the current state of creating DTs for robot systems. The various aspects of a robot system (geometry, kinematics, robot program, automation logic, etc.) can be modeled using different tools (e.g., Robot Operating System (ROS) [57], OpenPLC [58], Unreal Engine [59], etc.). However, exchanging data between these tools is often complicated and sometimes impossible. Additionally, commercial tools from one vendor may not exchange data with those from another. Consequently, there's a pressing need for standardized, exchangeable data formats for DT devices and assets, particularly their connectivity with open-source tools. Initial standards developments such as the IEC 63278—Asset Administration Shell [60] have been made. Still, they are at an early stage of development, and their widespread adoption needs to address their suitability for various scenarios.

Creating and operating a DT necessitates investment in software tools, human resources, data processing ability, and communication technologies. However, it is essential to recognize that not every robot installation requires a DT. Fundamentally, a DT benefits robotic systems as it facilitates flexibility, reconfigurability, seamless deployment, predictive maintenance, and safe human–robot interaction, among other benefits. Assessment methods must be available to identify the need and value of a DT in a given context and asses the ROI of creating and using a DT.

In manufacturing systems, hardware modularization involves incorporating parallel modules for capacity or capability adjustments. Combining flexible automation with human skills, this modular approach facilitates reconfigurability and agility. Achieving a high degree of customization involves adding, replacing, or eliminating modules. A similar modular approach is needed in DTs. Given that the current approach to developing DTs is time-consuming, the creation of libraries of modular DTs of assets and information blocks can pave the way for the formation of reusable and exchangeable DTs.

Flexible robot installations face the challenge of effective and dynamic task scheduling. Task scheduling for robots involves the creation of an optimal schedule within a system. This schedule specifies which robot is assigned to each task and how the tasks will be processed. Dynamic task allocation goes a step further, encompassing the planning of automation processes and allowing robots to adapt during operations. To address this challenge, a comprehensive approach that integrates event-based logical simulation, probabilistic analysis, and statistical data analysis into a DT model is needed. This integration can enhance the DT's capability to facilitate dynamic and efficient task scheduling, ensuring adaptability to changing operational needs.

## 11 Conclusion

A manufacturing system can be characterized as a network of subsystems, including equipment, machines, humans, and robots, working together to transform raw materials into finished products. Robots have become integral components of modern manufacturing facilities and are increasingly gaining traction. However, designing, developing, deploying, and reconfiguring robot cells is time-consuming, spanning weeks to months. The extensive involvement of a larger workforce in the project contributes to higher overall system costs and results in a prolonged ROI. This extended timeframe is influenced by various factors, such as sequential development processes, complex programming techniques, tool development, and the need to adhere to safety standards.

This chapter outlines the fundamental elements or modules constituting a robotic system's DT. The research arena has documented the advantages of employing DTs for robot systems. However, realizing the full potential of DTs for robot work cells requires the creation of a reliable and comprehensive digital representation that accurately models the elements and dynamics of the observed robot system. Flexibility remains critical in ensuring the DT can seamlessly adapt to changing circumstances. This chapter underscores the significance of DTs in manufacturing robotic cells and outlines the essential criteria for developing these DTs.

The interoperability challenge arises in creating DTs for robot systems, as exchanging data between different tools proves difficult. Standardized and exchangeable data formats are essential, emphasizing compatibility with open-source tools. While investing in DT tools and resources is crucial, it's important to note that not every robot system requires a DT. Modularization is critical in manufacturing and DT development, offering flexibility and reconfigurability. Flexible robot installations encounter challenges in dynamic task scheduling, which can be addressed by integrating logical simulation and statistical analysis into DT models, ensuring adaptability to changing operational needs.

**Disclaimer** This research was conducted through the support of a NIST cooperative agreement [60NANB23D234]. Specific commercial products and systems are identified in this paper to facilitate understanding. Such identification does not imply that these software systems are necessarily the best available for the purpose. No approval or endorsement of any commercial product by NIST is intended or implied.

## References

1. Malik AA, Brem A (2021) Digital twins for collaborative robots: a case study in human-robot interaction. Robot Comput Integr Manuf 68:102092. https://doi.org/10.1016/j.rcim.2020.102092
2. Rios J, Bernard A, Bouras A, Foufou S (2017) Product lifecycle management and the industry of the future. In: 14th IFIP WG 5.1 international conference. Springer, Seville, Spain
3. Morgan J, Halton M, Qiao Y, Breslin JG (2021) Industry 4.0 smart reconfigurable manufacturing machines. J Manuf Syst 59:481–506. https://doi.org/10.1016/j.jmsy.2021.03.001
4. Yazdani MA, Khezri A, Benyoucef L (2022) Process and production planning for sustainable reconfigurable manufacturing systems (SRMSs): multi-objective exact and heuristic-based approaches. Int J Adv Manuf Technol 119:4519–4540. https://doi.org/10.1007/s00170-021-08409-0
5. Shao G, Helu M (2020) Framework for a digital twin in manufacturing: scope and requirements. Manuf Lett 24:105–107. https://doi.org/10.1016/j.mfglet.2020.04.004
6. Advanced Manufacturing (2022) National strategy for advanced manufacturing
7. Malik AA (2023) Simulation based high fidelity digital twins of manufacturing systems: an application model and industrial use case. In: Winter simulation conference 2023
8. Masood T, Weston R (2011) An integrated modelling approach in support of next generation reconfigurable manufacturing systems. Int J Comput Aided Eng Technol 3:372–398
9. Bilberg A, Malik AA (2019) Digital twin driven human–robot collaborative assembly. CIRP Ann 68:499–502. https://doi.org/10.1016/j.cirp.2019.04.011
10. Maruyama T, Ueshiba T, Tada M, Toda H, Endo Y, Domae Y (2021) Digital twin-driven human robot collaboration using a digital human 1–46. https://doi.org/10.3390/s21248266
11. Liu S, Wang XV, Wang L (2022) Digital twin-enabled advance execution for human-robot collaborative assembly. CIRP Ann 71:25–28. https://doi.org/10.1016/j.cirp.2022.03.024

12. Yao B, Xu W, Shen T, Ye X, Tian S (2023) Digital twin-based multi-level task rescheduling for robotic assembly line. Sci Rep 13:1–20. https://doi.org/10.1038/s41598-023-28630-z
13. Masood T, Kern M, John Clarkson P (2021) Characteristics of changeable systems across value chains. Int J Prod Res 59:1626–1648. https://doi.org/10.1080/00207543.2020.1791997
14. Giones F, Brem A (2017) From toys to tools: the co-evolution of technological and entrepreneurial developments in the drone industry. Bus Horiz 60:875–884
15. Stadnicka D, Antonelli D (2019) Human-robot collaborative work cell implementation through lean thinking. Int J Comput Integr Manuf 32:580–595. https://doi.org/10.1080/0951192X.2019.1599437
16. Malik AA, Masood T, Kousar R (2020) Repurposing factories with robotics in the face of COVID-19. Sci Robot 5, (2020) https://doi.org/10.1126/scirobotics.abc2782
17. Malik AA, Masood T, Kousar R (2021) Reconfiguring and ramping-up ventilator production in the face of COVID-19: can robots help? J Manuf Syst 60:864–875. https://doi.org/10.1016/j.jmsy.2020.09.008
18. Anumbe N, Saidy C, Harik R (2022) A primer on the factories of the future
19. Nof SY (2009) Springer handbook of automation. Springer, Berlin
20. Boy GA (2018) From automation to interaction design. Handb Hum-Mach Interact 431–443. https://doi.org/10.1201/9781315557380-22
21. Malik AA, Masood T, Management E, Brem A (2024) Intelligent humanoid robots in manufacturing: addressing the worker shortage and skill gaps in assembly cells. https://doi.org/10.1145/3610978.3640765
22. Kurfess TR (2004) Robotics and automation handbook
23. Shneier MO, Messina ER, Schlenoff CI, Proctor FM, Kramer TR, Falco JA (2015) Measuring and representing the performance of manufacturing assembly robots
24. Andersson SKL, Granlund A, Hedelind M, Bruch J (2020) Exploring the capabilities of industrial collaborative robot applications. Adv Transdisciplinary Eng 13:109–118. https://doi.org/10.3233/ATDE200148
25. Pedersen MR, Nalpantidis L, Andersen RS, Schou C, Bogh S, Krüger V, Madsen O (2016) Robot skills for manufacturing: From concept to industrial deployment. Robot Comput Integr Manuf 37:282–291
26. Ji S, Lee S, Yoo S, Suh I, Kwon I, Park FC, Lee S, Kim H (2021) Learning-based automation of robotic assembly for smart manufacturing. Proc IEEE 109:423–440. https://doi.org/10.1109/JPROC.2021.3063154
27. Buerkle A, Eaton W, Al-Yacoub A, Zimmer M, Kinnell P, Henshaw M, Coombes M, Chen WH, Lohse N (2023) Towards industrial robots as a service (IRaaS): flexibility, usability, safety and business models. Robot Comput Integr Manuf 81:102484. https://doi.org/10.1016/j.rcim.2022.102484
28. Malik AA, Bilberg A (2019) Complexity-based task allocation in human-robot collaborative assembly. Ind Robot 46:471–480. https://doi.org/10.1108/IR-11-2018-0231
29. Malik AA, Bilberg A (2019) Developing a reference model for human–robot interaction. Int J Interact Des Manuf 13:1541–1547. https://doi.org/10.1007/s12008-019-00591-6
30. Rahman SMM, Sadrfaridpour B, Wang Y, Walker ID, Mears L, Pak R, Remy S (2020) Trust-triggered robot-human handovers using flexible manufacturing, 1–8
31. Edward J, Wannasuphoprasit CW, Peshkin MA (1996) Cobots: robots for collaboration with human operators. In: International mechanical engineering congress and exposition, Atlanta
32. Krüger J, Lien TK, Verl A (2009) Cooperation of human and machines in assembly lines. CIRP Ann Manuf Technol 58:628–646
33. Malik AA, Bilberg A (2019) Human centered lean automation in assembly. Procedia CIRP 81:659–664. https://doi.org/10.1016/j.procir.2019.03.172
34. Ajoudani A, Zanchettin AM, Ivaldi S, Albu-Schäffer A, Kosuge K, Khatib O (2018) Progress and prospects of the human–robot collaboration. Auton Robots 42:957–975. https://doi.org/10.1007/s10514-017-9677-2
35. Povlsen K (2023) A new generation of robots can help small manufacturers. https://hbr.org/2023/11/a-new-generation-of-robots-can-help-small-manufacturers

36. BMW (2017) HRC system in production at BMW Dingolfing
37. Bänziger T, Kunz A, Wegener K (2017) A library of skills and behaviors for smart mobile assistant robots in automotive assembly lines. In: Proceedings of the companion of the 2017 ACM/IEEE international conference on human-robot interaction, pp 77–78
38. Kirova I (2020) Give me a hand—*The* potential of mobile assistive robots in automotive logistics and assembly applications Stefanie. Educ Technol J 11:329–331. https://doi.org/10.26883/2010.202.2349
39. Melissa R (2023) Articulated robot market size. https://statzon.com/insights/global-articulated-robot-market
40. Malik AA, Masood T (2022) Mobile robot assistants in wind turbines manufacturing. SSRN Electron J. https://doi.org/10.2139/ssrn.4241697
41. ISO: ISO 19649_2017(en) Mobile robots—Vocabulary
42. Grieves M (2002) Completing the cycle: using PLM information in the sales and service functions, Troy
43. De Oliveira Hansen JP, Da Silva ER, Bilberg A, Bro C (2021) Design and development of automation equipment based on digital twins and virtual commissioning. Procedia CIRP. 104:1167–1172. https://doi.org/10.1016/j.procir.2021.11.196
44. Grieves M, Vickers J (2017) Digital twin: Mitigating unpredictable, undesirable emergent behavior in complex systems. In: Transdisciplinary perspectives on complex systems. Springer, pp 85–113
45. ISO: ISO 10303-242_2022—Industrial automation systems and integration—Product data representation and exchange—Part 242_Application protocol_Managed model-based 3D engineering
46. International Electrotechnical Commissioning (2003) Programmable controllers-Part 1: General information
47. Malik AA, Brem A (2021) Digital twins for collaborative robots: a case study in human-robot interaction. Robot Comput Integr Manuf. 68. https://doi.org/10.1016/j.rcim.2020.102092
48. Qi Q, Tao F, Hu T, Anwer N, Liu A, Wei Y, Wang L, Nee AYC (2021) Enabling technologies and tools for digital twin. J Manuf Syst 58:3–21. https://doi.org/10.1016/j.jmsy.2019.10.001
49. Collaborative Robots (2016) Eliminate the mystery around safety robot safety standards for collaborative operation ISO 10218 2016. Curr Saf Stan
50. ISO: ISO 13849-1_(2023)(en) Safety of machinery—Safety related parts of control systems, Part 1: General principles for design
51. Robots ISO (2016) Robotic devices--collaborative robots. ISO 15066: 2016
52. Bauer W, Bender M, Braun M, Rally P, Scholtz O. Lightweight robots in manual assembly—best to start simply! In: Examining companies' initial experiences with lightweight robots,. Frauenhofer-Institut für Arbeitswirtschaft und Organisation IAO, Stuttgart
53. Tan JTC, Duan F, Zhang Y, Watanabe K, Kato R, Arai T (2009) Human-robot collaboration in cellular manufacturing: design and development. In: intelligent robots and systems, 2009. IROS 2009. IEEE/RSJ International Conference on. pp. 29–34 (2009)
54. Francalanza E, Borg J, Constantinescu Deriving a systematic approach to changeable manufacturing system design. In: Procedia CIRP. pp 166–171. Elsevier B.V.
55. Malik AA, Bilberg A (2017) Framework to implement collaborative robots in manual assembly: a lean automation approach. In: Katalinic B. (Ed.) Proceedings of the 28th DAAAM international symposium, Published by DAAAM International, ISSN. pp 1726–9679
56. ElMaraghy H (2014) Managing variety in manufacturing. Procedia CIRP 1–2 (2014)
57. ROS: Robot Operating System, https://www.ros.org/
58. GitHub: OpenPLC_ an open source industrial controller. https://github.com/thiagoralves/OpenPLC
59. EPIC Games: Unreal Engine. https://www.unrealengine.com/en-US
60. IEC: IEC 63278-1 (2023) ED1 *Asset* Administration shell for industrial applications—Part 1: Asset administration shell structure. (2024)

# Mobiliti: A Digital Twin for Regional Transportation Network Design and Evaluation

**Jane F. Macfarlane and Ismaeel Babur**

**Abstract** Mobiliti is introduced as a foundation for a Digital Twin for managing transportation systems across metropolitan regions and for planning and designing urban networks. Regional transportation systems consist of interconnected subnetworks, each governed by different municipalities. Although localization simplifies analysis, transportation projects must be evaluated within the context of the larger regional network due to their potential impacts on overall network performance. The computational challenges of simulating metropolitan networks are addressed through parallel discrete event simulation, enhancing the predictive capabilities of the Digital Twin framework. This technology demonstrates significant potential to enhance urban quality of life, encourage shifts to multi-person vehicle use, and provide actionable feedback on infrastructure investments to ensure that urban environments are economically viable, safe, and healthy.

**Keywords** Regional-scale transportation dynamics · Parallel discrete event simulation, · Traffic assignment · Equity measures · Road networks · Real-time mobile device data · Predicitive traffic · Generative network design · Greenfield city design

## 1 Introduction

The US transportation sector is in the midst of a significant digital transformation [12]. An increasing number of connected vehicles are appearing on our roads as original equipment manufacturers (OEMs) integrate advanced sensors into cars, trucks, and public transportation. These sensors, which include cameras, lidar, GPS, and

J. F. Macfarlane (✉) · I. Babur
Smart Cities Research Center, Institute of Transportation Studies, University of California Berkeley, 115 McLaughlin Hall Berkeley, Berkeley, CA 94720, USA
e-mail: janemacfarlane@berkeley.edu

I. Babur
e-mail: ismaeel@berkeley.edu

 
M. Grieves and E. Hua (eds.), *Digital Twins, Simulation, and the Metaverse*, Simulation Foundations, Methods and Applications, https://doi.org/10.1007/978-3-031-69107-2_14

radio communication devices, are collecting vast amounts of data on vehicle location, speed, surrounding environments, and the health of components. This data is then analyzed to derive insights into vehicle dynamics, such as driving patterns, congestion levels, and critical braking events.

At the same time, the infrastructure itself is becoming smarter. Sensors and cameras installed across various points enable remote monitoring of traffic conditions, asset status, accidents, and weather changes. Traffic management centers are using this wealth of information to make operational decisions that enhance response times and overall efficiency. This shift is replacing older hardware systems with intelligent transportation systems (ITS), which include automated features like adaptive traffic signals, dynamic tolling, and smart parking meters, all designed to streamline transportation management.

Additionally, the surge in data is being captured and analyzed through big data platforms, hosted on the cloud. These platforms aim to provide a holistic view of transportation networks by integrating data from both vehicles and infrastructure sensors.

A transformation toward Mobility-as-a-Service (MaaS) is taking place. This model leverages the new digital and communication infrastructure to offer on-demand services such as ride-sharing, bike-sharing, and microtransit through consumer-friendly mobile apps. These services are becoming viable alternatives to private car ownership, particularly in urban areas. Alongside this shift, there is a growing governmental push toward electrification, including the development of charging networks and electric fleets, aimed at mitigating the environmental impacts of internal combustion engines. Digital systems are also playing a crucial role in facilitating the integration and monitoring of these new electric vehicle demands.

As a whole, this evolving ecosystem of digital technologies is not only enabling new transportation services and optimizing infrastructure usage, it is also enhancing the capacity for comprehensive monitoring and data-driven planning and operations. This transformation is set to accelerate and will reshape the future of transportation in the years to come.

## 2 Digital Twins for Regional Transportation Networks

Urban transportation networks are inherently complex, consisting of numerous interconnected systems and a variety of stakeholders, both public and private. To navigate and influence this intricate landscape effectively, all involved entities must deeply understand the dynamic interactions associated with the movement of people and goods. Traditional static models and simulations have often fallen short, limited by their inability to fully capture the intricacies of regional transportation dynamics.

Emerging and evolving sensors that monitor transportation systems are paving the way for innovative solutions like Digital Twins. These tools provide a virtual representation of transportation systems, aligning real-time data with system models to mirror the physical network's key components. This allows planners to simulate various scenarios, understand the cascading effects of network changes, and importantly, identify unintended consequences of these modifications. Enhanced computational power and advanced simulation technologies are amplifying the capabilities of Digital Twins, offering unprecedented insight into the dynamics of transportation.

The development of effective Digital Twins for regional planning hinges on several key research areas: defining clear fundamental domain taxonomies, advancing simulation methodologies, integrating data from multimodal transportation sources, and using artificial intelligence to codify urban complexities. Thoughtfully designed, Digital Twins have the potential to revolutionize the transportation planning process. They enable more precise analyses of investment options and policy decisions, offer the capability to evaluate alternatives swiftly, and provide rapid feedback, all of which contribute to more agile and effective decision-making processes that enhance mobility outcomes. Realizing the full potential of Digital Twins will necessitate strong partnerships among researchers, technology providers, planning agencies, and private industry.

## *2.1 Modeling Regional Transportation Systems*

Regional transportation systems consist of complex networks made up of interconnected subnetworks, each often governed independently by different municipalities. Although local governance can simplify some aspects of analysis, it is crucial for transportation projects to consider their role within the broader regional context due to significant impacts on network performance. This is particularly important as urbanization continues to increase vehicle densities in metropolitan areas. Historically, the capability to analyze extensive regional networks was hampered by computational limitations, necessitating substantial simplifications in network and demand modeling.

Metropolitan regions comprise multiple cities, each with its unique dynamics influenced by its geographical setting. Cities typically operate autonomously, and smaller cities may be particularly constrained by limited financial resources. This autonomy often results in a patchwork of traffic management strategies, linked together only by state Department of Transportation (DOT) managed interstates and highways.

Within cities, traffic management is handled through various mechanisms including traffic regulation tools like signals and stop signs, road network controls such as speed limits and turn restrictions, physical traffic calming measures like speed bumps and lane narrowing, and infrastructure investments focused on non-vehicular traffic, including protected bike lanes and intersection curb extensions. Meanwhile, state DOTs are responsible for managing broader infrastructure elements like interstates, state highways, and bridges, overseeing broader traffic sensing, incident management, and ramp metering.

Adding another layer to this complex system, navigation applications in vehicle dashboards and smartphones act as a third party influencing metropolitan traffic management. These apps can dynamically reroute travelers to faster routes as they become available, which, while while potentially helpful in reducing congestion, can also lead to unintended traffic disruptions in residential areas or lead to ill-informed behaviors in unusual circumstances, such as during wildfire evacuations.

The primary objective of regional transportation modeling is to codify these complexities of disjointed traffic management systems. The aim is to develop analytics that assist city planners in enhancing transportation options and dynamics across cities and to design strategies that encourage mode shifts which will be crucial for the future of traffic management in metropolitan areas.

## 2.2 *Mobiliti: Transformative for Transportation Planning*

Simulation tools for transportation systems come in various forms, typically categorized into microsimulation, mesosimulation, and macrosimulation. Our novel platform, Mobiliti, stands out by overcoming traditional computational limitations associated with the simulation of large regional road networks. It uses an agent-based, parallel discrete event simulation framework where each road link acts as an agent. These agents control the movement of vehicles across their geospatial locations, managing traffic flow according to constraints such as vehicle capacity, speed limits, and queue lengths. For instance, in the San Francisco Bay Area, Mobiliti can model 24 h of activity for approximately 7 million personal vehicles and 4 million trucks, translating to approximately 19 million trips across a network of 1 million road links, in less than eight minutes—a scale of simulation and computational speed unmatched by any current system.

Similar simulation tools limited by computational constraints are forced to reduce the scale of the simulation. This typically involves decreasing the number of trips or population size, reducing the size of the road network or simplifying its geometry, or any combination of these approaches. These reductions and simplifications fail to accurately represent the true dynamics of the metropolitan regions being studied. In practice, this limitation is particularly critical when considering dynamic rerouting, which often involves lower-class neighborhood roads that are often omitted in reduced network models.

In contrast, planning agencies sometimes use optimization algorithms to understand road network loading. These tools analyze trip demand profiles and assign trips to the network, aiming to optimize travel times. However, these tools also face computational challenges, and the traffic assignment models often operate with very low spatial and temporal fidelity—for example, running optimizations over a static four-hour period. Despite this, these tools can still require up to 24 h to process and frequently yield results that significantly underestimate congestion patterns.

### 2.3 The Need for Regional Models

The importance of regional transportation models stems from their ability to accurately represent the travel demand profiles of individual cities, a necessity for several reasons. First, economic and land use impacts are inherently regional. Activities such as employment, shopping, and recreation generate travel patterns that often begin in one city and conclude in another within the same region. Consequently, examining a single city in isolation does not capture critical origin–destination flows and requires assumptions about boundary conditions for trips that cross these boundaries.

Moreover, the transportation needs of a city are deeply influenced by regional developments. Public transportation options and road networks extend across city limits, creating an interconnected system where traffic conditions in one area can affect another downstream. Additionally, cities often share significant infrastructure such as ports, airports, and major state-run corridors, emphasizing the interdependence of urban areas.

Transportation policy and investment decisions, typically made at a regional level, require insights into travel behaviors across multiple cities. By focusing solely on individual cities, significant interactions and the use of shared assets may be overlooked. Regional models address these challenges by providing a comprehensive view of mobility patterns, capturing the intricacies of how people move within and between urban areas.

With projections indicating that 68% of the global population will reside in major cities by 2050, the relevance of regional models becomes even more pronounced. As urbanization continues to drive growth in major cities and expands them into future mega-regions, understanding the complex dynamics of regional transportation will be crucial for managing congestion and planning sustainable urban environments.

## 3 Digital Twin Versus Model Based Controller

When discussing Digital Twins, it's important to recognize that they are composed of three distinct components: a physical system operating in real-time, dynamic data about this system streamed at various frequencies, and a computationally manageable simulation of the system. This chapter explores a Digital Twin specifically designed to address the complex challenges of managing transportation systems across metropolitan regions.

The Digital Twin Consortium defines a Digital Twin as "a virtual representation of real-world entities and processes, synchronized at a specified frequency and fidelity". This emphasizes the Digital Twin's role in mirroring real-world systems and processes accurately and in real-time.

Contrasting with this is the model-based controller, a concept well-established in traditional control theory. Model-based controllers use knowledge of physical laws governing systems to estimate responses to external inputs. This knowledge is embedded within a feedback control system, allowing the controller to predict and optimize system behavior under various conditions. The challenge lies in developing accurate models that can sufficiently account for all relevant system behaviors.

In comparison, a Digital Twin goes beyond control. It serves as a comprehensive virtual counterpart of the system, not only monitoring and optimizing performance but also enabling scenario testing without impacting the physical system. Digital Twins thus offer substantial benefits, such as providing valuable insights, facilitating remote monitoring and control, and enhancing decision-making processes.

Key differences between a model-based controller and a Digital Twin include:

- A model-based controller uses a simplified model primarily for control, aiming to meet specific control requirements. It does not typically mirror the full complexity of the physical system. Conversely, a Digital Twin seeks to emulate the physical system as closely as possible, incorporating much greater detail and complexity.
- The model in a model-based controller is generally static, with updates based on new data being slow or non-existent. On the other hand, a Digital Twin continuously updates its virtual representation with real-time data from the physical system.
- The primary objective of a model-based controller is to calculate optimal control inputs to achieve desired performance. In contrast, the Digital Twin aims to understand, analyze, and potentially optimize the physical system, providing a predictive analytics capability for exploring future states.

While model-based controllers use simplified models designed specifically for control algorithms, Digital Twins aim to fully virtualize the physical system and maintain synchronization with real-time data. This allows Digital Twins to facilitate a broader exploration of scenarios in the context of the system's current state. It is likely that as Digital Twins evolve, model-based control techniques will be integrated within them, validated against these higher-fidelity representations to further enhance their utility and effectiveness.

## 4 Managing Computational Complexity

Transportation modeling can become exceedingly complex at larger scales. To manage this complexity, a key feature of any Digital Twin for transportation networks is its computationally manageable simulation capability. As the advancements predicted by Moore's law began to plateau, there was a significant shift in focus toward developing new approaches for traditional algorithms, particularly in the realm of parallel computation.

Parallel discrete event simulation [10] has emerged as a pivotal technology in this context. It offers a new level of scalability for modeling extensive network systems. Mobiliti uses this simulation approach to tackle the challenges posed by regional-scale transportation networks, which are increasingly common in urban metropolitan areas. This platform leverages parallel discrete event simulation to effectively manage and simulate the complex dynamics of these extensive networks, providing a valuable tool for understanding and optimizing transportation infrastructure.

## 4.1 Parallel Discrete Event Simulation

Parallel discrete event simulation (PDES) is a specialized simulation technique that represents processes as a sequence of time-stamped events occurring on predefined actors. These actors use local state information to schedule subsequent events. A prime example of PDES application is in vehicle transportation simulation, the focus of the Mobiliti development effort. In this scenario, the road network is divided into segments or links, which serve as the actors, while vehicles are modeled as the events that move between these links. The vehicle travel times are predicted based on delays caused by the presence of other vehicles on the same link.

One of the significant challenges in discrete event simulations is the effective parallelization of these processes. The core promise of a PDES engine is that by the simulation's conclusion, all events will have been processed in an order that respects the chronology of their timestamps. In context of the complexity of transportation modeling, where predicting event arrival times on any given actor is inherently challenging, the PDES community developed two main strategies for parallelization:

- Conservative Parallelization: This method uses model-specific information to ensure that no events will unexpectedly arrive within a certain time frame, thus guaranteeing that the simulation within that time interval is safe to execute. Although this approach often results in reduced parallelism, it can achieve good performance if sufficient detail about the problem is known to expose enough parallel opportunities.
- Optimistic Synchronization: Unlike conservative methods, optimistic synchronization does not require that events are initially executed in the correct order. Instead, it allows events that are processed out of order to be replayed to ensure that the final state of each actor aligns with what would have occurred had all events been executed correctly. However, this approach can face challenges if too many events are executed incorrectly, leading to significant runtime overheads. Additionally, making an event "reversible" often involves considerable coding effort and can present substantial engineering challenges, especially when third-party libraries are used.

Deciding between these two parallelization techniques involves understanding their performance trade-offs, which can vary widely depending on the specific characteristics of the problem at hand. Conservative methods generally offer more predictable performance but at the cost of limited scalability, while optimistic approaches provide greater flexibility but can lead to increased complexity and potential performance penalties.

## 4.2 Discrete Event Simulation

Discrete event simulation (DES) is a computational approach used to model systems that evolve over time through instantaneous state changes, known as events, which occur at specific timestamps. Each event not only modifies the state of the system but also has the potential to trigger additional future events. A DES application is characterized by its state space, an initial state, and an initial set of events. The framework responsible for DES ensures that these events are executed in the order of their timestamps, continuing until no further events remain, at which point the simulation concludes.

To manage the sequence of events effectively, DES typically employs a time-sorted queue. This queue continuously checks for the event with the earliest timestamp and processes it until the queue is empty. Building upon this, parallel discrete event simulation (PDES) extends DES by introducing mechanisms that allow for the parallel processing of events. This is achieved by dividing the application state into logical processes (LPs), with each LP handling events that only affect its specific portion of the state. This segmentation enables safe concurrent execution of events across different LPs.

However, parallelizing events introduces significant complexity, particularly in maintaining the correct order of event execution within each LP. A typical challenge arises when events from different LPs are executed concurrently, and an event from one LP triggers a new event in another LP with a timestamp that predates the currently processed event. This scenario leads to a violation of the time-ordering constraint and creates a conflict in the simulation's execution process.

## 4.3 Conservative Versus Optimistic Execution

To manage these complexities, two general strategies are employed: conservative and optimistic execution.

In the conservative approach to execution, events within a runtime are only processed when it is certain that no preceding events will be generated. Typically, this means that the only event deemed safe for execution is the one with the earliest global timestamp. This method, however, imposes a strict global linear order of event times, effectively limiting opportunities for parallel computation and thus constraining the computational scalability.

To optimize this approach, events can be distributed across multiple CPUs by dividing the master event queue into several per-LP (logical process) queues. The application must provide the minimum timestamp difference for any pair of events running in respective LPs, where the first event triggers the second. Events are considered safe for concurrent execution if their timestamps are within the bounds set by their respective LPs in this matrix. During execution, CPUs must frequently exchange their earliest timestamp values to calculate the most lenient execution bounds. However, due to the necessity of adhering to the most conservative case, this method often results in missed opportunities for parallel execution and requires aggressive synchronization.

### 4.3.1 Optimistic Execution

Optimistic execution, pioneered by Jefferson in his description of the Time Warp mechanism [7], adopts a more localized approach. It operates under the assumption that the earliest event in an LP's queue is likely safe to execute, despite the risk that an earlier event might arrive from another LP. To manage this risk, the runtime must be capable of 'rolling back'—reverting the state of the LP to the point before the out-of-order event was executed. This process requires that the application provides mechanisms to reverse one or more events, a complex challenge in itself. Additionally, the system must handle potentially invalid event orderings, which will need to be corrected through rollbacks. The key advantage of optimistic execution over conservative methods is that it allows for the execution of events without immediate synchronization, taking advantage of the frequent safety of event execution despite the lack of concrete proof of valid event ordering.

### 4.3.2 Asynchronous Global Virtual Time Computation

A crucial aspect of any Time Warp-based parallel discrete event simulation (PDES) is managing the computation of Global Virtual Time (GVT). GVT is the consensus timestamp up to which all events are confirmed as correctly processed, serving as a baseline for system integrity. Maintaining a queue that can revert the system to any prior state is essential for enabling rollbacks when events are processed out of order, though this can significantly increase memory requirements.

To enhance scalability and manage this complexity, a highly asynchronous version of Time Warp is implemented that eschews blocking collective operations. A primary challenge in optimizing Time Warp performance is to curb excessive optimism, which can lead to frequent and resource-intensive rollbacks. Rollbacks are necessitated when an event, once processed, must be reversed due to subsequent receipt of an earlier time-stamped event. This also involves rolling back any events that the initial event might have triggered on neighboring processors.

The system leveraged by Mobiliti, called Devastator, was developed to run on a supercomputer at Lawrence Berkeley National Laboratory by Cy Chan, Max Bremer, et al. It employs a strategy known as the moving time window. This method limits the execution interval to GVT plus a tunable parameter ($w$), thus containing the temporal range within which events can be executed. As GVT progresses, this window advances, progressively allowing more pending events to be processed. Devastator automatically adjusts the width of this window based on operational efficiency metrics.

In summary, while the conservative synchronization approach grapples with the scalability challenges posed by frequent collective synchronization, the optimistic synchronization method offers a solution. By reducing the latency associated with smaller, point-to-point messages, it decreases the likelihood of delays caused by 'straggler' events that necessitate rollbacks, thus enhancing the system's scalability and suitability for application needs.

## 5 Mobiliti—A Platform for Regional Traffic Modeling

Agent-based traffic simulation serves as a robust mechanism for understanding real-world transportation systems and facilitating the development of Digital Twins [16, 17]. By leveraging Devastor, Mobiliti is designed to effectively model extensive regional transportation networks. This is achieved by treating road links as parallel actors that manage the flow of vehicle events. Mobiliti is structured upon three primary software layers: GASNet-Ex, Devastator, and Mobiliti itself. GASNet-Ex [5] is an open-source software that facilitates high-performance inter-process communications across distributed memory systems, ideal for handling small active messages. Devastator operates above GASNet-Ex, implementing the Time Warp optimistic parallel discrete event protocol which manages the scheduling, execution, rollback, commit, and Global Virtual Time of events. At the top layer, Mobiliti introduces domain-specific logic that defines both the actors and events within the road network, and it orchestrates the parallel domain decomposition, mapping actors to processors.

Within Mobiliti, road links act as computational actors, calculating vehicle traversal times and enforcing timing and capacity constraints, which naturally lead to queue formation as congestion builds. The platform tracks congestion by monitoring vehicle flow rates on these links across specified time windows. Designed to run on the supercomputer at Lawrence Berkeley National Laboratory, Mobiliti demonstrates strong scalability, effectively running on up to 1024 cores. Performance tests reveal that the baseline simulation can process over 2.4 billion discrete events in less than 30 s using 1024 cores, handling massive networks efficiently [2].

Beyond simulating basic vehicle movement across a network, Mobiliti also supports dynamic rerouting, allowing a predetermined number of vehicles to request new routes based on real-time congestion data. This feature reflects the pervasive use of navigation systems in modern urban settings and offers adjustable controls for rerouting parameters and the distribution of rerouting permissions among vehicles.

Figure 1 outlines the components of Mobiliti illustrating how different data inputs interact and contribute to the overall modeling process. The model is broken down into three primary components, each colored differently to represent distinct aspects of the simulation:

1. **Road Network (Green)**: Represents the physical layout of the transportation infrastructure, including all roads and paths vehicles can traverse. It provides the foundation for the simulation where all other data will interact.

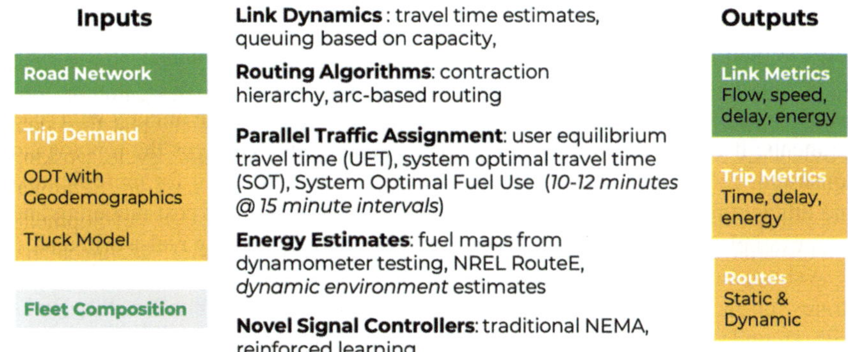


**Fig. 1** Overview of Mobiliti highlighting the interconnected components used to analyze and optimize urban mobility

2. **Trip Demand (Yellow)**:
   - **ODT with Geodemographics**: Origin–destination time data enhanced with geodemographic information from a travel demand model helps in understanding the demographic characteristics of travelers and their travel patterns.
   - **Truck Model**: Specific modeling considerations for truck movements, which might differ in terms of routes, regulations, or times of operation compared to other vehicles.
3. **Fleet Composition (Gray)**: Details the types of vehicles involved in the transportation network and is used for determining energy use.

The outputs of the model, based on the inputs from the three areas above, are divided into three types of metrics:

- **Link Metrics**: Measures related to individual road segments such as flow, speed, delay, and energy consumption.
- **Trip Metrics**: Focuses on the efficiency and impact of individual trips within the network, including travel time, delays, and energy usage.
- **Routes**: Involves both static and dynamic routing information. Static routes are predetermined, while dynamic routes can change in response to traffic conditions, blockages, or other real-time factors.

The model uses these inputs and outputs to simulate and analyze transportation scenarios, optimizing for various factors such as reduced travel times, decreased energy consumption, or improved safety.

Mobiliti uses road network graphs derived from HERE Technologies maps [6] or may use any detailed road network as long as essential attributes like free-flow speed, lane count, and road geometry are specified. An example network built using open-source data for Houston is detailed in [4]. The vehicle demand profile is typically generated from regional travel demand models and specifies the origin, destination, and timing of trips.

Mobiliti employs an arc-based routing algorithm that modifies Dijkstra's classic shortest path algorithm to operate directly on road segments rather than intersections. This approach simplifies the graph representation and improves cost modeling accuracy by associating costs like travel time or fuel consumption directly with road segments. It avoids the need for node preprocessing and sidesteps the topological constraints required by traditional Dijkstra's algorithm, allowing for more flexible and efficient routing. Connected car data was used to determine cost modeling and is an example of how the Digital Twin is enhanced by integrating real-world data.

A simulation of the Los Angeles area, visualized using kepler.gl in Fig. 2, demonstrates the platform's capabilities on a network of approximately 2 million links with around 40 million trips in a single day, completing the simulation in about 15 min. This simulation was based on data from the 2020 Southern California Area Government Group Travel Demand Model. Validation of simulation results incorporates comparisons against multiple real-world data sources such as PeMS, Uber Movement speeds, on-road tube counts, regional surveys, and products from large real-time data collection vendors. Mobiliti has recently surpassed the regional level by demonstrating its ability to handle simulations at an unprecedented scale. The California Statewide Travel Demand Model was used to simulate vehicle and truck movements across the entire state, accounting for over 102 million trips. This represents one of the largest transportation simulations ever created, offering new insights into statewide traffic dynamics and congestion patterns. These statewide insights are crucial, as regional and city-level simulations often struggle to adequately represent 'out-of-region' or 'out-of-city' trips, frequently resorting to assumptions or simplifications that fail to capture the full scope of traffic flows beyond their planning jurisdictions. Figure 3 visualizes the daily vehicle count over the entire California road network, based on results from the statewide simulation.

## 6 Enriching Digital Twins Through Advanced Data Integration

Developing a precise and operationally effective Digital Twin requires a detailed, high-quality depiction of the dynamics and behaviors of real-world transportation systems. This involves modeling and quantifying the complex interactions among various components and features throughout the regional network. The model's complexity is further enriched by integrating dedicated sensor data and ancillary datasets, enhancing its ability to account for unpredictable events, such as traffic incidents or unplanned construction on routing behaviors.

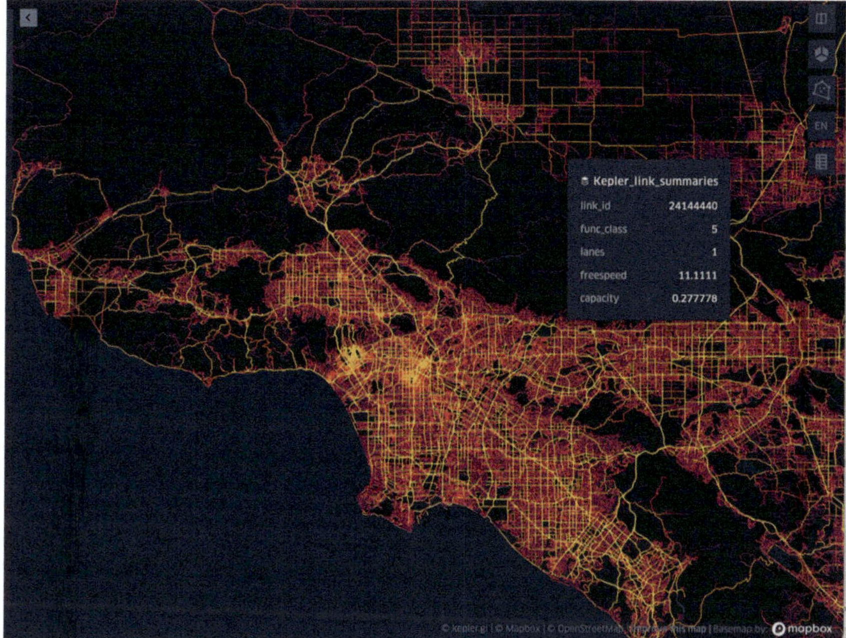


**Fig. 2** Mobiliti link speeds for the Los Angeles basin using the 2020 SCAG travel demand model, visualization by kepler.gl

## *6.1 Overcoming Data Fragmentation*

Traditionally, government entities have developed or acquired software solutions tailored to specific operational needs, resulting in a fragmented data landscape where information is optimized for narrow applications within each department. For instance, individual departments may often collect data primarily to analyze congestion patterns to support transit planning or infrastructure investments. However, this domain-specific data is not easily usable or accessible to other city stakeholders who might leverage this traffic analysis to identify logistics inefficiencies or plan economic development initiatives. This underscores the importance of an interoperable Digital Twin framework that dismantles data silos and fosters a shared understanding across various urban applications.

While some government agencies have begun to use emerging data sources and analytical methods, these ad hoc approaches frequently lead to gaps and inefficiencies. Specific engagements with outside services often generate costly static reports with limited analytical capabilities, and the data remains siloed. Despite substantial investments in data collection, the lack of interoperability restricts the broader applicational use of these datasets, leaving their full potential value untapped. By organizing heterogeneous data into a unified, interoperable knowledge graph, datasets

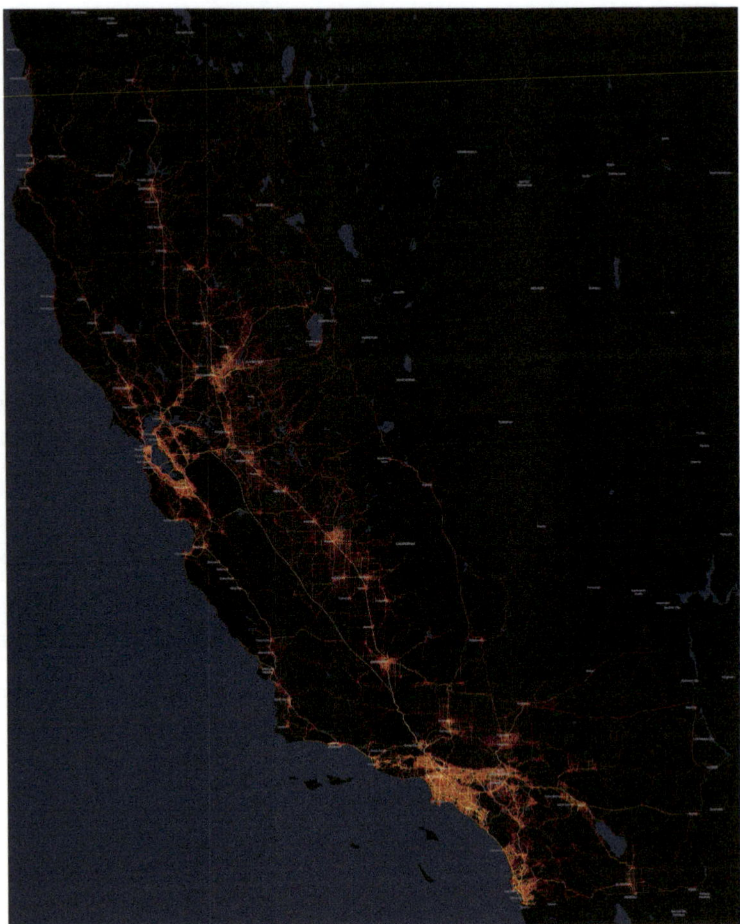

**Fig. 3** CA statewide daily vehicle count

originally intended for single purposes can be transformed into a comprehensive digital representation, supporting a broad spectrum of smart city functions.

A robust Digital Twin framework must include a common knowledge representation and adapters to align diverse data models into an extensible ontology. This facilitates the portability of analytics, allowing data science workflows and AI/ML models to be applied broadly within the Digital Twin, extracting reliable and reproducible insights from the integrated data set.

Larger municipal agencies might develop proprietary in-house solutions or engage technology firms to create custom systems. However, this approach often forces government bodies to assume the roles of software developers, taking on significant risks and costs without necessarily having the needed expertise. Conversely, the growing trend of using singular focused applications developed by tech companies offers proven solutions but still leads to siloed applications within municipalities.

To move beyond these challenges, a holistic Digital Twin framework is essential. It should provide a unified data infrastructure and an integrated environment for data and model integration, allowing seamless AI/ML workflows to operate across a consolidated, multi-source knowledge graph. This framework enables the development of reusable domain-adapted models that can be transferred across the Digital Twin's applications through shared service abstractions.

In contrast to custom, siloed applications, a model-driven Digital Twin architecture separates domain logic from data structures, enhancing portability and minimizing the need for rework. Versioned analytics and model management are essential for supporting robust tracking of assets and data lineage throughout the Digital Twin ecosystem, ensuring consistent and reliable decision support.

## 6.2 *Digital Twin Data Infrastructure*

The ultimate aim is to establish an explorative platform built upon unified data that represents a single source of truth while enabling Digital Twin scenario development capabilities. APIs that facilitate seamless access to this curated data and secure integration of third-party datasets will encourage users to contribute back to the collective pool. This fosters a beneficial cycle of data sharing and enhancement across domains, agencies, and jurisdictions.

Integrated data science tools embedded within the platform can extract reliable, reproducible insights from the combined data, positioning the Digital Twin as the authoritative source for interdisciplinary analysis and preventing the loss of institutional knowledge. By maximizing the value of multi-source data fusion, a shared data infrastructure enhances both data quality and reliability through scalable data science techniques.

For example, predictive models trained on extensive historical data can detect anomalies in incoming data streams that deviate from expected patterns. These conflicts among data from different sensor types can be automatically identified and flagged for review. This multi-source data fusion can potentially mitigate inherent biases in individual datasets, revealing biases related to underrepresented populations and usage patterns, thus supporting more equitable decision-making.

These capabilities, integrated into an automated data observatory that continuously monitors the Digital Twin's data ecosystem and identifies accuracy issues through AI/ML validation pipelines, are crucial to the Digital Twin's performance. Issues are prioritized based on their impact, with notifications sent to relevant experts for necessary adjustments. Engaging the relevant experts in the organization and involving them in model adjustments based on trending data will be essential to the long-term success of the Digital Twin.

Given baseline data quality is verified, completeness can be addressed. For the transportation domain, data coverage varies widely across different communities and may be sparse in certain areas or during specific time periods. Figure 4, for instance, illustrates traffic count data collected by contractors over several years in San Jose.

In situations where data gaps exist due to limitations in spatial or temporal coverage, or sensor malfunctions, the comprehensive data repository of the Digital Twin can be employed to create synthetic data to fill these gaps. Deep learning techniques like graph neural networks can assist in this case [14, 15].

However, it is critical to quantify the uncertainty associated with these synthesized data products. Real-time data can instill confidence by appearing to accurately reflect reality, yet it is very often subject to anomalies due to hardware issues, collection methods, and varying penetration rates. By establishing confidence intervals and error propagation models, downstream users can assess the reliability of the data and make informed decisions regarding its suitability for their workflows. This is illustrated in Figs. 5 and 6, which showcases the data validation processes using Mobiliti results.

These advanced algorithms provide a detailed and comprehensive view of regional behaviors, integrating various influencing factors such as time of day, seasonal variations, nearby events, weather conditions, population densities, the built environment,

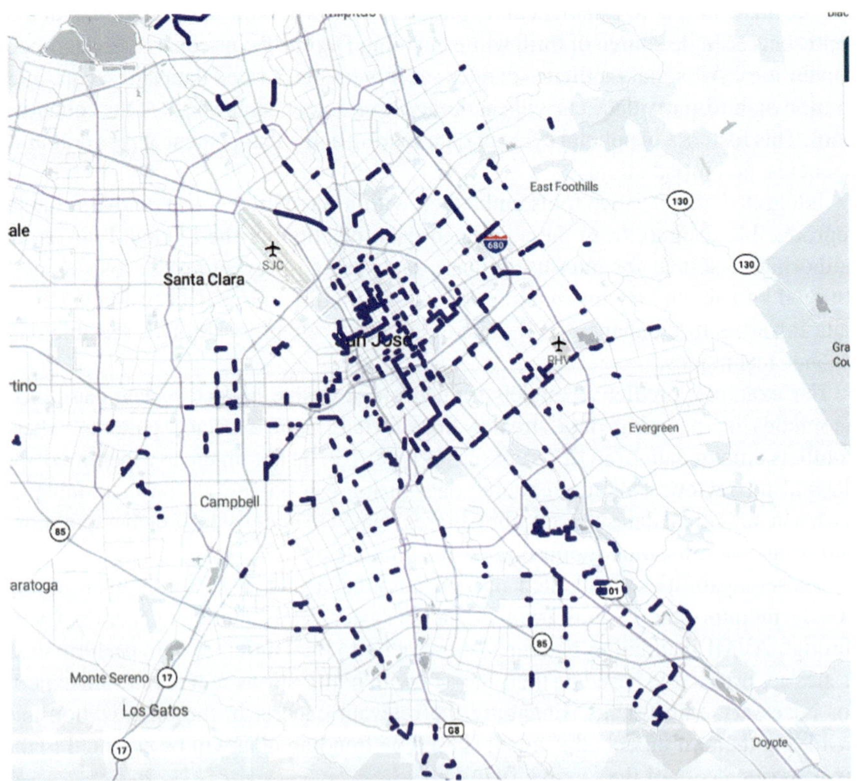


**Fig. 4** Sparsity of static sensor data

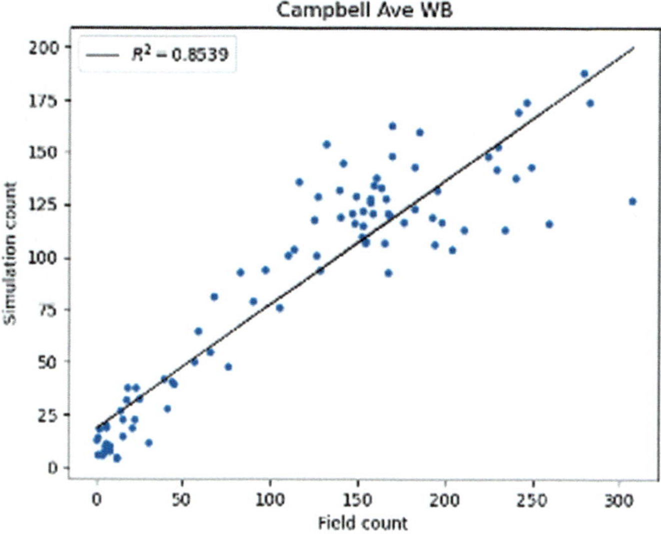


**Fig. 5** Validation of Mobiliti results with static sensor data

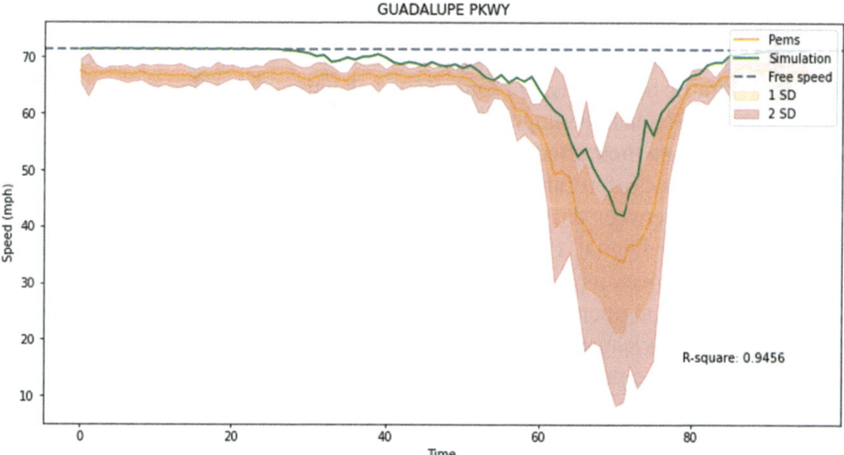


**Fig. 6** Validation with dynamic sensor data with uncertainty

and the economic landscape. These elements are combined within the Digital Twin to predict traffic flows, speeds, and congestion patterns at a granular level across the transportation network.

The simulation model allows for adjustments in individual parameters to gauge the sensitivity of output metrics to specific changes in the input conditions. The accuracy of the simulation engine is validated by comparing its predicted outputs with real-world ground truth data within the Digital Twin demonstrating a favorable

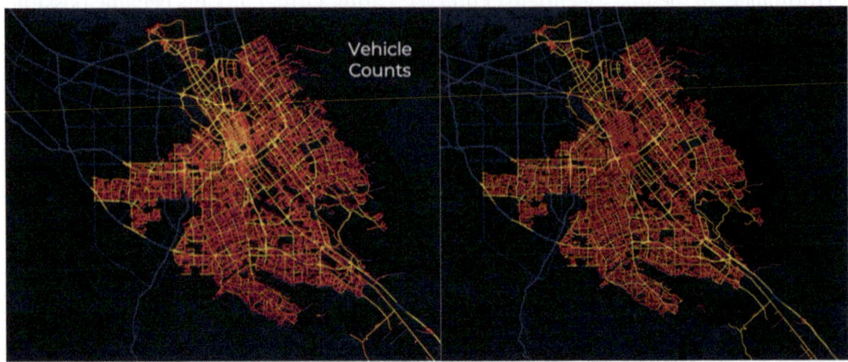


**Fig. 7** Vehicle count validation allows the Digital Twin to estimate vehicle counts (left) and speeds (right) across the full network—colored from red to yellow

alignment between simulated and observed data. Once the simulation fidelity is established through this rigorous benchmarking, the calibrated engine can be used to fill missing data gaps by generating synthetically simulated values, as shown in Fig. 7. This approach, corroborated against available estimates, ensures a continuous, holistic view of the transportation system's state.

For short-term predictions, the simulation engine can model the impacts of variables such as event attendance, weather forecasts, or disruptions in transit services on expected traffic conditions. For longer-term planning and policy analysis, it evaluates scenarios related to demographic shifts, changes in land use, or housing developments to estimate their effects on key performance indicators including mobility, accessibility, equity, safety, and environmental impacts.

The interplay between data-driven simulation and the central unified data repository enhances the Digital Twin framework's capabilities. The simulation engine continuously improves its accuracy by assimilating the latest real-world observations and can also augment the data repository in areas showing deficiencies. This closed-loop data enrichment ensures that the Digital Twin maintains a comprehensive, self-regulating representation of the physical system's state , enabling informed and reliable decision-making.

An integrated multi-source data environment facilitates the discovery of novel relationships and insights through advanced analytics and machine learning. By applying deep learning techniques to the consolidated data corpus, previously hidden patterns and complex correlations are uncovered, enabling more accurate modeling and prediction of system behaviors, as highlighted in Fig. 7, where the real-time data and learned behaviors tune the model and overcome data sparsity and penetration concerns.

Ultimately, by dismantling long-standing data silos and establishing an integrated Digital Twin environment, cities can unlock the immense potential of their data assets, deriving reliable, actionable insights to drive intelligent policies, resilient infrastructure investments, and equitable resource allocation across all communities.

### 6.3 Improving Travel Demand Estimates and Traffic Insights

Proper alignment of real-world data with simulation models is critical for building trust in the Digital Twin [11]. Travel demand models, which are intricate and costly to develop, are typically updated at five year intervals. One of the core advantages of Digital Twins is their ability to incorporate real-time data to refresh these models more frequently, reflecting the latest trends in demographics, economic conditions, and land use.

Moreover, as the Digital Twin generates a comprehensive virtual representation of traffic, new insights can emerge from the observed traffic patterns. Unlike traditional methods that may overlook fine-grained details, Mobiliti enables a detailed analysis across the entire network. For instance, critical link density, detailed in Fig. 8, is indicative of where vehicle accumulation peaks during the day and provides valuable insights that might not be feasible to capture through conventional methods.

Derived traffic insights highlight links that are consistently over-capacity, shedding light on potential congestion causes. Figure 9 identifies locations of saturation (left) and temporal congestion (right). This information, coupled with expert analysis, can lead to innovative infrastructure solutions or policy adjustments that alleviate traffic issues.

The ability to simulate and predict traffic under various conditions is another significant benefit of the Digital Twin. By adjusting simulation parameters, the model can assess the impact of changes in traffic flow, speed, and congestion across different times of day, seasons, or in response to specific events. These simulations are validated against real-world data, ensuring the model's accuracy and reliability.

Figure 10 compares Average Daily Traffic (ADT) from Mobiliti with actual observed data, demonstrating the model's alignment with real-world conditions. It also shows a comparison to a popular tool that relies solely on mobile device data. This benchmarking is crucial for confirming the simulation's accuracy before using it to fill data gaps or guide decision-making processes.

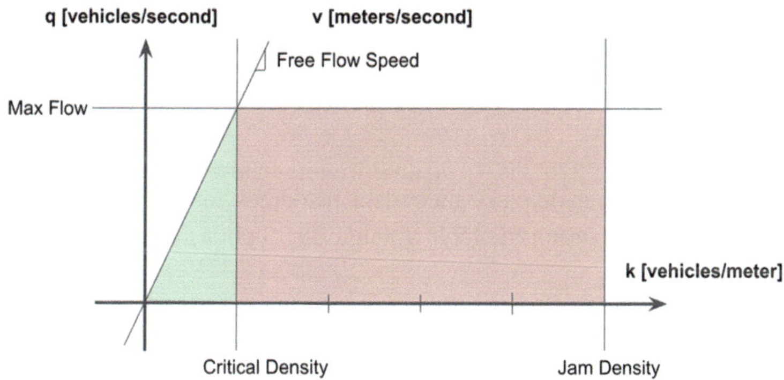


**Fig. 8** Road link level critical density indicates the point at which accumulation of vehicles occurs

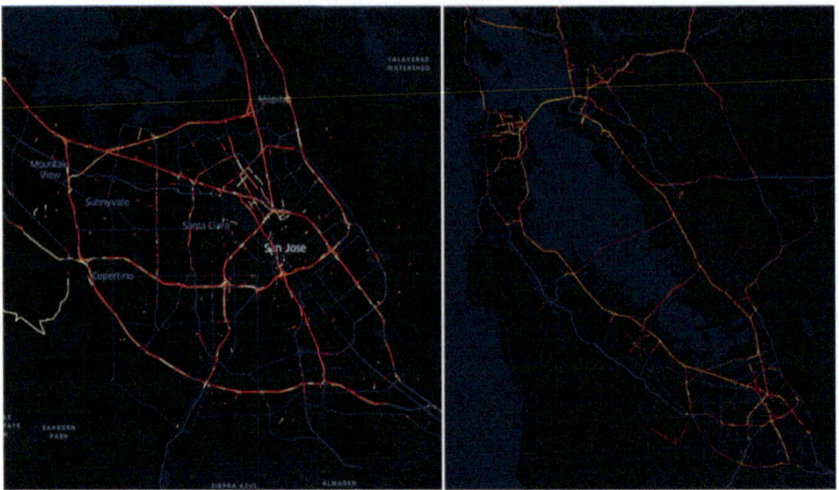


**Fig. 9** Saturation index for the Bay Area model (left); temporal congestion index for the Bay Area model (right)

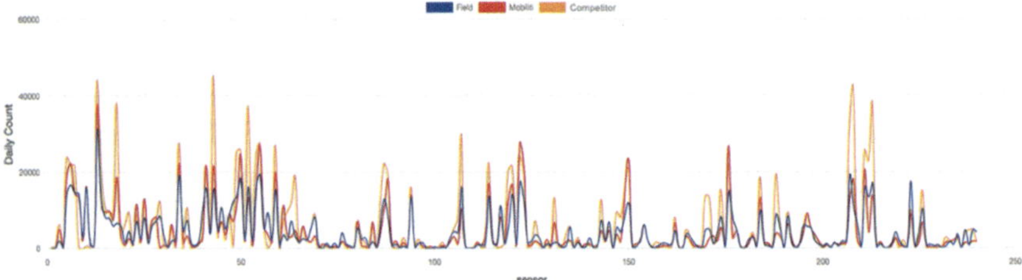


**Fig. 10** Real-world data validation with Mobiliti results

This dynamic interplay between data-driven simulation and a unified data repository empowers the Digital Twin to continuously learn and adapt, enhancing its ability to provide a detailed and accurate representation of the transportation system. This ongoing data enrichment ensures that the Digital Twin remains a vital, self-updating tool for decision support.

Ultimately, by integrating these capabilities into a cohesive Digital Twin framework, municipalities can harness their data resources more effectively, driving smarter policies, and ensuring equitable resource distribution across all communities.

## 7 Connecting City Dynamics to City Performance

While the Digital Twin capability provides powerful new computational assets for practitioners, it does not reduce the complexity of the network fundamentals. As a consequence, defining performance indices that relate to the core needs of a city's population remains an important endeavor. To this end, a new framework for evaluating transportation policies and investments that accounts for social equity impacts alongside traditional mobility performance metrics has been developed called Socially Aware Evaluation Framework for Transportation—SAEF [9]. Conventional evaluations often focus narrowly on mobility and fail to sufficiently consider social outcomes. As such, a socially aware framework can provide a post-simulation data analysis that incorporates quantitative measures of accessibility, exposure, and externalities experienced by different social groups, including low-income, minority, and disabled populations. These social impact measures are integrated with mobility metrics within a multi-criteria decision analysis to enable a more holistic assessment of transportation projects and policies.

### *7.1 Understanding the Impacts of Traffic Dynamics on City Level Performance Indices*

The complexity of city management can lead to various urban issues that significantly affect the quality of life for residents. The Mobiliti platform is currently geared toward understanding vehicle traffic dynamics, as traffic congestion is a prominent concern for future city planning, with direct impacts on public health, safety, and regional economic health. Accordingly, the platform investigates the city-level impacts of various vehicle routing strategies to enhance traffic management and mitigate negative consequences. Beyond traffic management, it can be used for a variety of analyses, including alternative mode planning.

The term 'socially-aware' encompasses four measurement themes: neighborhood, safety, mobility, and environment. To address socially relevant aspects of vehicle dynamics, themes and indicators are developed beyond simple congestion measurements. These themes, supported by city performance indicators from extensive literature reviews [8], identify critical factors such as accidents for safety and emissions for environmental quality. This approach has led to the creation of 24 indicators from an initial list of over 100, categorized into themes that reflect neighborhood quality, safety, mobility, and environmental quality.

The framework offers insights into the trade-offs between social and mobility objectives, aiding in decision-making processes. This methodology is designed to support more equitable transportation planning outcomes. Fifteen performance indices, as depicted in Fig. 11, range from positive (green) to negative (red) outcomes for four different traffic scenarios in the City of San Jose.

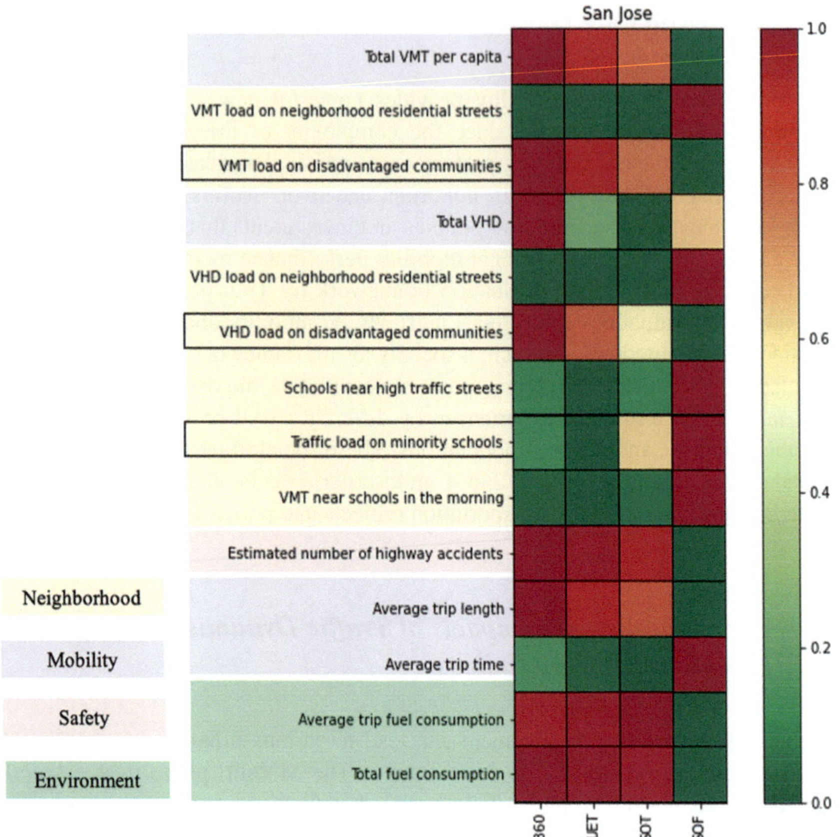


**Fig. 11** SAEF evaluation indices for the city of San Jose under four traffic scenarios (green indicates a positive outcome while red indicates a poor outcome in comparison across the four traffic scenarios). *VHD* vehicle hours delayed, *VMT* vehicle miles traveled

The Mobiliti platform includes traditional traffic assignment algorithms that have been parallelized [3]. Four different scenarios associated with different traffic outcomes were evaluated and detailed in Fig. 11. The scenarios include a Baseline 60 (B60) where routes are initially determined by the shortest path and then dynamically adjusted for 60% of the trips based on real-time congestion [1]. Other scenarios use a Frank Wolfe algorithm that optimizes trip assignments to achieve objectives like user equilibrium travel time, system optimal travel time, or system optimal fuel consumption—identified as UET, SOT, and SOF, respectively, in Fig. 11. Evaluation of the impacts of these scenarios reflects the complexities of traffic dynamics, indicating no single optimal approach but rather highlighting the importance of understanding trade-offs and exploring multiple objectives in transportation planning.

The SAEF evaluation creates indicators that can then be used to take a deeper look into the associated traffic dynamics. Vehicle Miles Traveled (VMT) and Vehicle

Hours Delayed (VHD) in Equity Priority Communities (EPC) can be extracted from the simulation results. Designations of disadvantaged communities vary across cities and area government groups and may consider a variety of factors such as income level, population age, and ethnicity. For these analyses, geospatial representations of these EPC communities were used to extract link level speed and flow data and provide VMT and VHD metrics. As observed in Fig. 12, three of the four scenarios have poor outcomes for VHD and VMT in EPCs. Further analytics, shown in Fig. 12, then demonstrate that a significant number of longer trips (arbitrarily selected as trips greater than 40 min) that start and end in the city centers of San Francisco and San Jose are using the EPC road links during their journey. These are considered "pass through" trips, in which the vehicle has no connection to the specific neighborhood and are short-cutting congestion by using neighborhood streets on their way to their destination. Further analysis indicates that these trips are mostly work-related trips or commute trips and that 60% of these trips were rerouted due to congestion, thus incurring undue safety and health impacts on lower income communities.

While the simulation may not exactly mirror reality, it provides critical insights into traffic phenomena, particularly those influenced by driver behaviors, which are often challenging for traffic planners to measure [13]. This demonstrates a key value of using Digital Twins in transportation planning.

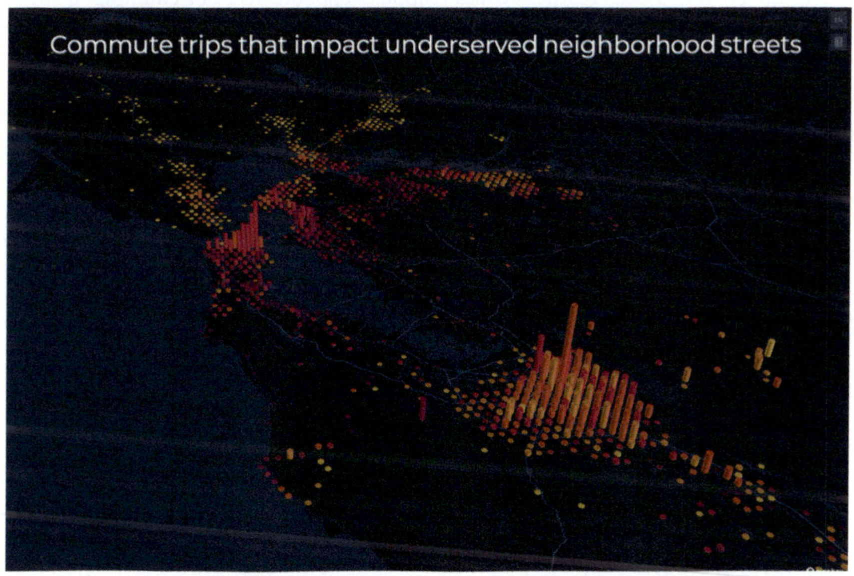


**Fig. 12** Origins and destinations of San Jose and San Francisco commute trips that pass through lower income communities thus impacting community health and safety

## 8 Iterative Exploration

A high-fidelity virtual representation that mimics the real system enables the safe and rapid evaluation of scenarios that are challenging or impossible to test in reality. Gained insights can inform investment plans and policy decisions. The ultimate aim is to equip practitioners with tools to configure the Digital Twin to reflect potential changes in infrastructure or technology and assess their impacts on system performance.

Examples include:

- Reducing the number of lanes on an arterial road to create space for protected bike lanes and observing the effects on congestion and traffic dynamics influenced by dynamic routing.
- Adjusting the Origin–Destination Time (ODT) dataset in the Digital Twin to reflect changing conditions such as population growth and demand levels and observing the resulting strain on network resources. For instance, in San Francisco, a simulation where current transit riders were instead assigned a personal vehicle showed the significant congestion that would result without existing transit systems (Fig. 13-left).
- Introducing disruptive events like natural disasters or major accidents into the Digital Twin to assess resilience and response capabilities. For example, a simulated earthquake in San Jose directed drivers within a certain radius to return

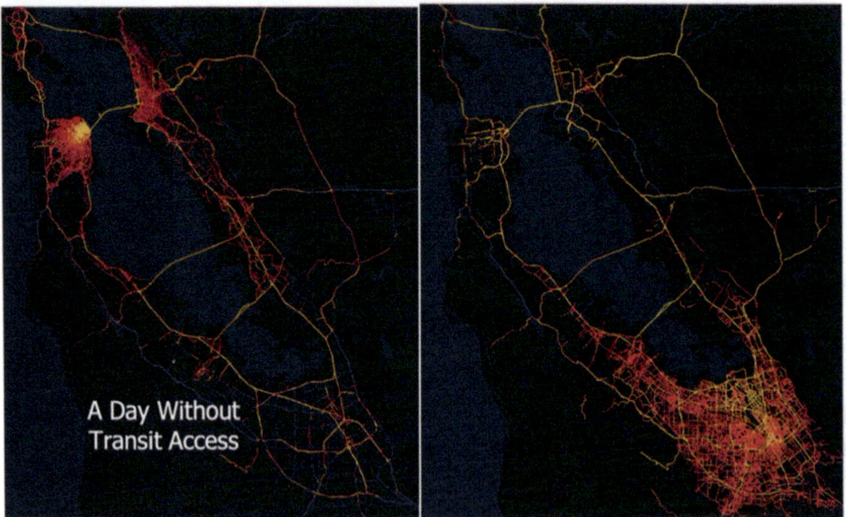


**Fig. 13** Example scenarios for iterative exploration: include transit riders as vehicle owners/drivers (left) and a simple earthquake model in which vehicle drivers return home (vehicle density is indicated from red to yellow)

home, dramatically affecting local traffic and also impacting main arterials far beyond neighboring cities (Fig. 12-right).
- Using machine learning within the Digital Twin to develop predictive models.
- Engaging stakeholders to define scenarios of interest and exploring consequences to foster an intuitive understanding of existing conditions.

While simulations within the Digital Twin may not always perfectly align with reality, the insights they provide are invaluable. They offer a safe and efficient means for traffic planners and other practitioners to evaluate scenarios that might be challenging or impossible to test in real life. This high-fidelity virtual representation helps inform investment plans and policy decisions, equipping practitioners with robust tools to adjust the Digital Twin according to potential changes in infrastructure or technology, thereby allowing them to effectively predict and analyze impacts on system performance.

## 9 Unlocking Generative Urban Network Design Through Digital Twins

Urban network design is entering a transformative era with the advent of generative modeling techniques. This is showcased in the development of a New Greenfield City network. By borrowing network components and the population's activity patterns from neighboring cities, planners can now simulate and visualize the intricate dynamics of a new urban development even before it materializes. This method leverages Mobiliti to create robust, predictive models that were previously unattainable.

The generative design process for a New Greenfield City network is particularly innovative because it synthesizes not just the physical infrastructure of the network but also the underlying human activity that drives traffic patterns. This approach includes borrowing and modifying both geometric elements and Origin–Destination Time (ODT) data, providing a rich dataset that informs the design and anticipates the needs of the urban area.

The effectiveness of this modeling approach is evident in the visualizations that detail the origins and destinations of all regional trips including a New Greenfield City network. These visualizations, Figs. 14 and 15, illustrate the density and spread of trips in the region.

By adjusting the network elements—links added, removed, or modified—the City Network Persona evolves. This persona reflects the changes in demand driven by population activities and commercial vehicle flows, enabling a responsive design process that adapts to predicted needs.

Initial model results highlight the existing Bay Area network with vehicle miles traveled (VMT) before the integration of the New Greenfield City is shown in Fig. 16. Figure 17 introduces the new city network in the upper right of the network, showing how it integrates within the existing framework and the accompanying travel demand.

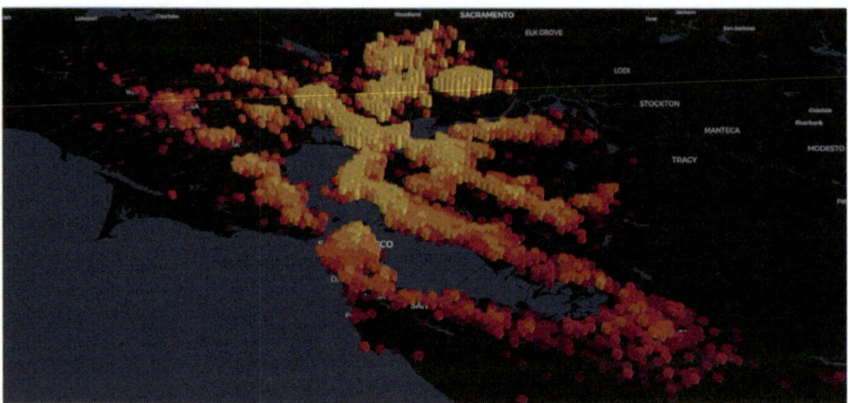


**Fig. 14** Origins of all trips taken by the population of a New Greenfield City network

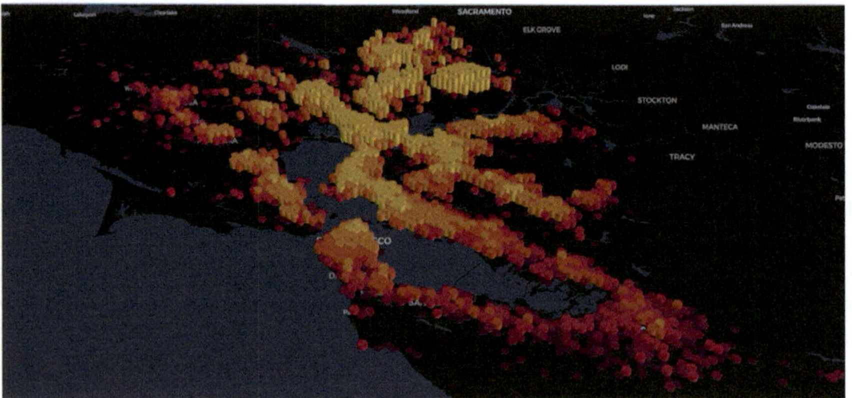


**Fig. 15** Destinations of all trips taken by the population of a New Greenfield City network

A zoomed-in view, Fig. 18, provides a detailed perspective of how the new network segments interact with the existing network.

The final visualization, Fig. 19, demonstrates the regional impact of the New Greenfield City on VMT. This includes not only the expected increase in VMT across the entire Bay Area but also specific impacts on highway and county roads in both the immediate and extended vicinities of Greenfield City. These areas show significant changes, highlighted in the visual outputs, indicating a substantial increase in traffic congestion and predicted pressure points. This analysis can be incorporated as a feedback process for an automated generative design framework that can tune the Greenfield City network and its associated demand to meet the design objectives for the Greenfield City, while also addressing regional concerns associated with the traffic impacts of the new travel demand.

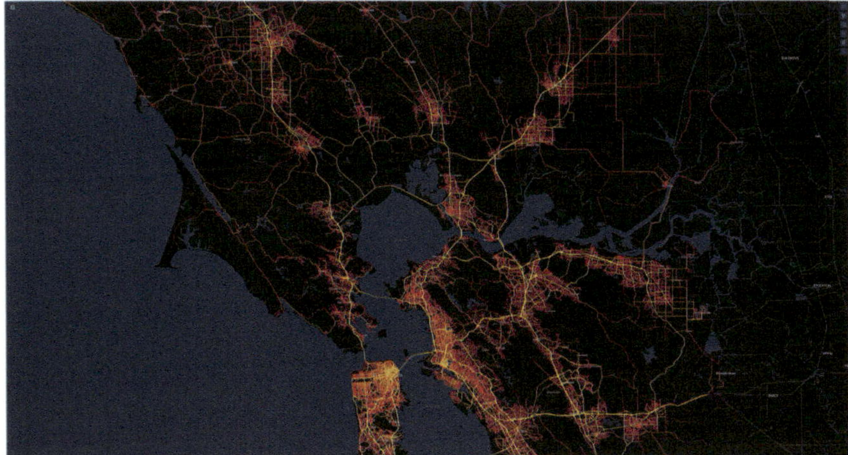

**Fig. 16** Daily vehicle count in the San Francisco Bay area before the addition of the New Greenfield City network

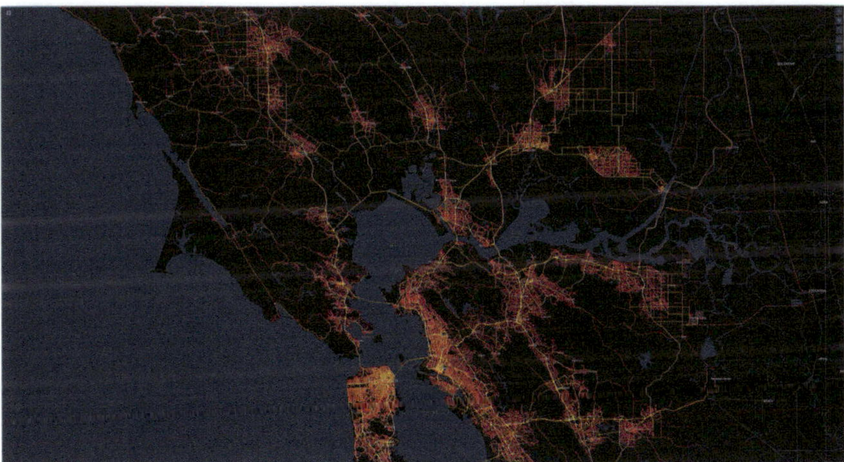

**Fig. 17** Daily vehicle count in the San Francisco Bay area after the addition of the New Greenfield City network

The ability to perform such detailed and predictive analyses through Digital Twins represents a significant leap forward in urban planning. Digital Twins enable planners to unlock insights that inform investment plans and policy decisions, facilitating a proactive approach to urban development. The generative design capabilities facilitated by Mobiliti ensure that planners and stakeholders can explore various scenarios and their consequences, enhancing understanding and preparation for real-world implementations. This comprehensive approach not only optimizes the

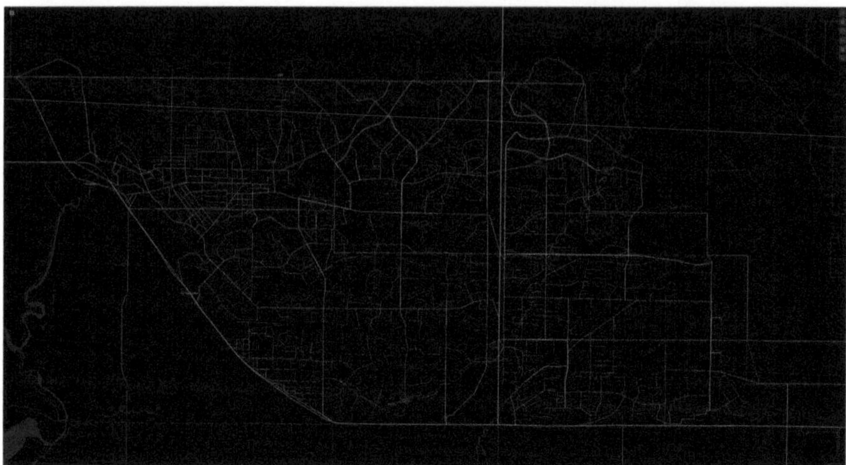

**Fig. 18** Daily vehicle count in the New Greenfield City network after integration with the San Francisco Bay area network

**Fig. 19** Impact of the integration of the New Greenfield City network with the San Francisco Bay area network in terms of daily vehicle count

network design but also ensures that new developments, such as the New Greenfield City, are integrated seamlessly into the larger urban fabric, supporting sustainable and efficient growth.

## 10 Future Research and Transportation Data Ecosystem Development

The exploration of Digital Twins in transportation and urban planning is expanding. The Mobiliti platform has prioritized vehicle traffic due to its significant impact on health, safety, and quality of life in cities. The application of Digital Twins in urban planning offers substantial potential for leveraging computational methods to tackle complex real-world urban challenges. Traditionally, urban designs are developed by human designers through extensive analysis, discussion, and iterations. However, emerging computational models can streamline this process by optimizing designs and generating multiple alternatives for designers to select from.

Computational AI methods, especially deep reinforcement learning (DRL) algorithms, are being proposed to solve intricate challenges in urban science. Urban design problems can be framed as sequential decision-making tasks with rewards linked to various metrics, such as the creation of 15 min cities. With high-performance computational support, the model can learn and gradually approach the optimal solution, achieving strategies that surpass human capabilities.

When integrated with Mobiliti, using a multi-agent reinforcement learning approach, a comprehensive urban planning, development, and management process can be facilitated. This includes enhancing spatial designs by optimizing them with respect to dynamic evolutionary outcomes like simulated traffic flow. Mobiliti is applicable not only to individual cities but can also efficiently plan multiple cities collectively, generating optimized plans that offer mutual benefits and yield more efficient collective outcomes.

Several developmental areas could further unlock the full potential of Digital Twins in transportation management:

- **Digital Twin Taxonomies**: There is a need to clarify the definition of Digital Twins to resolve confusion in the data ecosystem. This could involve establishing a clean definition or creating a taxonomy of Digital Twins with example instances, enhancing data alignment, codifying physical system complexities, and defining simulated models that capture primary components of the physical world but acknowledge inherent randomness.
- **Surrogate Models Trained by Simulation**: Deep learning could capture complex simulation aspects and offer model-based controller solutions.
- **Data Access and Analytics**: The transportation data ecosystem is rapidly evolving. Enhancements in data availability and integration are crucial, alongside developing common ontologies for network and land use to maximize value.
- **Addressing Urban Complexity**: As urbanization and the formation of mega-regions continue, managing transportation in the context of city values will grow in complexity. Accessibility to new computational approaches for urban planners and broader ontologies covering domains like health, safety, and quality of life will be essential for developing effective and equitable solutions and performance indices that reflect the intricate interactions within city environments.

**Acknowledgements** The Mobiliti Platform is the compilation of the work of a lot of excellent scientists: Cy Chan, Srinath Ravulaparthy, and Max Bremer at Lawrence Berkeley Laboratory, Anu Kuncheria, Dimitris Vlachogiannis, Colin Laurence, Anthony Patire, Rowland Awadagin Herbert-Faulkner, and LuLu Lui at University of California Berkeley, Prasanna Balaprakash, and Tanwi Mallick at Argonne National Laboratory. The work described was sponsored by the US Department of Energy (DOE) Vehicle Technologies Office (VTO) under the Big Data Solutions for Mobility Program, an initiative of the Energy Efficient Mobility Systems (EEMS) Program. The following DOE Office of Energy Efficiency and Renewable Energy (EERE) managers played important roles in establishing the project concept, advancing implementation, and providing ongoing guidance: David Anderson and Prasad Gupte.

## References

1. Chan C, Kuncheria A, Macfarlane J (2023) Simulating the impact of dynamic rerouting on metropolitan-scale traffic systems. ACM Trans Model Comput Simul 33(1–2):1–29
2. Chan C, Wang B, Bachan J, Macfarlane J (2018) Mobiliti: scalable transportation simulation using high-performance parallel computing. In: 2018 21st international conference on intelligent transportation systems (ITSC), pp 634–641
3. Chan C, Kuncheria A, Zhao B, Cabannes T, Keimer A, Wang B, Bayen A, Macfarlane J (2021) Quasi-dynamic traffic assignment using high performance computing. ArXiv:2104.12911 [Cs]
4. Faltesek A, Dakshinamoorthi B, Prabhala S, Thobani A, Kuncheria A, Macfarlane J (2021) Urban traffic simulation: network and demand representation impacts on congestion metrics
5. Gasnet. [online]. Available: http://gasnet.lbl.gov
6. HERE Technologies—The world's #1 location platform. en. url: https://wwwhere.com/(visited on 11/11/2021)
7. Jefferson D (1985) Virtual time. ACM Trans Program Lang Syst 7, 3 (July 1985), 404–425
8. Koglin T, et al (2011) Measuring sustainability of transport in the city—development of an indicator-set. https://doi.org/10.13140/RG.2.1.1451.0240
9. Kuncheria, A., Walker, J.L. and Macfarlane, J. (2023). Socially-aware evaluation framework for transportation, Transportation Letters, pp. 1–18.
10. Kunz G (2010) Parallel discrete event simulation. In: Wehrle K, Güneş M, Gross J (eds) Modeling and tools for network simulation. Springer, Berlin, Heidelberg. https://doi.org/10.1007/978-3-642-12331-3_8
11. Macfarlane J (2021) Mobile device data analytics for next-generation traffic management (Report)
12. Macfarlane J (2019) The transforming transportation ecosystem—a call to action
13. Macfarlane J (2019) When apps rule the road: the proliferation of navigation apps is causing traffic chaos. It's time to restore order. IEEE Spectr 56(10):22–27
14. Mallick T, Balaprakash P, Rask E, Macfarlane J (2021) Transfer learning with graph neural networks for short-term highway traffic forecasting. In: 2020 25th international conference on pattern recognition (ICPR), pp 10367–10374
15. Mallick T, Balaprakash P, Rask E, Macfarlane J (2020) Graph-partitioning-based diffusion convolutional recurrent neural network for large-scale traffic forecasting. Transp Res Rec 2674(9):473–488
16. *MATSim.org*. [online]. Available: https://www.matsim.org/
17. *Polaris*. [online]. Available: https://polaris.es.anl.gov/

# Digital Twin in the Battery Industry

**Soumya Singh, Michael Oberle, Daniel Schel, Julian Grimm, Olga Meyer, and Kai Peter Birke**

**Abstract** The increase in global demand for lithium-ion battery cells has driven the establishment of numerous manufacturing facilities. However, the complex design and manufacturing of Li-ion battery cells pose several challenges, such as ensuring high cell quality while maintaining high process stability and efficiency. During the usage phase, factors such as operating conditions, and charging/discharging profiles can significantly impact the performance and longevity of battery cells. Furthermore, addressing recycling and proper disposal in the end-of-life phase presents substantial challenges, given the risks associated with hazardous materials and the scarcity of resources. Hence, effective management of these phases is essential for achieving sustainable and efficient energy storage systems. Digital twins applied in various industries have shown promising results in improving product lifecycle management and smart manufacturing processes. However, it is important to scientifically test and validate its effectiveness within the specific context of the battery industry. The main question is whether digital twins can realistically address the growing challenges of the battery industry, such as degradation evaluation, usage optimization, manufacturing inconsistencies, second-life application possibilities, etc. Without a thorough understanding of the benefits of digital twins, it is difficult to claim that they are the best solution for the battery industry's challenges. This chapter explores the potential applications and use cases of digital twins in the battery lifecycle, outlining the requirements and concepts necessary for implementing digital twins effectively. This approach ensures a comprehensive understanding of the system's benefits and limitations, thereby enabling stakeholders to make informed implementation decisions.

**Keywords** Digital Twin · Battery lifecycle · Manufacturing · Li-ion battery · AAS

---

S. Singh (✉) · M. Oberle · D. Schel · J. Grimm · O. Meyer · K. P. Birke
Fraunhofer Institute for Manufacturing Engineering and Automation IPA, Nobelstr. 12, 70569 Stuttgart, Germany
e-mail: soumya.singh@ipa.fraunhofer.de

K. P. Birke
Institute for Photovoltaics, Electrical Energy Storage Systems, University of Stuttgart, Pfaffenwaldring 47, 70569 Stuttgart, Germany



M. Grieves and E. Hua (eds.), *Digital Twins, Simulation, and the Metaverse*, Simulation Foundations, Methods and Applications, https://doi.org/10.1007/978-3-031-69107-2_15

## 1 Introduction

Lithium-ion batteries (LiBs) are widely favored energy storage systems due to their promising attributes such as high energy density, low self-discharge property and long cycle life. The European Battery Alliance foresees a potential market worth up to 250 billion euros for automotive batteries in Europe from 2025 onwards. Presently, European production meets only 6% of global LiB demand, prompting a political target to achieve 30% by 2030 [1]. To support this goal, the Federal Ministry for Economic Affairs and Energy has allocated nearly three billion euros for battery cell production, attracting proposals from over 40 German companies [2]. This funding prioritizes sustainability and digitalization, aiming for efficient material usage, energy conservation, and innovative recycling approaches. In order to meet such requirements and the rising global demand, the challenges across the battery lifecycle need to be addressed through cutting-edge technologies [3].

Driven by the rising market demand for battery electric vehicles (BEVs), consumer electronics, and battery-driven grid storage, stakeholders across the battery value chain encounter numerous challenges throughout the battery's design, manufacturing, usage, and end-of-life (EoL) phases. This section will subsequently explore the challenges within the different lifecycle stages.

During battery design, challenges can arise from market demands, such as customers along the supply chain, including original equipment manufacturers (OEMs), or end-users. Regarding BEVs, users demand high energy density, a long driving range, long calendar and cycle life, fast and efficient charging, safety, and low costs [4]. On the other hand, OEMs face design limitations because they aim for application-specific batteries. The variety of applications potentially influences the challenges in the design stage [5]. Furthermore, the battery market is dynamic at the moment, with heterogeneous manufacturers, different packs/cell designs and formats, electrode designs and formats, different active and inactive material combinations with various levels of readiness, as well as promising future technologies [6, 7]. Bringing all this together, battery design has multiple and fast-changing objectives, conditions, and circumstances, which, as a result, leads to tradeoffs and batteries that cannot meet all aspects.

The demands within the battery market generate technical challenges for battery design and development. It is crucial to reach good performance measures with a specific battery design. Reaching a high energy density is said to be the overarching goal [8]. Besides this, safety, thermal stability and temperature ranges, and durability are further influencing factors.

The aim of achieving a high energy density and fast charging rates, results in challenging cooling efforts on the battery system for dissipating and supplying heat. Furthermore, increasing the size and thickness of individual cells to meet the target values changes the electrical and thermal conditions at a cell level in a negative manner [8]. Thus, reducing the amount of inactive material in either a battery cell or a battery system may lead to undesired conditions. In contrast, the system's thermal management requirements can be reduced by increasing the cell's thermal operation window [5]. In addition to thermal requirements, there is also a tradeoff between general safety and energy density. In a battery system, there are electrical, thermal, and mechanical safety protection measures on the cell, module, pack, and vehicle levels [5]. This results in an effective safety system but ends up in a lower energy density. Furthermore, battery durability is a critical parameter due to warranty issues of the battery manufacturer and OEMs. Therefore, in the design phase, selecting suitable tests and confirming test results for possible future degradation of the battery during usage under operating conditions is challenging [5].

Aside from the design challenges due to technical requirements, there are also issues arising from scarce and expensive raw materials. Since mining and refining operations are concentrated in a few countries, future resource supply could become challenging [9]. Efforts are underway to reduce reliance on scarce resources like cobalt and nickel in battery production [5]. With large-scale battery facilities being developed, it is crucial to understand how changes in battery design might affect manufacturing processes and supply chains. Overall, battery design encounters challenges due to tradeoffs in target values and evolving conditions [7].

In battery manufacturing, the main objectives are cost savings, quality improvement and fostering sustainability. However, achieving these objectives present challenges such as scaling up production, enhancing productivity, minimizing scrap, and optimizing resource and energy efficiency [10]. These challenges can be addressed at different levels, such as manufacturing operations or process levels.

As elaborated in [10], certain manufacturing processes have prolonged processing times, consequently reducing the yield rate. Notably, the electrolyte filling process stands out as one of the most time-consuming steps. With an increase in the cell's size, the reduction of the electrolyte filling process time becomes even more challenging. Another example related to large cell formats with higher energy density involves the necessity of adjusting their tab designs according to the specific format, resulting in the need for greater positioning accuracy and posing handling challenges [8]. Last but not least, modification in cell designs, as well as the integration of new battery designs into existing production lines, can be very challenging because they require significant adjustments to manufacturing processes, machinery, and workflow, often leading to disruptions, increased costs, and potential delays in production schedules [7].

During battery usage, challenges may arise due to user expectations, particularly concerning the total costs of ownership (TCO), including purchasing costs, electricity costs, and maintenance costs. To mitigate these expenses, efforts must be made across the design, manufacturing, and EoL stages, involving process improvements, material innovations, and implementing circular economy principles [5].

A major challenge for LiBs is that different operating conditions and load profiles create individual life cycles that greatly affect the aging of a cell. Consequently, the aging of each LiB cell within a battery pack differs from one another, leading to internal inconsistency between the battery cells within one module or pack [11]. Lastly, challenges such as monitoring performance, ensuring safety, managing thermal conditions, access to charging infrastructure, and accurately predicting battery life are crucial for maintaining efficient battery operation and extending longevity.

Generally, the EoL of batteries involves processes such as recollection, sorting, dismantling, and material recovery. Currently, the return rate of used batteries is low [11]. The methods used for processing batteries at EoL can be as unique as the design of each individual cell. Overall, research in the field of EoL batteries is still ongoing and there are mainly no standardized processes, making it very challenging [12, 13]. It is unclear which batteries should be processed for which circular value creation solution, e.g., reuse, remanufacturing, repurposing, repair, and recycling. Safety is a major issue for all these solutions. Once a used battery is returned, the battery's and cell's condition is unclear, but they are still electrochemically active and contain chemically stored energy and flammable materials. During dismantling, the batteries are treated mechanically and are thus exposed to further potential damages. For reuse, batteries should generally be available at a high remaining capacity and low cost [12].

Currently, second-life applications are in development and testing, and there are not yet proven technologies or processing standards. Additionally, ongoing research on recycling [5, 13] faces the challenge of separating tightly coupled materials within battery systems (dismantling) and battery cells (recycling). Standardized procedures for dismantling and recycling are currently lacking, and there are various process routes for handling different battery chemistries [13].

Besides challenges related to standardization, safety, and future market developments, new regulatory regulations also influence activities in the EoL of batteries. This includes, for example, the EU Battery Regulation called Battery Passport, which mandates the use of digital passports for EVs and industrial batteries over 2 kWh sold into the EU market starting from February 2027. These passports must contain essential information accessible to the general public, regulatory bodies, and battery service providers. The Battery Passport aims to enhance sustainability and circularity in battery value chains by promoting transparency, facilitating efficient recycling, and encouraging sustainable practices among manufacturers.

In summary, the battery life cycle presents significant challenges in design, manufacturing, usage, and EoL stages. This chapter explored the applications of the concept of Digital Twin (DT) as a promising solution to address these challenges. By aligning DT use cases with industry challenges, the potential benefits of a DT implementation become more evident. The discussion delves into specific use cases within different lifecycle stages, supported by an implementation framework and a case study example.

### *1.1 Digital Twin Definition*

In 2002, the basic concept of DTs was first presented as "The Conceptual Ideal of PLM" and later in 2005 as the Mirrored Spaces Model by Dr. Michael Grieves. This was the first time that the idea of virtual representation of real objects was applied in the context of industrial production. Dr. Grieves, who was the Co-director of a research center at the University of Michigan at the time, is considered a pioneer who put the concept of DT into practice [14, 15].

DT technology enables the recreation of real objects or systems in a virtual environment. Data is collected from sensors and other sources to create a replica of the real object. The DT can then be used for various purposes, such as simulating operating procedures, optimizing processes, or predicting malfunctions.

The term "Digital Twin" was introduced by John Vickers from NASA in 2010 [15, 16]. It serves as a metaphor illustrating the idea that the DT is a virtual copy of a real object. Similar to real-life twins, the real object and the DT share certain characteristics and features. Since then, the number of publications on the subject has increased enormously [17].

Jones et al. in [18] recorded a sharp increase in publications on the topic of DTs as early as 2014. Despite the significant increase in the interest in the DTs in recent years, research is still at a very early stage, with initial practical applications remaining tentative and not fully developed. As evidenced by various publications on DTs in [18, 19], there was already a pressing need to consolidate research and industry in the early 2020s to establish a common understanding of the topic and ensure that future research efforts are based on solid foundations for each specific area.

The term "Digital Twin" has become ambiguous due to the abundance of scientific publications and industrial application scenarios. Consequently, organizations have been formed at the national and international levels with the goal of promoting and disseminating the concept of DT. However, there exists a range of definitions, graphic models, and complementary concepts. Nevertheless, the lowest common denominator in the context of DT remains the Mirrored Spaces Model by Grieves [20, 21].

The conceptual process is significantly delayed by the need to define terminology and concepts for the DT. Consequently, numerous definitions have been broadly used in both industry and academia for some time now, posing a major challenge for users to find the right definition and base their work on useful practical specifications. The lack of consistency and a number of non-consensual solutions, based on a variety of different definitions and frameworks can potentially risk the benefits that the DT was originally intended to provide [18].

For several years now, standardization has been actively pursuing a coordinated approach to the topic of DTs. The forerunner in this area is the Joint Technical Committee of ISO and IEC (JTC 1), established to develop global standards for information and communications technology (ICT) for business and consumer applications. Within JTC 1, there is a special subcommittee, SC 41, dedicated to standards for the Internet of Things. Since the beginning of 2021, this subcommittee has expanded its scope and terms of reference by establishing a new working group (WG 6) dedicated to DT applications across multiple sectors, including smart manufacturing, utilities, smart grids, smart buildings, and smart cities [22].

Unfortunately, due to the high interest and participation of numerous experts from many countries at the international level, coordination has slowed down and the full description of the concept for different areas has become a lengthy process. Nevertheless, the first basic standards have already been published in initial drafts or are officially due for publication, which may lead to initial successes in addressing definitional confusion.

At this stage, we are rely on the definition of the Industry 4.0 platform, which forms the basis for the rest of this chapter:

> A digital twin is a digital representation, information in the digital information world that represents characteristics and behaviors of an entity, sufficient to meet the requirements [23].

It took some time to agree on the definition and formulation of the basic terms, however, the situation is different in practical research. Here, developers started a bit earlier and came up with the first solutions. A glimmer of hope is coming from the field of manufacturing. Due to the advancing digitization in the industrial environment, the development of concepts and closer practical solutions for the DT is being tackled more quickly than in other areas.

One of these areas is battery cell production. In the following sections, we will take a closer look at the product, the battery cell, or rather the life cycle of battery cells and the possible applications for DTs in this context.

## *1.2 Digital Twin Concept*

As previously mentioned, Grieves' Mirrored Spaces Model is the lowest common conceptual denominator in the context of DT. It describes how the real space is connected to one or more virtual spaces (VS) through a linking mechanism based on data, information, and processes. The various virtual spaces show different representations of the same physical object, facilitating the exploration of alternative designs, among other purposes. Grieves emphasizes that the flow of data and information in its original state is regularly processed and forwarded by humans. To realize its full potential, a "robust, accurate, and timely [14]" link is essential. While numerous other concepts exist, some build on this framework while others offer different perspectives on DT [21].

However, such a definition alone is not sufficient to describe or classify the various application scenarios of DT. Grieves addresses this by introducing three types of DT along the product lifecycle.

The Digital Twin Prototype (DTP) includes all potential products that can be created. It represents the product and its various versions as it progresses step by step, from the initial idea to the production of the first manufactured item.

The Digital Twin Instance (DTI) is created at the start of production. Each product that is created can have a DTI, which represents the collection of different types of information for the respective product instance, such as serial numbers, service information, part measurements, assembly process information, etc. Much of the information is fed from the product's DTP, which is linked to the DTIs.

The Digital Twin Aggregate (DTA) is the aggregate of all manufactured products. It represents the collection and aggregation of data from the product population and thus forms the summary of all DTIs, always with the aim of creating value [15]. These DT types are of a basic nature, meaning the complexity regarding the product lifecycle, the entities involved in it, and their relationships to each other are greatly reduced, allowing only a rudimentary classification of the application scenarios of DT.

Another interesting concept that supports the DT classification is the Digital Twin Structure Model (DTSM), developed despite the lack of a coherent DT definition and based on the information and knowledge about DT available by 2020. This model incorporates various key elements, including an asset lifecycle perspective integrated with the Product, Process, and Resource (PPR) model. It facilitates the structuring, categorization, and comparison of DT use cases [20].

We will only discuss the most important aspects of the DTSM and relate them to the Grieves DT types. Fig. 1 provides a coherent description scheme for application scenarios of DT. The DTSM essentially consists of three dimensions: Lifecycle, PPR, and Cyber-Physical.

1. The first dimension, "Lifecycle", consists of two phases: type and instance. These phases are identical to the Grieves DT types. The DTP is relevant in the type phase, while in the instance phase, the DTI and DTA represent multiple DTIs.
2. The second dimension, the "PPR" model, describes the distinction between product, process, and resource and is widely assumed in production and manufacturing (see [24, 25]). It is a helpful model to describe the relationships and entities involved in the product creation process in sufficient detail. According to the DTSM, products, processes, and resource have their own lifecycle and can therefore be represented by their own DTs. We make use of this in the description of the application scenarios, e.g., to differentiate between the DTI of a product and a resource that generates it.
3. The third dimension, "Cyber-Physical", describes the relationship between the physical world and the virtual world in the context of the DT. An asset (i.e., product, process, resource) exists in the physical world and is represented by the cyber part (i.e., DT) in the virtual world. A further distinction is made in the cyber part between two levels: Descriptive and Executive. These describe the static (i.e., descriptive) and dynamic (i.e., executive) aspects of the DT and are therefore very well suited to classify the behavior of a DT.

This shows that the two concepts, the DTSM and the Grieves DT types, can complement each other. The combination of these two concepts serves as a foundation for exploring DT in the battery industry, as outlined in this chapter. To this end, the applicability of DT in the battery industry is described based on identified use cases along the battery life cycle with the aid of both the concepts.

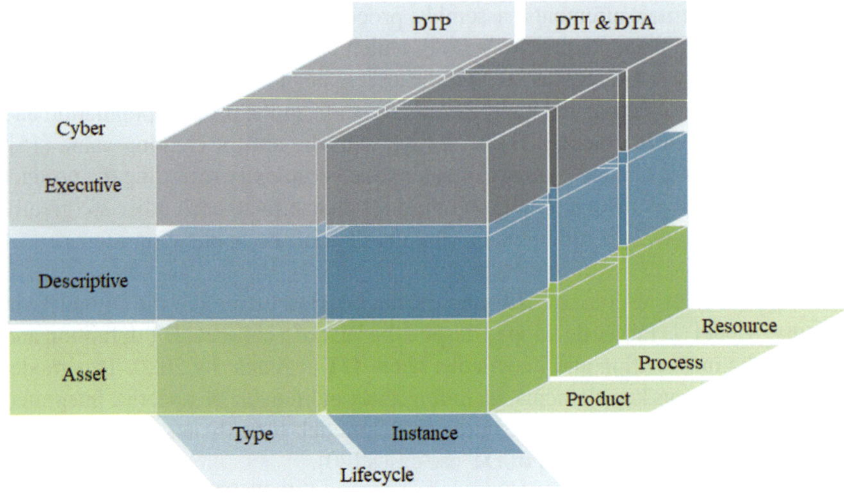


**Fig. 1** Classification of DTs based on [20] and [15]

## *1.3 Battery Lifecycle*

Product Lifecycle Management (PLM) is the business activity of managing a product throughout its lifecycle, from the initial idea to its retirement and disposal [26]. The concept of DTs originally emerged to address PLM but has since been extended to include battery lifecycle management [27]. There isn't a universally recognized PLM standard dedicated solely to the battery lifecycle. Instead, PLM standards are typically used to manage the lifecycle of products and systems overall, which may include LiBs and their components. The concept of product lifecycle includes the following major stages: conceptualization, organization of resources, design, analysis, simulation, and verification, manufacturing and distribution, in-use performance and customer feedback, and eventual disposal [28]. In this context, a battery lifecycle can be described according to the nodes and loops represented in Fig. 2.

LiB lifecycle primarily starts with raw material extraction and its supply to electrode manufacturers. Battery systems can be divided into three primary components and their respective production stages: electrode manufacturing, cell manufacturing, and module/pack manufacturing. According to PLM standards, each of these components can undergo individual management stages such as product ideation and organization design, conceptual design, detailed design, design evaluation, production planning, production, distribution, usage/service life, second life, and EoL. The service life of a battery system is connected with the usage of the entire battery module or pack within its respective application. Lastly, the EoL stage can involve direct recycling, remanufacturing, reuse, or repurposing at both the cell and system levels.

DTs applied in various industries have shown promising results in improving PLM and smart manufacturing processes. However, it is important to scientifically test and validate their effectiveness in the specific context of the battery industry. Without a thorough understanding and substantial efforts in quantifying the benefits of DT, it is challenging to assert that it is a suitable solution to the battery industry's challenges [29].

However, DT implementations present their own set of challenges, such as ensuring data accuracy and integrity, managing the complexity of integrating various data sources, addressing cybersecurity risks, and ensuring scalability and interoperability. Additionally, there may be challenges associated with the cost of implementation, organizational resistance to change, and the need for specialized expertise in areas such as data analytics and machine learning.

This brings us to the research questions addressed in this chapter. Firstly, what are the potential applications of DTs across various stages of the battery lifecycle? We identify specific use cases where strategic implementation of DTs, as hypothesized, could alleviate the challenges currently faced. Secondly, how can the development and implementation of a DT be optimized for a battery-specific industrial context? By delving into both the theoretical and practical dimensions of these inquiries, we aim to contribute insights into leveraging DTs for the battery industry.

## 2 Potential Use Cases of Digital Twins in the Battery Lifecycle

While it is essential to identify use cases and understand the potential business advantages and functional benefits of DT applications, accurately quantifying these benefits remains challenging due to the current technological landscape and the ambiguity surrounding DT classification. Existing literature [3, 27, 30] highlights the theoretical analyses of DT applications in the battery lifecycle; however, translating these theoretical insights into tangible business advantages necessitates actual implementation. Recent studies emphasize the need to move beyond theoretical discussions to quantitatively assess the real-world impacts of DT utilization. In this context, Fig. 3 summarizes the potential challenges within the battery lifecycle that could be mitigated through the practical implementation of DT.

In the following sub-sections, we explore the potential use cases of DTs within each lifecycle phase to address these challenges.

### *2.1 Digital Twin for Battery Design*

#### 2.1.1 Objective

With the continuous evolution of battery materials, the pursuit of an optimal battery remains ongoing, requiring the introduction of new designs and technologies at both cell and pack levels. This process is both costly and time-intensive. A crucial use case of DTs is in providing decision-support for battery system design by leveraging real-world data to reflect the battery's behavior and performance in various operational and environmental conditions.

### 2.1.2 Overview

The trial and error approach in developing new battery designs often results in significant consumption and wastage of critical raw materials, such as nickel and cobalt. The motivation behind using a DT during the design phase is to provide stakeholders with a decision-supporting tool for streamlining the design processes, thereby minimizing material waste and optimizing resource utilization. It aims to facilitate the exchange of data and information across different lifecycle stages as well as the prediction of battery behavior and performances using data from various operating conditions and data from previous product generations, hence, aiding in the design of more durable and efficient battery system. Additionally, the operational data of the battery can be used as feedback to adjust the design and optimization processes.

During the standard design process of LiB systems, several key features need consideration. For cell design, aspects such as nominal capacity, voltage, internal resistance, energy and power requirements, cell format (cylindrical, prismatic, pouch), and electrode material (e.g., lithium cobalt oxide, lithium iron phosphate) play a crucial role. Safety features, including prevention of overcharging, discharging, and thermal runaway, along with cooling systems and electronic control through Battery Management System (BMS), are vital considerations [31].

Recent studies [32, 33] have proposed various DT frameworks for smart product design. Applying DT in the product design stages allows for providing a quantitative design tool for efficient and optimal design decisions using data from previous product generations [34], as well as operational data from the usage phase. In the design process, battery models, such as electrochemical-thermal models or other simulation-based and data-driven methods, play a crucial role in scenario simulation. These tools are instrumental for enhancing battery design, reducing long design iterations, and expediting the time-to-market.

### 2.1.3 Implementation Framework

This use case falls under the category of a DTP of the product. According to Fig. 1, there is a physical component and a cyber-component of the product DTP. Aligning the use cases according to the classification system in Fig. 1 helps establish the system boundary for implementation. This process of defining the system boundary will be applied to all subsequent use cases outlined in the following sections.

The battery DTP primarily comprises a combination of data, models, and algorithms already employed for battery simulations. In the design phase, the battery

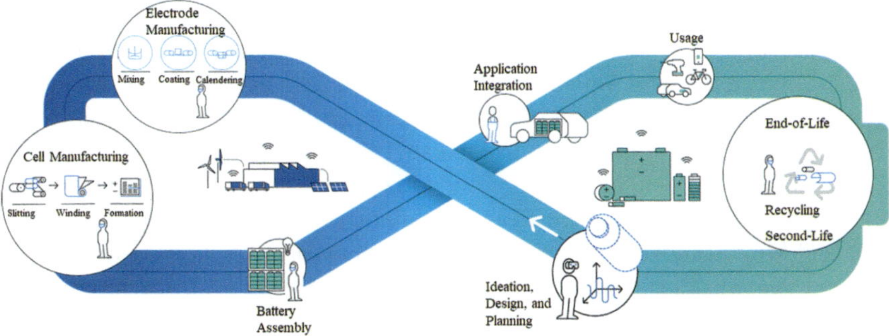


**Fig. 2** Battery lifecycle

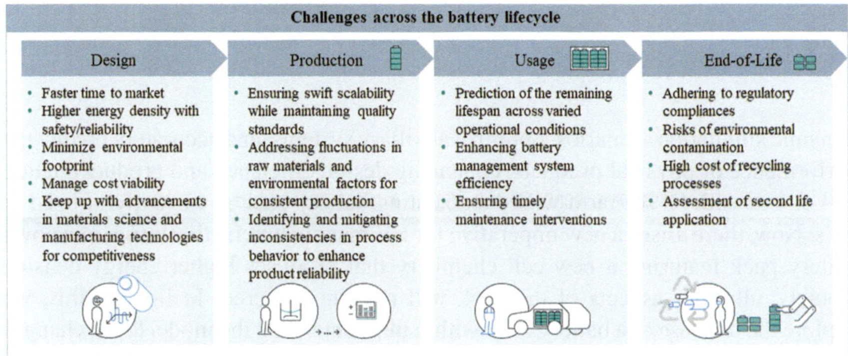


**Fig. 3** Challenges across the battery lifecycle, potentially addressed by DTs

DTP can determine the impact of various parameters on system behavior, identify problems within the system, and simulate deterioration deterrence options [35].

Figure 4 depicts the DTP for battery design. On the left side of the figure, each life cycle phase is associated with the types of DTs based on their definitions (in Sect. 1.2), namely DTP, DTI, and DTA. The battery DTP streamlines the design processes by utilizing the design data, experimental data, real-time and environmental-impacted data to feed the battery models. Various modeling approaches are available for simulating the battery states (such as state of health, state of charge, etc.), performance, aging and safety of LiBs. These include electrochemical models (like pseudo-2D-P2D model, single particle model SPM, etc.), electrical models (equivalent circuit model ECM), thermal models, mechanical models, empirical models, data-driven approaches, and hybrid approaches. These models are coupled with algorithms, listed in Fig. 4, to streamline the scenario simulations.

The DTP serves as a virtual prototype, simulating design scenarios and facilitating efficient testing and redesigning. This reduces costs by identifying flaws before actual production. By harnessing the capabilities of the DTP, battery designers can create

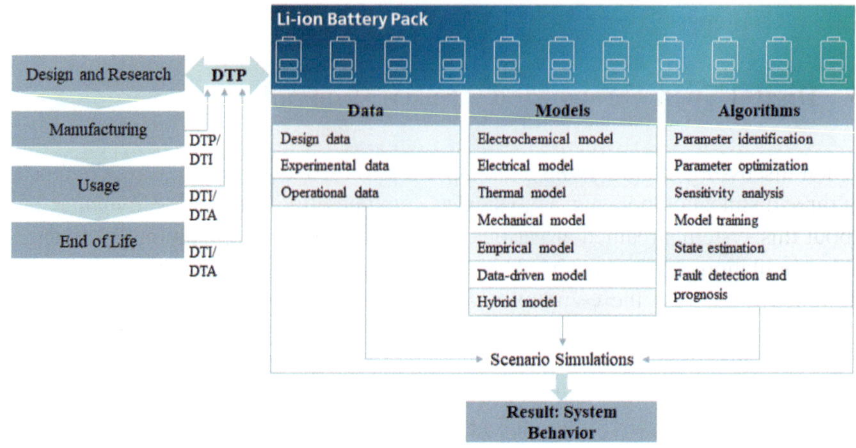


**Fig. 4** DT for battery design

dynamic simulation scenarios, optimize auxiliary systems, and accurately predict the performance of physical products, enhancing design efficiency and product quality.

Let's consider a scenario where an existing battery pack is currently utilized in EVs. Now, there arises a new imperative for EV manufacturers: the design of a novel battery pack featuring a new cell chemistry that promises higher energy density. Notably, all other aspects of the pack will remain unaltered. In light of this, we explore the use case of a battery DTP within this context. At the model level, changes in cell chemistry affect a wide range of electrochemical parameters, such as particle radius, electrode thickness, conductivity, and diffusivity of the active materials. These affected parameters can be provided by the material supplier, measured through test experiments, or identified through various parameter identification algorithms. The existing pack design also has DTI and DTA from the manufacturing, usage, and EoL phases. Using the information on the new design requirements and the DTI/DTAs of the previous product version, a DTP of the new product prototype is developed.

The DTP of the new design offers a transformative solution for battery designers by integrating data from various lifecycle phases, facilitating simulation, testing, and optimization of battery performance in a virtual environment. Consequently, before initiating the entire value chain, the compatibility of the new design at both the module/pack and system levels can be tested virtually. When designing or optimizing the current battery design to align with operational needs, the real-time data provided by the DTIs offer insights into the battery pack's operating cycle, working environment, and other critical information not easily obtained through traditional laboratory experiments. As a result, the DTP promotes battery design, development, and research, ushering in a more efficient and informed approach to advancing battery technology.

#### 2.1.4 Challenges

- The models used for simulation the design scenarios are the core element of the DTP. Ideally, the battery model is a high-fidelity multiscale that can incorporate both temporal and spatial dynamic features. One of the challenges is to quantify and manage the uncertainty inherent in these scenario simulations and subsequence choices based on them. Furthermore, the verification of such models is complex and computationally intensive.
- The challenge associated with the effective integration of data from various life cycle phases and technical domains during the implementation of a DTP for battery design optimization lies in ensuring seamless communication and compatibility between disparate datasets. Achieving this integration requires overcoming technical barriers, standardizing data formats, addressing interoperability issues, and ensuring data consistency and accuracy across different sources.
- The aggregation of multiple datasets used to feed the models is associated with multidisciplinary challenges concerning the sharing methods, privacy, security, and governance systems.

Overcoming these challenges requires robust data integration mechanisms, advanced modeling techniques, and seamless communication between the battery systems and the DTP. The concept of DTI plays an integral role here, potentially serving as a bridge between the diverse data sources and the DTP, facilitating real-time data exchange and enhancing the accuracy and reliability of the simulation and optimization processes.

### 2.2 *Digital Twin During Battery Manufacturing*

#### 2.2.1 Objective

Batteries are required to adhere to strict performance and quality standards. However, manufacturing processes include non-deterministic factors like material inconsistencies, environmental conditions, human error, and equipment wear. These factors contribute to variability in the outcomes of the manufacturing process. The implementation of DTs in battery manufacturing is aimed at mitigating these inconsistencies in order to minimize scrap, increase throughput, and enhance reliability of the battery cell during use.

#### 2.2.2 Overview

To achieve the mentioned objectives, different types of DTs are required, each serving specific purposes. If the aim is to reduce scrap and ensure cell reliability, the DT of the process becomes paramount. Conversely, if the objective is to increase throughput,

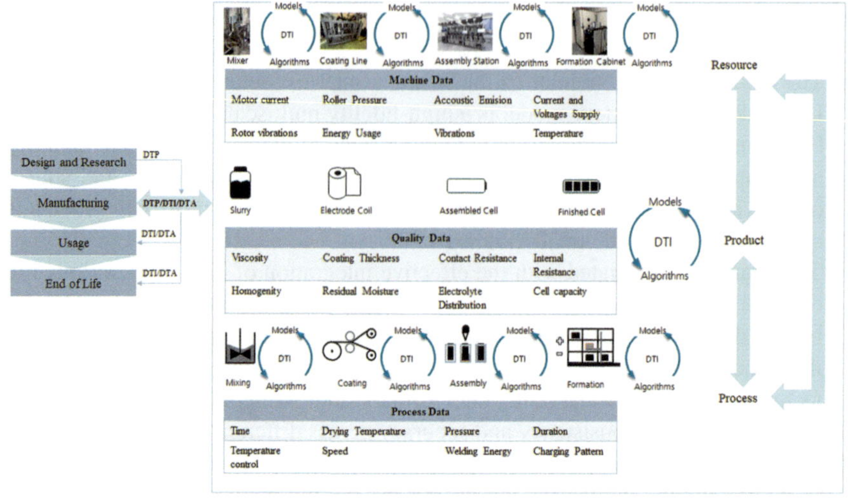


**Fig. 5** DT during battery manufacturing

then the DT of the resource is key. Additionally, the DTI of the product also plays a crucial role in supporting both objectives.

While the DTP of the product is critical during product design and provides the baseline structure and data for the DTI, the DTI of the product becomes key during the manufacturing life cycle. There exists an essential link between the DTI of the product and the DTI of the process and resource. On one hand, it collects data throughout the entire manufacturing process, while on the other, it identifies if the asset is within specifications. DTAs of the product enable statistical analysis and machine learning approaches to derive models and rules that can be leveraged for the creation of DTP of the processes.

In a use case aimed at ensuring cell quality by maintaining manufacturing process outcomes consistently within specifications, the DTI of the process utilizes information from the product DTI regarding process results and product specifications for its calculations. For example, this can allow for the dynamic calculation of a recipe for adjusting the dryer during coating based on the viscosity of the slurry. Similarly, the DTI of the product supports the DTI of resources in the battery manufacturing process. Information on the result of each process step is essential information for the DTI model in terms of calculating the wear level. Based on this information, the DTI can determine when maintenance procedures need to take place to minimize resource downtime (Fig. 5).

### 2.2.3 Implementation Framework

Considering the DTSM, both the DTI of the product and the DTI of the process and resource have descriptive and executive components. Figure 6 shows a framework for

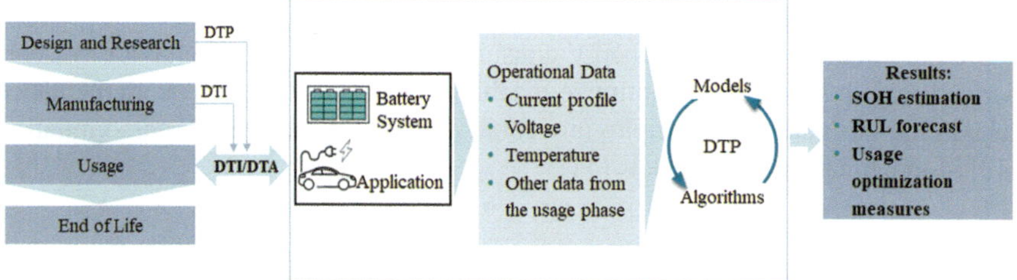


**Fig. 6** DT during operation

DTs in battery manufacturing for PPR and illustrates their interrelations, including the primary data required.

For the DTI of the product, quality data is collected at each transformative stage of the manufacturing process. This data is continuously enriched from the mixing of slurry to the manufacturing of electrodes, cell assembly, and finally, cell formation. The collection of quality data is particularly crucial during manufacturing, not only for compliance with customer standards but also for internal processes.

The product's DTI includes an executive component for evaluating the current state of its quality, employing either statistical or machine learning models. Alternatively, it can utilize the DTP and its models developed during battery design. Quality data collected up to the current work in progress is utilized for this purpose. For example, in slurry mixing, quality data encompasses viscosity and homogeneity. Quality models can assess whether these values fall within acceptable parameters and predict their impact on future manufacturing steps. Typically, these models are constructed using machine learning [36, 37].

On the manufacturing shop floor, each process has a separate DTI due to its distinct nature. For example, the mixing and coating processes are continual, whereas assembly and formation are discrete. It is evident that each manufacturing step can differ significantly from another, even without considering the specific techniques involved. The process DTI comprises descriptive and executive components. The descriptive parts include all parameters (e.g., time or temperature for mixing) that define the process. The executive part models the process behavior with the aim of providing actionable information for process control. Typically, this involves a recipe that includes process parameters for the current run of the process.

Information of the product DTI is crucial for the process DTIs. On one hand, it facilitates adapting the process to the current product by retrieving the current characteristics achieved by preceding processes. On the other hand, it can provide indications of effectiveness by retrieving the outcomes of subsequent processes to adapt for future process runs [38]. These dynamic short-term adjustments enabled by the DTAs of the product allow for more fundamental process adjustments by conducting deeper analyses on hundreds, if not millions, of occurrences [39].

The resource DTI functions similarly to the process DTI in interactions and capabilities but focuses on machine data rather than process data. For instance, in a battery manufacturing sub-process like tab welding, executed using different machines such as laser and resistance welders, the process DTIs control the process, while resource DTIs handle energy consumption and machine behavior. Models derived from resource DTIs optimize machine-related factors such as maintenance scheduling or cycle times, employing techniques such as discrete event simulation [40] or machine learning models [41].

### 2.2.4 Challenges

- As described in the previous paragraphs, the interplay of the PPR DTIs plays a crucial role. However, besides communicating with each other, this requires gathering data from various systems, such as ERP, MES, machines, and tools. Integrating this data from diverse sources and ensuring its accuracy and timeliness is a significant challenge. This is exacerbated by the vast amount of data that needs to be aggregated. Considering that several hundred thousand battery cells are produced each day, the data volume that needs to be processed and stored is immense. Hence, careful planning of the system architecture is required.
- In addition, the creation and maintenance of the DTIs for process and resource, particularly the executive models, can be challenging. For both, accurately modeling the dynamic behavior of the real asset is essential. This in particularly challenging in battery manufacturing processes, which are complex to model. A prime example is the coating process. While the fundamental steps required for coating are well understood, modeling becomes difficult due to numerous influencing factors (e.g., slot control, dryer control) that affect the behavior of the model. Furthermore, the relationship between processing conditions (e.g., slurry composition, coating speed, drying temperature) and the resulting microstructure is an area of active research.

Addressing these challenges requires strategic integration and a robust system architecture. One potential approach is to implement standardized description formats and interfaces that enable uniform models and access to diverse data sources. This approach could also facilitate abstracting physical data storage, separating representation from big data high architecture design. Furthermore, the creation and maintenance of executive models could be streamlined by integrating simulation models and machine techniques. This integration can help mitigate situations where a lack of data impedes machine learning efforts, or validate data-driven models with simulations.

## 2.3 *Digital Twin During Battery Usage*

### 2.3.1 Objective

The objective of a DT during the battery usage phase is to predict and investigate the degradation mechanisms and potential failures, thereby optimizing performance and extending the service life. Through the integration of battery operation data with the models, a DT enables proactive management of battery health and performance.

### 2.3.2 Overview

The ever-increasing demand for efficient energy storage solutions necessitates a deeper understanding of battery behavior throughout its lifecycle. This involves ensuring safety during operation and unlocking its full potential in terms of performance and service life. A battery DTI can potentially serve as a powerful tool, to comprehensively investigate degradation mechanisms and enhance the performance and longevity of batteries during usage.

The primary goal of implementing a DTI in this context is twofold. Firstly, it aims to advance our capability to foresee and comprehend the degradation mechanisms and potential failures that batteries may undergo during usage. Secondly, the DTI serves as a predictive and optimization tool, enabling a proactive approach to enhance battery behavior. By seamlessly integrating real-time data and simulations, the DTI facilitates not only a better understanding of battery dynamics but also empowers strategies for optimizing their overall performance and extending longevity. This pursuit is pivotal in ensuring the reliability and effectiveness of battery systems in various applications, from portable electronics to electric vehicles and renewable energy storage.

The potential functionalities of a battery DTI during usage and to some extent, also for EoL are identified in [29], namely: product design and optimization; behavioral integration in other life cycle phases; Prediction of Remaining-Useful-Life (RUL) prediction; evaluation of battery ageing indicators; charging optimization; adaptive control of thermal management; fault diagnosis. In a technical context, the results that such a DTI can offer are as follows:

1. Estimation and prediction of State of Charge (SOC) and State of Health (SOH)
2. Prediction of RUL of the battery cell/module
3. Prediction of voltage response
4. Measures for usage optimization

Holistic approaches for implementing DTs for the battery operation phase have been sparingly explored in the literature except in [29, 42]. In general, most of these approaches are a combination of the following components: physics-based model, data-driven model or a hybrid model; model parameterization approach; a data management module; model evolution approach; and result calculation algorithm.

Together, these components act as a backbone intelligence that can be uploaded on the BMS.

### 2.3.3 Implementation Framework

This use case falls under the category of DTI of a product, specifically the battery system. According to Fig. 1, there are both physical and cyber components of the product DTI. In the context of this use case, a DTI represents a collection of data generated when the battery system is under operation in a certain application, such as the data from the BMS, service information, serial number, etc.

The DTI of the individual cells, modules, and battery systems are already created during the manufacturing phase. Battery systems are complex products composed of multiple modules and a significant number of battery cells. The DTI of a battery system is composed of the DTAs of the cells and the DTAs of the modules. During the usage phase, the DTI of the battery system is connected to the specific physical twin and remains connected throughout the entire life of that physical twin (i.e., the battery system).

Figure 6 represents a battery system DTI. The behavioral profile of the battery is recorded as operational data during its actual usage. This typically includes data acquired from on-board BMSs, consisting of information from current, voltage, and temperature sensors. In most cases, BMS data is stored locally. However, recent research efforts have focused on mapping real-time battery data from BMS to cloud-enabled systems. This approach can potentially minimize local computational needs and aggregate large datasets to improve the performance of ML-based algorithms.

The DTI can continue to utilize the models and algorithms in the DTP created during the design phase. By feeding operational data into the DTP developed during the design phase, the battery model needs to evolve to align with battery usage profiles. This ability for model evolution is a vital component of the algorithm, providing the battery DTI with essential "intelligence" during the usage phase. The exploration of model evolution represents an intriguing research gap within battery modeling and is the focus of extensive investigation. This concept is further elaborated upon and applied in a case study presented in Sect. 3.

The DTI's result for this use case involves degradation assessment in terms of SOH estimation and RUL forecast, along with optimization measures based on predicting voltage response. These capabilities enable the derivation of maintenance measures during servicing, facilitating efficient troubleshooting, improving customer satisfaction, and providing advanced monitoring capabilities. While individual users may not delve deeply into battery degradation mechanisms, organizations utilizing battery systems for applications such as EVs, smart grids, or industrial power require a comprehensive understanding of how the usage patterns influence system aging. These users benefit from the results of a battery DTI in terms of optimizing charging patterns, accurate range estimation, estimating system reliability, and responsible energy management.

#### 2.3.4 Challenges

- The DTI of a battery system ultimately constitutes a DTA of multiple modules. Such an architecture can be challenging to develop and maintain, particularly in the absence of standardized reference DTIs at the cell and module levels.
- Development and verification of a LiB model that efficiently emulates the real-time behavior of the system is complex and requires massive amounts of data. Moreover, the method of model evolution is not well-established and still needs further research. Additionally, accurately estimating the changing parameters for the battery model is challenging due to inherent cell-to-cell variations arising from subtle manufacturing differences, resulting in non-linear degradation over time.
- Integrating operational data into the model can be challenging. While cloud services have been used with onboard BMSs, standardizing data integration methods for battery DTs is crucial, despite requiring initial investment and implementation time.

To address these challenges, standardizing reference DTIs at both the cell and module levels is crucial. This will simplify the development and maintenance of the architecture. Additionally, progress in Li-ion battery modeling techniques, along with the establishment of standardized data integration methods, and further research in model evolution methods are essential steps. These endeavors will enhance architecture development, improve model accuracy, streamline data integration, and refine battery models, ultimately optimizing system performance and fostering innovation in the energy storage sector.

### *2.4 Digital Twin for End-of-Life Applications*

#### 2.4.1 Objective

The objective of a DT at battery EoL is to enable informed decision-making for optimal second-life applications and to ensure effective and safe remanufacturing, recycling, and repurposing of batteries. It serves as a bridge, filling the informational gap that exists at EoL concerning other life cycle phases.

#### 2.4.2 Overview

With the increase in demand for LiBs, prolonging the overall utility of the batteries is becoming increasingly important. Due to the presence of critical materials, such as cobalt and nickel, and the high manufacturing costs of battery packs, there is economic, political, and environmental interest in utilizing existing battery system in second-life applications. Currently, there is insufficient information available for choosing a second-life application. Examples of second-life applications include power grid storage systems, automated guided vehicles like forklifts, and home storage systems for storing photovoltaic electricity temporarily. In such applications, the battery system experiences lower demands in terms of external temperatures, charge/discharge currents, and SOH.

DTs provide an approach to tackle this challenge by creating a digital representation of the battery´s past life, including design details, manufacturing details, usage history, and environmental interactions. This provides a comprehensive insight into both the structural and functional aspects of the batteries. Such information is crucial for designing, planning, and executing the processes of remanufacturing, recycling, and repurposing. Additionally, since the disassembly of batteries is safety–critical, this information also enables process automation.

The concept of a DT at battery EoL aligns with that of a Digital Battery Passport (DBP). According to the European Commission legislation, all industrial batteries and EVBs should implement a DBP by 2026 [43]. A DBP, as defined by the European Commission, is an electronic record (e.g., QR code) of an individual battery containing crucial information about the sustainability characteristics of batteries, such as capacity, performance, durability, and chemical composition. It aims to support the scaling of circular economy strategies such as disassembly, repurposing, remanufacturing, and recycling.

### 2.4.3 Implementation Framework

This use case falls under the category of DTI/DTA of a product, specifically the battery system. When data from a population of battery systems is collected and aggregated, it is referred to as a DTA. DTI/DTAs provide a method to aggregate the entire lifecycle data of battery systems.

As illustrated in Fig. 7, data relevant to EoL decision-support is collected from all other lifecycle phases. Design information from the DTP provides insights into the original specifications and performance characteristics of the battery. Operational data from the battery DTI/DTA is used to estimate SOH and predict RUL of the battery in a second-life application. Promising methods exist for classifying end-of-life batteries using physics-based models or data-driven models and algorithms. Manufacturing data from the DTI offers details about the assembly format, materials, and quality control processes used in producing the battery. The design-support tool can use such information to identify inherent limitations or weaknesses in the battery design that may impact its performance or safety during a second life.

The DTI/DTA at EoL facilitates the decision-making process to identify the most suitable second-life applications. Similarly, the data can also be used to determine other suitable pathways such as remanufacturing, repurposing, or recycling.

We propose the integration of DTIs as a solution for the DBP, providing a readily available implementation. The technical design and operation of a DBP involve structured product-related data with a predefined scope, agreed data ownership, and access rights conveyed through a unique identifier. By linking DTIs with DBP, stakeholders

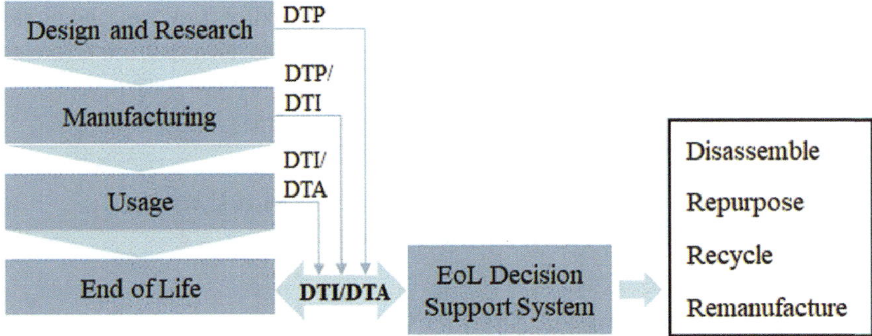


**Fig. 7** DT at EoL

can access a wealth of information regarding the origin, usage, and disposal considerations of batteries, thereby enhancing decision-making processes and promoting sustainable practices.

DTIs contribute to a closed-loop lifecycle for batteries, supporting a circular economy and responsible resource management. Their application promises significant environmental and economic benefits, transforming battery lifecycle management toward sustainability and efficiency.

### 2.4.4 Challenges

- Similar to the usage phase, implementing DT at EoL also requires accurate modeling of battery degradation, real-time data integration, and ensuring model compatibility with diverse battery chemistries and applications. However, such sophisticated modeling techniques are not commercially available.
- Managing uncertainties in predicting battery performance and health is another obstacle, as uncertainties in degradation mechanisms, environmental factors, and usage patterns can impact the accuracy of the predictions.
- Despite the potential benefits of integrating DTIs with DBP, the criteria for EoL decisions remain heterogeneous and context-dependent. Standardizing these criteria is essential for ensuring consistency and reliability in determining the EoL status of batteries. This standardization process requires collaboration among industry stakeholders, regulatory bodies, and academic researchers to establish universally accepted guidelines for assessing the suitability of batteries for repurposing, recycling, or disposal.

To overcome these challenges, standardizing reference DTIs linked with DBP can play a critical role. This approach ensures consistency and interoperability across different stages of a battery's life cycle. These reference models provide a common framework for data collection, sharing, and analysis, enhancing transparency and efficiency in tracking battery information. Standardization in the DBP can help overcome issues related to data silos, lack of interoperability, reliability concerns, and clarity regarding responsibilities at various EoL stages. Ultimately, standardized reference models play a crucial role in streamlining.

## 3 Case Study: Implementing Digital Twin of a Battery System During the Operation Phase

### *3.1 Asset Administration Shell as Cornerstone for Implementation of Digital Twin*

In order to establish a general framework for implementing a DT covering the entire lifecycle of a battery, as in our case, a standards-based approach is essential. This is necessary to establish syntactic and semantic interoperability between different actors or systems involved in the battery lifecycle. Such a standardized approach can be found in the Asset Administration Shell (AAS), which we will closely examine in this section.

Numerous terms and definitions introduced as part of Industry 4.0, such as Asset Administration Shell (AAS) or DT, lacked a uniform understanding, even among experts. Therefore, the focus of Wagner et al. [44] is not on defining additional terms, but on explaining and substantiating existing terms with the aim of resolving apparent contradictions. This takes the form of a classification of the AAS and the DT in the lifecycle of a factory.

In their work, Tantik et al. [45] propose an approach to harmonize the specifications of the World Wide Web Consortium (W3C) with the guidelines of the Plattform Industrie 4.0 (I4.0), thereby establishing a uniform structure for industrial Cyber-Physical Systems (CPS). Using the recommended AAS for I4.0 components, they identify and allocate required functionality to different segments.

The Plug-and-Produce concept necessitates the exchange of configuration data when connecting a new module to a system. Lang et al. [46] suggest a novel concept for secure Plug-and-Produce functionality, leveraging the combination of the AAS and Blockchain technology.

In their work, Platenius-Mohr et al. [47] outline requirements and a solution for enabling interoperable DTs by flexibly transforming their information models. They illustrate the application of this solution in an industrial context by transforming IIoT system-based DTs into the AAS format.

The AAS provides a standardized electronic representation of industrial assets that act as DTs and enable interoperability between automated industrial systems and CPS. Inigo et al. [48] present a case study on the application of AAS in an industrial context, where the integration of a machine tool ecosystem with a robotic arm was implemented to validate the AAS in a manufacturing scenario.

The AAS can also integrate existing standards from different domains or product lifecycle phases. For example, Lüder et al. [49] discuss the capture and representation of engineering data as part of AAS. They present the AAS completion method based on the AutomationML standard to facilitate the capture of the AAS engineering dataset and its export to an AAS serialization.

The AAS can be considered from two perspectives: the virtual representation as an asset description and the technical functionality as a smart manufacturing service. In order to enable the effective provision of information for DT applications, Park et al. [50] propose Virtual REpresentation for a DIgital twin application (VREDI), which is an asset description for the operational procedures of a work-center-level DT application.

These descriptions are aligned with the intended use of a DT, reflecting the perspectives of involved professionals such as process designers, control specialists, and managers. Cimino et al. [51] introduce a paradigm, the Digital Multiverse, to encompass major DT interpretations, not only in terms of data integration but also by establishing and enforcing consistency rules involving both data and models.

As defined in the Reference Architectural Model for Industry 4.0 (RAMI 4.0), an AAS represents a practical embodiment of the DT concept. Its realization involves the integration of operational technologies and information and communication technologies. Ye et al. [52] aim to present the current status of AAS development, design an intuitive method for implementing AASs, and develop an AAS-enabled digital solution for cyber-physical applications in the manufacturing sector.

The survey by Jacoby et al. [53] of open-source implementations of the AAS involved a structured search on GitHub and Google Scholar, which led to the selection of four implementations for in-depth analysis. This paper represents a first attempt to provide a comprehensive comparison of AAS implementations: ASX Server, Eclipse BaSyx, FA3ST Service and NOVAAS.

In conclusion, it can be summarized that the AAS is now gaining traction after many years of development and testing. Additionally, the AAS is open to integrating existing standards and has the flexibility to cover all life cycle phases of an asset. It can also be extended to design DTs for various application areas and domains. Furthermore, a growing open-source community is also contributing to its widespread adoption.

## *3.2 Digital Twin Concept Implementation with Asset Administration Shell*

As previously mentioned, the consideration of DT in the battery lifecycle is based on use cases along this lifecycle, which are described using two DT concepts: the DTSM and the DT types. How does the proposed DT implementation approach based on the AAS relate to these two DT description concepts?

An AAS represents a DT implementation of an asset. An asset can be represented by one or more AAS in virtual space, with each AAS specifically representing one asset. An asset can encompass anything that holds value for a company, whether tangible or intangible. References to other AAS can also be managed within an AAS in order to describe certain relationships or dependencies. Due to this nature of the AAS, the DT types – DTP, DTI, DTA – can be implemented with the help of the AAS.

The two dimensions of the DTSM, Lifecycle and PPR Model, can also be implemented based on these characteristics of the AAS. Part of the third dimension of the DTMS is already represented by the relationship between the AAS and the asset. For the remaining cyber part (Descriptive and Executive), a deeper consideration of the AAS metamodel is necessary. The AAS describes the aspects of a specific DT of an asset using sub-models. These sub-models represent the semantic description for such a DT aspect and consist of a collection of properties that describe certain characteristics of the asset, such as the length, height, and width of a product or the current temperature. These properties form the descriptive part. The executive part consists of operations that can be defined as part of a sub-model. Operations take properties as input parameters and return properties as results. In this way, executable functions and services can be described and called.

The actual execution of an operation is implementation specific; for instance, a simple arithmetic operation or a simulation model can be executed.

For this reason, the AAS is also ideal for implementing a DT for the battery lifecycle, as discussed in more detail in the following sections through implementation examples.

### 3.3 Case Study: DT Implementation for State Estimation of LiB Cells

Building upon the understanding of the AAS, we will now examine how it serves as an ideal framework for implementing a battery DT during the operation phase to estimate the state of the battery. In this case study, a battery DT software application is implemented, which can be installed and accessed by various stakeholders, such as battery specialists seeking insights into the cell behavior in actual applications during usage. Models and battery data are associated with the assets, thereby serving as the backend of the software application, which is then used for practical use cases.

There is no one-size-fits-all approach for implementing this software application because its usage heavily depends on the IT infrastructure of the application landscape. Therefore, a tailored implementation strategy is necessary to accommodate the unique characteristics and requirements of each application environment.

Drawing inspiration from the foundational principles outlined in [54–56], a DTI for state estimation of LiBs is implemented, and the architecture of the implementation is presented in Fig. 8.

The DTI components are as follows:

1. Asset: This refers to the real-world entity or system that the DT aims to replicate. For this implementation, the asset is a battery pack. Please note that an asset of

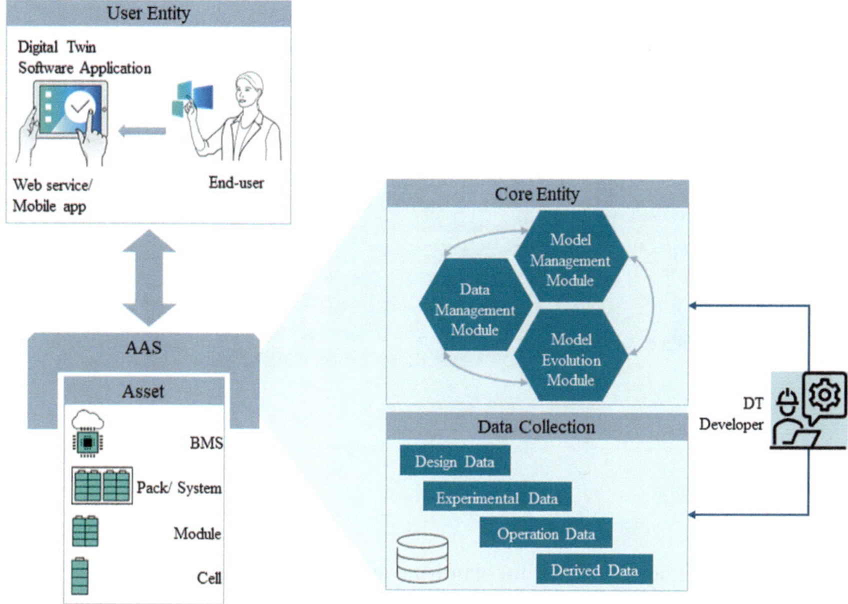


**Fig. 8** DTI architecture for state estimation of LiBs using AAS

a cell or module can be referenced to the AAS of the battery pack. This also implies that the DTI of a battery pack is a DTA of the battery modules.

The AAS, implemented through the AASX package explorer, is shown in Fig. 9. It consists of a number of sub-models in which all the information and functionalities of the battery – including its features, characteristics, properties, statuses, parameters, measurement data, and capabilities are stored. The granularity of operation data fed into the DTI depends on the interface responsible for real-time data transfer from the BMS.

By creating an AAS for the battery pack, we have established a standardized framework for accessing, managing, and exchanging data related to the battery's operation, maintenance, and performance. This AAS enables seamless integration of operation data with the models, facilitating holistic monitoring, analysis, and optimization of the battery pack's usage and health. Through the AAS, stakeholders can gain valuable insights into the battery's state, make informed decisions, and drive continuous improvement in its performance and reliability.

2. Data Collection and Device Control Entity: The AAS refers to this entity, which comprises two sub-entities. The first sub-entity is responsible for collecting data from different sources. Battery-related data such as design data, experimental data, operation data, data derived from simulations, and expert knowledge about battery behavior form the basis data for battery DT. These data are used for model development, verification, evolution, and execution. It is important to note that

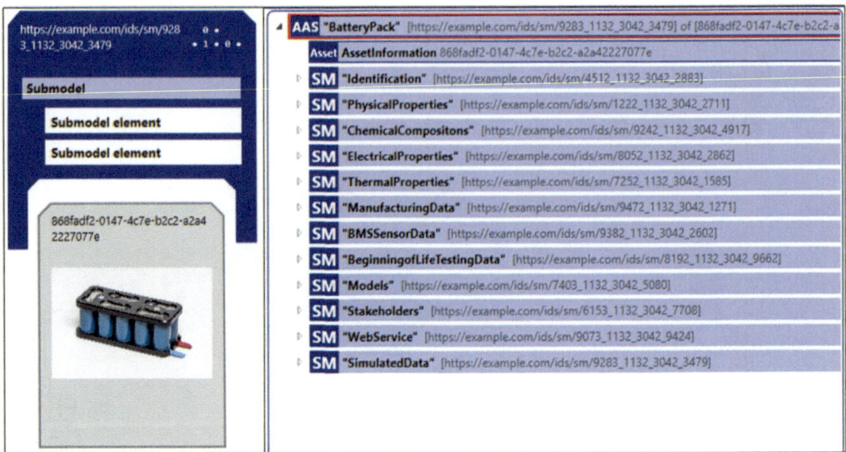


**Fig. 9** AAS of a battery

the methods/interfaces used for acquiring and storing these diverse data sets may vary. The second sub-entity is the control unit, like the BMS, which sends control programs to the battery system when adjustments are needed.

3. Core entity: The core entity acts as the "brain" of the DTI, serving as a cognitive center. The functions performed by the core entity include data management, data analytics, model development and management, model verification, result generation, and model evolution. Additionally, the interfaces that connect the software components are also part of the core entity.

The **data management module** processes and manages all types of data acquired in the data collection entity. In the **model management module**, all models can be stored, used, managed, and updated. The DT battery models include physics-based electrochemical models, ageing models, data-driven model lifecycle prediction models, etc. The algorithms responsible for generating results (estimations, predictions, or forecasts, and battery behavior evaluation) from the models or the data are also integrated in this module. As the battery is used the DT accumulates data from the usage phase such as charging/discharging profiles, working environments sensor acquisitions, and maintenance data. By combining historical and real-time data, the DTI parameters can be updated and evolved using methods such as neural networks, Bayesian regression, maximum likelihood parameter estimation, and other parameter estimation algorithms. This **model evolution module** works continuously to enhance the accuracy of the DTI model.

4. User Entity: The DT software application is a web service accessed by the user. This application should include some type of visualization module for users to access the results of the DTI. For this implementation, we utilized the AAS Web GUI [57]. AAS Web GUI is a user interface provided by Basyx for interacting with the AAS Server and AAS Registry through a web browser. For further details

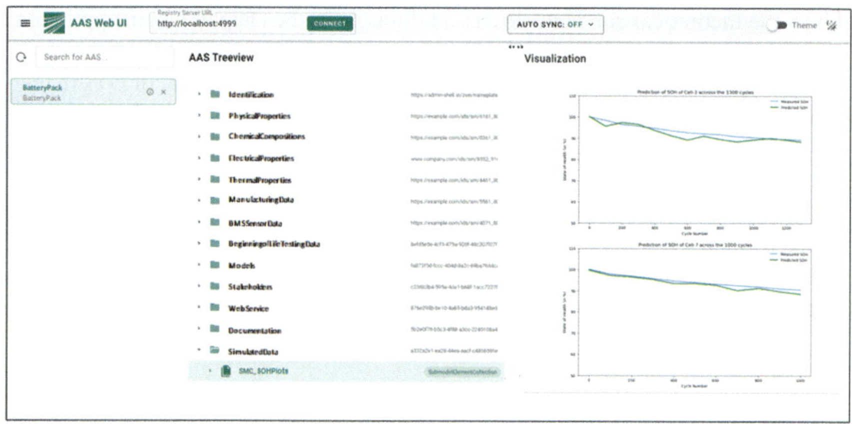


**Fig. 10** AAS GUI of the battery DTI

on asset integration through Basyx, readers are referred to [58]. Figure 10 shows a sample view of the AAS GUI for the battery pack.

The end user of the DTI is not necessarily the end user of the battery system as well. The DT developer is responsible for accessing the core entity and its connection with the other entities.

Implementing a DT for state estimation of LiBs cells holds significant promise in enhancing our understanding of its degradation mechanisms and in optimizing its usage in applications. The AAS framework enables an interface for bi-directional data exchange between the physical and the virtual space. However, a limitation remains that the accuracy of the DT heavily relies on the quality and granularity of data fed into the battery models. This challenge can be mitigated by employing reduced-order and adaptive models, which offer possible ways to improve the model's robustness. Despite these challenges, a DT implementation represents a significant step forward in advancing battery management practices. By continually refining methodologies and overcoming limitations, we can unlock the full potential of DTs in revolutionizing the way we monitor, analyze, and optimize LiBs for improved performance and reliability in real-world settings.

## 4 Conclusion

Across the sections of this chapter, we discussed the use cases in different lifecycle phases of a battery, illustrating the implementation framework and the corresponding challenges encountered during implementation. We shared insights into the practical considerations and strategies necessary for the implementation. Moreover, we described an example of our own implementation, offering a tangible demonstration of how the theoretical concepts discussed throughout the chapter can in applied in a in real-world battery applications.

In conclusion, this chapter highlights the transformative potential of DTs in revolutionizing decision-making and operations throughout the battery lifecycle. From real-time monitoring to predictive maintenance and continuous data-driven insights, digital twins emerge as powerful catalysts for efficiency, innovation, and sustainability in the dynamic landscape of battery technology. The chapter underscores the profound impact of the DT concept on advancing capabilities and environmental considerations associated with batteries across various applications, setting the stage for a paradigm shift in how we perceive and interact with battery systems.

## References

1. VDI/VDE Innovation + Technik GmbH: Projektträger und Dienstleister für Innovationen (2024) Battery cell manufacturing ecosystem in Europe. https://vdivde-it.de/en/publication/battery-cell-manufacturing-ecosystem-europe. Accessed 25 Mar 2024
2. Wirtschaft und Klimaschutz, BMWK—Bundesministerium für (2024) Batterien für die Mobilität von morgen. https://www.bmwk.de/Redaktion/DE/Artikel/Industrie/batteriezellfertigung.html. Accessed 25 Mar 2024
3. Naseri F, Gil S, Barbu C et al (2023) Digital Twin of electric vehicle battery systems: comprehensive review of the use cases, requirements, and platforms. Renew Sustain Energy Rev 179:113280. https://doi.org/10.1016/j.rser.2023.113280
4. Deng J, Bae C, Denlinger A et al (2020) Electric vehicles batteries: requirements and challenges. Joule 4:511–515. https://doi.org/10.1016/j.joule.2020.01.013
5. Masias A, Marcicki J, Paxton WA (2021) Opportunities and challenges of lithium ion batteries in automotive applications. ACS Energy Lett 6:621–630. https://doi.org/10.1021/acsenergylett.0c02584
6. Link S, Neef C, Wicke T (2023) Trends in automotive battery cell design: a statistical analysis of empirical data. Batteries 9:261. https://doi.org/10.3390/batteries9050261
7. Stephan A, Hettesheimer T, Neef C, et al (2023) Alternative battery technologies roadmap 2030+. Fraunhofer-Gesellschaft
8. Link S, Neef C, Wicke T, et al (2022) Development perspectives for lithium-ion battery cell formats. Fraunhofer ISI
9. Baars J, Domenech T, Bleischwitz R et al (2021) Circular economy strategies for electric vehicle batteries reduce reliance on raw materials. Nat Sustain 4:71–79. https://doi.org/10.1038/s41893-020-00607-0
10. Michaelis S, Schütrumpf J, Kampker A, et al (2023) Roadmap battery production equipment 2030. Update 2023. VDMA Verlag
11. Kampker A, Heimes HH, Offermanns C et al (2023) Prediction of battery return volumes for 3R: remanufacturing, reuse, and recycling. Energies 16:6873. https://doi.org/10.3390/en16196873
12. Thielmann A, Wietschel M, Funke S, et al (2020) Batterien für Elektroautos: Faktencheck und Handlungsbedarf. Sind Batterien für Elektroautos der Schlüssel für eine nachhaltige Mobilität der Zukunft?[Batteries for electric cars: Fact check and need for action. Are batteries for electric cars the key to sustainable mobility in the future?]. Perspectives—Policy Briefs
13. Doose S, Mayer JK, Michalowski P et al (2021) Challenges in ecofriendly battery recycling and closed material cycles: a perspective on future lithium battery generations. Metals 11:291. https://doi.org/10.3390/met11020291

14. Grieves MW (2005) Product lifecycle management: the new paradigm for enterprises. IJPD 2:71. https://doi.org/10.1504/IJPD.2005.006669
15. Grieves MW (2023) Digital Twins: Past, Present, and Future. In: Crespi N, Drobot AT, Minerva R (eds) The Digital Twin. Springer International Publishing, Cham, pp 97–121
16. Piascik R, Vickers J, Lowry D, et al (2010) Technology area 12: materials, structures, mechanical systems, and manufacturing road map. NASA Office of Chief Technologist:15–88
17. Kerber F (2022) Der digitale zwilling im industriellen kontext: eine regionale perspektive. https://www.tha.de/Binaries/Binary55211/Whitepaper.pdf. Accessed 17 Nov 2023
18. Jones D, Snider C, Nassehi A et al (2020) Characterising the Digital Twin: a systematic literature review. CIRP J Manuf Sci Technol 29:36–52. https://doi.org/10.1016/j.cirpj.2020.02.002
19. Krauß J, Schmetz A, Fitzner A, et al (2023) Der digitale zwilling in der batteriezellfertigung. Fraunhofer-Gesellschaft
20. Lechler T, Fuchs J, Sjarov M, et al (2020) Introduction of a comprehensive structure model for the Digital Twin in manufacturing. In: 2020 25th IEEE international conference on emerging technologies and factory automation (ETFA). IEEE, pp 1773–1780
21. Sjarov M, Lechler T, Fuchs J, et al (2020) The digital twin concept in industry—a review and systematization. In: 2020 25th IEEE international conference on emerging technologies and factory automation (ETFA). IEEE, pp 1789–1796
22. ISO/IEC JTC 1 (2021) SC 41 new scope includes Digital Twin. https://jtc1info.org/sc-41-new-scope-includes-digital-twin/. Accessed 17 Nov 2023
23. (2022) Industrie 4.0 Begriffe: VDI-Statusreport Juni 2022 = Industrie 4.0 Terms and definitions. Blaue Papiere. VDI e.V, Düsseldorf
24. Cutting-Decelle AF, Young R, Michel JJ et al (2007) ISO 15531 mandate: a product-process-resource based approach for managing modularity in production management. Concurr Eng 15:217–235. https://doi.org/10.1177/1063293X07079329
25. Maropoulos PG, McKay KR, Bramall DG (2002) Resource-aware aggregate planning for the distributed manufacturing enterprise. CIRP Ann 51:363–366. https://doi.org/10.1016/S0007-8506(07)61537-6
26. Asiedu Y, Gu P (1998) Product life cycle cost analysis: state of the art review. Int J Prod Res 36:883–908. https://doi.org/10.1080/002075498193444
27. Anandavel S, Li W, Garg A et al (2021) Application of digital twins to the product lifecycle management of battery packs of electric vehicles. IET Collab Intel Manufact 3:356–366. https://doi.org/10.1049/cim2.12028
28. Subrahmanian E, Rachuri S, Bouras A, et al (2006) The role of standards in product lifecycle management support. National Institute of Standards and Technology, Gaithersburg, MD
29. Singh S, Weeber M, Birke KP (2021) Implementation of battery Digital Twin: approach, functionalities and benefits. Batteries 7:78. https://doi.org/10.3390/batteries7040078
30. Wang W, Wang J, Tian J et al (2021) Application of digital twin in smart battery management systems. Chin J Mech Eng 34:1–19. https://doi.org/10.1186/s10033-021-00577-0
31. U.S. Agency for International Development (2023) Lithium-ion battery standards | Energy | U.S. agency for international development. https://www.usaid.gov/energy/powering-health/technical-standards/lithium-ion-batteries. Accessed 24 Nov 2023
32. Tao F, Cheng J, Qi Q et al (2018) Digital Twin-driven product design, manufacturing and service with big data. Int J Adv Manuf Technol 94:3563–3576. https://doi.org/10.1007/s00170-017-0233-1
33. Zhang M, Sui F, Liu A, et al (2020) Chapter 1—digital twin driven smart product design framework. In: Nee AYC, Hu T, Liu A, et al (eds) Digital twin driven smart design. Academic Press, pp 3–32
34. Wagner R, Schleich B, Haefner B et al (2019) Challenges and potentials of digital twins and industry 4.0 in product design and production for high performance products. Proc CIRP 84:88–93. https://doi.org/10.1016/j.procir.2019.04.219
35. Semeraro C, Aljaghoub H, Abdelkareem MA et al (2023) Guidelines for designing a digital twin for Li-ion battery: a reference methodology. Energy 284:128699. https://doi.org/10.1016/j.energy.2023.128699

36. Turetskyy A, Wessel J, Herrmann C, Thiede S (2020) Data-driven cyber-physical system for quality gates in lithium-ion battery cell manufacturing. Proc CIRP 93:168–173. https://doi.org/10.1016/j.procir.2020.03.077
37. Stock S, Pohlmann S, Günter FJ, Hille L, Hagemeister J, Reinhart G (2022) Early quality classification and prediction of battery cycle life in production using machine learning. J Energy Storage 50:104144. https://doi.org/10.1016/j.est.2022.104144
38. Meiners J, Fröhlich A, Dröder K (2022) Potential of a machine learning based cross-process control in lithium-ion battery production. Proc CIRP 112:525–530. https://doi.org/10.1016/j.procir.2022.09.093
39. Thiede S, Turetskyy A, Kwade A et al (2019) Data mining in battery production chains towards multi-criterial quality prediction. CIRP Ann 68:463–466. https://doi.org/10.1016/j.cirp.2019.04.066
40. Titmarsh R, Assad F, Harrison R (2022) Energy saving in lithium-ion battery manufacturing through the implementation of predictive maintenance. In: 2022 international conference on computing, electronics & communications engineering (iCCECE). IEEE
41. Giner J, Lamprecht R, Gallina V, et al (2021) Demonstrating reinforcement learning for maintenance scheduling in a production environment. In: 2021 26th IEEE international conference on emerging technologies and factory automation (ETFA). IEEE
42. Dubarry M, Howey D, Wu B (2023) Enabling battery digital twins at the industrial scale. Joule 7:1134–1144. https://doi.org/10.1016/j.joule.2023.05.005
43. (2024) Regulation—2023/1542—EN—EUR-Lex. https://eur-lex.europa.eu/eli/reg/2023/1542/oj#document1. Accessed 25 Mar 2024
44. Wagner C, Grothoff J, Epple U, et al (2017) The role of the Industry 4.0 asset administration shell and the digital twin during the life cycle of a plant. In: IEEE international conference on emerging technologies and factory automation limassol C2:((ed) ETFA 2017: 2017 22nd IEEE international conference on emerging technologies and factory automation : September 12–15, 2017, Limassol, Cyprus. IEEE, pp 1–8
45. Tantik E, Anderl R (2017) Integrated data model and structure for the asset administration shell in industrie 4.0. Proc CIRP 60:86–91. https://doi.org/10.1016/j.procir.2017.01.048
46. Lang D, Friesen M, Ehrlich M. et al (2018) Pursuing the vision of industrie 4.0: secure plug-and-produce by means of the asset administration shell and blockchain technology. In: IEEE 16th international conference on industrial informatics (INDIN): Proceedings : Faculty of engineering of the university of Porto, Porto, Portugal, 18–20 July, 2018. IEEE, Piscataway, NJ, pp 1092–1097
47. Platenius-Mohr M, Malakuti S, Grüner S, et al (2019) Interoperable digital twins in IIoT systems by transformation of information models. In: Proceedings of the 9th international conference on the internet of things. association for computing machinery, New York,NY,United States, pp 1–8
48. Inigo MA, Porto A, Kremer B, et al (2020) Towards an asset administration shell scenario: a use case for interoperability and standardization in industry 4.0. In: NOMS 2020—2020 IEEE/IFIP network operations and management symposium. IEEE, [S.l.], pp 1–6
49. Lüder A, Behnert A-K, Rinker F, et al (2020) Generating Industry 4.0 asset administration shells with data from engineering data logistics. In: 2020 25th IEEE international conference on emerging technologies and factory automation (ETFA). IEEE, pp 867–874
50. Park KT, Yang J, Noh SD (2021) VREDI: virtual representation for a digital twin application in a work-center-level asset administration shell. J Intell Manuf 32:501–544. https://doi.org/10.1007/s10845-020-01586-x
51. Cimino C, Ferretti G, Leva A (2021) Harmonising and integrating the digital twins multiverse: a paradigm and a toolset proposal. Comput Ind 132:103501. https://doi.org/10.1016/j.compind.2021.103501
52. Ye X, Hong SH, Song WS et al (2021) An Industry 4.0 asset administration shell-enabled digital solution for robot-based manufacturing systems. IEEE Access 9:154448–154459. https://doi.org/10.1109/ACCESS.2021.3128580

53. Jacoby M, Baumann M, Bischoff T et al (2023) Open-source implementations of the reactive asset administration shell: a survey. Sensors (Basel). https://doi.org/10.3390/s23115229
54. Singh S, Weeber M, Birke K-P (2021) Advancing Digital Twin implementation: a toolbox for modelling and simulation. Proc CIRP 99:567–572. https://doi.org/10.1016/j.procir.2021.03.078
55. Yang D, Cui Y, Xia Q et al (2022) A digital twin-driven life prediction method of lithium-ion batteries based on adaptive model evolution. Materials (Basel). https://doi.org/10.3390/ma15093331
56. Issa R, Badr MM, Shalash O et al (2023) A data-driven digital twin of electric vehicle Li-ion battery state-of-charge estimation enabled by driving behavior application programming interfaces. Batteries 9:521. https://doi.org/10.3390/batteries9100521
57. GitHub (2024) Eclipse-basyx/basyx-applications: applications. https://github.com/eclipse-basyx/basyx-applications/tree/main. Accessed 25 Mar 2024
58. (2024) BaSyx How-To—Asset Integration. https://eclipse.dev/basyx/asset_integration/. Accessed 25 Mar 2024

# The Impact of Japanese Anime Games and AI on the Development of the Metaverse

**Hiromi Komuro, Oussouby Sacko, Yu Yuan, Qiongzhao Ellen. Schicktanz, Ramesh Ramadoss, and Miyuki Komuro**

**Abstract** The concept of the Metaverse was initially introduced in 1992 by Neal Stephenson in his science fiction novel "Snow Crash". The book portrays the Metaverse as a three-dimensional digital environment where the boundaries between the physical and virtual realms blur. In this space, individuals interact and amuse themselves using digital representations known as "avatars".

Japanese anime, manga, games, and other content are widely esteemed globally for their distinctive qualities and exceptional standards. As a prominent content producer with a rich heritage, Japan's content creation is underpinned by professional know-how, cultural richness, and top-tier talent, endowing it with an allure that sets it apart. The Metaverse, a novel platform for expression and interaction, aligns well with Japanese content and presents even greater possibilities.

It is crucial to harness the strengths of Japanese content while infusing global elements. We can engage audiences worldwide using 3DCG technology to deliver more captivating visual presentations and immersive experiences. By tapping into

H. Komuro (✉)
International Metaverse Association of Japan, 35-5, Kitanonokami-Cho Shimogamo Sakyo-Ku, Kyoto 606-0846, Japan

O. Sacko
Kyoto Seika University, Japan, 137 Kino-Cho, Iwakura, Sakyo-Ku, Kyoto 606-8588, Japan

Y. Yuan
Metaverse Acceleration and Sustainability Association (MASA), Scientific Research Building Suite 107H, Tsinghua High-Tech Park, Shenzhen 518057, China

Q. Ellen. Schicktanz
Rodin International Artists Foundation (U.S.A.), 1001 Chestnut Street#303E, Philadelphia, PA 19107, USA

R. Ramadoss
IEEE Blockchain Technical Community, Piscataway, NJ, USA

M. Komuro
Collège du Léman International School Geneva, Collège du Léman Route de Sauverny, 74 Case Postale 156 Versoix, 1290 Versoix, Switzerland

 
M. Grieves and E. Hua (eds.), *Digital Twins, Simulation, and the Metaverse*, Simulation Foundations, Methods and Applications, https://doi.org/10.1007/978-3-031-69107-2_16

our creativity and technical prowess, we can provide compelling experiences for users worldwide.

Japanese companies actively expand and participate in the Metaverse by utilizing their unique technologies and intellectual property (IP), content creation capabilities, and game development techniques. This chapter will introduce the current status and future trends of Japan's development in the metaverse field and explore how Japan's unique cultural background influences its role in developing game technology and the Metaverse.

## 1 Introduction

The Metaverse is a virtual world derived from the development of digital technology and the Internet. It blends the physical and virtual worlds, creating an extended reality where individuals can interact with each other in an immersive digital space [1]. The Metaverse supports social interactions, developable relationships, and transferable economic values [2]. Japan has technological advantages and rich content creation experience in the metaverse field, laying the foundation for the Metaverse. Japan's advantages in technology, culture, and other aspects and the companies' emphasis on user experience have attracted many users. In the development of the Metaverse, AI technology has improved user experience and interactivity, providing a steady stream of content materials for the virtual world. The support of the government and the International Metaverse Association for Metaverse is crucial, promoting the healthy development of the industry and global cooperation. Strengthening international cooperation and exchanges and jointly formulating standards and norms will help promote the prosperity of the worldwide metaverse industry. Integrating the metaverses of various countries and the Japanese Metaverse through international collaboration will bring a richer and more immersive virtual experience, benefiting users worldwide.

Japan has long had a deep cultural heritage and advanced technology in the fields of anime, games, and technology, and these factors provide Japan with a unique advantage in constructing the Metaverse. Anime and manga culture has spawned many globally recognized virtual idols and characters that can serve as virtual inhabitants or interactive elements in the Metaverse, attracting users from different cultural backgrounds worldwide. The Japanese game industry is known for its innovative and high-quality game design, and these design concepts and technological prowess are being utilized to build more immersive and interactive metaverse experiences. Japanese game developers specialize in creating elaborate storylines and complex characterizations, skills that help create engaging virtual worlds in the Metaverse.

As AI technology advances, Japanese companies also explore incorporating cutting-edge technologies such as augmented reality (AR) and virtual reality (VR) into metaverses to provide a more realistic and three-dimensional user experience. By integrating these technologies, users can interact with virtual spaces in ways never before possible, which not only changes the way entertainment and social interaction are done but also opens up transformative possibilities in education, work, and several other fields. Japan's Metaverse development has benefited from its rich cultural resources, technological innovation, and industrial base. In the future, it is expected to continue to play a vital role in developing the global Metaverse, especially with its unique contributions to content creation and game technology, which will bring users more prosperous and more diverse virtual world experiences worldwide.

## 2 History of the Metaverse

The term metaverse is a combination of "meta" (meaning virtual and transcendent) and "verse" (meaning universe) [3]. It refers to a set of virtual spaces where one user can interact with other users in the same or multiple physical spaces in real time [4]. The term metaverse is widely discussed in academic circles [5]. The concept of the Metaverse first appeared in 1992, proposed by Neal Stephenson in his science fiction novel "Snow Crash". The book describes the Metaverse as a three-dimensional digital space where the natural and virtual worlds are closely connected, where people can communicate and entertain themselves through their respective "avatars". As a virtual world, the Metaverse enables people to create, share, collaborate, and build communities and share experiences across the globe. It is a place where people can work, learn, play, and even live in a more accessible and customizable manner than the physical world. The Metaverse could be the future virtual tourist destination, a retail store, classroom, office, or a gaming platform where one can participate through digital self [6–8].

In the 1990s, the term metaverse was coined in science fiction and among researchers, rooted in early research into virtual reality (VR) and the development of online games. During this period, with the development of computer graphics technology and virtual reality technology, discussions were underway about more immersive virtual spaces and virtual worlds. Technologies such as VRML (Virtual Reality Modeling Language) were used to create 3D virtual spaces. It can be said that it was derived from theoretical ideas and philosophical reflections.

The concept of the Metaverse in the 2000s reflects the changing times toward more specific technologies and applications. During this period, digital spaces such as online games, virtual worlds, and social media developed rapidly, where multiple users can participate interactively. With this, it can be said that the traditional concept of the Metaverse has been embodied more concretely and practically. The early to mid-2000s saw the rise of popular online games and virtual worlds such as Second Life and World of Warcraft. These platforms allowed users to interact and engage in activities through interaction with other players and content within a virtual space. In addition, the spread of social media has expanded communication and business activities in the virtual space. Against the backdrop of the development of the Internet and digital technology, the Metaverse in this era quickly became popular as a new social and economic space concept that is different from the real world. This is considered an early example of the Metaverse.

In 2003, many companies were looking for business opportunities, such as "Second Life", which created a virtual space on the Internet which converted money from a virtual space into currency in a physical space and bought and sold land in a virtual space. This virtual space service, which incorporates elements such as online games, SNS, and online meeting services, has been attracting attention as a metaverse since the second half of 2010. Virtual space services have been growing in response to these changes, and services such as VR-Chat, Roblox, and Horizon Worlds have a variety of characteristics, such as VR-supporting, gaming, and business-oriented. It is spread all over the world as a "metaverse".

The Metaverse in the 2010s was when technology evolved, and digital communication developed further. During this era, technologies such as virtual reality (VR), augmented reality (AR), and mixed reality (MR) advanced rapidly. They began to be leveraged by businesses and individuals to deliver new metaverse experiences. Virtual spaces have become more realistic and immersive and are used in various fields, such as entertainment, education, and business [9]. VR and AR-powered games, simulations, educational platforms, and commercial facilities have emerged to provide users with new experiences. Social VR platforms and online collaboration tools have also emerged, allowing users to communicate and collaborate virtually. These platforms have enabled interaction and collaboration across distances, enabling new forms of cooperation and communication. It can be seen as an era in which the Metaverse will become more diverse and even more pervasive in daily life. The Metaverse in the 2020s is more widely recognized, with many companies exploring business and entertainment opportunities within the virtual space. The Metaverse does have a potential impact on business value [10]. Cryptocurrencies and NFTs (non-fungible tokens) are also emerging as relevant elements of economic activity within the Metaverse. Metaverse unifies disruptive technologies like IoT, AI, AR, VR, XR, Web 3.0, and blockchain (BCT). It integrates them to create a real-time [11] massively scaled and completely immersive environment capable of providing real-time experiences in three-dimensional virtual space [12].

These factors make the Metaverse a critical concept that explores the possibilities of a future where the digital and physical worlds merge. The potential revolutionary impact of the Metaverse on the way we work, shop, socialize, and play [13]. With the rapid growth of the Metaverse and the growing importance of remote work and digital communication in the wake of the COVID-19 pandemic, many companies, including large technology companies, are rolling out metaverse-related projects. We are seeing the expansion of the Metaverse in a wide range of fields, such as digital asset trading using NFTs and the emergence of new business moles in virtual space. As a result, the Metaverse is expected to evolve beyond mere entertainment to link with new economic zones and social structures. In 2024, OpenAI's Sora technology and Apple's Vision Pro VR goggles are expected to bring innovative developments to the Metaverse.

## 3 Overview of Metaverse

The Metaverse is a term that refers to a virtual space that is different from the physical universe and the real world. In general, it refers to a digital space like a virtual world. The Metaverse can leverage virtual reality technology and artificial intelligence to create an environment allowing real-world experiences and activities. This will not only allow people to have new opportunities for communication and interaction, but it will also create new possibilities in creativity and business. There are cultural, social, and ethical issues to consider when designing and developing the Metaverse[14] (Fig. 1).

Metaverse services that general users widely use include game-type metaverses such as "Fortnite" and "Roblox" and Social Networking Service (SNS) based metaverses such as "VR Chat" and "Cluster." While these services can be accessed from a PC or smartphone, accessing them from a head-mounted display like the Meta Quest 2 allows for a more immersive virtual experience.

In addition, at the base of the Metaverse is XR, an advanced technology that fuses the real and the virtual. The Metaverse based on VR (virtual reality) is called the VR metaverse, and it is based on the virtual world and is mainly accessed using head-mounted displays (HMDs) such as Metaverse (Figs. 2 and 3).

The definition of the Metaverse refers to a digital environment in a virtual world or augmented reality space. Built primarily on the Internet, it is a platform that allows users to interact and engage in activities with other users and physical spaces within a virtual space. The Metaverse offers realistic experiences and interactions, including various forms of digital representations, agents, avatars, objects, and more. Some

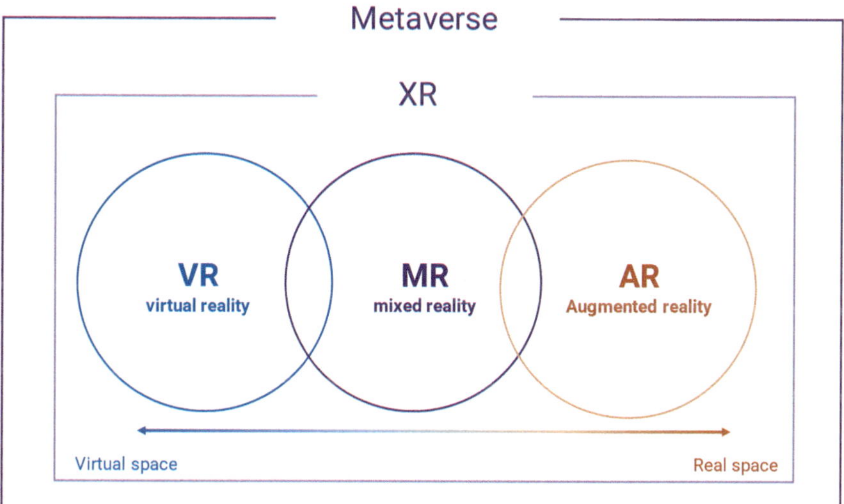


**Fig. 1** Metaverse architecture diagram. *Source* https://metaversesouken.com/metaverse/market/# [15]

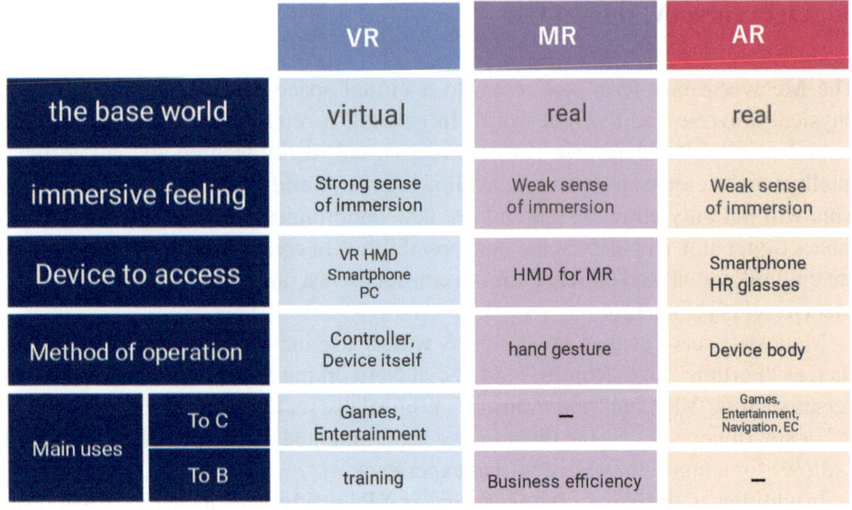

| | | VR | MR | AR |
|---|---|---|---|---|
| the base world | | virtual | real | real |
| immersive feeling | | Strong sense of immersion | Weak sense of immersion | Weak sense of immersion |
| Device to access | | VR HMD Smartphone PC | HMD for MR | Smartphone HR glasses |
| Method of operation | | Controller, Device itself | hand gesture | Device body |
| Main uses | To C | Games, Entertainment | – | Games, Entertainment, Navigation, EC |
| | To B | training | Business efficiency | – |

**Fig. 2** Metaverse architecture diagram. *Source* Metaverse Research Institute [16]

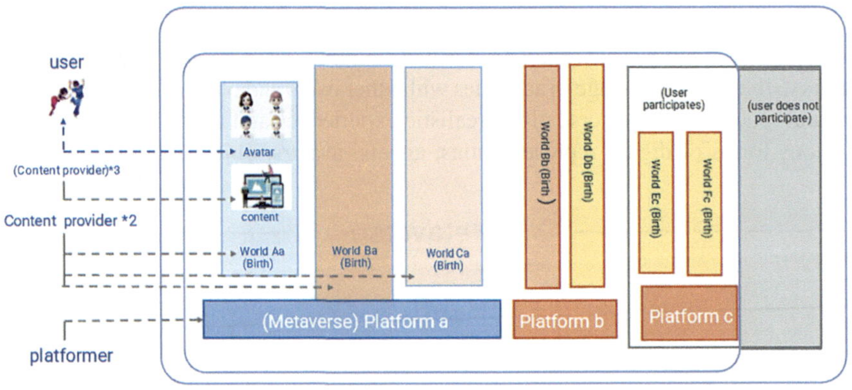


**Fig. 3** Conceptual diagram of "metaverse" and related terms. *Source* Ministry of Internal Affairs and Communications "Study Group on the Utilization of Metaverse, etc. for the Web3 Era" Report 2023.7 [17]

metaverses also incorporate in-game fantasy elements in which users can engage in multiple activities and creativity. The Metaverse aims to push the boundaries of the natural world and provide new experiences and opportunities for communication. It is a broad concept encompassing many elements, including virtual reality, augmented reality, online gaming, social media, virtual worlds, and digital assets. The Metaverse is used in various sectors, including entertainment, education, business, and communications.

## 4 Metaverse Market Size

The worldwide metaverse market is forecasted to reach USD 678.8 billion in 2030. This is a 17-fold increase in comparison to 2021. Japan's domestic market is predicted to be about 4 trillion yen in 2025 and 24 trillion in 2030. This is the Japanese market share of about 10% of the world, estimated by the Mitsubishi Research Institute's trial exchange rate standard. The most significant segments of the Japanese domestic market are expected to be gaming, entertainment, and medical. Health, manufacturing plants, and office work are expected to expand and develop one after another (Fig. 4).

According to the Yano Research Institute, the metaverse market size in Japan is expected to grow to about 1 trillion yen by fiscal 2026. It will likely increase like the global market, with an annual growth rate of about 170% (Fig. 5).

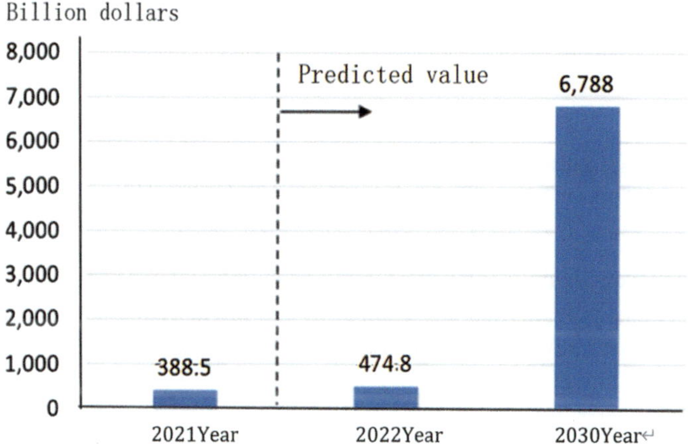


**Fig. 4** Global Metaverse market size (Sales) forecast. *Source* Ministry of Internal Affairs and Communications "Study Group on the Utilization of Metaverse, etc. for the Web3 Era" Report 2023.7 [17]

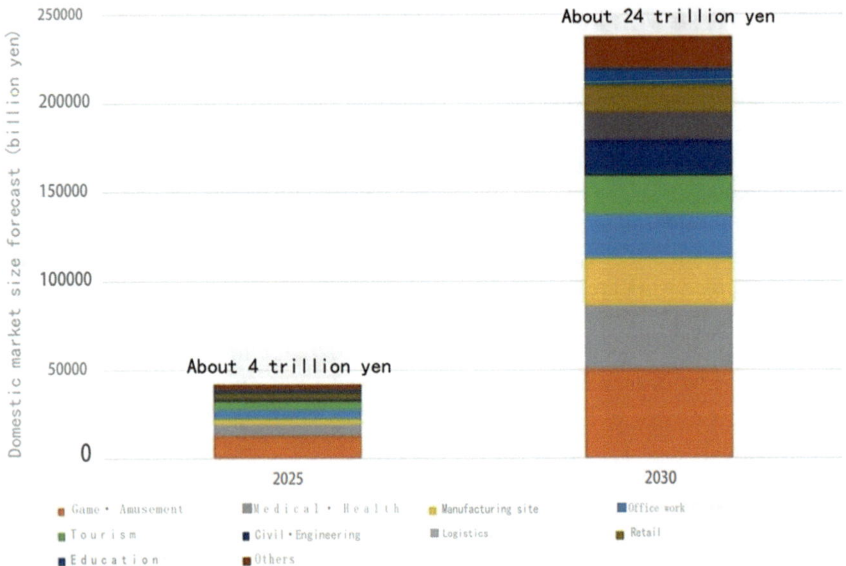


**Fig. 5** Japan metadata market size (Sales) forecast. *Source* Mitsubishi Research Institute, CX2030 Metaverse in a Wide Context as a Place to Utilize Virtual Technology [18]

Various players, mainly companies involved in online games and their infrastructure, have entered the market, and competition is intensifying. Companies such as Sony and Nintendo have entered the market in Japan, and several ventures operating metaverse platforms have emerged.

Some immersive metaverse services that use HMDs are beginning to exceed 20,000 simultaneous users.

## 5 Metaverse Usage by Age, Region, etc.

According to a study by the Mitsubishi Research Institute, the percentage of respondents who have experience using Metaverse is higher among younger age groups, with more than 10% of those in their 20 s and younger. The level of awareness was also higher among younger age groups, with more than 20% of respondents in their teens saying they know it well enough to explain it to others. Comparing urban areas (three major metropolitan areas) and rural areas (other regions), the understanding rate was generally higher among respondents in urban areas for the same age group. Still, the difference by age group was more pronounced (Fig. 6).

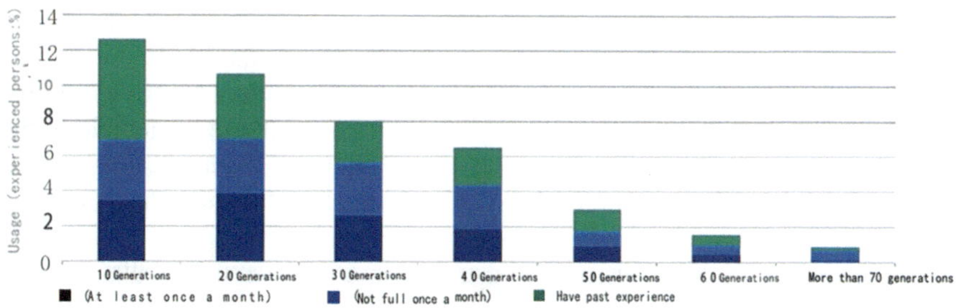


**Fig. 6** Comparison of metaverse usage by age group. [19] *Source* Mitsubishi Research Institute, Results of a questionnaire on metaverse cognition and use

## 6 Japan Companies Are Entering the Metaverse Market

The current state of the metaverse market and users is expected to accelerate as the Metaverse is widely used globally. Various corporate and individual players will continue to enter the market, and the market size is expected to expand at an accelerated pace. Moreover, the Metaverse is becoming more widely recognized in the country as participants in metaverse-related games are aware of the Metaverse.

(1) **Spread of remote communication due to Corona**

Remote communication has become widespread due to Corona. Due to the spread of the coronavirus infection, people's communication opportunities have shifted from face-to-face to remote, and it has become commonplace for private and professional communication to be conducted through chat, such as Slack and video conversations, such as ZOOM.

Not only is the demand for digital communication expanding, but people must also use digital communication without resistance.

(2) **Popularization of the Metaverse among Young People**

Today, the Metaverse is experiencing a rapid increase in young users, especially for online gaming applications. This is because young people have been using smartphones daily since childhood, and there are more and more situations where they use communication functions in online games instead of SNS as a means of communication.

Fortnite and Roblox boast an overwhelming number of users, about 350 million and 200 million, respectively, as representative services of the game-type Metaverse.

(3) **Growing interest in Web3, including NFTs**

The term Web3 refers to new capabilities enabled by blockchain and distributed ledger technologies, while Web 3.0 refers to the next iteration of the World Wide Web. Sometimes, the terms Web3 and Web 3.0 are used interchangeably. However, there is no widely accepted definition for these terms.

With the establishment of communication through the Metaverse, it is essential to determine the extent to which communication with other users in the form of an avatar, which is one's alter ego on the Metaverse, will be established to spread the Metaverse. It is believed that if many users get the feeling of talking to other users in avatars as if they were talking face-to-face, many remote communications will be replaced by communication through the Metaverse. It can be said that the foundation for the establishment of communication through avatars has already begun to be prepared, such as the popularity of Vtubers in the world and the installation of a function called "Memoji" on the iPhone that allows you to send facial expressions with your original avatar.

It is a legal development related to the Metaverse and NFTs. Currently, the Japanese government has identified Web3 as one of the growth industries in Japan. It is actively moving to develop the industry, such as setting up the Web 3.0 Policy Promotion Office. On the other hand, laws related to the Metaverse and NFTs have not yet been developed, and there is a risk that it will lead to troubles between companies and individuals. For example, there is currently a lack of progress in various legal developments, such as the taxation system for NFT-related businesses, the ownership of digital assets, the handling of harassment and slander, and the issue of inclusion.

(4) **Improvement of Performance and UX of VR/AR Devices**

It is an advancement in metaverse-related technologies. The Metaverse is a service made up of technologies in various fields, but the development of technology has dramatically improved the value of experience in recent years. Specifically, improvements in communication technology, computer processing performance, and the emergence of devices for the Metaverse have made the experience in the Metaverse space smoother (Fig. 7).

(5) **High growth potential of the Metaverse market**

The growth of the metaverse market requires a variety of human resources, such as engineers who support the Metaverse from behind the scenes, creators who create content on the Metaverse, and business developers who promote the business use of the Metaverse. One of the most Important aspects is creators who create content in the Metaverse. It is no exaggeration to say that the value of the metaverse experience is primarily determined by the number of users who gather in the Metaverse and the number of attractive contents that attracts them. Therefore, the extent to which we can cultivate creators who can create captivating 3D content on the Metaverse is an essential key to the spread of the Metaverse. For example, Roblox, one of the leading metaverses, is made up of user-created game titles, as it is called the game version of YouTube, and has successfully utilized UGC (User Generated Contents) and has grown to the point where about 200 million people use it. Companies such as Meta and Microsoft overseas and Psychic VRLab in Japan are already investing in creator development and operating programs, and it is believed that the development of many creators will progress in the future.

It is monetization by companies. Many companies, from big tech companies such as Meta to leading companies in a wide range of industries in Japan, are actively

| Company Name | PICO Technology Company | Meta Platforms Company | Sony Interative Entertainment Company | Meta Plaforms Company | HTC Company | Apple Company |
|---|---|---|---|---|---|---|
| Model Name | PICO 4 128GB | Quest 2 128GB | PlayStation VR2 | Quest Pro | VIVE XR Elite | Apple Vision Pro 256GB |
| Appearance | 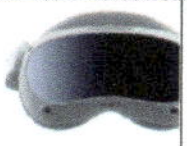 |  |  |  |  | 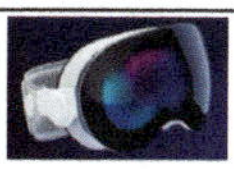 |
| Price range | Around $350 | Around $350 | Around $470 | Around $1100 | Around $1200 | Around $3499 |

**Fig. 7** Examples of major HMDs (Until May 2024). *Source* PICO 4 128 GB: https://www.picoxr.com/jp/products/pico4 [20], Quest2128GB: https://www.meta.com/jp/quest/products/quest-2/ [21], PlayStation VR2: https://www.playstation.com/ja-jp/ps-vr2/ [22], Quest Pro: https://www.meta.com/jp/quest/quest-pro/ [23], VIVE XR Elite: https://htcvive.jp/item/99HATS004-00.html [24], Apple Vision Pro: https://www.apple.com/apple-vision-pro/ [25]

investing in the business utilization of the Metaverse. This is because various applications that lead to improved profits are being considered, such as business creation in the metaverse area, efficiency of work sites, and use for multiple simulations.

On the other hand, the Metaverse is still in its infancy as a technology, and not many companies have linked it to significant monetization. Therefore, every time a company succeeds in establishing a business model or ecosystem that leads to medium- to long-term earnings, each company's investment in the Metaverse will accelerate, and it will make a significant contribution to the development of the market.

## 7 Japan's Special Position and Potential Influence in Metaverse Development

Japan holds a unique position and significant potential influence in developing the Metaverse—the economic role of NFT technology in Japanese games.

NFT-backed innovations possess unique identities and statuses [26] NFTs can manifest in various guises, encompassing avatars, digital accessories, and exclusive digital compilations that highly value product uniqueness, unattainability, and exclusivity [27]. These tokens bestow a sense of digital rarity upon digital possessions or assets, simultaneously verifying their authenticity, authorship, and ownership credentials [28] In the Metaverse, Japanese games leverage the NFT technology to bolster user economic activities, as demonstrated by several vital aspects. Firstly, establishing digital asset ownership through NFTs ensures the uniqueness and authenticity of in-game items, allowing players to own and trade these assets, such as exclusive skins or characters. This ownership extends to cross-game asset interoperability, where NFTs enable items to be exchanged and used across various games and platforms, amplifying their value and players' economic engagement. Creating virtual markets and transaction systems within Japanese games facilitates player interaction and introduces new revenue streams. These markets are underpinned by incentive and reward mechanisms, where NFTs motivate players to complete tasks and advance through the game, enhancing engagement and loyalty. Additionally, the monetization of art and collectables through NFTs transforms digital creations into tradable assets, further expanding the economic scope of the Metaverse.

NFTs act as a bridge between the game economy and the real economy, enabling players to liquidate their virtual assets for real-world currency and thus increasing the assets' economic value. The immutable nature of NFTs also offers robust intellectual property protection for game developers, fostering innovation and investment. Moreover, integrating NFTs with decentralized and game finance introduces new financial products and services, adding depth to the Metaverse's economic framework.

Japanese games utilize NFT technology to enhance user economic activities in the Metaverse, driving the growth of the virtual economy and generating new opportunities and value for players and developers alike. NFTs reshape the value of luxury [29]. Luxury brands have launched different NFT marketing campaigns since late 2020 [30, 31].

## 8 Current Status of the Japan Metaverse

The current state of Japan's Metaverse is rapidly evolving but not yet fully established. Here are some points about the current state of the Japan metaverse: First, regarding the spread of virtual reality (VR) and augmented reality (AR) technologies are spreading in Japan, and metaverse spaces utilizing these technologies are gradually spreading. With the increasing use of VR headsets and AR applications, businesses and individuals are exploring the Metaverse for various purposes. The Japanese entertainment industry also uses the Metaverse to provide new content and experiences. Content such as anime and games have been integrated into the Metaverse, enabling interactive communication with fans.

Japanese companies are also making efforts to utilize the Metaverse for business. From remote work to online events and virtual shops, the Metaverse is helping to expand and improve the efficiency of business activities in various fields. The Japan metaverse market faces multiple challenges, including privacy and security issues, improving user experience, and developing interactive communication tools. In response to these issues, efforts are being made to enhance technological innovation, establish laws and regulations, and educate users.

The metaverse market in Japan is currently witnessing vigorous growth, fueled by increasing interest from both businesses and consumers. Looking ahead, as technology continues to advance and new services come to the forefront, there is anticipation that the metaverse market in Japan will undergo even greater expansion.

## 9 Metaverse Business in Japan Companies

### *9.1 Sony*

Sony, which develops various businesses such as music, games, and anime, emphasized its entry into the metaverse business at its management policy briefing in May 2022. It will be combined with game technology, the core of Sony's business, to create a new entertainment experience and build a live network space where games, IP, and music intersect.

Sony plans to acquire a company that provides live services, works with sports teams to make stadiums virtual, and develops a next-generation VR system for PlayStation 5. In addition, a service will be provided that uses a camera system that supports sports judgment to reproduce movements captured in live video in virtual space in real time.

Sony is a leading electronics and video game company, Japan's top tech and media conglomerate, and a major film and TV producer with a vast cash reserve and iconic IPs like Spider-Man and Men in Black. In August 2021, Sony acquired the animation streaming company Crunchyroll, which added an extra bargaining chip to the layout of its Metaverse; combined with a considerable number of IPs backed by a vast library of high-quality content resources, Sony is expected to play an essential role in the metaverse era of content supply, while this business will become the focus of the company's potential growth.

Sony has extended its gaming business into metaverse development, investing $1.45 billion in Epic in 2020 and 2021 and invested $1 billion in the Fortnite game, with a combined contribution of 4.9%. Epic Games, which has more than 300 million users, will be used as a base to build the foundation of the Internet virtual space "metaverse". Sony seeks to become a platform provider in the entertainment field, just like the American IT giants. In addition to games, it has begun to function as a multifunctional space by hosting concerts by famous artists. Over the past few decades, Sony has also been working on developing technologies and products such as VR and developing related content. The accumulated reserves of related capabilities are considered an advantage for Sony's metaverse business. Sony has put forward a long-term vision of increasing the number of users directly contacted through the sale of games, movies, music, etc., to one billion people. The company plans to make games and other proprietary content the center piece of its efforts to win users in the Metaverse.

Epic Games develops its games and provides computer graphics (CG) production technology from outside. Sony, specializing in CG production in software development for PlayStation, believes combining the two companies' technologies will give it an advantage in technologies such as spatial reproduction, which is required for the Metaverse.

There are five directions for Sony's game technology to be extended to the metaverse development:

1 Launch of the latest VR head-mounted display, PlayStation VR.
2 Launch of mobile motion capture copy for 2 Metaverse.
3 Implementation of state-of-the-art XR real-time projects in virtual space.
4 Recreating Manchester City Stadium on VR.
5 The development of Location, a voice-enabled AR app, and having fun in the streets and facilities. Metaverse technology upgrades will lead to better immersive experiences and applications.

## *9.2 Ricoh*

Ricoh, which develops and manufactures multifunction printers and printers for offices, has created a "Ricoh Virtual Workplace" service to realize a world where people can work more conveniently than in real life. Ricoh Virtual Workplace is a service that reproduces any space in VR and allows everyone to gather in the same space using a VR headset. By recreating the office workspace in VR, you can communicate as if you were in the same office, even when working from home. In addition, there are examples of use in the construction industry. There is also a way to use 3D data that you usually use to appear in a virtual space, and people with different companies and departments, such as orders, designers, and builders, can discuss it in the same virtual space. Ricoh's services can be said to be a metaverse business with BtoB business in mind.

## *9.3 Panasonic*

Panasonic, which develops, manufactures, and sells various types of electrical equipment such as home appliances and air conditioners, announced in January 2022 three products that symbolize a full-scale entry into the metaverse business from Shiftall Inc., a 100% subsidiary of Panasonic. The MeganeX VR headset is an ultra-high-resolution and ultra-lightweight VR headset. It is lightweight so that heavy users, who spend more than 2000 h a year in the Metaverse, can wear it comfortably. In addition, Panasonic announced the wearable cooling and heating device "Pebble Feel" and the metaverse-compatible microphone "Mutalk" with a sound leakage prevention function. In the future, Panasonic will utilize its advanced technological capabilities to develop further new products related to the Metaverse.

### 9.4 Cluster

Cluster is a metaverse platform operated by Cluster Inc. As of July 2022, the number of app downloads has exceeded 1 million, and the cumulative attendance at the event is about to exceed 10 million. Cluster is characterized by the fact that you can easily enjoy it from the app even if you don't have a VR headset. In clusters, users can create virtual spaces called "worlds" by themselves, and 44,000 original worlds have been created. Some heavy users who log in 6–7 days a week have an average stay time of more than 4 h per day. Another prominent feature is that the number of users who spend their leisure time in the virtual space is increasing daily.

### 9.5 SATCHX

SATCHX is an application that allows you to enjoy AR/VR content provided by KDDI. In addition to the conventional AR function that displays characters on magazines and cards, you can experience 3D content appearing in space. The app has exceeded 5 million downloads. It also functions as a QR code reader, and it is also unique in that it has uses other than metaverse platforms. It also gives you easy access to over 10,000 AR/VR content created by artists and creators worldwide. In the future, we plan to implement "Spatial Awareness XR Content" that recognizes the surrounding space, such as the cityscape, displays content, and allows users to experience the same content from real and virtual perspectives.

### 9.6 Reality

REALITY is a virtual live-streaming app operated by REALITY Inc., a subsidiary of GREE. The number of app downloads is about 8.5 million. Anyone can transform into their favorite avatar with a single smartphone and enjoy live streaming, games, and communication without showing their face. It is also possible to read facial expressions with the in-camera of a smartphone, link the facial expressions of avatars, and collaborate with multiple people. Communicating by sending gifts and comments during live streams and participating in virtual communities among users have also become popular. Another prominent feature is that you can collaborate with the cluster and share your REALITY avatar with the cluster for distribution (Fig. 8).

**Fig. 8** REALITY. *Source* https://www.capturingreality.com/ [35]

## 10 Virtual Space and Games Have a Strong Affinity

In a broad sense, the Metaverse refers to virtual space itself, which is said to be compatible with the entertainment business of creating one's own alter ego. In particular, game content is loved by people of all ages, and it is not unusual for it to be the first opportunity for people to come into contact with the Metaverse. Through the experience of game content, it is possible to reduce resistance to virtual space and draw people directly into the world of the Metaverse. In recent years, games that incorporate elements of the Metaverse are gaining momentum, and there have been cases of collaboration events with famous artists. Events are being held to attract users with core game content and utilize the virtual space.

Nintendo values immersive experiences in games. Nintendo's Assemble! (Animal Crossing: New Horizons) is not designed explicitly for the Metaverse, but it plays a role in the concept of the Metaverse. Some observers and players have recognized the game as relevant to the metaverse concept due to its open world, social interaction, and high degree of freedom, and it has over 15 million players (Fig. 9).

Toppan Printing is actively promoting its business in the metaverse field by utilizing its deep know-how and expertise accumulated in the printing industry to create social value and solve social problems. Toppan Printing is a leading Japanese printing company and one of the first to enter the metaverse industry, mainly because of the availability of many metaverse-related technologies and the valued attention on new social platforms.

It mainly carries out the following businesses in the Metaverse:

1. Metaverse shopping center.
2. VR culture and tourism metaverse.

**Fig. 9** Animal crossing: New horizons. *Source* https://www.nintendo.com.hk/switch/animal_crossing_new_horizons/ [36]

3. Enterprise-oriented metaverse platform construction.
4. Digital human automatic generation service.

With the metaverse service platform "Mira Verse" and avatar management platform "AVATECT" as the core, the company aims to build an intelligent society without boundaries by providing safe, reliable, and realistic services. The company strives to create new lifestyles and business models in this society, allowing everyone to realize their talents fully.

The company's service combines the recreation of a real-time illuminated virtual space with a unique experience that can only be realized in a virtual space. Remote users can enter a virtual illuminated space anytime, anywhere, and move and interact freely as avatar characters. In addition, by integrating location information from fixed cameras, guides wearing "IoANeck," and doppelganger robots, users can synchronize their interactions with virtual representations of these entities in the virtual space, realizing an experience that feels like they are there.

Toppan Printing also participates in the "Digital Twins" world travel program, part of the "LINKSPARK Osaka" co-creation space developed by Nishi-Nippon Telegraph and Telephone Corporation. Starting in April 2023, the project will begin empirical experiments to verify the effectiveness of Digital Twin technology in providing remote face-to-face guidance. Compared to video distribution and web conferencing systems, Digital Twin technology can reflect real-world events into virtual space in real time, providing users a new remote experience.

The "Digital Twin" is a technology and concept for reproducing the natural world in cyberspace, so named because it mirrors the natural world in virtual space.IoANeck is a wearable communication device that provides a live experience from the wearer's point of view, enabling immersive interactions regardless of whether people are off the beaten path or far away. Interaction.

Through these innovations, Toppan Printing is driving the development of metaverse services that offer users a new way of communicating and experiencing. Among them are "Digital Twin World Trip," "IoANeck," "Mira Verse," and "AVATECT. "AVATECT" is a registered trademark of Toppan Printing Co.

Hakuhodo, a leading Japanese advertising company, has successfully expanded into the Metaverse based on its deep experience in advertising. In May 2022, Hakuhodo's Digital Advertising Alliance was the first to launch its business in the Japanese domestic market, realizing a breakthrough in metaverse publicity through selling advertisement slots on the U.S.-based online gaming platform "Roblox". By advertising sale slots on Roblox, an American online gaming platform, Hakuhodo made a breakthrough in metaverse promotion. This innovative advertising sales model, the first of its kind in Japan, not only breaks the constraints of traditional advertising in terms of cost and physical limitations but also opens up unlimited creative possibilities in the Metaverse.

In the Metaverse, the production and display of advertisements are no longer limited by the physical constraints of the natural world, making the creation and implementation of advertising content more flexible and cost-effective. In addition, to further enhance the impact and accuracy of the advertisements, BoBaoTang adopts advanced tracking technology, which is not only able to monitor the user's behavioral trajectory in the virtual environment but also accurately capture the user's perspective movement while wearing VR glasses, thus providing data support for the optimization of the advertisement effect. Hakuhodo has demonstrated its industry-leading insight and innovation in Metaverse's advertising business through these innovative attempts.

Dentsu, a well-known Japanese advertising company, has gained insight into the strengths and weaknesses of advertising in the development of Metaverse. The company is working to apply its extensive advertising experience to the virtual world to realize more efficient campaigns, although this attempt is still in progress. Dentsu has established a cross-functional organization, "XRXSTUDIO," which focuses on using advanced technologies such as the Metaverse to support the growth and development of businesses.

The head of XRXSTUDIO, Jin Kim, profoundly understands the uniqueness of Metaverse advertising. The director pointed out that the effectiveness of traditional billboards may be limited in the Metaverse due to factors such as viewing angle, resolution, and focus adjustment that are unique to the Metaverse environment. These factors may cause the user's attention to be focused on objects that are not in the core field of view. Therefore, to improve the effectiveness of advertisements, it is necessary to design innovative advertising strategies that can attract users and thus achieve better marketing results in the emerging field of Metaverse.

## 11 Specific Applications of AI in the Business of Enterprise Metaverse Development

Japanese social networking giant GREE's investment and development plans in the metaverse space demonstrate its importance in this emerging market. In August 2021, GREE planned to invest 10 billion yen in developing its metaverse business and set a goal of securing hundreds of millions of users globally in two to three years. This demonstrates GREE's optimism about the growth prospects of the Metaverse and its willingness to invest significant financial resources in it. GREE utilizes AI technology to create a more engaging and immersive user experience in the Metaverse while driving social interaction and user engagement.

GREE's metaverse business consists of three main parts: the platform business (REALITY), the enterprise metaverse business (REALITY XR Cloud), and the VTuber business (REALITY Studios). AI technology is critical in providing personalized virtual experiences and services in these business areas. For example, in the platform business (REALITY), GREE realizes the integration of the Metaverse with daily life scenarios by creating different virtual spaces and incorporating elements such as personal rooms and pets, which requires AI technology to enhance the personalization and interactivity of the user experience. In addition, GREE's VTuber live-streaming platform, REALITY, utilizes AI technology to generate avatars and provide customized live-streaming content. The application of AI in this area may include motion capture and real-time speech synthesis to enhance the interactivity and realism of live streaming; REALITY's platform emphasizes real-time voice communication based on avatars, where AI technology may be used for speech recognition and natural language processing to provide a more natural and smooth communication experience; GREE's metaverse operation and development emphasizes the combination of 3D space and user-generated content (UGC).AI may play a role in content review, recommendation systems, and content creation tools to encourage user participation and creativity; to reach a broader range of users, REALITY has opted to develop lightweight metaverse applications for smartphones. AI technology may be used to optimize the user experience in this process, including interface design and performance adaptation, to ensure that a wide range of users can easily access the Metaverse; REALITY proposes the concept of a "Creator Economy", whereby AI technology can help users to earn financial income by participating in games and selling original virtual goods, such as through smart contracts and virtual currencies. REALITY proposes the concept of a "creator economy", where AI technology can help users generate economic income by participating in games and selling original virtual goods, for example, through smart contracts and virtual currency trading support; GREE's metaverse platform emphasizes the influence of Japanese anime culture, and AI may play a key role in content creation, image generation, and stylistic adaptation to appeal to the target user group.

## 12 Status of Overseas Companies Investing in Japan's AI Industry

Microsoft has declared a substantial investment of $2.9 billion, approximately 440 billion yen, into its Japanese division to address the escalating demand for generative artificial intelligence. This capital will be instrumental in reinforcing the company's data centers, which are vital for AI development, and setting up a new research hub, marking Microsoft's largest-ever investment in Japan. Over the next two years, the funds will enhance Microsoft's data center capabilities in Tokyo and Osaka, thereby increasing their computational power and integrating cutting-edge AI semiconductors such as GPUs. This initiative also involves establishing a Tokyo-based research center that addresses societal challenges through AI and robotics research.

Microsoft's commitment extends to educational initiatives, with plans to train engineers proficient in AI and to support three million individuals through training programs for re-learners, including non-traditional employees, women, and AI developers. The company also aims to bolster collaboration with the Japanese government on cybersecurity and threat intelligence sharing. During a meeting, Brad Smith, Microsoft's President, discussed this strategic investment with Prime Minister Kishida, highlighting the importance of leveraging AI to overcome demographic challenges and build the technological infrastructure necessary for Japan's future. Prime Minister Kishida expressed his appreciation for Microsoft's investment and the mutual benefits of collaboration in the digital space. In an exclusive NHK interview, Smith emphasized the rapid progression of AI in Japan and the significant skills gap that needs to be addressed. He sees the investment as essential for Japan's future and an opportunity to democratize AI skills nationwide.

The investment by Microsoft is part of a broader trend where U.S. IT giants are expanding their presence in Japan in anticipation of increased demand for AI services. AWS, for instance, plans a $14.9 billion investment in Japan by 2027 to bolster its data centers and generative AI capabilities. Google has already established a data center in China. Japanese companies like NTT and NEC focus on developing AI systems tailored for the Japanese market, emphasizing language processing and specialized AI applications. According to JEITA, the demand for generative AI services in Japan is projected to grow exponentially, reaching 1.77 trillion yen by 2030, indicating a continued trend of investment in this sector by various companies [32].

U.S. software giant Oracle has announced plans to invest $8 billion in its Japanese business over the next ten years. With the popularity of AI, the demand for data centers has grown, and in this area, large U.S. companies have been investing in Japan, one after the other [33].

## 13 U.S.-Based OpenAI Opens the Japan Office

OpenAI's Japan office is relevant to the Metaverse, as it positions the company to leverage Japan's technological prowess and cultural creativity. By integrating advanced AI capabilities into the Metaverse, OpenAI can enhance user experiences through personalized content, sophisticated interactions, and immersive simulations. The Japan office also serves as a strategic hub for research and development, fostering collaborations with local tech industries and contributing to creating a diverse and inclusive virtual world. This presence in Japan underscores OpenAI's commitment to shaping the future of the Metaverse with cutting-edge AI technologies that resonate with a global audience. OpenAI, a developer of interactive artificial intelligence (AI) such as "ChatGPT," announced on April 14, 2024 that it has opened "OpenAI Japan" in Tokyo as its first Asian base. With the center's opening, the company aims to develop secure AI that meets Japan's specific needs and creates new opportunities in cooperation with the Japanese government, companies, and research institutions. OpenAI CEO Sam Altman met with Prime Minister Fumio Kishida in 2023 to express the company's intention to open a Japanese base.

Tadao Nagasaki, who served as president of Amazon Web Services Japan, was appointed president of OpenAI Japan. As a first step, the company has launched enterprise early access to the ChatGPT-4 custom model, which runs up to three times faster than its predecessor and is optimized specifically for the Japanese language. The custom model will be released more widely in the coming months.

TechCrunch (April 15, 2024), an industry publication, reported that while ChatGPT has been available in multiple languages for some time, optimizing the model specifically for the Japanese language should allow for a better understanding of Japanese nuances, including cultural backgrounds, making it more effective for customer service and content creation, among others. The report says this should make it more effective, especially in customer service and content creation. OpenAI Japan has also announced that its efforts in Japan will bring it closer to major companies such as Daikin, Rakuten, and Toyota Connected, which have already deployed the ChatGPT Enterprise, as well as help accelerate local governments' efforts to use technology to improve the efficiency of public services. The company has also announced that it will help local governments accelerate their efforts to use technology to enhance the efficiency of public services. As an example of a local government, Yokosuka City in Kanagawa Prefecture reported that nearly all of its employees had access to ChatGPT, and 80% of its employees had seen an increase in productivity.

This is just the first step in a long-term partnership with the Japanese people, government leaders, businesses, and research institutions, Altman said in a video message at the opening of the Tokyo office. This comes shortly after Microsoft, OpenAI's largest partner, announced on April 9, 2024 its cooperation with Japan in strengthening its AI and cloud infrastructure, and investment in Japan in the AI field continues (see April 16, 2024, article).

## 14 Japanese Influence in Anime, Games, and AI

Japan's anime is highly regarded worldwide, and its production techniques and storytelling are born from its unique cultural background. Japan's anime encompasses various genres and target audiences, and its deep storytelling and character designs have influenced fans and creators. Anime conventions and film festivals are held worldwide, and Japanese anime is recognized as a unique cultural soft power.

Japan game developers have unique designs, rafixes, music, storytelling, etc. In particular, Japan's role-playing games (RPGs) and arcade games are globally famous, and Japan's gaming culture is highly regarded worldwide.

Japan is also one of the world leaders in the research and development of AI technology. Japan companies and research institutes are making advanced efforts in robotics, autonomous driving, natural language processing, and image recognition. Japan's AI technology has brought about innovation in the industrial and service fields, and its results are attracting worldwide attention. Japan is also actively working on AI ethics and law and demonstrating leadership for properly using global AI technology.

Japan's strength and global influence in these areas are backed by advanced technological capabilities, unique cultural backgrounds, creative approaches, and years of research and development. As a world leader in these fields, Japan is expected to continue its innovative activities and expand its influence.

## 15 Centralized Metaverse Versus Decentralized Metaverse

The centralized Metaverse and the decentralized Metaverse represent two distinct approaches to creating and managing virtual worlds and digital ecosystems.

In a centralized metaverse, a single entity or corporation typically controls and owns the virtual environment. This entity governs the rules, infrastructure, and economy of the Metaverse, acting as a central authority. Users interact within this controlled space, often subject to the terms and conditions set by the central authority. The controlling entity unilaterally makes decisions regarding content moderation, virtual asset ownership, and overall governance. Examples of centralized metaverse platforms include proprietary virtual worlds and online gaming environments where the company behind the platform retains significant control over user experiences (e.g., Meta's Horizon Worlds, Roblox, Minecraft, Fortnite).

On the other hand, the decentralized Metaverse operates on principles of decentralization, blockchain technology, and distributed consensus. In a decentralized metaverse, there is no single controlling authority. Instead, the Metaverse is built on decentralized networks and protocols, where governance is distributed among a community of users and developers. This model prioritizes principles such as transparency, interoperability, and user empowerment. Decentralized autonomous

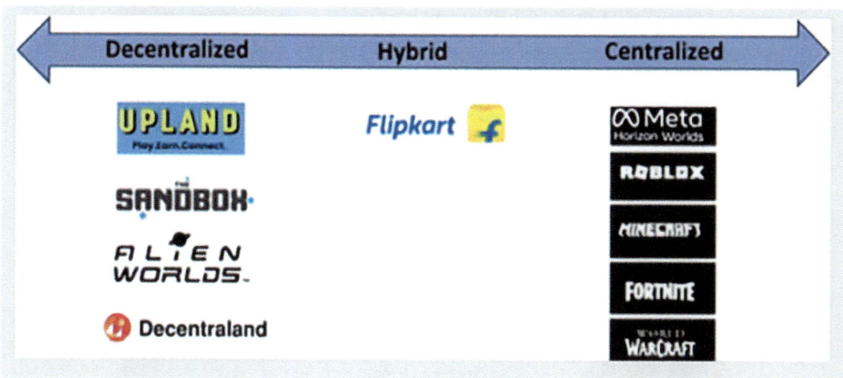


**Fig. 10** Decentralized Metaverse

organizations (DAOs), smart contracts, and community-driven consensus mechanisms often determine content creation, asset ownership, and governance decisions. Users have greater autonomy and ownership over their digital assets and experiences. Decentralized metaverse platforms aim to create open, interconnected virtual environments that transcend individual platforms and enable seamless interactions across different applications and ecosystems. Some examples of decentralized metaverses include Upland, Sandbox, Alien Worlds, Decentraland, etc.

While centralized metaverse platforms offer curated experiences and centralized control, decentralized metaverse platforms emphasize decentralization, user ownership, and interoperability. Both approaches have their advantages and challenges, and the ongoing development of the Metaverse will likely involve a hybrid metaverse, which is a combination of centralized and decentralized elements as the concept continues to evolve. For example, FlipVerse is a hybrid metaverse platform created and managed by Flipkart with NFTs issued on Polygon blockchain (Fig. 10).

## 16 The Future of the Metaverse in Japan

The metaverse market is still in its infancy as a three-phase period and is expected to develop through three phases: the "dawn period," "popularization period," and "establishment period" around 2040. Let's look at each of the phases in an easy-to-understand manner (Fig. 11).

| phase name | period | outline |
|---|---|---|
| dawn period | ~2025 | With the evolution of technological elements and increasing social needs, many general users are showing interest in Metaverse. |
| Popularization period | 2025～2030 | Due to the enrichment of services provided onMetabass, Metaverse will suddenly spread into people's lives. |
| Settling period | 2030～ | Elemental technologies have fully matured, and many people, regardless of age or gender, will be working in theMetaverse space as a matter of course. |

**Fig. 11** The Metaverse Market. Metaverse Research Institute. *Source* https://metaversesouken.com/metaverse/market/#:~:text= [34]

## 17 Japan Organizations And Metaverse Professional Societies Supporting Metaverse Businesses

Many organizations have been launched in Japan to support the development of the metaverse business, including the International Metaverse Association and Metaverse Japan.

The International Metaverse Association was established in May 2022, with the Metaverse's position being "a space full of possibilities that can realize things that cannot be done in the real world depending on the idea". "To become a supporter of the metaverse-related businesses that will be born in the future, the association continues its activities such as study sessions, business matching, joint research with universities, regular international metaverse competitions, and forums where you can talk with experts from around the world. On the other hand, the International Metaverse Association" will serve as a hub for widely sharing cutting-edge information and worldviews across industries and companies and will serve as a foundation for discussing the new concept of the Metaverse.

By carrying out various activities to unleash the power of Japan's IP and artists and the power of new entrepreneurs in the metaverse era, we aim to make the power of Japan shine in the global market and user community.

## 18 Japan's Metaverse Business Has the Potential for Rapid Growth

Japan's metaverse business has only room for growth. The number of users who use the Metaverse is also minimal, and only a few companies and platforms are involved in the Metaverse business. However, Japan has advanced technological capabilities that support high economic growth and Japan's unique anime and manga culture. If we can brand original content using Japanese anime and manga characters overseas, it will not be long before Japan's metaverse business shines again. As the Metaverse rapidly penetrates the world, the iterative development of AI generation significantly impacts the Japanese Metaverse industry.

## 19 Problems in the Development of the Metaverse

The infrastructure development of cloud-based databases also plays a crucial role in influencing the efficiency and effectiveness of usage. Ensuring data security and compliance with relevant standards are paramount considerations in the development of the metaverse.

Regarding the key factors driving the development of the metaverse market, device evolution and compelling content remain central.

Advancements in device technology, exemplified by products like VR goggles and AR glasses led by industry pioneers such as Meta Quest, are instrumental in expanding the metaverse market. While current devices like the Meta Quest have made significant strides in terms of size and weight reduction compared to a decade ago, the challenge of prolonged wear discomfort persists due to their bulkiness. A future where these devices achieve a weight and form factor akin to sunglasses holds the promise of bringing us closer to a sci-fi realm where the boundaries between reality and the virtual blur, seamlessly integrating the device into our daily lives.

Equally critical is the availability of hit content that extends beyond gaming experiences, catering to a broader audience encompassing communication, entertainment, and productivity. Presently, the metaverse landscape is predominantly saturated with VR games, which predominantly cater to gamers. To drive widespread adoption among the general populace, the introduction of captivating content in various domains is essential. The emergence of hit content that resonates with diverse user segments is expected to trigger a rapid expansion of the metaverse. Drawing from the success of gaming platforms like Nintendo DS and Wii, where titles such as Mario, Pokemon, and Super Smash Bros. played a pivotal role in popularizing the devices, underscores the pivotal role of engaging content in driving metaverse adoption across different demographics and interests.

Vision Pro, a visionary tech company at the forefront of immersive experiences, recognizes the significance of infrastructure and content in shaping the future of the metaverse. By leveraging cutting-edge technologies and collaborating with creators to develop diverse, captivating content, Vision Pro is dedicated to creating a compelling metaverse experience that appeals to a wide audience. Moreover, Vision Pro places a strong emphasis on data security and compliance with industry standards to ensure a safe and reliable metaverse environment. Through Vision Pro's commitment to technological innovation, content diversity, and data integrity, the metaverse is poised to evolve into a vibrant digital ecosystem that revolutionizes user interactions with virtual worlds.

## 20 Conclusion

The Metaverse stands at the forefront of digital innovation, with vast potential and significant challenges. While it's true that not everyone is currently participating in the Metaverse, the landscape is swiftly changing. Investments from corporations are surging, and legislative frameworks are beginning to take shape, signaling a solid belief in the Metaverse's future impact. These developments suggest that the current barriers and uncertainties surrounding the Metaverse are not insurmountable but temporary hurdles that will likely be overcome as the sector matures.

Corporate investment in the Metaverse is not just about financial influx but also about validating the Metaverse as a viable field for innovation and economic activity. Companies are exploring new business models, customer engagement strategies, and operational efficiencies within this digital space. This investment is a testament to the belief in the Metaverse's potential to redefine industries from retail and entertainment to education and healthcare. As more companies enter the Metaverse, their collective efforts will likely accelerate the development of technologies, standards, and best practices, making the Metaverse more accessible and valuable to a broader audience.

The role of domestic and international legislation cannot be overstated. Financial risks are introduced by the speculative nature of investments in metaverse-related assets, including NFTs. As the Japanese government's initiative illustrates, recognizing and fostering digital innovation sectors like Web3 and the Metaverse is crucial. However, the development of specific laws governing the Metaverse and NFTs is still in its infancy. The challenge for lawmakers is to create regulations that protect users and promote fairness without innovation. This includes addressing complex issues such as digital asset ownership, privacy, cybersecurity, and the ethical use of technology. As legal frameworks evolve, they will provide the necessary stability and security for individuals and businesses to invest in and explore the Metaverse confidently.

Discussing and experimenting with the Metaverse's business applications today is more than an exercise in speculation; it's a foundational step toward building the future. These early explorations and implementations are critical for understanding what works, what doesn't, and what's possible. Moreover, they serve as a crucial learning phase for businesses, helping them adapt to and shape the emerging digital era. Inclusion remains an essential concern, ensuring that the Metaverse becomes a space where diversity is celebrated, and all participants have equal opportunities. The development of the Metaverse should be guided by accessibility, inclusivity, and ethical considerations to realize its potential as a transformative force for society. In the future, the primary focus should be on guiding young people's learning and interest in AI, as well as assisting them in developing correct cognition and values regarding the metaverse, AI, and virtual currency. Guide them to enjoy the convenience of technology, establish a proper sense of morality, and contribute to constructing a harmonious and beautiful cyberspace. Simultaneously, various countries around the world will host youth metaverse forums. As we stand on the brink of the "new era of the metaverse," it's clear that the journey ahead is filled with promise and challenges. The collective efforts of businesses, governments, and communities are essential to unlock the full potential of the Metaverse. By fostering innovation, establishing robust legal frameworks, and committing to inclusivity, we can ensure that the Metaverse becomes a vibrant, dynamic space that benefits all. Taking proactive steps today is vital for navigating the complexities of the Metaverse and seizing the opportunities of this new digital frontier.

Japan's technological development in the field of Metaverse is expected to accelerate with significant advancements in network and cloud computing technologies and the increased ability of AI technologies to generate realistic images, audio, video, and text. For example, through AI technology, more realistic virtual experiences can be created, including simulation of various sensory experiences in the real world; the global metaverse market is expected to continue to expand, and its domestic market size is also likely to grow significantly, and it is anticipated that the size of Japan's metaverse market will reach tens of trillions of yen by 2030, which will attract more companies to participate in metaverse-related development and investment; the Japanese public's Awareness of the Metaverse is gradually increasing, especially among the younger generation. As users' understanding of the Metaverse deepens, it can be expected that more users will participate in the Metaverse experience, which will drive market development and innovation. The Japanese government has begun to pay attention to the development of the metaverse industry and may introduce relevant policies to support the industry's healthy growth. For example, by formulating data-related laws and standards and promoting international cooperation, Japan hopes to occupy a dominant position in the global virtual space industry; with the rise of the concept of Metaverse, Japanese companies such as GREE and Sony have begun to invest in the field of Metaverse and layout, and it is expected that in the future there will be more companies to promote the development of the Metaverse through cooperation and innovation; the application scenarios of the Metaverse will be more and more diversified, from entertainment, education, business to industrial applications, etc., all of them may become the application fields of metaverse technology. For example, telecommuting, education and training, product demonstration, etc., through the Metaverse; with the development of the Metaverse, its impact on society will also become an important topic. Protecting personal privacy, the legal status of virtual identity, and moral and ethical issues in the Metaverse need to be discussed and solved by all sectors of society. Japan will also pay attention to international developments in the development of the Metaverse and seek cooperation with other countries and international organizations. By participating in developing and exchanging the future of the Metaverse in Japan's international standards, Japan can play a more significant role in developing the global Metaverse.

To summarize, the future outlook of Japan's Metaverse is multifaceted and involves many dimensions, such as technology, market, law, and society. With technological advancement and social acceptance, the Metaverse is expected to become a new engine for economic growth and innovation in Japan. At the same time, attention needs to be paid to the legal, ethical, and social issues that accompany its development to ensure the healthy development of the Metaverse.

## References

1. Giang Barrera K, Shah D (2023) Marketing in the Metaverse: Conceptual understanding, framework, and research agenda. J Bus Res 155. https://doi.org/10.1016/j.jbusres.2022.113420
2. Dwivedi YK, Jebabli I, Wamba SF (2022) Metaverse beyond the hype: multidisciplinary perspectives on emerging challenges, opportunities, and agenda for research, practice, and policy. Int J Inf Manag 66. https://doi.org/10.1016/j.ijinfomgt.2022.102542
3. Park S, Kim S (2022) Identifying world types to deliver Gameful experiences for sustainable learning in the Metaverse. Sustainability 14
4. Hollensen S, Kotler P, Opresnik MO (2023) Metaverse—The new marketing universe. J Bus Strategy 44(3):119–125. https://doi.org/10.1108/JBS-01-2022-0014
5. Almoqbel MY, Naderi A, Wohn DY, et al (2022) The Metaverse: a systematic literature review to map scholarly definitions. In: Companion publication of the 2022 conference on computer supported cooperative work and social computing, pp 80–84
6. McKinsey (2022) Value creation in the Metaverse. McKinsey & Company. https://wwwmckinsey.com/capabilities/growth-marketing-and-sales/our-insights/value-createon-in-the-metaverse. Accessed 12 July 2023
7. Judy J (2017) Information bodies: computational anxiety in Neal Stephenson's snow crash. Interdisciplinary Literary Studies
8. Lee LH, Braud T, Zhou P et al. All one needs to know about Metaverse: a complete survey on technological singularity, virtual ecosystem, and research agenda
9. Johri A, Sayal A, Chaithra N et al (2024) Crafting the techno-functional blocks for Metaverse-A review and research agenda. Int J Inf Manage Data Insights 4(1):100213
10. Polyviou A, Pappas IO (2022) Chasing Metaverses: reflecting on existing literature to understand the business value of metaverses. Inf Syst Front J Res Innovation 1–22. https://doi.org/10.1007/s10796-022-10364-4
11. Cao W, Cai Z, Yao X et al (2023) Digital transformation to help carbon neutrality and green sustainable development based on the Metaverse. Sustainability 15(9):7132
12. Lee LH, Braud T, Zhou P et al (2021) All one needs to know about Metaverse: a complete survey on technological singularity, virtual ecosystem, and research agenda. arxiv preprint arxiv: 2110.05352
13. Trunfio M, Rossi S (2022) Advances in metaverse investigation: streams of research and future agenda[C]//Virtual Worlds. MDPI 1(2):103–129
14. Hohendanner M, Ullstein C, Miyamoto D et al (2024) Metaverse perspectives from Japan: a participatory speculative design case study. arxiv preprint arxiv: 2401.17428
15. Source: https://metaversesouken.com/metaverse/market/#:~:text=
16. Source: Metaverse Research Institute
17. Source: Ministry of Internal Affairs and Communications "Study Group on the Utilization of Metaverse, etc. for the Web3 Era" Report 2023.7

18. Source: Mitsubishi Research Institute, CX2030 Metaverse in a Wide Context as a Place to Utilize Virtual Technology
19. Source: Mitsubishi Research Institute, Results of a questionnaire on metaverse cognition and use
20. PICO 4 128GB: https://www.picoxr.com/jp/products/pico4
21. Quest2 128GB: https://www.meta.com/jp/quest/products/quest-2/
22. PlayStation VR2: https://www.playstation.com/ja-jp/ps-vr2/
23. Quest Pro: https://www.meta.com/jp/quest/quest-pro/
24. VIVE XR Elite: https://htcvive.jp/item/99HATS004-00.html
25. Apple Vision Pro: https://www.apple.com/apple-vision-pro/
26. Belk R, Humayun M, Brouard M (2022) Money, possessions, and ownership in the Metaverse: NFTs, cryptocurrencies, Web3 and Wild Markets. J Bus Res 153:198–205
27. Kapferer J-N (2015) Kapferer on luxury: How luxury brands can grow yet remain rare. Kogan Page Publishers
28. Chohan R, Paschen J (2021) NFT marketing: how marketers can use non-fungible tokens in their campaigns. Bus Horiz
29. Bao W, Hudders L, Yu S et al (2024) Virtual luxury in the Metaverse: NFT-enabled value recreation in luxury brands. Int J Res Market
30. Joy A, Zhu Y, Peña C, Brouard M (2022) Digital future of luxury brands: Metaverse, digital fashion, and non-fungible tokens. Strateg Chang 31(3):337–343
31. Alexander B, Bellandi N (2022) Limited or limitless? Exploring the potential of NFTs on value creation in luxury fashion. Fash Pract 14(3):376–400
32. https://www3.nhk.or.jp/news/html/20240409/k10014416831000.html
33. https://www3.nhk.or.jp/news/html/20240418/k10014425941000.html#:~:text=
34. Metaverse Research Institute https://metaversesouken.com/metaverse/market/#:~:text=
35. Source: https://www.capturingreality.com/
36. Source: https://www.nintendo.com.hk/switch/animal_crossing_new_horizons/

# The Latest Developments in China's Metaverse

**Caichun Gong and Jin Qiu**

**Abstract** This chapter presents different understandings of the metaverse among major Chinese scholars, as well as the approval of the China National Committee for Terminology in Science and Technology on the final definition of the metaverse. The chapter then provides the attitudes and policies of governments at different levels in China toward the metaverse in different periods. Subsequently, this chapter describes famous white papers and monographs concerning the metaverse in China. Finally, it introduces major academic institutions and commercial organizations related to the metaverse in China.

**Keywords** China's metaverse · Latest development · Definition of metaverse · Attitudes toward metaverse · Action plans

## 1 Introduction

On March 10, 2021, Roblox successfully listed on the New York Stock Exchange. Its favorable capital performance ignited a strong interest in the metaverse among Chinese investors. The rebrand of Facebook as Meta on October 28, 2021, further captivated countless Chinese, sparking debates about whether the metaverse is a new round of technological scams as a mere fantasy or the future trend of Internet development and the direction for technological advancement. A plethora of scholars, experts, entrepreneurs, jurists, and economists in China started their journey of exploration, laying the foundation for the development of the metaverse in the country. On August 29, 2023, the General Office of the Ministry of Industry and Information

C. Gong (✉)
Guilin University Of Electronic Technology, #1 Jinji Road, Qixing, Guilin, Guangxi Zhuang Autonomous Region, P.R. China
e-mail: gongcaichun@honestcareer.com

J. Qiu
Dalian University of Technology, #2 Linggong Road, Ganjingzi, Dalian, Liaoning Province, P.R. China
e-mail: qiujin@dlut.edu.cn



M. Grieves and E. Hua (eds.), *Digital Twins, Simulation, and the Metaverse*, Simulation Foundations, Methods and Applications, https://doi.org/10.1007/978-3-031-69107-2_17

Technology of the People's Republic of China (MIIT) and other four government departments jointly issued the "Metaverse Industry Innovation and Development Three-Year Action Plan (2023–2025)," officially refuting the assertion that the metaverse is a technological scam. This chapter narrates the history and current state of the metaverse's development in China.

## 2 China's Understanding of the Metaverse

Since the term "metaverse" experienced the transition from a science fiction concept to a technological one, various sectors in China, including capital, media, academia, industry, and government, have explored and studied it. Some, in pursuit of sensationalism, have presented absurd and bizarre viewpoints, suggesting that "there are a thousand versions of the metaverse in a thousand people's mind, just as there are a thousand Hamlets in a thousand people's eyes [1]." The definition of "metaverse" varies, influenced by different interests, stances, educational backgrounds, knowledge levels, and work experiences.

In Chinese, the metaverse is translated as "元宇宙 (Yuan Yu Zhou)." Some interpret these three characters individually: "元" (Yuan) signifies transcendence, "宇" (Yu) space, and "宙" (Zhou) time [2, 3]. This approach resembles how some explained blockchain (区块链, Qu Kuai Lian) technology character by character in its early days. There are diverse views of the metaverse: some see it as a 3D Internet [4–6], others regard it as a decentralized new social relation [6], and still others take it as a new phase in human civilization [7].

To determine a precise definition and the conceptual boundaries for the metaverse, over 140 Chinese experts started to carry out relevant research at the end of 2021. In early July 2022, the China National Committee for Terminology in Science and Technology (CNTERM) initiated drafting the definition of "metaverse." After more than two months of research, 19 experts from academia and industry achieved 19 different definitions. On September 13, 2022, the CNTERM, with experts including academicians Li Deren and Zhuang Songlin, and Dr. Gong Caichun, reviewed and approved the core terminologies for the metaverse, thus forming the current draft for public consultation [8].

> The metaverse is a virtual world built by humans using digital technologies, which either mirrors or transcends the real world and can interact with it.

## 3 The Attitudes of China's Central and Local Governments Toward the Metaverse

Over nearly three years, the metaverse in China has gone through a transition from being misunderstood and underrated to being actively supported and promoted. The attitude of Chinese governments at all levels toward the metaverse can be divided into four phases as follows.

(1) **Phase One**: **Negative Attitudes from Both Central and Local Governments**

The first phase lasted from March 2021 to December 2021. Given the adverse impact blockchain technology had on China's real economy and the fact that blockchain would serve as one of the core technologies of the metaverse, both central and local governments were extremely cautious and generally negative toward the metaverse. By the end of 2021, with heightened enthusiasm for the metaverse in media, capital, industry, academia, and legal circles, there was a widespread anticipation for a clear stance from the central government.

In this context, the "People's Daily," the official newspaper of the Communist Party of China (CPC), published its first commentary on the metaverse on November 17, 2021, entitled "Is Everything 'Metaverse'?" [9] The commentary stated:

> The discussion about the metaverse continues, with some filled with optimism and yearning, and many others skeptical. Whether it is a fleeting illusion or a tangible future, a capital speculation or a new race track, a rehash of old ideas in a new bottle or a breakthrough in technology, it's worth the patience of "letting the bullets fly for a while" before any conclusions are drawn. However, it is clear that some new concepts have the power of affecting people's confidence in technological development and their expectations for a better future life. More importantly, it takes time for new concepts and the related industries to mature and lay a solid foundation toward an enticing technological future. Just as virtual reality, augmented reality, or mixed reality all center around "reality," without the support of reality, any concept would end up being a mirage only, or a tree without roots. "A shaky foundation will cause the mountains to tremble"—this principle applies in both the real universe and the metaverse.

The public's desire for central ministries to express their views on the metaverse grew. Surprisingly, the first national ministry to speak out about the metaverse was the Central Commission for Discipline Inspection (CCDI). On December 23, 2021, an article titled "How the Metaverse Could Rewrite Human Social Life" was published on CCDI's official web portal. It argued:

> There is no single technology known as the "metaverse"; it is a combination and upgrade of existing technologies, akin to a "3D version of the Internet." We suggest a rational view of the metaverse's role in the new round of technological revolution and its impact on society. The market's skepticism about this new concept mainly stems from the immaturity and incompatibility of technological implementation in terms of hardware conditions and network environment. Nevertheless, the ultimate version of the metaverse and its key technologies like computing power, platforms, and network construction still require a long time to be perfected. Currently, some companies' hype around the metaverse concept differs significantly from the reality of the metaverse, necessitating discernment between truth and

falsehood as well as the need for cautious judgment. From the perspectives of the participants in the industry, the metaverse will remain a goal for the next generation of Internet development for some time, depending on core technological breakthroughs and changes at the level of underlying technology and computing power.

It can be seen that, although neither "People's Daily" nor the CCDI's website explicitly defined a stance on the metaverse, the diplomatic rhetoric used in the expressions conveyed obviously negative sentiments to those familiar with Chinese language nuances. As a consequence, scholars who originally intended to conduct more research on the metaverse abandoned their plans, and many entrepreneurs started to withdraw from the field, causing a decline in the stock prices of many metaverse-related companies.

(2) **Phase Two: Supportive Views on the Metaverse Industry from Local Governments**

The second phase spanned from January to June 2022. Despite the central government's skepticism toward the metaverse and the lack of a clear stance, it is an undeniable fact that the country was suffering from a severe national epidemic and the real economy was confronted with great challenges. Starting from January 2022, local governments across China began to introduce policies to develop the metaverse industry. It is generally believed that during this phase, local governments did not have a deep understanding of what the metaverse really was or how its industry could develop. However, given the metaverse's popularity and its widespread promotion across various sectors, some daring local governments simply decided to give their support. Perhaps it is in line with former Chinese president Deng Xiaoping's famous saying, "It doesn't matter whether a cat is black or white, as long as it catches mice."

On January 5, 2022, the Office of the Digital Economy Development Leadership Group of Zhejiang Province issued the "Guiding Opinions on the Construction of Zhejiang Province's Future Industry Pilot Zones" [10], identifying a wealth of key future industrial technologies, including artificial intelligence, blockchain, third-generation semiconductors, quantum information, flexible electronics, future networks, aerospace integration, bioengineering, cutting-edge new materials, advanced equipment manufacturing, advanced energy, and the metaverse. This was the first time a local government included the metaverse in its future work plan, bringing a glimmer of hope to scholars and entrepreneurs still committed to researching the metaverse.

On January 11, 2022, Cheng Yongwen, the Mayor of Wuhan, proposed in the "2022 Wuhan City Government Work Report" to promote the integration of the metaverse and the real economy [11]. Subsequently, local governments in Beijing, Shanghai, Hangzhou, Nanchang, Wuxi, Chengdu, Guangzhou, Shenzhen, Jinan, Hefei, and Nanning successively introduced metaverse industry policies. The positive stance of local governments seemed to stand in stark contrast with the central government's negative attitude.

(3) **Phase Three: Positive Voices from Ministries on the Metaverse**

The third phase lasted from June 2022 to February 2023. During this period, central ministries began to express positive views on the metaverse. The dissemination of various white papers, such as the "China Metaverse White Paper" [4], played a significant role in scientifically popularizing the concept; stakeholders in academia, industry, government, and law developed more mature considerations of the metaverse than before.

On June 25, 2022, Wan Gang, Chairman of the China Association for Science and Technology(CAST), said when having a talk with college students at Central South University: "I hope to find a development path for the metaverse with Chinese characteristics." This marked the first time a ministry leader openly discussed the metaverse in a public setting, signaling the emergence of positive voices from central ministries regarding the metaverse.

Later, on September 13, 2022, the CNTERM held the "First Symposium on Metaverse and Core Concept Terminology," where 26 experts approved and released the national standards for the definitions of metaverse terminology. This event marked the first time the academic community positively voiced its views on the metaverse, leading to a flourishing period of metaverse-related research [8].

On February 18, 2022, the Development Research Center of the State Council (DRC) published a series of books under the theme "New Journey of Modernization," edited by Long Guoqiang, the Deputy Director of the Center. "Metaverse: New Opportunities for Digital China," [12] written by Gong Caichun and Jiang Ming, was part of this series, focusing on how local governments should develop the metaverse industry. The book received widespread acclaim at the National People's Congress (NPC) and the Chinese People's Political Consultative Conference (CPPCC), indicating that the central government's policy on the metaverse industry was urgently needed.

(4) **Phase Four: Metaverse Action Plans Formulated by Central Ministries**

The fourth phase extends from March 2023 to the present. During the 2023 NPC and CPPCC, leaders of relevant ministries and attending representatives positively addressed concerns about the development of the metaverse industry. Since then, various ministries have begun formulating strategic plans, action plans, and implementation schemes for the metaverse, prompting even those local governments that are still hesitant to introduce their metaverse industry development action plans.

On March 1, 2023, at a press conference held by the State Council Information Office under the theme "Authoritative Ministries Speak about the Start," Minister of MIIT Jin Zhuanglong stated the need to accelerate the pace of building the framework in fields like humanoid robots, the metaverse, and quantum technology. This was the first time a ministry leader expressed his support in public for the metaverse industry development, marking a significant shift in China's metaverse landscape.

On April 2, 2023, the DRC's "Economic Reference" initiated a major research project. Thereafter, experts from the DRC, the Research Office of the State Council (ROSC), the National Development and Reform Commission (NDRC), the CAST, the Chinese Academy of Management Sciences (CAMS), and the Beijing Information Industry Association(BIIA) began to carry out research on "Major Issues in the Development of China's Metaverse Industry," [13] submitting their findings as internal references to the State Council. On August 22, 2023, four departments including the MIIT, the Ministry of Science and Technology (MST), the National Energy Administration(NEA), and the National Standardization Administration Committee (SAC) jointly released the "New Industry Standardization Pilot Project Implementation Plan (2023–2035)," [14] officially designating the metaverse and brain-computer interfaces as future industries.

On August 29, 2023, several offices including the MIIT, the Ministry of Education (ME), the Ministry of Culture and Tourism(MCT), the State-owned Assets Supervision and Administration Commission of the State Council (SASAC), and the National Radio and Television Administration (NRTA) jointly issued the "Metaverse Industry Innovation and Development Three-Year Action Plan (2023–2025)" [15] which sets out the development goals for China's metaverse:

> By 2025, we should achieve breakthroughs in metaverse technology, industry, applications, and governance, establishing a significant growth pole in the digital economy. Meanwhile, we aim to expand and rationalize the industry scale, perfect the technological system, and strengthen the industry's technical foundation so as to elevate the comprehensive strength to a world-leading level. The goal is to cultivate 3–5 globally influential ecosystem companies and several specialized and innovative small and medium enterprises while creating 3–5 industrial development clusters. Initial success in industrial metaverse development, the creation of typical applications, and the formation of benchmark production lines, factories, and parks are also envisioned. Typical metaverse software and hardware products are expected to be widely applied in areas like lifestyle consumption and public services, forming a range of new businesses, models, and formats.

The introduction of the "Action Plan" signifies the official inclusion of the metaverse as a key industry in national development, just a step away from being determined as a national strategy. It is foreseeable that in the future, universities will establish numerous metaverse research institutes and colleges, and the Ministry of Science and Technology (MST) will also issue major R&D plans for the metaverse.

## 4 China's Metaverse-Related Action Plans and Implementation Strategies

Since the term "metaverse" evolved from a science fiction concept into a technological term, staunch supporters have been convinced that it would inevitably become part of national strategy, although the process might be lengthy.

In August 2021, the MIIT and the Office of the Central Cyberspace Affairs Commission (CCAC) jointly released the "Guidelines for Accelerating the Application and Industrial Development of Blockchain Technology," [16] proposing to cultivate outstanding blockchain products, enterprises, and industrial parks, to build an open-source ecosystem, and to speed up the development of a complete blockchain industry chain by addressing weaknesses and enhancing strengths. The release of these guidelines coincided with the peak of the metaverse's popularity in China.

Although the guidelines did not mention the metaverse specifically, it was widely understood that blockchain technology is one of the core technologies of the metaverse. Without doubt, these guidelines raised hopes for the development of the metaverse industry at the national level.

Shanghai, as China's largest city and pioneer in reform and opening up, also played a leading role in the development of the metaverse. As early as December 30, 2021, the Shanghai Municipal Commission of Economy and Informatization (SMCEI) officially released the "14th Five-Year Plan for the Development of the Electronic Information Industry in Shanghai," which for the first time included the metaverse in a local government's industrial plan [17]. On June 24, 2022, the Shanghai Municipal Government's Office issued the "Action Plan for Cultivating the Metaverse New Track (2022–2025) [18]," aiming to achieve a scale of 350 billion CNY(about 48.6 billion USD) in metaverse-related industries by 2025, so as to drive the city's software and information service industry to exceed 1.5 trillion CNY (about 208.4 billion USD), and the electronic information manufacturing industry to break through 550 billion CNY (about 76.4 billion USD).

As China's capital, cultural center, national hub for international interactions, and center for technological innovation, Beijing played a pivotal role in the development of the metaverse, influencing many cities' metaverse policies. On February 23, 2022, the government of Beijing's Tongzhou District issued "Several Measures for Accelerating the Innovative and Leading Development of the Metaverse in Beijing's Sub-Center." [19] The measures proposed the creation of a batch of metaverse demonstration projects and support for metaverse application scenarios, relying on the Tongzhou Industrial Guidance Fund and adopting a "fund of funds plus direct investment" approach to create a fund covering the metaverse industry. On August 22, 2022, the Tongzhou District government issued the "Metaverse Innovation Development Action Plan for Beijing's Sub-Center (2022–2024)." [20] The plan outlined the cultivation of four major metaverse industry chains and the creation of four major metaverse-empowered application scenarios, with a view to promoting the innovative development of the metaverse industry in the sub-center (i.e., Tongzhou District) and fostering deep integration of digital technology with the physical economy. The goal was to make the sub-center a demonstration area for metaverse applications featuring cultural and tourist themes and contents within three years by cultivating and introducing over 100 metaverse ecosystem companies, completing more than 30 metaverse-empowered typical application projects, and promoting the formulation of a batch of metaverse-related standards.

On October 28, 2022, the MIIT, the ME, and other departments issued a notice for the "Action Plan for the Integrated Development of Virtual Reality and Industry Applications (2022–2026)." [21] The plan emphasized the combination of virtual reality with industrial production in key vertical industries. Specific goals included promoting the deep integration of virtual reality and the industrial Internet, supporting the application and promotion of virtual reality technology in key stages of product life cycles such as design, manufacturing, operation, maintenance, and training, enhancing compatibility with digital twin models and data, and advancing the integration and intelligence of the entire industrial production process. Although the action plan did not mention the metaverse directly, virtual reality, as a source of immersive experience in the metaverse, is one of its most important core technologies. It was believed that the MIIT's documents on the metaverse were already in the making. On the same day, the Industrial Metaverse Collaborative Development Organization (IMCDO) led by the Industrial Culture Development Center (ICDC) of the MIIT was established and released the "Three-Year Action Plan for Industrial Metaverse Innovation and Development (2022–2025)." As this organization is the unit that pays the most attention to the metaverse in the MIIT, which is one of the most important national ministries, it was widely assumed that the ministry's issuance of metaverse-related documents could be in the works.

On August 29, 2023, the MIIT, the ME, and other three departments issued the "Metaverse Industry Innovation and Development Three-Year Action Plan (2023–2025)" [22], which set forth the goals for the development of China's metaverse industry. The release of the action plan marked the official inclusion of the metaverse as a key future industry in national development, indicating that significant human, material, and financial resources would be attracted to the field. The metaverse finally landed on solid ground, eliciting cheers from the media, capital, academic, and industrial sectors.

## 5 Key White Papers and Monographs on the Metaverse

On August 1, 2021, Zhao Guodong, Yi Huanhuan, and Xu Yuanchong published "Metaverse: The Future of the Internet" [23]. This was the first book related to the metaverse in China, playing a significant role in publicizing the metaverse within the country.

On November 22, 2021, Gong Caichun organized 137 well-known experts and scholars from academic, media, capital, industry, and legal sectors to draft the "China Metaverse White Paper" [4]. The whitepaper was released on January 22, 2022, after three months of preparation, causing a stir in the industry. Major media outlets like Xinhua, People's Daily, and China Network reported on the release ceremony. The "China Metaverse White Paper" comprehensively expounded on the metaverse, covering a variety of topics including concepts, origins, current status, industry, technology, law, ethics, economics, and talents. On May 27, 2022, the updated "China Metaverse White Paper (2022)" was released, with prefaces by academicians such as Zhuang Songlin, Wu Zhiqiang, Ma Jianzhang, and Chen Xiaohong. As this white paper established the main research categories of the metaverse, subsequent metaverse white papers and monographs more or less received its influence.

The exploration of core technologies of the metaverse has never ceased. "Metaverse: The Future of the Internet" posits the idea that the metaverse consists of six major technology categories: blockchain, interactive technology, game engines, artificial intelligence, network technology, and Internet of Things technology, collectively referred to as BIGANT [23]. Yan Yang, in "Metaverse Science and Technology Industry," argues that the metaverse comprises a combination of blockchain, gaming, artificial intelligence, spatial computing, cybersecurity, digital twin and digital native technologies, interactive technology, brain-computer interfaces, the Internet of Things, integration of virtual reality (VR), augmented reality (AR), mixed reality (MR), and extended reality (XR), cloud computing, chips, communication networks, edge computing, and energy reconstruction, known as the 1B, 2A, 3S, 4I, 5C, or BASIC technology system [24]. Following BIGANT, there has been a trend of creating various catchy acronyms with significant meanings, such as the BAND technology system proposed by Song Jiaji, the We Aimed to Begin system by Shen Yang, and the Transformer system by Meng Hong.

In July 2022, Ye Yurui and his co-authors published "Top Ten Technologies of the Metaverse." The book categorizes the core technologies of the metaverse into two major classes: foundational technologies, including computing, storage, network, system security, as well as artificial intelligence technologies; and pillar technologies, comprising interaction and display, digital twin and digital native, identity and economic system creation, content creation, and governance technologies [25]. In October 2022, Gong Caichun's "Metaverse: the Eve of Great Revolution" draws on the Open System Internetwork Reference Model to propose a hierarchical reference model for the metaverse. Gong divides the metaverse's technologies into the infrastructure layer, interface technology layer, basic software layer, data governance layer, digital creation layer, and application development layer (see Fig. 1) [26].

In February 2023, China Development Press published the "New Journey of Modernization" series. Mr. Long Guoqiang, Deputy Director of the DRC, served as the chief editor of the series which is also regarded as twelve recommendations proposed by the DRC to the State Council for China's economic development. "The Metaverse: New Opportunities for Digital China" is one of the books in this series [12]. Targeting leaders of counties and cities across China as its main readership, the book discusses how these administrative regions and areas can leverage their industrial advantages to develop the metaverse industry. The book was later brought to the NPC and CPPCC by Deputy Director Long and was well-received by participants.

In March 2023, the China Academy of Information and Communications Technology (CAICT) released the "Industrial Metaverse White Paper" [27]. The white

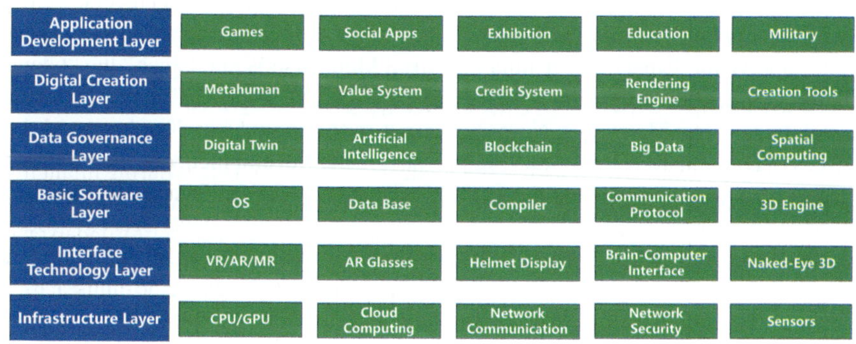


**Fig. 1** A hierarchical reference model for the Metaverse By Gong Caichun

paper emphasizes that the industrial metaverse is an important application field of the metaverse, representing a new type of industrial system, ecology, and model serving the industrial economy, based on the core infrastructure and application concepts of the metaverse. The "Industrial Metaverse White Paper" has led researchers of the metaverse to focus on industrial scenarios to address real economic issues.

On November 3, 2023, the ICDC and the BIIA jointly issued the "China Industrial Metaverse White Paper" [28]. This white paper summarizes the definitions of the metaverse and industrial metaverse, reviews the historical processes and related policies of industrial development in major industrial countries, and suggests that the world is at a crossroads of a new industrial revolution. Asserting that the industrial metaverse is a manifestation of the next generation of industry, the "China Industrial Metaverse White Paper" presents the technological framework of the industrial metaverse, describes its main application scenarios, and analyzes potential legal, ethical, and moral issues that may arise during the development of the industrial metaverse. As the book was released after the MIIT had already issued the "Metaverse Industry Innovation and Development Three-Year Action Plan (2023–2025)" [15], it has garnered widespread attention from various sectors.

## 6 Major Metaverse Research Institutions in China

In recent years, China has established over 40 metaverse-related associations, special committees, research institutes, and academies, which have played an active role in the promotion, publicization, and research of the metaverse in the country.

On June 29, 2022, the Wuhan Metaverse Research Institute was established in the Dongxihu District of Wuhan. The institute, committed to the overall planning and design of the Chinese metaverse track, is primarily engaged in the research on the metaverse's economy, social governance, ethics and morals, law, industry, and technology. Its responsibilities involve assisting in formulating China's metaverse development strategies and policy standards, organizing the drafting of metaverse-related industry standards, national standards, and international standards, writing technical monographs on the metaverse, and drafting white papers for various metaverse fields. So far this institute has issued a wealth of documents and books (including the "China Metaverse White Paper," the "China Industrial Metaverse White Paper," the "China Digital Collectibles Regulatory White Paper," "The Metaverse: The Eve of Great Revolution," "The Metaverse: New Opportunities for Digital Chin"), completed the DRC's "Economic Reference" on "Major Issues in the Development of China's Metaverse Industry," and organized the CNTERM's meeting for the review and approval of metaverse terminology.

On November 7, 2022, the MIIT approved the establishment of the National Virtual Reality Innovation Center, which is based on the Nanchang Virtual Reality Research Institute Co., Ltd. The center focuses on "industrializable key common technology R&D," addressing the key common technological challenges that constrain the development of China's virtual reality industry. It aims to build platforms for key common technology R&D, pilot verification, testing, technical services, talent training, and international cooperation, gradually constructing an industrial innovation ecosystem that covers the entire virtual reality industry chain and promotes the high-quality development of China's virtual reality industry.

On April 16, 2022, the Metaverse Culture Laboratory was officially launched at the School of Journalism and Communication of Tsinghua University. The laboratory will carry out studies in the fields of future media technology development, metaverse cultural and creative industries, metaverse index, virtual digital human index, etc., through Industry-University-Research collaboration in the hope of building itself into a domestic industry-leading, theoretically pioneering, and research and innovation-oriented metaverse scientific research institution. The establishment of the Metaverse Culture Laboratory will further open up the entire value chain of scientific and technological achievements operation, provide content support and reference for accelerating industrial layout, and ultimately empower the vigorous development of the emerging industry of metaverse.

On September 23, 2022, the School of Artificial Intelligence at Nanjing University of Information Science & Technology renamed the Department of Information Engineering to the Department of Metaverse Engineering, becoming the first department in a domestic university to be named after the Metaverse. Professor Pan Zhigeng, Director of the Department of Metaverse Engineering at Nanjing University of Information Science & Technology, has also set up the Jiangsu Metaverse Engineering Research Center and the China Metaverse Technology and Application Innovation Platform, making Nanjing University of Information Science & Technology an important hub for metaverse research, teaching, and technology transformation in domestic universities. It is hoped that Nanjing can be developed into a national innovation center for the metaverse.

In March 2023, the Metaverse Research Institute of Guilin University of Electronic Science and Technology was established, with Professor Jiang Ming serving as the dean. The Metaverse Research Institute is mainly engaged in the industrial development, technological progress, interdisciplinary research, achievement transformation and industrialization, application scenarios and case studies, and ethical research of the metaverse, helping to promote the high-quality development of Guangxi's real economy. On December 1, 2023, the Guangxi Zhuang Autonomous Region Metaverse Application Scenario Innovation Engineering Research Center was officially established in Nanning. In the future, the center will leverage the rich cultural, tourism, and vocational education resources of Guangxi, combined with the geographical advantages and characteristics of China ASEAN (Association of Southeast Asian Nations), to explore typical application scenarios and attract high-level investment and talents.

Later, Fudan University, Renmin University of China, Beijing University of Posts and Telecommunications, Guangzhou University and other institutions have also set up research centers related to the metaverse, hoping to contribute their own efforts to conducting relevant studies in this new field.

## 7 The Metaverse Progress in China's Industrial Sector

Since Roblox went public in 2021, Chinese Internet giants have been actively deploying resources in the metaverse, and a multitude of innovative enterprises related to the metaverse have emerged. At the same time, traditional enterprises have also started to explore the metaverse sector.

(1) **Tencent**

In September 2021, Tencent collaborated with Roblox to launch "QQ Music Star Town," an immersive entertainment game where players could experience virtual concerts, new song releases, music festivals, and other activities. Tencent's ZPLAN project started public recruitment in September 2021, offering up to 97 positions across technology, design, and product categories. This project is considered by the public to be Tencent's first metaverse project.

(2) **Baidu**

Baidu has not only increased its investment in the metaverse but also continuously introduced various products based on metaverse technology, such as virtual reality games and virtual idols. Xi Rang, Baidu's immersive virtual space parallel to the physical world, was officially launched at the company's AI Developer Conference on December 27, 2021, providing a virtual space for people to freely explore, create, and socialize. Baidu's layout in the metaverse field includes several aspects: the "Baidu Brain," an AI-based virtual human system for virtual idols, virtual anchors, etc.; "Xi Rang," a virtual world construction platform based on WebGL technology, which allows users to create high-quality 3D scenes through simple operations; a VR-based smart driving assistance system, developed in collaboration with the virtual reality company Alt12 through Baidu's investment; and "Baidu Map," a smart map application based on AR technology that combines the real world with the virtual world to provide more intelligent navigation services.

(3) **NetEase**

NetEase is also actively engaged in the metaverse field through investment and independent research and development. In 2018, NetEase launched a game called "Justice" based on virtual reality technology, with plans to release more related games in the future. In addition, NetEase has positioned itself in the virtual reality hardware and content production sectors through acquisitions and investments. NetEase Yaotai, a metaverse immersive digital event platform, has become the go-to for many large-scale conferences with its metaverse meeting rooms.

(4) **ByteDance**

ByteDance has launched several products based on virtual reality technology through investment and independent research and development, such as the virtual hosts on Douyin. Furthermore, ByteDance has also engaged itself in the virtual reality hardware and content production sectors through acquisitions and investments. In September 2021, ByteDance acquired PICO (a Chinese VR manufacturer founded in 2015 and known for producing high-quality VR equipment and content, including VR headsets, controllers, and software) for 9 billion CNY (about 1.25 billion USD), which was seen as a move to obtain a "ticket" to the metaverse. Highly regarded by ByteDance, PICO received investments of tens of billions of CNY from the company over the next two years.

In February 2021, ByteDance launched its game brand "Morning and Evening Light Years" and thereafter, it has not only spent several billions of dollars to fully acquire mobile game developers such as Mutong Technology and Youai Interactive Entertainment, but also heavily invested in the game engine developer Code Qiankun. On March 11, 2022, ByteDance acquired Particle, which was considered an important addition to its metaverse content ecosystem. In July 2022, ByteDance introduced an online virtual community "Party Island" in the Chinese market. Prior to this, ByteDance had already launched "Pixsoul," a social app with a metaverse concept featuring avatar customization, in Southeast Asia in September 2021.

(5) **Huawei**

Huawei has various initiatives in the metaverse domain. To begin with, it holds patents for XR-specific chips and game controllers. Moreover, the company not only has released XR hardware products, but also continues to enrich the HarmonyOS content ecosystem through independent R&D, supporting developers, and collaborating with game manufacturers. Huawei Vision Glass, an AR glasses product, offers features such as smart viewing, AR space, and surround display. Additionally, Huawei has launched the "Shougang Park Metaverse" project based on AR capabilities, combining the virtual world with the real world and allowing users to experience virtual performances, light shows, and other multimedia interactions.

To promote the development of the metaverse, Huawei has introduced the content creation platform "Huaban," which supports users in creating, rendering, and editing 3D models. The Huaban platform also provides creators such as designers and artists with creation tools and a resource library, thus further promoting the rapid development of the metaverse.

(6) **Other Innovative Enterprises**

Rayneo Innovation has a high profile in the AR glasses market for products like Air 1S, which are known for their good display effect and high clarity.

Rokid is a well-known brand in China's AR glasses field, with application cases in smart cities, industry, culture and tourism, architecture, etc., used in more than 80 countries worldwide.

Cocafe is the metaverse marketing pioneer for most MNCs in China, and the leading 3D digital assets producer in e-commerce. It utilizes proprietary 3D, AR, Web3, and other digital technologies, along with generative AI capabilities, to provide one-stop services from digital asset content production and technology development to digital marketing.

MiHoYo is a company committed to technical R&D and the exploration of cutting-edge technologies, possessing leading capabilities in cartoon rendering, artificial intelligence, cloud gaming technology, etc. It has created products like "Genshin

Impact," "Honkai Impact," "Nova Desktop," "MiHoYo Community," and more, building a multifaceted product ecosystem around their IPs including animations, comics, music, novels, and merchandise.

Inmyshow launched the virtual social software "Rainbow Universe," which creates a virtual social platform based on real social systems by integrating social media platforms like Bilibili and Weibo, with a focus on "identity, social interaction, content, and consumption."

Hengxin Shambala has devoted itself to the layout of related VR businesses by drawing on high-quality resources in the VR field, covering VR content production, high-quality image collection and post-processing, VR content aggregation, VR education, VR interactive equipment, and other core links.

Fengyu Zhu has gradually formed core business segments including digital content production, augmented reality technology applications, and cultural tourism scene operations. In digital content production, the company uses WebGL technology to achieve 3D graphic rendering, providing customers with comprehensive digital scene solutions including creative planning, design and production, and development execution.

## 8 Conclusion

The metaverse has been a hot topic in China for nearly three years, evolving from a "confusing and overlooked" technological term to a thriving future industry that is just one step away from becoming a national strategy of China. The academic sector is intensifying theoretical research, the industrial sector is expanding application scenario exploration, the capital sector is increasing investment efforts, the media sector is amplifying promotional activities, and the legal sector is devising risk response strategies. It is believed that in the near future, Chinese universities will establish metaverse majors or even colleges. The metaverse is also expected to further empower industries such as agriculture, industry, construction, healthcare, military, and education, thus promoting the transformation and upgrading of traditional industries with its immense power.

## References

1. Kong R et al (2021) Exploring the metaverse framework: The third revolution of productivity. Tianfeng securities. Available from: https://xueqiu.com/3966435964/197725500
2. Deloitte Consulting (2022) A comprehensive view of the metaverse—vision, technology, and response. Available from: https://www2.deloitte.com/cn/zh/pages/technology-media-and-telecommunications/articles/metaverse-report.html
3. Shen Y (2022) Metaverse development research report 2.0. Tsinghua University School of Journalism and Communication New Media Research Center. Available from: https://www.docin.com/p-4525195689.html

4. Gong C (2022) China metaverse white paper (2022). Beijing Information Industry Association. Available from: https://wenku.baidu.com/view/8470598a2179168884868762caaedd3382c4b56c.html
5. Tencent News (2022) Fudan University School of Journalism and communication: 2021–2022 metaverse report. 2022. Available from: https://www.mayiwenku.com/p-33608040.html
6. Guan X, Li Y (2021) How the metaverse rewrites human social life. In: Central commission for discipline inspection of the CPC and national supervisory commission website. Available from: https://www.ccdi.gov.cn/toutiaon/202112/t20211223_160087.html
7. Ding J (2022) Metaverse: Or redefining human civilization 4.0. Big Data Digest. Available from: https://m.163.com/dy/article/GTS453V20511831M.html
8. Gong C et al (2023) Research on the terminology definition of "Metaverse" and related issues. China Sci Technol Terminol 25(01):27–35
9. Zhang J (2021) Can everything be "metaverse"? People's Daily Commentary 2021. Available from: https://mp.weixin.qq.com/s/HS5NKh30_vTILjyyzdSrSA
10. Office of the Zhejiang Province Digital Economy Development Leadership Group (2022) Guidance opinions on the construction of future industry pilot zones in Zhejiang Province. Available from: http://www.myjjzx.cn/cj/view.php?aid=484
11. Cheng Y (2022) Wuhan municipal government work report. In: The first session of the fifteenth Wuhan municipal people's congress. Available from: http://www.whrd.gov.cn/html/rdzl/dbdhwj/1501/2022/0121/20932.html
12. Yu L, Che H (2023) The Metaverse: new opportunities for digital China. China Development Press
13. Shen Y, Gong C (2023) Major issues in the development of China's metaverse industry. Econ Ref 11:1–58
14. Ministry of Industry and Information Technology (2023) Ministry of science and technology, national energy administration, national standardization administration. In: Implementation plan for the new industry standardization pilot project (2023–2035). Available from: https://www.gov.cn/zhengce/zhengceku/202308/content_6899527.htm
15. Offices of the Ministry of Industry and Information Technology (2023) Ministry of Education, Ministry of Culture and Tourism, State-owned Assets Supervision and Administration Commission of the State Council, State Administration of Radio and Television. In: Metaverse industry innovation and development Three-Year Action Plan (2023–2025). Available from: https://www.gov.cn/zhengce/zhengceku/202309/content_6903023.htm
16. Ministry of Industry and Information Technology, Office of the Central Cyberspace Affairs Commission (2021) Guidelines on accelerating the application and industrial development of blockchain technology. Available from: https://www.cac.gov.cn/2021-06/07/c_1624629407537785.htm
17. Shanghai Municipal Commission of Economy and Informatization (2021) 14th Five-Year Plan for the development of the electronic information industry in Shanghai. Available from: https://www.shanghai.gov.cn/gwk/search/content/99677f56ada245ac834e12bb3dd214a9
18. Office of the Shanghai Municipal People's Government (2022) Action plan for cultivating the "metaverse" new track (2022–2025). Available from: https://www.shanghai.gov.cn/202214bgtwj/20220720/90aa73b046464b9c8799ef2339026d7d.html
19. Beijing Tongzhou District People's Government (2022) Several measures for accelerating the innovative and leading development of the metaverse in Beijing's sub-center. Available from: https://www.bjtzh.gov.cn/bjtz/jdhy/202203/1515478.shtml?eqid=dcadfed100006bfc000000036479149f
20. Beijing Tongzhou District People's Government (2022) Metaverse innovation development action plan for Beijing's sub-center (2022–2024). Available from: https://www.bjtzh.gov.cn/bjtz/jdhy/202208/1612429.shtml
21. Ministry of Industry and Information Technology, Ministry of Education et al (2022) Action plan for the integrated development of virtual reality and industry applications (2022–2026). Available from: https://www.gov.cn/zhengce/zhengceku/2022-11/01/5723273/files/23f1b69dcf8b4923a20bd6743022a56f.pdf?eqid=d573637b0016c746000000056485dc71

22. Ministry of Industry and Information Technology, Ministry of Education et al (2023) Three-year action plan for the innovative development of the metaverse industry (2023–2025). Available from: https://www.gov.cn/zhengce/zhengceku/202309/content_6903023.htm
23. Zhao G et al (2021) The Metaverse: the future of the internet. China Translation & Publishing House.
24. Yan Y (2022) Metaverse science and technology industry. Party School of Central Committee of CPC Press
25. Ye Y et al (2022) Top ten technologies of the Metaverse. China Translation & Publishing House.
26. Gong C (2022) The Metaverse: The eve of great revolution. Central Party School Press.
27. China Academy of Information and Communications Technology, Industrial Internet Industry Alliance (2023) Industrial Metaverse white paper. Available from: https://aii-alliance.org/uploads/1/20230324/8c43ae95c9991a94d3b71e61f6d6ba8a.pdf
28. Industrial Culture Development Center of the Ministry of Industry and Information Technology, Beijing Information Industry Association (2023) China industrial Metaverse white paper. Available from: https://wenku.baidu.com/view/423794dd862458fb770bf78a6529647d26283427.html

# Index




M. Grieves and E. Hua (eds.), *Digital Twins, Simulation, and the Metaverse*, Simulation Foundations, Methods and Applications, https://doi.org/10.1007/978-3-031-69107-2

**D**



**E**

**W**



**X**

GPSR Compliance

*The European Union's (EU) General Product Safety Regulation (GPSR) is a set of rules that requires consumer products to be safe and our obligations to ensure this.*

*If you have any concerns about our products, you can contact us on ProductSafety@springernature.com*

In case Publisher is established outside the EU, the EU authorized representative is:

Springer Nature Customer Service Center GmbH
Europaplatz 3
69115 Heidelberg, Germany

**Batch number: 10091901**

Printed by Printforce, the Netherlands